Antibody Drug Discovery

Molecular Medicine and Medicinal Chemistry

Published:

MicroRNAs in Development and Cancer
edited by Frank J. Slack (Yale University, USA)

Merkel Cell Carcinoma: A Multidisciplinary Approach
edited by Vernon K. Sondak, Jane L. Messina, Jonathan S. Zager, and Ronald C. DeConti (H Lee Moffitt Cancer Center & Research Institute, USA)

DNA Deamination and the Immune System: AID in Health and Disease
edited by Sebastian Fugmann (National Institutes of Health, USA), Marilyn Diaz (National Institutes of Health, USA) and Nina Papavasiliou (Rockefeller University, USA)

Antibody Drug Discovery
edited by Clive R. Wood (Bayer HealthCare, Germany)

Forthcoming:

Molecular Exploitation of Apoptosis Pathways in Prostate Cancer
by Natasha Kyprianou (University of Kentucky, USA)

Fluorine in Pharmaceutical and Medicinal Chemistry: From Biophysical Aspects to Clinical Applications
edited by Véronique Gouverneur (University of Oxford, UK) and Klaus Müller (F Hoffmann-La Roche AG, Switzerland)

Volume 4 Molecular Medicine and Medicinal Chemistry

Antibody Drug Discovery

Clive R Wood
Bayer HealthCare, Germany
Editor

ICP
Imperial College Press

Published by

Imperial College Press
57 Shelton Street
Covent Garden
London WC2H 9HE

Distributed by

World Scientific Publishing Co. Pte. Ltd.
5 Toh Tuck Link, Singapore 596224
USA office: 27 Warren Street, Suite 401-402, Hackensack, NJ 07601
UK office: 57 Shelton Street, Covent Garden, London WC2H 9HE

British Library Cataloguing-in-Publication Data
A catalogue record for this book is available from the British Library.

Molecular Medicine and Medicinal Chemistry — Vol. 4
ANTIBODY DRUG DISCOVERY

ISBN-13 978-1-84816-628-8
ISBN-10 1-84816-628-1

Typeset by Stallion Press
Email: enquiries@stallionpress.com

Printed in Singapore by Mainland Press Pte Ltd.

Contents

Contributors

Foreword

John McCafferty, University of Cambridge, Department of Biochemistry, Tennis Court Road, Cambridge, CB2 1QW, UK

Editor

Clive R. Wood, Bayer HealthCare AG, GDD-Global Biologics, Aprather Weg 18a, 42096 Wuppertal, Germany

Authors

David R. Buckler, Dyax Corp., 300 Technology Square, Cambridge, MA 02139, USA

Jie Chen, Dyax Corp., 300 Technology Square, Cambridge, MA 02139, USA

Helen Dooley, Pfizer, Inc., Foresterhill, Aberdeen, AB25 2ZS, UK

William J. J. Finlay, Pfizer, Inc., Grange Castle Business Park, Clondalkin, Dublin 22, Ireland

Davinder Gill, Pfizer, Inc., Global Biotherapeutic Technologies, 87 Cambridgepark Drive, Cambridge, MA 02140, USA

Olivier Léger, Merck Serono International S.A., Protein Engineering & Antibody Technologies, Geneva Research Center, Merck Serono International S.A., 9, chemin des Mines, 1202 Geneva, Switzerland

Hans-Georg Lerchen, Bayer HealthCare AG, Medicinal Chemistry, Aprather Weg 18a, 42096 Wuppertal, Germany

David Lowe, MedImmune Ltd., Milstein Building, Granta Park, Cambridge, CB21 6GH, UK

Beate Müller-Tiemann, Bayer HealthCare AG, Global Biologics, Aprather Weg 18a, 42096 Wuppertal, Germany

Andrew Nixon, Dyax Corp., 300 Technology Square, Cambridge, MA 02139, USA

Stephane Olland, Pfizer, Inc., Global Biotherapeutic Technologies, 87 Cambridgepark Drive, Cambridge, MA 02140, USA

José W. Saldanha, National Institute for Medical Research, Division of Mathematical Biology, The Ridgeway, Mill Hill, London NW7 1AA, UK

Darren Schofield, MedImmune Ltd., Milstein Building, Granta Park, Cambridge, CB21 6GH, UK

Daniel J. Sexton, Dyax Corp., 300 Technology Square, Cambridge, MA 02139, USA

Sherif El Sheikh, Bayer HealthCare AG, Medicinal Chemistry, Aprather Weg 18a, 42096 Wuppertal, Germany

Paul Stephens, UCB Celltech, Discovery Biology, 216 Bath Road, Slough, Berkshire SL1 4EN, UK

Sean Stevens, Ablexis LLC, 409 Illinois Street, San Francisco, CA 94158, USA

William R. Strohl, Centocor R&D, Inc., Biologics Research, 145 King of Prussia Road, Radnor, PA 19087, USA

Berni M. Sweeney, UCB Celltech, Discovery Biology, 216 Bath Road, Slough, Berkshire, SL1 4EN, UK

Lioudmila Tchistiakova, Pfizer, Inc., Global Biotherapeutic Technologies, 87 Cambridgepark Drive, Cambridge, MA 02140, USA

Tristan J. Vaughan, MedImmune Ltd., Milstein Building, Granta Park, Cambridge, CB21 6GH, UK

Trevor Wilkinson, MedImmune Ltd., Milstein Building, Granta Park, Cambridge, CB21 6GH, UK

Jörg Willuda, Bayer HealthCare AG, Therapeutic Research Group Oncology, Muellerstrasse 178, 13353 Berlin, Germany

Zhenping Zhu, Kadman Pharmaceuticals, 450 East 29th Street, New York, NY 10016, USA

Foreword

*John McCafferty**

In 1996 only two antibodies were approved for therapeutic use in humans. In the years that have followed, however, 26 more antibody products have been approved with hundreds more entering clinical trials[1]. Antibodies are the fastest growing class of drugs, and the factors behind this success are clear. Their long half-lives allow relatively infrequent dosing, and they are well-tolerated and highly specific for their targets. Their basic properties can even be improved by engineering binding or effector functions. Simply stated, antibodies make good drugs, and they have become accepted by drug developers, regulators and users alike.

Monoclonal antibodies were first described in 1975, so why did it take so long for their potential to be realized in the clinic? Initially, there was great optimism around the clinical use of antibodies as "magic bullets" for targeting tumors or for modifying immunological function. However, this initial optimism soon gave way to pessimism when murine antibodies were found to induce unwanted immune responses in human recipients. Additionally, poor engagement with human effector function limited their potency. The answer to the problem was to either develop antibodies directly from a human source or make murine antibodies more "human-like". Over the last two decades a number of solutions have emerged to allow generation of antibody drugs which are accepted by the patients' immune system. It was these technological developments that unlocked the potential of monoclonal antibodies and drove the "antibody revolution", leading to the expansion of antibody drugs in the world's dispensary.

Advanced molecular biology techniques allowed the creation of chimeric antibodies by grafting variable domains from mouse monoclonal antibodies onto human constant regions. The reduction in immunogenicity was taken a step

* University of Cambridge, UK.

further by "CDR grafting" (also called "humanization") where the individual binding loops from murine variable domains were grafted onto human variable region frameworks. These approaches circumvented the problems associated with using murine antibodies in humans, and many of the clinical candidates from the 1990s were generated by this route. The current profile of approved drugs reflects this pedigree, with chimeric or CDR grafted antibodies accounting for two thirds of all antibody-based drugs currently marketed in the USA. Other technology developments during the early 1990s enabled direct isolation of human antibodies. The demonstration 20 years ago that functional antibodies could be displayed on the surface of filamentous bacteriophage opened the door to a powerful method for isolating or improving human antibodies using large antibody display libraries. Fully human antibodies could also be directly isolated from transgenic mice in which murine antibody genes were replaced with human antibody genes, allowing the generation of high affinity human antibodies by immunization. Over the last decade an increasing proportion of antibodies in development have been derived from phage display or transgenic technologies, with 10 such antibodies already approved or undergoing review by the FDA and 88 more currently undergoing clinical trials[1].

Technological developments have not only driven our ability to generate desired binding specificities but have also helped enhance clinical effects and outcomes by allowing flexibility in the choice of antibody class, isotype and mode of action. In particular, engineering of antibody constant domains has played an important role over recent years in increasing half-life or controlling engagement with effector functions.

These combined developments over the last two decades have moved antibody drugs from being on the fringes of medical practice to becoming a major weapon in combating disease, a trend that is likely to continue in the coming years. In *Antibody Drug Discovery* Clive Wood has assembled a number of excellent overviews and expositions of the transforming technologies that have driven this revolution. This book is a valuable resource for those seeking to understand and benefit from these mature technologies. Through their efforts, the contributors have distilled extensive "real world" knowledge and experiences to create a volume that will hopefully inform and inspire the next generation of drug developers.

John McCafferty, Cambridge, 2011

[1] Nelson A.L., Dhimolea E., Reichert J.M. (2010). Development trends for human monoclonal antibody therapeutics. *Nat Rev Drug Discov* **9**: 767–74.

Preface

*Clive R. Wood**

The recombinant DNA-based antibody technologies crafted during the last few decades have allowed us to exploit the natural defense mechanisms that antibodies provide. We can identify antibodies that bind almost any target with high affinity and exquisite specificity, and in an IgG scaffold antibodies have the benefits of bivalent binding, effector mechanisms and a long plasma half-life. Since the market launch of two humanized antibodies in 1997, these drugs have found a wide range of clinical uses. In 2009, only fourteen years later, four IgG1antibodies and one IgG1 fusion protein were represented in the list of prescription drugs with the twenty largest worldwide markets. This productive trend is forecast to continue with the market for antibody drugs having a compound adjusted growth rate (2008–2014) of 10.3% (PharmaVitae: Monoclonal Antibodies: 2009 Update, Datamonitor), dramatically outpacing that of small molecule drugs. This success is based on delivering value to patients that is not provided by other means. Antibodies are firmly embedded in our armory of approaches against disease.

This book was developed as a place to capture an overview of the foundations of antibody drug discovery technology, with each chapter written by leaders in the field. The guiding principle for these authors was that they should clarify how each area has developed and emphasize key points for readers working in this area to consider. These foundations are well established and will serve us well for years to come. Their general accessibility is increasing today as patents covering some areas have expired or are close to expiring. Many of us in the field are excited by emerging new areas of antibody technology and novel scaffolds that may have antibody-like properties. Nonetheless, the relative ease of use of

* Bayer HealthCare AG, Germany.

existing technologies and their increasing availability with patent expiration will increase the demands on new approaches to deliver benefits over and above those of the foundation technologies.

Three core technologies are used for identifying antibody leads: humanization of mouse monoclonal antibodies (mAbs), human immunoglobulin transgenic mice, and *in vitro* display. Once a murine mAb has been isolated with the desired properties, the fastest route to a drug candidate can be by grafting the murine mAb CDR sequences into a human scaffold. The 'Humanization of Antibodies' chapter (Olivier Léger and José Saldanha) provides a succinct overview of the different humanization technologies that are currently used. The two other core technologies do not use a murine mAb as a starting point. In the transgenic approach, a number of mouse strains have been engineered that block the ability of the endogenous murine immunoglobulin genes to contribute to an antibody response, while introducing human immunoglobulin genes that can rearrange, be used in the murine B-cell response and isolated via conventional hybridoma technology. The chapter on 'Transgenic Mice Rearranging Human Antibodies' (Sean Stevens) addresses key advances in this field. Additionally, *in vitro* display with combinatorial libraries of antibody fragments has proven very successful. Display systems based on phage particles and yeast cells are the most extensively employed. Among other advantages, they allow the identification of antibodies with epitope specificities that would be silenced by immunologic tolerance *in vivo*. Therefore, with well-designed selection strategies, there is a higher probability of finding leads with unique specificities, including the ability to bind both the human target antigen and its orthologs in preclinical species, thereby accelerating pharmacological and toxicological testing. I am particularly pleased that the chapter on 'Selection and Screening of Antibody Phage Display Libraries' (David Buckler *et al.*) is a unique collaboration between two of the pioneering organizations in phage display and represents a summation of many decades of experience. It is followed by a chapter on how antibodies from different approaches, especially from phage display, have been affinity optimized (David Lowe *et al.*). Most antibody phage display libraries have used immunoglobulin sequences from naïve donors, so the resulting leads have not been through affinity maturation *in vivo*. Thus, the leads may have low affinities, and in some cases affinity maturation is necessary to match antibodies from *in vivo* immunization approaches.

Once an antibody lead is identified, the structure of the Fc must be selected, expressed and then purified to provide supply to the research team for further characterization. The majority of marketed antibody therapeutics are of the human IgG1 isotype, which has proven to be a very successful scaffold for a wide variety of antigen specificities and clinical uses. This success justifies treating it as a preferred isotype unless the effector properties of IgG1 are verified as being unsuitable for a particular use. In recent years, a number of antibodies with other isotypes have been approved. The 'Isotype Selection and Fc Engineering' chapter (William Strohl) provides a comprehensive review of Fc isotypes and engineering strategies. The 'Antibody Expression from Bacteria to Transgenic Animals' chapter (Paul Stephens and Bernie Sweeney) provides an overview of the current systems used for antibody expression and an evaluation of some of the alternative systems that are being developed. In addition to expressing the antibody, a purification process is needed for research supply and an evaluation of the feasibility of a development process. 'Current Trends in Antibody Purification and Lead Selection' (Jie Chen and Andrew Nixon) provides an overview of key purification approaches and describes how the feasibility of development processes is increasingly playing a role in selecting lead candidates.

Two chapters address distinct and important themes in antibody therapeutic discovery. 'Design and Application of Immunoconjugates for Cancer Therapy' (Sherif El Sheikh *et al.*) provides an update on the field of conjugating antibodies to cytotoxic agents. This exciting concept developed in the early days of antibody technology and has had a relatively slow development since. However, many pivotal clinical studies are now being reported and are anticipated to be completed in the next few years. "Likewise, the development of multispecific antibodies has been much anticipated but slow to develop momentum. In 'Dual-Targeting Bispecific Antibodies as New Therapeutic Modalities for Cancer' (Zhenping Zhu) an update on the progress and options in this area is described." This book does not aim to cover in depth the wide range of alternative technologies designed to supplement or replace what is possible with antibodies. However, in 'Antibody Fragments and Alternate Protein Scaffolds,' (Lioudmila Tchistiakova *et al.*) there is a valuable perspective on how one biologics discovery organization has approached these areas.

I am very grateful to all of the authors of this volume for their willingness to share their expertise and perspectives. In addition, I thank Britta Werthwein for her excellcnt administrative assistance; the Sarkis Science Editing team for their superb editorial support; and many colleagues for their reviews and input, especially René Hoet, Ulrich Haupts, Wayne Coco, Jörg Willuda, Heiner Apeler, John Murphy and Rick Harkins. The advances described in this book are based on the contributions of a wide range of scientists and inventors. I salute these pioneers who have made antibody therapeutics possible. They can take pride in both having delivered drugs to patients today and having provided the basic tools to build new drugs as we gain insights into disease and identify innovative targets.

Clive R. Wood, Gloucester, MA, 2011

A number of antibody drugs have been approved for marketing during the preparation of this volume. Any statements on the exact number are bound to be outdated quickly in this fast-moving field. For current information, Janice Reichert's list at http://www.landesbioscience.com/journals/mabs/about/#background is recommended as a regularly revised catalog.

1

Humanization of Antibodies

Olivier Léger and José W. Saldanha*†

1.1 Introduction

Antibody humanization utilizes proven strategies for lessening the immunogenicity of well-characterized monoclonal antibodies (mAbs) from animal sources (commonly mice) and for ameliorating their activation of the human immune system, thus producing clinical diagnostics and therapeutics. At the time of writing, there were thirty approved antibodies listed at the Food and Drug Administration (FDA) website and over a hundred humanized antibodies in clinical trials.

Antibodies produced from rudimentary forms of inoculation are likely to have been used for centuries before the English scientist Edward Jenner pioneered vaccination in the late-eighteenth century. However, the concept of using antibodies as drugs stems from their discovery in 1890, when von Behring and Kitasato reported that small doses of diphtheria toxin injected into animals yielded sera containing an active ingredient called "antitoxin," which we now know as "antibody." This antiserum could protect against lethal doses of toxins in humans, and since the late 1800s antisera have been used to treat acute disease and in prophylaxis, as well as being used *in vitro* as diagnostic tools to establish and monitor disease. Because antiserum is a crude extract containing not only animal antibodies but also other animal proteins, it invariably induces complications, particularly upon multiple administrations. The resulting immune complex disease and anaphylaxis were termed "serum sickness." The active animal antibody apparently needed to be produced separately from the other animal proteins. This came about when Köhler and Milstein (1975) at the Medical Research Council

* Merck Serono International S.A., Switzerland; † National Institute for Medical Research, UK.

Laboratory in Cambridge (UK) published their method of producing custom-built antibodies *in vitro*. They fused antibody producing spleen cells from an immunized rodent with immortal tumor cells (myelomas) from the bone marrow of mice to produce hybridomas. A hybridoma combines the cancer cell's ability to reproduce almost indefinitely with the immune cell's ability to produce antibodies. Once screened to isolate the hybridoma yielding antibodies of the required antigen specificity and affinity, a hybridoma will grow and divide, mass-producing antibodies of a single type (monoclonal). Nearly a century previously, the German scientist Paul Ehrlich envisaged that such entities could be used as "magic bullets" to target and destroy human diseases, and hybridomas were the factories for producing these magic bullets.

An early clinical success of mAb use was in the treatment of organ transplant rejection with a mouse IgG2a antibody reactive against the cell–surface receptor CD3 (Goldstein, 1987). However, exposing humans to mouse antibodies provoked a response similar to the serum sickness of antisera therapy, namely that when mouse mAbs are administered in multiple doses, the patient almost invariably raised an immune response to the mAbs. This human anti-mouse antibody response (HAMA) (Schroff *et al.*, 1985) can develop shortly after initiation of treatment and precludes long-term therapy. Antibodies produced by the HAMA response can neutralize or result in clearance of the therapeutic antibody in immune complexes, or can sensitize the patient to allergic reactions on re-administration, thus compromising the efficacy, biological half-life or safety of the mAb. Moreover, mouse mAbs are inefficient activators of the human effector functions that result in a therapeutic effect. Although they may bind an antigen, it may not be targeted for clearance. As such, although mAbs from hybridoma technology are immensely useful as scientific research and diagnostic reagents, they have not completely fulfilled the possibilities inherent in Ehrlich's vision. Ideally, human antibodies should be used in human therapy. However, immortalization of human antibody-producing cells has had limited success, and besides being difficult to prepare, human hybridomas are unstable and generally secrete low levels of mAbs.

1.1.1 *First step: chimerization*

An alternative approach to mouse antibodies is chimeric antibodies, where animal variable domains are joined to human constant domains (Boulianne *et al.*, 1984; Morrison *et al.*, 1984; Neuberger *et al.*, 1985). This can be achieved by linking the animal and human genes coding for each domain and then expressing the engineered, recombinant antibody gene in rodent myeloma cell lines. In mouse/human chimeric antibodies, the immunogenic mouse constant domains are replaced with their human counterparts to reduce the likelihood of a HAMA response, while retaining the mouse variable domains that allows the antigen-binding affinity and specificity in the engineered molecule to remain unaltered. Effector functions of chimeric antibodies may also be changed by choosing the constant domains of different human immunoglobulin isotypes (Dyer *et al.*, 1989), since the isotypes have different capabilities in complement and cell-mediated clearance. While chimerization reduced the HAMA response of some murine antibodies, others showed immunogenicity upon repeated administration (Khazaeli *et al.*, 1994) — the so-called human anti-chimeric antibody (HACA) response — due in part to the presence of immunogenic epitopes in the mouse variable domains.

1.2 CDR Grafting: Standard Technology

Another approach to reducing immunogenicity even further is to reshape a human antibody so it contains only the complementarity determining regions (CDRs) from an animal monoclonal antibody of the desired affinity and specificity. This approach was pioneered by Greg Winter and colleagues at the Medical Research Council, Cambridge. (Jones *et al.*, 1986) The reshaped, humanized antibody retains only the six loops essential for binding from the mouse antibody, while the rest of the molecule, including the β-sheet framework of the variable domains is human. The goal is to maintain the antibody combining site (paratope) that forms the basis for antigen binding, by minimizing any alterations to the structural conformation of the CDRs. The choices required in protein engineering the mouse antibody are (1) the extent of the mouse CDRs to graft, (2) the human framework acceptor regions into which the CDRs should be grafted, (3) the backmutations required to restore the original affinity (4) and the isotype required to provide the desired biological function. The isotype choice is governed as for chimeric antibodies and the other choices are discussed below.

1.2.1 *Choice of CDR region to graft*

Humanization by CDR grafting requires that the antigen-binding residues from the mouse antibody be retained in the humanized antibody; thus, the identification of these residues obviously plays an important role in the protocol. The CDRs were originally defined by Wu and Kabat (1970) as the regions of the variable domain with the highest variability in the protein sequence: three in the light chain variable domain (V_L) and three in the heavy chain variable domain (V_H). These were predicted to contain the antigen-binding site which was subsequently verified by the X-ray structures of antibody/antigen complexes. Consequently, the CDRs are retained in CDR-grafted molecules, although they are not all necessary since not all six CDRs might interact with an antigen in a particular antibody. Chothia *et al.* (1989) defined CDRs on the basis of known antibody structures, considering only the positional variability of residues in the structural loops.

Their definition is more restricted than that of Kabat (1991), and therefore the CDR-grafted molecule based on this definition in theory should have fewer immunogenic non-human epitopes. This, in fact, has not proven true, and many CDR-grafted antibodies actually contain the Kabat definition, except for CDR-H1 where the Kabat and Chothia definitions are often combined (Presta *et al.*, 1997), effectively increasing the possible immunogenic non-human residues. This may be minimized by changing some non-antigen contacting mouse residues in the CDRs back to the human residues (Presta *et al.*, 1993).

1.2.2 *Choice of human framework acceptor regions*

The selection of human framework acceptor regions into which the mouse CDR regions are grafted is generally accomplished by searching protein sequence databases with the mouse variable region sequences as input. The major strategies are described below.

1.2.2.1 *Fixed frameworks*

In the fixed frameworks strategy, no database search is employed since the human framework acceptor regions are fixed regardless of the source antibody or its sequence similarity. Usually the human myeloma antibodies REI (for the light chain) and NEW (for the heavy chain) are used, since their X-ray three-dimensional structures are freely available for scrutiny, and they are well characterized in terms of stability and expression. NEW and REI human framework acceptors were used for the first therapeutically CDR-grafted antibody, the anti-cancer Campath antibody (alemtuzumab) (Riechmann *et al.*, 1988), and Actemra (tocilizumab) (Sato *et al.*, 1993) which was approved for Castleman's disease and rheumatoid arthritis. The advantages of this fixed frameworks approach are having the structures of the human frameworks and the wealth of data collected on these antibodies. However, any difference in sequence identity between the source antibody and these fixed human frameworks will inevitably result in considerable structural differences. This may be offset by the inherent flexibility of the antigen-binding site, which can allow for these structural differences by induced fit to the antigen, but greater sequence difference will also lead to the need for a greater number of backmutations and a greater possibility of immunogenicity.

1.2.2.2 *Best fit*

In the best fit approach, human framework acceptor regions are selected based on shared sequence similarities (hence, the approach is also called "homology matching") to the source antibody's variable regions. Homologous human frameworks should have a structure that will minimize distortions to the conformations of the grafted CDR loops, requiring fewer backmutations, thereby recreating the binding of the original antibody with less potential for immunogenicity. This was the approach taken by Queen *et al.* (1989) for the V_H of the anti-TAC antibody, now marketed as Zenapax (daclizumab) for the prophylaxis of acute organ rejection in patients receiving renal transplants. In this case, the V_L was chosen to match the V_H (i.e. the same human antibody for both chains), while others have used the most similar V_L and V_H from different human antibodies. The advantage of using V_L and V_H from the same clone is that the packing of the interface between the domains is less likely to be interrupted, but this is offset by a lower sequence similarity between at least one domain and the source antibody. Both the fixed frameworks and best fit approaches to human framework selection can be limited to the mature variable regions found in protein sequence databases, which are usually characterized in terms of solubility, stability and expression. However, this runs the risk of somatic hypermutation in these variable regions (V-regions), creating potentially immunogenic epitopes.

1.2.2.3 *Consensus*

A way around the problem of somatic hypermutation is to select human framework acceptor sequences which are the consensus of the subgroups in the Kabat database (Kabat *et al.*, 1991). The Kabat database classifies the subgroups using a scheme that is based on analysis of the N-terminal 35 residues in the sequence. The consensus sequence of each Kabat subgroup is composed of the most frequent amino acid at each framework position, thus ridding the frameworks of unusual mutations. In this approach, consensus sequences for the V_H and V_L most similar to the mouse sequences are chosen for CDR grafting. However, consensus sequences are not natural and may contain unusual (and immunogenic) sequence motifs, even though they may not contain unusual residues. Comparisons of fixed framework, best fit, and consensus selection strategies can be found in the following publications (Maeda *et al.*, 1991; Kolbinger *et al.*, 1993; Sato *et al.*, 1994), but no method is generally the best.

1.2.2.4 *Germline*

Since consensus sequences may contain unnatural and immunogenic sequence motifs, an alternative strategy is to use the human germline sequences (originally suggested by Shearman *et al.*, 1991) that are natural human sequences not containing the somatic hypermutations found in the protein- and cDNA-derived sequences. The complete human immunoglobulin germline gene loci are available in the Immunogenetics Database (IMGT) (www.imgt.org), and genes most similar to the mouse V_L and V_H sequences are usually selected. Consensus and germline sequences can also be selected in a fixed frameworks type of approach with the understanding that the most abundant human subgroups are V_L subgroup kappa I and V_H subgroup III, or have high sequence similarity to the original mouse variable region in a best fit approach.

1.2.3 *Backmutations*

It must be stressed that simply grafting CDRs from a mouse antibody into human framework acceptor sequences does not necessarily re-create a functional antibody binding site. In fact, it often results in reduced affinity and on occasion may abolish it. The conformation of the CDRs and thus the ability to bind antigen can depend on interactions with the framework regions. Residues in the framework sequences can also sometimes directly interact with the antigen. Moreover, the packing and orientation of the β-sheets of the variable domains must not be altered, since they can alter the topography of the binding site and impact antigen binding. In most cases, it is necessary to alter the human framework acceptor sequences by introducing backmutations which revert certain human residues to mouse in order to restore the affinity of the original antibody, and in some cases even increase it (Kolbinger *et al.*, 1993). Of course, if too many backmutations are made, the result is a hyper-chimeric and not a CDR-grafted antibody, so some limitation must be placed on the number of backmutations introduced to keep the molecule "human" and to reduce its immunogenicity while still recreating antigen binding.

CDR grafting is made possible by the limited repertoire of conformations (called canonical classes) for five of the six CDRs. These were first identified by Chothia and Lesk (1987) and depend on a few critical residues in each class that must be conserved in order to retain the CDR conformation. Once the canonical class is elucidated for the mouse CDR, then backmutation of these critical residues in the human frameworks to their murine counterparts ensure that the conformation of the CDR can be maintained. It should be noted that sometimes backmutation of a canonical residue has no effect and may be even detrimental to the binding of the humanized antibody. These observations might be explained by the CDR not contacting the antigen, or that the human residue at that position alters the CDR conformation for better binding than the mouse residue.

It is common in CDR grafting to select human framework acceptors from two different antibodies for the variable light and heavy regions. This increases the possibility that these domains may pack differently to the source antibody and the different orientations will disturb the topography of the antigen binding site. For this reason, backmutations are often introduced to conserve the residues at the V_L/V_H domain packing interface. The residues at the packing interface were defined by Chothia *et al.*, (1985) and their importance in retaining binding affinity is exemplified in the humanization of antibody 1B4 (Singer *et al.*, 1993).

Interestingly, their importance might also be functional since in the humanization of antibody KM966 (Nakamura *et al.*, 2000), retention of murine residues at the packing interface had little effect on binding, but improved the complement-dependent cytolysis of the humanized antibody.

Residues in the framework sequences of the murine antibody that interact with the antigen can be readily identified if the X-ray structure of the antibody-antigen complex is known. However, this is rarely the case and usually a three-dimensional molecular model is built using either web tools such as PIGS (www.biocomputing.it/pigs) or proprietary software such as from Accelrys Inc. (accelrys.com). The framework is built on structural templates chosen for their sequence similarity to the mouse variable domain. The computer programs can superimpose V_L and V_H domains so that the correct packing orientation is obtained. Non-identical residues in the structural template are mutated to those in the mouse, and their side chain angles are optimized to minimize clashes. CDR conformations can be built on canonical templates or if these are not applicable, then conformational search methods can be used. CDR-H3 is usually the longest and most variable CDR loop that generally makes the most contact with antigen; however, despite its importance in the combining site, no canonical conformations are known for it. Shirai *et al.* (1996, 1999) described both "kinked" or "extended" conformations for the base of the loop, which can be predicted by sequence-based rules. Further rules can be applied to determine additional features of CDR-H3 and improve the modeling of its conformation. Accurate modeling, particularly of CDR-H3 is important for the identification and selection of residues for backmutation.

Somatic hypermutations in the murine sequences are the first residues to consider when analyzing the model, but analysis of the model for any unusual residues compared to the Kabat mouse subgroup in the vicinity of the CDRs gives an indication of whether they require backmutation in the resulting CDR-grafted molecule. Somatic hypermutations in the human framework acceptor sequences, which are also rare residues in comparison with that position in the human subgroup, can also be mutated to the consensus human residue in an effort to eliminate any immunogenic epitopes. N-terminal residues can also influence binding, either directly or indirectly, particularly in the V_L where they form a continuous surface with the combining site (Kolbinger *et al.*, 1993). Carbohydrates can also influence the binding of antigen, and therefore putative N- and O-glycosylation sites are predicted in the sequence and analyzed on the structural model.

However, the identification of backmutations is still, to some extent, an art and is the most difficult and unpredictable procedure in humanization, sometimes requiring many different versions of the CDR-grafted molecule to achieve success. Some data are available on the Humanization bY Design website (www.cryst.bbk.ac.uk/~ubcg07s), which could help in identifying strategic alterations, but each new source antibody can present new challenges. Since the backmutations of one antibody are not necessarily transferable to another, it is difficult to generalize. Therefore, it is usually necessary to analyze an accurately built three-dimensional model of the source antibody (and perhaps its CDR-grafted counterpart), closely observing critical amino acids that are buried in close proximity to the CDRs or those on the surface of the molecule that are likely to bind antigen.

1.3 Affinity versus Stability

For many of the early CDR-grafted antibodies, biophysical properties were not usually considered. When they were examined, problems appeared to occur frequently, clearly limiting the use of such recombinant antibodies even if they had very promising binding properties. Today, CDR grafting has the dual purpose of increasing the stability as well as reducing the immunogenicity of the source antibody. In addition, the biophysical properties of the therapeutic agent, such as its thermodynamic stability and its aggregation propensity, can critically affect the pharmacokinetics of such molecules (Willuda *et al.*, 1999).

Stability is a requirement for the development of modern antibody-based therapeutics. The hypervariable antigen-binding region (Fv) is responsible for stability variation between IgGs of identical subclass. In order to determine the range of stabilities that may be expected for humanized antibodies, Garber and Demarest (2007) performed differential scanning calorimetry on 17 humanized antibodies. They found that the Fabs of these antibodies exhibited thermal unfolding transitions with midpoints (TMS) varying from 57 to 82°C. Antibodies with very low Fab stabilities were found to aggregate and to be poorly expressed.

Solvent-exposed loops of CDRs and humanization-related mutations have a significant impact on the stability of the Fab fragment, and the impact can be much larger than the difference in stability between different human Fab frameworks. According to Ionescu *et al.*, (2008), the stability of an antibody can be affected by CDR grafting and the humanization process by a combination of any of the following factors: (1) selection of human variable framework acceptors, (2) impact of CDRs and additional mutations in the human framework on the intrinsic stability of each variable domain (V_H or V_L), and (3) impact of CDRs and additional mutations in the human framework on the energy of interaction between the domains. Understanding the stability of mAbs is important in the development of new therapeutic agents, and it should be monitored during the humanization process, which may also improve the selection process between similar human germline frameworks during an antibody humanization campaign.

In view of the large evolutionary distance between the different antibody germline families within the same species, it is not surprising that the biophysical properties of the proteins derived from these sequences vary considerably. Ewert *et al.*, (2003) systematically compared the biophysical properties of isolated human antibody variable domains representing the consensus sequences of the different germline families and of the single-chain Fv (scFv) fragments composed of these consensus domains. Human V_H3 was found to be the most stable of the human frameworks, followed by human V_H1 and V_H5, whereas human V_H2, human V_H4 and particularly human V_H6 showed very poor biophysical properties. One would therefore anticipate that CDR grafting into human V_H3 and V_H1 would offer acceptable biophysical properties, but grafts to non-optimized human V_H2, V_H4, and V_H6 frameworks are very likely to yield antibodies with suboptimal properties. This is the approach that was followed in a recent publication by Kügler *et al.* (2009). They designed a CDR graft to human frameworks of a CD19-scFv fragment derived from the murine antibody 4G7. The murine CDRs were grafted into the human acceptor frameworks V_K3-V_H3. Eleven additional point mutations had to be introduced to obtain a better match between CDR graft and acceptor framework. The CDR-grafted version of 4G7 maintained the comparatively high stability of the murine loop donor but did not improve it beyond this. Their analysis indicated that this was due to subtle strain introduced between CDRs and the framework that mitigated the otherwise highly favorable properties of the human acceptor framework. The CDR-grafted version of 4G7 exhibited a three- to four-fold reduced affinity compared to the source scFv. This is probably due to steric conflict between the framework and CDR residues that could potentially alter the CDR conformation.

This elegant piece of work brings one back to the difficult and critical issue of the choice of the human framework acceptor regions in the humanization design. One often has to make a choice. On the one hand a graft to the human framework most closely related to the framework of the CDR donor could minimize the probability of losing binding affinity, but it may also lead to a construct of equally insufficient stability as the donor. By contrast, grafting to the most stable of the human V_L and V_H frameworks reduces the risk of producing a construct with insufficient stability, but increases the risk of reducing antigen-binding affinity. To further complicate matters, the most stable framework is not necessarily the one that gives the best production yields in a given expression system. A set of rules (Honegger, 2008; Honegger *et al.*, 2009) have been tested in many instances that will usually yield a loop graft that retains most, if not all, of the antigen-binding affinity. However, this approach usually necessitates that a number of additional point mutations are introduced, which means that the issue of immunogenicity will need to be taken into account.

1.4 Alternative Approaches

1.4.1 *Veneering or resurfacing*

Padlan (1991) proposed that murine variable domains could be veneered in order to reduce their antigenicity. Veneering, also known as resurfacing, is a technique whereby the exposed framework residues are replaced by amino acids found at corresponding positions in human antibody sequences. This technique is based on the observation that the antigenicity of a protein is determined by the accessible and protruding residues (Novotny *et al.*, 1986). Since nothing is changed on the CDR-framework interactions at the interior, and only subtle changes are made to the exterior, this usually has only minor effects on the conformation and the activity of the resurfaced antibody (Delagrave *et al.*, 1999; Roguska *et al.*, 1994; 1996). Classically, the surface residues are identified according to Pedersen *et al.* (1994), who defined a set of positions in heavy and light chain variable region sequences that describe the exposed framework surface of the Fv region, and those that are non-human-like are subsequently mutated.

Staelens *et al.* (2006), working on the humanization of monoclonal antibody 82D6A3 directed against the von Willebrand factor (VWF) A3-domain, introduced a new and accurate approach for the identification of non-human-like framework surface residues by evaluation of the surface accessibility within the context of the modeled Fv. The resurfacing procedure was not accompanied by alterations of the binding affinity and inhibition properties of the scFv, thereby fulfilling a primary goal of any humanization attempt. Although many antibodies have been humanized by variable domain resurfacing, to our knowledge no clinical studies with these antibodies have been published until now.

1.4.2 *Grafting of abbreviated CDRs containing specificity-determining residues*

CDR-grafted humanized antibodies carry xenogeneic CDRs which may evoke anti-variable region, including anti-idiotypic, responses. Not all CDR residues of a murine antibody are essential for antigen binding. This rationale comes from a comprehensive analysis of the three-dimensional structures of the antibody-combining site where it was suggested that only 20–33% of the CDR residues are critical in the antigen–antibody interaction (Padlan, 1994). These residues, located in the regions of high variability and most likely to be unique to each antibody, are designated as specificity-determining residues (SDRs). The SDRs are mainly in CDR-H1, in the N-terminal and middle regions of CDR-H2, in CDR-H3 (but not in the terminal region), in the C-terminal region of CDR-L1, in the first and (sometimes) the middle parts of CDR-L2, and in the middle region of CDR-L3 (Padlan *et al.*, 1995). A new approach to the humanization of antibodies, therefore, was based on grafting only the SDRs of a xenogeneic antibody into the human antibody frameworks. Tamura *et al.*, (2000) humanized the anti-tumor-associated glycoprotein-72 mAb CC49 by grafting only its SDRs onto the human antibody frameworks. The humanized CC49 variant retained the antigen-binding properties of the source mAb. In *in vitro* binding analyses using patients' sera, the SDR-grafted humanized antibodies resulted in lower immunogenic potential compared to antibodies constructed by CDR grafting (Tamura *et al.*, 2000; Yoon *et al.*, 2006). The immunogenicity of an antibody therefore could be reduced by transplanting only those parts of the CDRs that contain the SDRs. The "abbreviated" CDRs have been defined (Padlan *et al.*, 1995) as the boundaries of the potential SDRs in various antigen-combining sites.

1.4.3 *Superhumanization*

This humanization strategy is based on structural homologies between mouse and human CDRs and essentially ignores the frameworks. The first step of this CDR-grafting method is to identify human germline V genes that, in combination, have the same canonical structure class as that of the mouse antibody to be humanized. Within that matching subset, typically a half dozen genes from 44 functional V_H or 41 functional V_L genes in the human genome, H, and L chain gene segments are selected whose CDRs have the best residue-to-residue homology to the mouse antibody. In the selected sequences, the remaining non-homologous CDR residues are simply converted to the mouse antibody sequence. mAbs constructed by this strategy retain the ability to bind antigen, and because they are CDR-grafted in a way that minimizes deviation from human sequences, such Abs are called "superhumanized." Tan *et al.* (2002) described the application of this method to humanizing the mouse anti-human CD28 antibody 9.3: the fully superhumanized antibody lost 30-fold affinity but retained biological activity. As a second case study, Hwang *et al.* (2005) published a detailed description of the superhumanization method and used this method to humanize anti-lysozyme antibody D1.3. The affinity loss of superhumanized D1.3 was six-fold. In a third published example, Hu *et al.* (2007) humanized a murine anti-Venezuelan equine encephalitis virus antibody 1A4A1, which was demonstrated to retain antigen-binding specificity and neutralizing activity, but no affinity measurement comparison between the humanized and source antibody 1A4A1 was reported.

Because the choice of the human frameworks is driven by the sequence and structure of the CDRs, this strategy has the potential to generate humanized antibodies that retain good binding to their cognate antigen. However, the general applicability of such a framework-ignoring strategy is uncertain since (1) favorable conformation(s) of a given CDR can be uniquely dependent on some specific source framework residues (canonical residues), and (2) particular source framework residues may in some cases actively participate in direct interactions with the antigen, as has been observed in certain antigen/antibody complexes.

1.4.4 *Human string content optimization*

Recently, Lazar *et al.* (2007) introduced a new method of humanization based on a novel and immunologically relevant metric of antibody humanness, termed "human string content" (HSC), that quantifies a sequence at the level of potential MHC/ T cell epitopes. Human identity is defined as the number of total "human 9-mers," which is an exact count of 9-mer stretches in the Fv that perfectly match any one of the corresponding stretches of nine amino acids in their set of functional human germline sequences. Human 9-mer content is used as a measure of T cell tolerance (i.e. if a stretch of nine consecutive amino acids matches one of the human germline segments, it is more likely that T cells will not recognize it due to developmental tolerance for the original human 9-mer). This approach utilizes the homology present in human germline sequences to make murine-to-human substitutions that increase the human sequence content of the Fv. The humanness of the resulting Fv is derived from several discrete germline sequences. Positions that are not within or proximal to the CDRs and V_H/V_L interface are optimized in the process, and the relative humanness of the resulting Fv is comparable to mAbs humanized using CDR grafting and other humanization techniques. This method was applied to the humanization of four antibodies with different antigen specificities showing better binding signals than or comparable to those of the source non-human antibodies.

1.4.5 *Framework shuffling and human framework adaptation*

Dall' Acqua *et al.* (2005) described a new humanization approach called "framework shuffling," which does not require any rational design or structural information and for which there is no need to design backmutations. In this method, a given non-human monoclonal antibody is humanized by synthesizing a combinatorial library comprised of its six CDRs fused in frame to a pool of mixed and matched human germline frameworks. These human frameworks encompass all known heavy and light chain human germline genes. Libraries are cloned into a M13-based phage expression vector. The primary screen consists of a single point ELISA using periplasmic extracts prepared from individual recombinant M13 clones. Using this technology, the group humanized the mouse monoclonal antibody B233 raised against the potential cancer target human receptor tyrosine kinase EphA2. Following a two-step selection process, in which the light and heavy chains of the source antibody were successively humanized, they identified several humanized variants that retained binding to EphA2. Following conversion to human IgG1, the best framework-shuffled, humanized binder exhibited a relative affinity five- to eight-fold lower than source mAb B233. In a subsequent study, a significant improvement to this technology was carried out: a one-step humanization process, in which the heavy and light chains of a given antibody are humanized simultaneously (Damschroder *et al.*, 2007). In the later study, they humanized another anti-EphA2 mouse antibody, EA2 and an anti-EphB4 mouse antibody, 47. They characterized in detail these two newly humanized antibodies as well as the previously humanized mAb B233, in terms of binding affinity to their respective antigens, functional activity, thermostability, electric charge, and expression yields. They showed that several of these parameters were either maintained or improved in all three humanized antibodies when compared with their respective chimeric counterpart.

Recently, an alternative to framework (FR) shuffling, human framework selection (HFS) was described by Fransson *et al.* (2010) as the first step of a new humanization method called human framework adaptation (HFA). In this method, the human germline genes are selected based on sequence and structural considerations. The first step of the HFA protocol entails generating a relatively small library of HFS variants by combining the binding site of the non-human antibody with human germline genes. Generation of the HFS library differs in several respects from other humanization methods, including CDR grafting and FR shuffling. First, the definition of the binding site, and thus regions used to transfer specificity from the non-human antibody in the human FRs, includes not only CDRs (Kabat definition) but also the hypervariable loops (Chothia definition). Second, a relatively small library of HFS variants is generated by taking into account sequence and structural considerations. Third, FRs are not mixed to generate the HFS library; only FRs coming from the same human IGV_H or IGV_K genes are used as building blocks of the library. Using this method, Fransson and co-workers describe the humanization of a mouse anti-human IL-13 antibody, 836. HFS of mouse 836 resulted in a five-fold loss of affinity. In a second step, SDRs were diversified in one of the best HFS variants, and clones with improved affinity, up to 100-fold compared to the HPS antibody, were selected using phage display (SDR optimization).

1.4.6 *Combinatorial library approaches*

1.4.6.1 *Guided selection*

An alternative method, guided selection, has been developed to convert murine antibodies into completely human antibodies with similar binding characteristics. This methodology was originally developed by Jespers *et al.*, (1994). Mouse V_H and V_L domains are sequentially or in parallel replaced by human V_H and V_L domains, respectively, using phage selection to derive human antibodies with best affinity. For example, in this pioneering study, using the mouse antibody mAb32 against TNF-α, the original mouse V_H Fab domain was combined with V_L repertoire-derived human B cells, and the resulting human V_L-shuffled antibody Fab library was displayed on the phage surface followed by selection on the target antigen. The selected V_Ls were then used for construction of a human V_H-shuffled antibody Fab library followed by selection on the target antigen, which led to the isolation of human Fab fragments binding to TNF-α. One of the selected Fab fragments, P3A2, when expressed as a complete antibody, was found to recognize the same epitope and have a binding affinity similar to the original mouse mAb32. Using a parallel rather than a sequential replacement of V_H and V_L domains, Humira (adalimumab), the first human antibody to gain marketing authorization approval, was eventually generated (Osbourn *et al.*, 2005). In this pivotal study, they used another murine anti-TNF-α mAb, MAK195, to guide the selection. As a result, a fully human antibody was obtained but a stepwise CDR mutagenesis was then necessary to generate the high affinity and high potency antibody known as Humira.

Guided selection has also been used successfully to humanize rodent monoclonal antibodies directed against cell surface antigens. For example, recently Figini *et al.* (1998 and 2009) generated completely human Fab fragments by conversion of murine anti-folate receptor (a specific epithelial ovarian carcinoma antigen) antibodies to human antibodies, through the combined use of human patients' antibody phage display libraries biased toward an anti-tumor response, guided selection, and chain shuffling. A potential disadvantage of the guided selection approach is that shuffling one or both antibody chains can result in epitope drift (Kang *et al.*, 1991; Ohlin *et al.*, 1996). This can, however, be seen as an advantage for introducing minor but essential changes in specificity, such as removing cross-reactivities to unwanted antigens (Christensen *et al.*, 2009). In order to maintain the epitope recognized by the source non-human antibody, CDRs can be conserved. In this alternative method, one or both non-human CDR3s are commonly retained since they usually play a critical role in the recognition of the antigen. For example, Rader *et al.* (1998) kept both murine CDR3 V_H and V_L in the construction of the library when humanizing murine LM609, an anti-human integrin α5β3. Binding specificity and affinity of the isolated human Fab were conserved.

1.4.6.2 *Framework libraries*

Rosok *et al.* (1996) described a general antibody engineering strategy that combines elements of structure-based approaches with *in vitro* evolution strategies to address the difficulty of maintaining antibody-binding activity following humanization. They used this alternative approach, based on the use of a phage-displayed combinatorial library, to humanize anti-carcinoma antibody BR96, which recognizes a tumor-associated antigen expressing a Lewis Y (Le^y)-related carbohydrate on the surface of many human carcinomas. Based on this strategy, the constructed library contained all possible combinations of source and human template amino acids at key framework positions, and tighter-binding antibodies were selected by panning against antigen. By using this strategy the activity of a set of humanized variants can be determined while simultaneously selecting tighter-binding variants, thus reducing or eliminating the trial-and-error process of CDR grafting. By using an M13-derived phage vector and plaque-lift assay as the selection method, several frameworks variants with antigen binding within two-fold of the source antibody were isolated. Subsequently, Baca *et al.* (1997) described a phage display method for optimizing the framework of humanized antibodies by random mutagenesis of important framework residues. They applied this method to the humanization of the anti-vascular endothelial growth factor (VEGF) murine monoclonal mAb A4.6.1. Affinity panning of a library of humanized A4.6.1 antibody mutants led to the selection of one variant with greater than 125-fold enhanced affinity for antigen relative to the initial humanized antibody with no framework changes. A single additional mutation gave a further six-fold improvement in binding. The affinity of this variant was six-fold weaker than that of a murine/human chimera of A4.6.1. Son *et al.* (2004) used a similar approach to the strategy described by Rosok (above) except that they used phage display as in Baca (above) rather than colony lift as the selection method. They described the humanization of agonistic anti-human 4-1BB mAb, BBK-4. Six humanized scFv fragments were selected from the combinatorial library expressing phage-displayed humanized scFv. All retained the epitope specificity of the original mAb but had affinities of lower than 1/10 of the original. In spite of their lower affinities, they behaved similarly to the original mAb in mixed lymphocyte reaction assays.

More recently, also using the combinatorial approach and phage display as the selection method, Rader *et al.* (2000) and Nishibori *et al.* (2006) humanized a rabbit and chicken antibody, respectively. All six CDRs were first grafted into the most sequence-similar human germline frameworks and fine-tuned using a phage-displayed combinatorial approach with permutation of important framework residues. After panning of the humanized rabbit and chicken library against the human A33 colon cancer antigen and recombinant mouse prion protein, respectively, the most humanized variants were selected and analyzed for antigen-binding activity. The best clones retained affinity comparable to the source rabbit or chicken antibody.

The processes of antibody humanization and *in vitro* affinity maturation are often performed in discrete steps. However, certain framework and CDR residues are spatially close and are likely to interact with one another. Therefore, sequential framework and CDR optimization approaches overlook potentially beneficial combinations of interacting framework and CDR residues. To address the limitations associated with sequential optimization strategies, Wu *et al.* (1999) performed a focused mutagenesis of the third CDR of the H and L chains concomitantly with expression of a combinatorial framework library. By using an M13-derived phage vector and plaque-lift assay as the selection method, simultaneous optimization of framework and CDR residues resulted in the identification of multiple humanized versions of the murine anti-CD40 mAb 40.2., displaying affinities better than the chimeric Fab. Moreover, at least two of the higher affinity variants, clones CW43 and 2B8, demonstrated a beneficial combination of framework and CDR residues that could not have been identified in a stepwise manner. Thus, simultaneous optimization of framework and CDR residues of mAb 4.20.220 resulted in the identification of variants that contained fewer murine framework residues yet displayed higher affinity.

1.4.7 *Humaneering™ technology*

KaloBios Pharmaceuticals has developed a novel proprietary technology to generate human therapeutic antibodies with good biophysical properties (Flynn *et al.*, 2005). The Humaneering™ process identifies the minimum sequence information required to determine antigen-binding specificity from the variable region of a reference source antibody. That information is then transferred to a library of human partial V-region gene sequences that are generally close to germline or completely germline to generate an epitope-focused library of human antibody V-regions. A novel microbial-based, signal-less secretion system is used to express members of the library as antibody Fab fragments. The library is then screened for antigen-binding Fab using a colony lift screened in solution as the binding assay. The resultant engineered human Fabs retain the binding specificity of the reference antibody, typically have equivalent or higher affinity for antigen than the reference antibody, and have V-regions with a high degree of sequence identity compared with human germline antibody genes. The minimum specificity determinant (MSD), which is required to generate the epitope-focused library, is represented by a short sequence within a CDR of the heavy chain (usually CDR-H3) and a short sequence within a CDR of the light chain (frequently CDR-L3). It should be noted that MSDs may be contained within any CDR of a heavy or light chain, and may also be spread among several CDRs. Paired heavy- and light-chain CDR segments, containing specificity determinants from the reference antibody, are used to constrain the binding specificity so that antigen-binders obtained from the library retain the epitope-specificity of the starting antibody. This process for engineering human antibodies results in sequences very close, in some cases identical, to the human germlines. Humaneering™ technology yields large panels of antibodies with the identical binding specificity; therefore, clones can be chosen that fulfill not only classic issues of antibody engineering (e.g. affinity and specificity), but also down-stream processing issues (e.g. expression level and physicochemical properties). KB001, which targets the *Pseudomonas aeruginosa* bacteria, and KB003, which targets human GM-CSF, are KaloBios' most advanced Humaneered antibodies, both being tested in phase I/II trials.

1.5 Concluding Remarks

As described in this chapter, many humanized forms of mAbs are possible. The most widely used methods involve grafting the CDRs of the non-human antibody into human frameworks typically selected for their sequence homology to the non-human framework. A compromise between humanizing a non-human framework residue to reduce its potential immunogenicity and retaining a non-human residue critical for antigen binding and/or specificity is often necessary. To mitigate the shortcomings of the traditional CDR-grafting approaches, various hybrid selection approaches have been tried in which portions of the non-human antibody have been combined with libraries of complementary human antibody sequences in successive rounds of selection for antigen binding. In the course of these multiple rounds of selection, most of the non-human sequences are gradually replaced with human sequences. These approaches have generally not fared any better than CDR grafting.

Other methodologies include chain replacement techniques where the non-human CDR3s are retained and only the remainder of the V-regions, including the frameworks and CDR 1 and 2, are individually replaced in steps performed sequentially. However, because inter-species compatibility imposes a high demand for structural homology in the selected human V-regions, only those most homologous to the non-human V-regions are generally selected. In this regard, the result is quite similar to that of the CDR-grafting approach, except that CDR 1 and 2 are initially human in the selected V-regions.

Although effective at making framework sequences more human, due to the perceived risk of antigen-affinity loss, few of these methods substantially engineer the CDRs. In essence, CDRs need not be considered untouchable. Humanization approaches such as SDR grafting have attempted to address this issue by reducing the CDR boundaries rather than using the strict CDR definition. Whatever the methodology used, future humanization protocols should include experimental probing of selected CDR residues in order to reduce the mouse content but also to evaluate its impact on affinity. Indeed, in many instances, increased antigen affinity has been reported when mutating single CDR residues. However, some humanized and even fully human sequence-derived antibody molecules still carry immunological risk. Harding *et al.* (2010) analyzed $CD4^+$ helper T cell epitopes in a set of eight humanized antibodies. Interestingly, all $CD4^+$ helper T cell epitopes were found only in CDR-sequence containing regions. They demonstrated that amino acid modifications to CDR regions can be designed that reduce the immunogenic potential while maintaining the bioactivity of the antibody molecule. Future humanization protocols could include a systematic search for $CD4^+$ helper T cell epitopes in order to reduce further the immunogenic potential of the therapeutic molecule.

The method of choice for humanizing a non-human antibody will depend on the immunogenicity of the molecule and the ease of implementing the method, weighed against the potential clinical utility of the antibody. There is no simple correlation between sequence humanness and immune response, as illustrated by the virtual absence of immunogenicity for the CDR-grafted humanized antibody trastuzumab (Herceptin®) and the 5–89% neutralizing response rate in a subset of patients, that varies depending on the disease and the therapy, to the fully human mAb adalimumab (Humira®) (Radstake *et al.*, 2009; Bender *et al.*, 2007; West *et al.*, 2008). Many factors in addition to protein sequence can affect immunogenicity, such as dose and administration parameters (route, duration and frequencies as well as target specific effects), but one of the most critical factors is the biophysical properties of the final molecule (De Groot *et al.*, 2007). These properties should be carefully monitored during the humanization process.

To conclude, we propose that a mixing and matching of techniques could be used in future protocols to take into consideration all the different aspects of humanization that we consider above, with the ultimate goal of engineering fully human mAbs. Such an attempt has recently been reported by Bernett *et al.* (2010). Starting from the source murine variable regions of three currently marketed mAbs targeting CD25 (daclizumab), VEGF (bevacizumab), and tumor necrosis factor alpha (TNF-α) (infliximab), they generated "fully human" antibodies. All three engineered mAbs had levels of sequence humanness comparable to other fully human mAbs and yet maintained antigen binding and *in vitro* activity comparable to the three marketed drugs.

References

Baca M., Presta L.G., O'Connor S.J., Wells J.A. (1997). Antibody humanization using monovalent phage display. *J Biol Chem* **272**: 10678–84.

Bender N.K., Heilig C.E., Droll B., Wohlgemuth J., Armbruster F.P., Heilig B. (2007). Immunogenicity, efficacy and adverse events of adilimumab in RA patients. *Rheumatol Int* **27**: 269–74.

Bernett M.J., Karki S., Moore G.L., Leung I.W., Chen H., Pong E., Nguyen D.H., Jacinto J., Zalevsky J., Muchhal U.S., Desjarlais J.R., Lazar G.A. (2010). Engineering fully human monoclonal antibodies from murine variable regions. *J Mol Biol* **396**: 1474–90.

Boulianne G.L., Hozumi N., Shulman M.J. (1984). Production of functional chimaeric mouse/human antibody. *Nature* **312**: 643–6.

Chothia C., Lesk A.M. (1987). Canonical structures for the hypervariable regions of immunoglobulins. *J Mol Biol* **196**: 901–17.

Chothia C., Lesk A.M., Tramontano A., Levitt M., Smith-Gill S.J., Air G., Sheriff S., Padlan E.A., Davies D., Tulip W.R., Colman P.M., Spinelli S., Alzari P.M., Poljak R.J. (1989). Conformations of immunoglobulin hypervariable regions. *Nature* **342**: 877–83.

Chothia C., Novotny J., Bruccoleri R., Karplus M. (1985). Domain association in immunoglobulin molecules. The packing of variable domains. *J Mol Biol* **186**: 651–63.

Christensen P.A., Danielczyk A., Ravn P., Larsen M., Stahn R., Karsten U., Goletz S. (2009). Modifying antibody specificity by chain shuffling of V_H/ V_L between antibodies with related specificities. *Scand J Immunol* **69**: 1–10.

Dall'Acqua W.F., Damschroder M.M., Zhang J., Woods R.M., Widjaja L., Yu J., Wu H. (2005). Antibody humanization by framework shuffling. *Methods* **36**: 43–60.

Damschroder M.M., Widjaja L., Gill P.S., Krasnoperov V., Jiang W., Dall'Acqua W.F., Wu H. (2007). Framework shuffling of antibodies to reduce immunogenicity and manipulate functional and biophysical properties. *Mol Immunol* **44**: 3049–60.

De Groot A.S., Scott D.W. (2007). Immunogenicity of protein therapeutics. *Trends Immunol* **28**: 482–90.

Delagrave S., Catalan J., Sweet C., Drabik G., Henry A., Rees A., Monath T.P., Guirakhoo F. (1999). Effects of humanization by variable domain resurfacing on the antiviral activity of a single-chain antibody against respiratory syncytial virus. *Protein Eng* **12**: 357–62.

Dyer M.J., Hale G., Hayhoe F.G., Waldmann H. (1989). Effects of CAMPATH-1 antibodies in vivo in patients with lymphoid malignancies: influence of antibody isotype. *Blood* **73**: 1431–39.

Ewert S., Huber T., Honegger A., Plückthun A. (2003). Biophysical properties of human antibody variable domains. *J Mol Biol* **325**: 531–53.

Figni M., Martin F., Ferri R., Luison E., Ripamonti E., Zacchetti A., Mortarino M., Di Cioccio V., Maurizi G., Allegretti M., Canevari S. (2009). Conversion of murine antibodies to human antibodies and their optimization for ovarian cancer therapy targeted to the folate receptor. *Cancer Immunol Immunother* **58**: 531–46.

Figni M., Obici L., Mezzanzanica D., Griffiths A., Colnaghi M.I., Winter G., Canevari S. (1998). Panning phage antibody libraries on cells: isolation of human Fab fragments against ovarian carcinoma using guided selection. *Cancer Res* **58**: 991–6.

Flynn P., Luehrsen K., Balint R.F., Her J.-H., Beddington C.R., Yarranton G.T. (2005). Antibody specificity transfer using minimal essential binding determinants. *US Patent Application No.* 20050255552.

Fransson J., Teplyakov A., Raghunathan G., Chi E., Cordier W., Dinh T., Feng Y., Giles-Komar J., Gilliland G., Lollo B., Malia T.J., Nishioka W., Obmolova G., Zhao S., Zhao Y., Swanson R.V., Almagro J.C. (2010). Human framework adaptation of a mouse anti-human IL-13 antibody. *J Mol Biol* **398**: 214–31.

Garber E., Demarest S.J. (2007). A broad range of Fab stabilities within a host of therapeutic IgGs. *Biochem Biophys Res Commun* **355**: 751–7.

Goldstein G. (1987). Monoclonal antibody specificity: Orthoclone OKT3 T-cell blocker. *Nephron* **46**: 5–11.

Harding F.A., Stickler M.M., Razo J., DuBridge R.B. (2010). The immunogenicity of humanized and fully human antibodies. *MAbs* **2**: 256–65.

Honegger A. (2008). Engineering antibodies for stability and efficient folding. *Handbook Exp Pharmacol* **181**: 47–68.

Honegger A., Malebranche A.D., Röthlisberger D., Plückthun A. (2009). The influence of the framework core residues on the biophysical properties of immunoglobulin heavy chain variable domains. *Protein Eng Des Sel* **22**: 121–34.

Hu W.G., Chau D., Wu J., Jager S., Nagata L.P. (2007). Humanization and mammalian expression of a murine monoclonal antibody against Venezuelan equine encephalitis virus. *Vaccine* **25**: 3210–4.

Hwang W.Y., Almagro J.C., Buss T.N., Tan P., Foote J. (2005). Use of human germline genes in a CDR homology-based approach to antibody humanization. *Methods* **36**: 35–42.

Ionescu R.M., Vlasak J., Price C., Kirchmeier M. (2008). Contribution of variable domains to the stability of humanized IgG1 monoclonal antibodies. *J Pharm Sci* **97**: 1414–26.

Jespers L.S., Roberts A., Mahler S.M., Winter G., Hoogenboom H.R. (1994). Guiding the selection of human antibodies from phage display repertoires to a single epitope of an antigen. *Biotechnology (N Y)* **12**: 899–903.

Jones P.T., Dear P.H., Foote J., Neuberger M.S., Winter G. (1986). Replacing the complementarity-determining regions in a human antibody with those from a mouse. *Nature* **321**: 522–5.

Kabat E.A., Wu T.T., Perry H., Gottesman K., Foeller C. (1991). Sequences of Proteins of Immunological Interest, Fifth Edition. *NIH Publication No.* **91–3242**.

Kang A.S., Jones T.M., Burton D.R. (1991). Antibody redesign by chain shuffling from random combinatorial immunoglobulin libraries. *Proc Natl Acad Sci USA* **88**: 11120–3.

Khazaeli M.B., Conry R.M., LoBuglio A.F. (1994). Human immune response to monoclonal antibodies. *J Immunother Emphasis Tumor Immunol* **15**: 42–52.

Köhler G., Milstein C. (1975). Continuous cultures of fused cells secreting antibody of predefined specificity. *Nature* **256**: 495–7.

Kolbinger F., Saldanha J., Hardman N., Bendig M.M. (1993). Humanization of a mouse anti-human IgE antibody: a potential therapeutic for IgE-mediated allergies. *Protein Eng* **6**: 971–80.

Kügler M., Stein C., Schwenkert M., Saul D., Vockentanz L., Huber T., Wetzel S.K., Scholz O., Plückthun A., Honegger A., Fey G.H. (2009). Stabilization and humanization of a single-chain Fv antibody fragment specific for human lymphocyte antigen CD19 by designed point mutations and CDR-grafting onto a human framework. *Protein Eng Des Sel* **22**: 135–47.

Lazar G.A., Desjarlais J.R., Jacinto J., Karki S., Hammond P.W. (2007). A molecular immunology approach to antibody humanization and functional optimization. *Mol Immunol* **44**: 1986–98.

Maeda H., Matsushita S., Eda Y., Kimachi K., Tokiyoshi S., Bendig M.M. (1991). Construction of reshaped human antibodies with HIV-neutralizing activity. *Hum Antibodies Hybridomas* **2**: 124–34.

Morrison S.L., Johnson M.J., Herzenberg L.A., Oi V.T. (1984). Chimeric human antibody molecules: mouse antigen-binding domains with human constant region domains. *Proc Natl Acad Sci USA* **81**: 6851–5.

Nakamura K., Tanaka Y., Fujino I., Hirayama N., Shitara K., Hanai N. (2000). Dissection and optimization of immune effector functions of humanized anti-ganglioside GM2 monoclonal antibody. *Mol Immunol* **37**: 1035–46.

Neuberger M.S., Williams G.T., Mitchell E.B., Jouhal S.S., Flanagan J.G., Rabbitts T.H. (1985). A hapten-specific chimaeric IgE antibody with human physiological effector function. *Nature* **314**: 268–70.

Nishibori N., Horiuchi H., Furusawa S., Matsuda H. (2006). Humanization of chicken monoclonal antibody using phage-display system. *Mol Immunol* **43**: 634–42.

Novotny J., Handschumacher M., Haber E., Bruccoleri R.E., Carlson W.B., Fanning D.W., Smith J.A., Rose G.D. (1986). Antigenic determinants in proteins coincide with surface regions accessible to large probes (antibody domains). *Proc Natl Acad Sci USA* **83**: 226–30.

Ohlin M., Owman H., Mach M., Borrebaeck C.A. (1996). Light chain shuffling of a high affinity antibody results in a drift in epitope recognition. *Mol Immunol* **33**: 47–56.

Osbourn J., Groves M., Vaughan T. (2005). From rodent reagents to human therapeutics using antibody guided selection. *Methods* **36**: 61–8.

Padlan E.A. (1991). A possible procedure for reducing the immunogenicity of antibody variable domains while preserving their ligand-binding properties. *Mol Immunol* **28**: 489–98.

Padlan E.A. (1994). Anatomy of the antibody molecule. *Mol Immunol* **31**: 169–217.

Padlan E.A., Abergel C., Tipper J.P. (1995). Identification of specificity-determining residues in antibodies. *FASEB J* **9**: 133–9.

Pedersen J.T., Henry A.H., Searle S.J., Guild B.C., Roguska M., Rees A.R. (1994). Comparison of surface accessible residues in human and murine immunoglobulin Fv domains. Implication for humanization of murine antibodies. *J Mol Biol* **235**: 959–73.

Presta L.G., Chen H., O'Connor S.J., Chisholm V., Meng Y.G., Krummen L., Winkler M., Ferrara N. (1997). Humanization of an anti-vascular endothelial growth factor monoclonal antibody for the therapy of solid tumors and other disorders. *Cancer Res* **57**: 4593–9.

Presta L.G., Lahr S.J., Shields R.L., Porter J.P., Gorman C.M., Fendly B.M., Jardieu P.M. (1993). Humanization of an antibody directed against IgE. *J Immunol* **151**: 2623–32.

Queen C., Schneider W.P., Selick H.E., Payne P.W., Landolfi N.F., Duncan J.F., Avdalovic N.M., Levitt M., Junghans R.P., Waldmann T.A. (1989). A humanized antibody that binds to the interleukin 2 receptor. *Proc Natl Acad Sci USA* **86**: 10029–33.

Rader C., Cheresh D.A., Barbas C.F. 3rd. (1998). A phage display approach for rapid antibody humanization: designed combinatorial V gene libraries. *Proc Natl Acad Sci USA* **95**: 8910–5.

Rader C., Ritter G., Nathan S., Elia M., Gout I., Jungbluth A.A., Choen L.S., Welt S., Old L.J., Barbas C.F. 3rd. (2000). The rabbit antibody repertoire as a novel source for the generation of therapeutic human antibodies. *J Biol Chem* **275**: 13668–76.

Radstake T.R, Svenson M., Eijsbouts A.M., van den Hoogen F.H., Enevold C., van Riel P.L. *et al.* (2009). Formation of antibodies against infliximab and adalimumab strongly correlates with functional drug levels and clinical responses in rheumatoid arthritis. *Ann Rheum Dis* **68**: 1739–45.

Riechmann L., Clark M., Waldmann H., Winter G. (1988). Reshaping human antibodies for therapy. *Nature* **332**: 323–7.

Roguska M.A., Pedersen J.T., Henry A.H., Searle S.M., Roja C.M., Avery B., Hoffee M., Cook S., Lambert J.M., Blättler W.A., Rees A.R., Guild B.C. (1996). A comparison of two murine monoclonal antibodies humanized by CDR-grafting and variable domain resurfacing. *Protein Eng* **9**: 895–904.

Roguska M.A., Pedersen J.T., Keddy C.A., Henry A.H, Searle S.J., Lambert J.M., Goldmacher V.S., Blättler W.A., Rees A.R., Guild B.C. (1994). Humanization of murine monoclonal antibodies through variable domain resurfacing. *Proc Natl Acad Sci USA* **91**: 969–73.

Rosok M.J., Yelton D.E., Harris L.J., Bajorath J., Hellström K.E., Hellström I., Cruz G.A., Kristensson K., Lin H., Huse W.D., Glaser S.M. (1996). A combinatorial library strategy for the rapid humanization of anticarcinoma BR96 Fab. *J Biol Chem* **271**: 22611–8.

Sato K., Tsuchiya M., Saldanha J., Koishihara Y., Ohsugi Y., Kishimoto T., Bendig M.M. (1993). Reshaping a human antibody to inhibit the interleukin 6-dependent tumor cell growth. *Cancer Res* **53**: 851–6.

Sato K., Tsuchiya M., Saldanha J., Koishihara Y., Ohsugi Y., Kishimoto T., Bendig M.M. (1994). Humanization of a mouse anti-human interleukin-6 receptor antibody comparing two methods for selecting human framework regions. *Mol Immunol* **31**: 371–81.

Schroff R.W., Foon K.A., Beatty S.M., Oldham R.K., Morgan A.C. Jr. (1985). Human anti-murine immunoglobulin responses in patients receiving monoclonal antibody therapy. *Cancer Res* **45**: 879–85.

Shearman C.W., Pollock D., White G., Hehir K., Moore G.P., Kanzy E.J., Kurrle R. (1991). Construction, expression and characterization of humanized antibodies directed against the human alpha/beta T cell receptor. *J Immunol* **147**: 4366–73.

Shirai H., Kidera A., Nakamura H. (1996). Structural classification of CDR-H3 in antibodies. *FEBS Lett* **399**: 1–8.

Shirai H., Kidera A., Nakamura H. (1999). H3-rules: identification of CDR-H3 structures in antibodies. *FEBS Lett* **455**: 188–97.

Singer II, Kawka DW, DeMartino JA, Daugherty BL, Elliston KO, Alves K, Bush BL, Cameron PM, Cuca GC, Davies P, *et al.* (1993). Optimal humanization of 1B4, an anti-CD18 murine monoclonal antibody, is achieved by correct choice of human V-region framework sequences. *J Immunol* **150**: 2844–57.

Son J.H., Lee U.H., Lee J.J., Kwon B., Kwon B.S., Park J.W. (2004). Humanization of agonistic anti-human 4–1BB monoclonal antibody using a phage-displayed combinatorial library. *J Immunol Methods* **286**: 187–201.

Staelens S., Desmet J., Ngo T.H., Vauterin S., Pareyn I., Barbeaux P., Van Rompaey I., Stassen J.M., Deckmyn H., Vanhoorelbeke K. (2006). Humanization by variable domain resurfacing and grafting on a human IgG4, using a new approach for determination of non-human like surface accessible framework residues based on homology modelling of variable domains. *Mol Immunol* **43**: 1243–57.

Tamura M., Milenic D.E., Iwahashi M., Padlan E., Schlom J., Kashmiri S.V. (2000). Structural correlates of an anticarcinoma antibody: identification of specificity-determining residues (SDRs) and development of a minimally immunogenic antibody variant by retention of SDRs only. *J Immunol* **164**: 1432–41.

Tan P., Mitchell D.A., Buss T.N., Holmes M.A., Anasetti C., Foote J. (2002). "Superhumanized" antibodies: reduction of immunogenic potential by complementarity-determining region grafting with human germline sequences: application to an anti-CD28. *J Immunol* **169**: 1119–25.

West R.L., Zelinkova Z., Wlobink G.J., Kuipers E.J., Stokkers P.C., van der Woude C.J. (2008). Immunogenicity negatively influences the outcome of adalimumab treatment in Crohn's disease. *Aliment Pharmacol Ther* **28**: 1122–6.

Willuda J., Honegger A., Waibel R., Schubiger P.A., Stahel R., Zangemeister-Wittke U., Plückthun A. (1999). High thermal stability is essential for tumor targeting of antibody fragments: engineering of a humanized anti-epithelial glycoprotein-2 (epithelial cell adhesion molecule) single-chain Fv fragment. *Cancer Res* **59**: 5758–67.

Wu H., Nie Y., Huse W.D., Watkins J.D. (1999). Humanization of a murine monoclonal antibody by simultaneous optimization of framework and CDR residues. *J Mol Biol* **294**: 151–62.

Wu T.T., Kabat E.A. (1970). An analysis of the sequences of the variable regions of Bence Jones proteins and myeloma light chains and their implications for antibody complementarity. *J Exp Med* **132**: 211–50.

Yoon S.O., Lee T.S., Kim S.J., Jang M.H., Kang Y.J., Park J.H., Kim K.S., Lee H.S., Ryu C.J., Gonzales N.R., Kashmiri S.V., Lim S.M., Choi C.W., Hong H.J. (2006). Construction, affinity maturation, and biological characterization of an anti-tumor-associated glycoprotein-72 humanized antibody. *J Biol Chem* **281**: 6985–92.

2

Selection and Screening of Antibody Phage Display Libraries

David R. Buckler, Darren Schofield†, Daniel J. Sexton*, David Lowe† and Tristan J. Vaughan†*

2.1 Introduction

The pioneering work of Kohler and Milstein to develop hybridoma cell lines allowed the production of highly specific monoclonal antibodies (mAbs) against a wide range of antigens and haptens (Kohler and Milstein, 1975). However, antibodies derived from murine hybridomas proved to be immunogenic in humans, prompting the need for partial or fully human mAbs for therapeutic development. Three different approaches rapidly advanced starting in the 1980s to address this need: humanization of mAbs through protein engineering and recombinant DNA technologies (Presta, 2008); generation of transgenic mammals capable of producing human antibodies after immunization (Lonberg, 2005); and *in vitro* display and selection of human antibody fragment libraries (Hoogenboom, 2005). This chapter will review the construction and use of human antibody fragment libraries in Ff filamentous bacteriophage, the most widely used of the display systems.

Constructing antibody phage display libraries involves the fusion of human V-region sequences to bacteriophage coat protein genes. A single antibody sequence is displayed on the surface of each phage particle, and the displayed antibody fragment can be selected from a highly diverse library by affinity-based methods. Clonal isolates that show the desired binding or functional properties

* Dyax Corp., USA; † MedImmune, UK.

are then decoded by sequencing the relevant region of the engineered phage-derived genome. This overall progression from a highly diverse library of immunoglobulin sequences to a small number with desired binding properties mirrors clonal selection and expansion that constitutes the mechanistic hallmark of the vertebrate adaptive immune system itself.

Generation of appropriate combinatorial antibody diversity requires an in-depth understanding of sequence diversity provided by the natural immune repertoire as well as insights into how antibody sequence features encode specific and high-affinity antigen binding. This chapter will therefore begin with an overview of genetic aspects of antibody sequence generation and methods to extract this sequence diversity to produce highly diverse libraries in phage display vectors. Relevant aspects of phage function and biology will be covered, including design considerations in constructing phage display libraries and highlights of landmark human antibody libraries reported in the literature. Finally, an overview of selection methods and approaches to effective screening of antibody phage display libraries will be provided.

2.2 Antibody Gene Rearrangement and Variable Region Sequence Diversity

2.2.1 *Antibody architecture and germline gene segment recombination*

Antibodies are highly modular structures made up entirely of subdomains of the immunoglobulin fold family, with roughly 110 amino acid residues per subdomain. The heavy chain is composed of a variable subdomain (V_H) at the N-terminus, followed by a series of three to four constant subdomains depending on the antibody isotype: A (α), D (δ), E (ε), G (γ), or M (μ). The light chain, divided into two genetically defined subgroups, kappa and lambda, is also composed of a variable subdomain (V_L) followed by a constant domain (C_L).

An expressed and folded IgG antibody pairs two heavy and two light chains to form a Y-shaped structure with the combined "fragment antigen binding" (Fab) arms and the constant region (Fc) fragment with each exhibiting pseudo two-fold symmetry (Fig. 2.1) (Valentine and Green, 1967; Harris *et al.*, 1998; Harris *et al.*, 1998). V_H and V_L subdomains dimerize through a noncovalent interface forming the F_V fragment, and the C_H1 and C_L domains form an interface that is further stabilized by an inter-chain disulfide bond between the heavy and light chains. A flexible linker region extends from the C-terminus of the C_H1 domain and includes two half cysteines that form disulfide bridges between the heavy chains. The C_H2 domain includes sites for glycosylation, and the final C_H3 domain forms a tight interface between paired heavy chain partners in the assembled IgG (Fig. 2.1).

The F_V, C_H1, and C_L subdomains together form the Fab fragment, providing the upper arms of the "Y" (Fig. 2.1), and the heavy-chain portion of the Fab

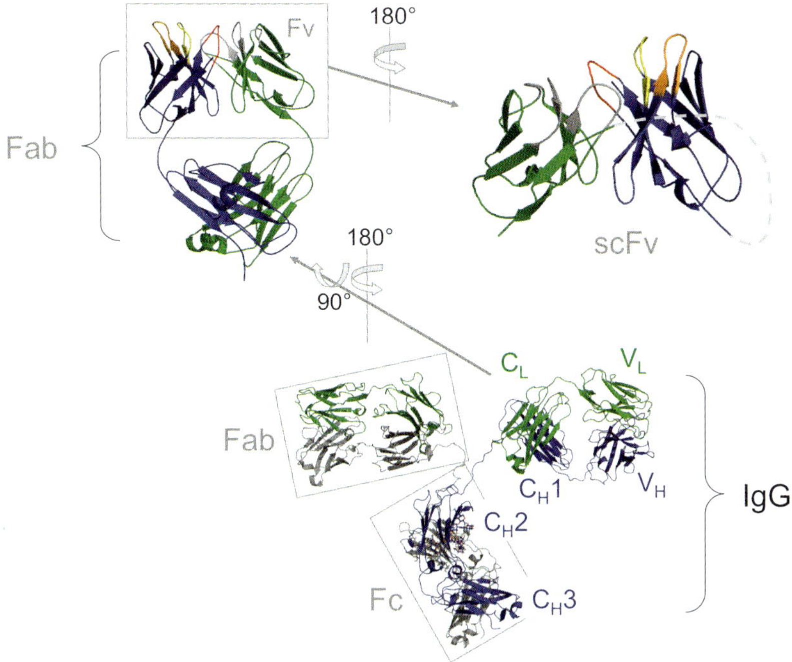

Figure 2.1 IgG and antibody fragment structures

The full IgG includes light (green) and heavy (grey and blue) chains paired through intra-chain disulfides and extensive inter-domain contacts. The hypervariable loops within the Fv fragment are rendered grey in the light chain V_L segment and red, yellow, and orange in the heavy chain V_H segment for CDR1, CDR2, and CDR3 loops, respectively. Fab and scFv fragment display has been most extensively used in antibody phage display.

(V_H-C_H1) is referred to as the Fd fragment. Both the F_V, typically engineered as a single-chain construct through a flexible linker (Bird *et al.*, 1988; Huston *et al.*, 1988), and Fab antibody fragments have been used extensively in display technologies because they provide the principal (and monovalent) antigen-binding unit of the antibody, focused at six regions of the hypervariable sequence defined as antigen-binding loops. The antibody constant regions, e.g. the Fc domain for IgG (Fig. 1), confer functional properties to the various antibody classes, such as recruitment of effector functions and half-life prolongation through Fc-receptor or complement C1q binding to the C_H2-C_H3 subdomains.

Genes encoding expressed antibodies in immature B cells result from combinatorial rearrangements between individual gene segments within the V, D, J, and C gene families organized on specific chromosomal loci in hematopoietic stem cells (HSCs). The number of gene segments in each family ranges from five to about fifty members for each of the V, D, J, and C gene segment families. The chromosomal rearrangements are enabled by recombination signal sequences (RSS) flanking the individual gene segments, and these germline segments in HSCs are organized into three distinct loci: IGH, IGK, and IGL (Lefranc and Lefranc, 2001).

The IGH locus contains the germline segments used for heavy chain recombination, which include V_H, D, J_H, and the heavy chain constant gene family segments. Single D and J_H gene segments first combine to generate a D-J_H intermediate. The D-J_H segment then combines with one of the V_H segments to generate the rearranged V_H-D-J_H variable region gene. Additional diversity arises from the possibility of random retention of small segments of the RSS at recombination joints between all rearranged gene segments (P-addition). Furthermore, introduction of up to 15 random nucleotides (N-addition), mediated by deoxynucleotidyl transferase, can occur at recombination joints involving D gene segments (Alt and Baltimore, 1982; Landau *et al.*, 1984). For the IGK and IGL light chain loci, only V, J, and C gene family members are present; therefore, D-segment diversity is not possible, providing the genetic rationale for the dominant role played by the heavy chain in antigen-binding specificity and affinity given its much greater potential to generate diversity.

In addition to the V, D, J (heavy chain), and V and J (light chain) gene rearrangements, described above, that form the F_V region of a mature antibody, one of the constant region segments (M, D, G, A, E) is joined to the 3′ terminus of the heavy chain V-region message to produce a full-length antibody. This segment is added depending on the stage of antibody maturation and on B cell function; it also determines functional class and whether the expressed polypeptide is directed to the B cell membrane or to extracellular secretion.

After completion of the chromosomal gene rearrangements at the pre-B cell stage, a signal is produced that precludes further variable-region gene rearrangement. This locks the newly formed variable-encoding gene arrangement into a fixed configuration. B cell clones expressing rearranged genes that do not produce functional antibodies undergo apoptosis. Additionally, rearrangements which produce binders to self-antigens are either clonally deleted or undergo a tolerance mechanism to prevent self-reactions (Burnet, 1976; Rajewsky, 1996).

At the immature pre-B cell stage, immunoglobulin gene rearrangement results in transcription of the IgM isotype, resulting in a surface-expressed antibody anchored to the outer cell membrane. In germinal centers and in the presence of antigen, the B cell antibody sequence can then undergo further maturation through somatic hypermutation (Berek and Milstein, 1988; Rajewsky, 1996) to increase binding affinity, mostly through point mutations (Tomlinson *et al.*, 1996; Wagner and Neuberger, 1996) but also through insertions and deletions (de Wildt *et al.*, 1999) within the antibody variable region. These mutations occur predominantly (Berek and Milstein, 1987; Tomlinson *et al.*, 1996) but not exclusively (Clark *et al.*, 2006) within the hypervariable regions. Resulting clonal mutants with increased antigen-binding affinity are amplified in preference to those with lower affinities and result in memory B cells or plasma cells expressing an affinity matured IgG antibody. In view of this sequence evolution from IgM to IgG, somatic hypermutations are typically found in antibody sequences from the IgG pool, whereas those from the IgM pool are close to the germline sequences.

The construction of highly effective recombinant antibody display libraries for the development of human therapeutics benefits from sequence considerations in several respects. To minimize immunogenicity, the antibody libraries should retain sequences as close to germline as possible, since these sequences implicitly evade a self-immune response. In addition, since sequence variability encoded by differences between members of the same gene segment family has evolved to promote binding specificity against a wide range of antigens, this information can provide valuable input to library design. The following section summarizes details of the immune gene families that help define the functional immune repertoire and therefore the extent of sequence diversity found in expressed antibodies. Structural insights on the antigen-binding sites that have been used to instruct library design will also be discussed.

2.2.2 *Variable region sequence diversity and structure guiding library design*

Sequence variability analyzed by Kabat and colleagues revealed three localized hypervariable regions within the N-terminal segments of both light chain (Wu and Kabat, 1970) and heavy chain (Kabat and Wu, 1971; Kabat *et al.*, 1991) antibody sequences that were shown experimentally to contact antigen (Amit *et al.*, 1986). These complementary determining regions (CDRs) are bounded by conserved framework (FR) sequences, and the linear arrangement of both heavy and light chain variable sequence segments is therefore denoted FR1-CDR1-FR2-CDR2-FR3-CDR3-FR4. CDRs located within heavy and light chains are denoted H-CDR and L-CDR, respectively. Details provided below highlight how specific sequence diversity within CDRs originates in germline gene segments and recombination events.

The multiple gene segments within each of the immunoglobulin gene families (V, D, J, C) arose from gene duplication during evolution. As a result, a large number of pseudogenes are also found within each of the immunoglobulin chromosomal loci that do not code for functional antibody sequences. Sequence data for mature antibodies have been accumulated from both direct sequencing of antibody proteins and from sequencing of the cDNA of rearranged antibody transcripts (Kabat *et al.*, 1991). When the sequences of mature antibodies were compared with those of germline gene segments, it became apparent that the CDR1 and CDR2 loops in both heavy and light chains are completely encoded by

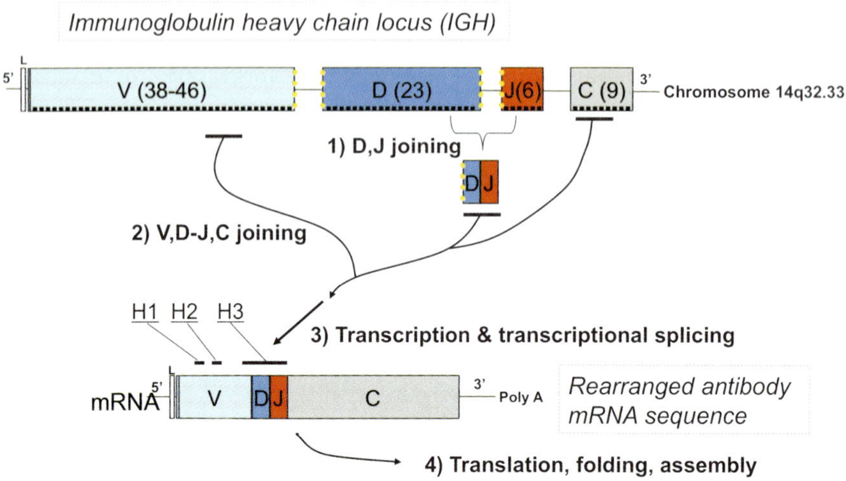

Figure 2.2 Gene segment rearrangements leading to the heavy chain transcript

The extensive sequence diversity of rearranged antibody sequences derives from sequential recombination of V, D, and J gene segments for the heavy chain. Rearrangements involving D-segments include single- or multiple-base additions or deletions at the joining regions. The H-CDR1 and H-CDR2 loops are encoded entirely within V-segments, whereas H-CDR3 loops arise from the combination of V-, D-, and J-segments. The light chain rearrangement mechanism is similar, except that no D-segment pool is available. Capture of "natural" immunoglobulin sequence diversity for phage display libraries generally involves PCR amplification of the rearranged antibody transcript from human immune cells.

the V family gene segments, while the CDR3 loops are encoded by the joining regions of the V, D, and J gene family segments for the heavy chain and at the V and J joining regions for the light chain (Fig. 2.2).

For IGH (located on chromosome 14; locus 14q32.33) the sources of variable region diversity are 38–46 functional V_H genes, 23 D genes, and 6 J_H genes; in addition, 9 C genes define the antibody isotype (Fig. 2.2) (Tomlinson *et al.*, 1992; Matsuda *et al.*, 1993; Cook and Tomlinson, 1995; Matsuda and Honjo, 1996; Matsuda *et al.*, 1998; Lefranc and Lefranc, 2001; Lefranc *et al.*, 2009). Hence, the diversity before somatic hypermutation for H-CDR1 and H-CDR2 is limited to the individual sequences encoded in the 38–46 V_H gene segments sequences. Mature human light chain genes originate from two distinct loci, kappa and lambda, each containing V and J (but no D) genes. The kappa locus, IGK, occurs on chromosome 2 (2p.11.2) and includes 31–35 functional V-kappa, 5 J-kappa genes, and a single C-kappa gene (Hieter *et al.*, 1980; Hieter *et al.*, 1982; Huber *et al.*, 1993; Schable and Zachau, 1993; Zachau, 1993; Cox *et al.*, 1994; Frippiat *et al.*,

1995; Tomlinson *et al.*, 1995; Williams *et al.*, 1996; Barbie and Lefranc, 1998; Lefranc *et al.*, 2009). The lambda locus, IGL, occurs on chromosome 22 (22q.11.2) and includes 29–33 functional V-lambda, 4–5 J-lambda, and 4–5 C-lambda genes (Frippiat *et al.*, 1995; Williams *et al.*, 1996; Kawasaki *et al.*, 1997; Pallares *et al.*, 1998; Scaviner *et al.*, 1999; Lefranc *et al.*, 2009). Hence, initial diversity for L-CDR1 and L-CDR2 originates from the approximately 60 light chain V family segments.

From these germline gene segment components alone, rearrangements of V_H, D, and J_H ($46 \times 23 \times 6 = 6{,}348$) and V_L and J_L ($35 \times 5 + 33 \times 5 = 340$) can produce about $6{,}348 \times 340$ combinations, i.e. $>10^6$ heavy/light gene segment combinations. Insertions and deletions around D-genes; joining mechanisms at V, D, and J junctions; and somatic hypermutation produce vast, additional diversity. The theoretical potential sequence diversity for antibody V-regions therefore greatly exceeds even the $\sim 10^{12}$ total B lymphocytes in the human body (Griffiths *et al.*, 1994). The nature of diversity generation for H-CDR3, in particular, results in complicated sequence patterns relating to D-segments (Abergel and Claverie, 1991) and widely-ranging lengths, from less than 3 to greater than 30 residues, with average length about 13 (Wu *et al.*, 1993), making these regions difficult to mimic synthetically in generating recombinant antibody libraries (Zemlin *et al.*, 2003). On the other hand, the limited diversity for the CDR1 and CDR2, originating from germline V_H and V_L gene segments, suggests that evolutionary pressures have selected a set of sequences with great potential for supporting binding to broad ranges of antigens.

As an introduction to a brief discussion on CDR sequences and structures, it should be noted that different antibody numbering schemes have been devised (Martin, 2007), and the Kabat numbering scheme is adopted here. It was recognized from early structural and sequence data that CDR loops (except H-CDR3) adopt a limited set of canonical conformations (Chothia and Lesk, 1987). For H-CDR1 (Chothia *et al.*, 1992), three major canonical structures were identified that could be correlated with overall loop size and residue type at specific loop positions. Canonical structure 1 was found to be the predominant structure for germline sequences. This structure includes an invariant Gly at position 26 to allow a backbone turn at that position to initiate the loop. In addition, loop position 29 was shown to be hydrophobic, forming buried contacts with positions 24 and 34, at which conserved, hydrophobic residues are also found. Two other canonical structures were observed for sequences with one or two additional residue insertions within the loop. Additional analyses of crystal structures indicate that H-CDR1 residues 33, 30, 31, and 35 often form antigen contacts (MacCallum *et al.*, 1996). Alignments of 3,500 sequences from the Kabat database showed that positions 29 and 32 are most often conserved, hydrophobic, and buried, while 28, 30, and 31, and 33 are variable and solvent-exposed (Sidhu *et al.*, 2004).

For H-CDR2, four canonical structures were observed (Chothia *et al.*, 1992), with loop sizes of five, six (for two canonical structures), or eight residues. For all canonical forms, position 51 was highly invariant as hydrophobic and buried while positions 52, 56, and 60 were highly variable and appeared to be surface-exposed in protein structures. Sequence analysis from 3,500 sequences from the Kabat database for H-CDR2 showed that positions 50, 52, 53, 54, 56, and 58 are variable, hydrophilic, and surface-exposed, while position 55 prefers glycine to allow a β-turn conformation (Sidhu *et al.*, 2004).

Similar analyses of canonical structures were completed for CDR loops of V-kappa (Tomlinson *et al.*, 1995) and V-lambda (Chothia and Lesk, 1987; Wu and Cygler, 1993; Ignatovich *et al.*, 1997), and results for V_H, V-kappa, and V-lambda were compiled (Al-Lazikani *et al.*, 1997). From these studies six and four canonical structures were identified for CDR1 of V-kappa and V-lambda, respectively, while single canonical structures were found for CDR2 of both V-kappa and V-lambda.

Loop structure analysis of CDR3 is more complex, given more extensive combinatorial diversity, especially for H-CDR3. Exhaustive sequence analysis for 382 rearranged light chain CDR3 sequences (Knappik *et al.*, 2000) showed that the kappa CDR3 (residues 91–96 from Kabat) most frequently exhibits a structural loop length of eight residues. Structural and sequence analysis showed antigen contact and sequence variation occurs at positions 91–94 and 96, whereas residue usage is highly conserved at positions 90 (Gln), 95 (Pro), and 89 (Gln). A similar analysis for lambda CDR3 (residues 89–96 from Kabat) for 147 lambda light chain sequences (Knappik *et al.*, 2000) showed over 92% with loop length of 8–10 residues. Sequence variability was observed for residues 93–96, and structural data showed frequent antigen contacts for positions 91, 94, and 96, and highly conserved residues were found at positions 89 (Gln) and 90 (Ser) (Knappik *et al.*, 2000).

2.2.3 *Amplification of variable region sequence repertoires*

The elucidation of sequences of the germline immunoglobulin loci as well as rearranged antibody sequences has provided the basis for amplifying V-gene repertoires for generating recombinant antibody display libraries from either chromosomal DNA (germline) or cDNA (rearranged) antibody sequences. As revealed from chromosomal mapping, the overall configuration for germline IGH gene segments includes a 5′ leader exon interrupted by an intron, the V_H exon, a conserved heptamer recombination signal, a 23 bp spacer region, and a conserved nonamer recombination signal (Lefranc and Lefranc, 2001). Winter's lab used this sequence information to design V_H-flanking primers that were family-specific against both leader sequence regions and the conserved nonamer and J-region sequences (Tomlinson *et al.*, 1992). This allowed PCR amplification and DNA sequencing of an exhaustive set of germline V_H family gene segments from a single individual, providing a nearly complete database of germline V_H segments. Based on sequence similarity of approximately 80%, the V_H gene segments have been grouped into six (Tomlinson *et al.*, 1992; Matsuda and Honjo, 1996) or seven (van Dijk *et al.*, 1993) subfamilies, VH1–VH7 (VH1 and VH7 are highly similar and sometimes grouped together). Similar families based on sequence comparison have also been defined through the Kabat database for rearranged antibody sequences.

For amplification by RT-PCR of rearranged immunoglobulin transcripts for cloning of V-region repertoires, the earliest approaches relied on the Kabat sequence database to construct oligonucleotide primers directed to conserved sequence regions. Early V_H amplification examples used mixtures of hybridoma clones as the source of RNA, and mixed primers were designed to anneal to either the N-terminal conserved regions in FR1 (Orlandi *et al.*, 1989) or the leader sequences preceding FR1 (Larrick *et al.*, 1989) for upstream priming. For the downstream regions, reverse primers were directed to the J-gene encoded regions at the CDR3/FR4 junction (Orlandi *et al.*, 1989) or to highly conserved sequences in the constant domain (Larrick *et al.*, 1989). It was soon after demonstrated that a similar approach could be successfully applied when the RNA was obtained from immunized mouse spleen, following some protocol modifications (Sastry *et al.*, 1989; Ward *et al.*, 1989). Using FR1/J-gene primer pairs, Huse and co-workers amplified both heavy and light chain variable region sequences from the RNA of immunized mouse spleen and randomly combined these in a lambda phage vector for expression of Fab in *Escherichia coli* (Huse *et al.*, 1989).

The initial advances for repertoire amplification described above were extended more broadly to human V-genes for the construction of antibody phage display libraries. Winter's lab (Marks *et al.*, 1991) completed a systematic analysis of 66 human V_H and 42 V-kappa sequences available from Kabat and other public databases after grouping by family to generate a series of family-specific primers. The upstream primers targeted the 5′ region of the V exon (FR1), while the downstream primers targeted family-specific regions of the J/C junction, and these could be used to amplify all known families of V_H and V-kappa human genes from human peripheral blood lymphocytes (PBLs) (Marks *et al.*, 1991). Similarly, Burton's lab used Kabat sequences to design V_H and V-kappa family primers targeting the 5′ end of the V exon together with reverse primers annealing to the hinge region (between the C_H1 and C_H2 constant domains of IgM or IgG) or to the 3′ end of C-kappa for heavy and light chain amplification, respectively (Persson *et al.*, 1991). Hoogenboom's lab later reported an updated set of primers based on available sequence data from V-Base (see Section 2.3.2) with upstream primers directed to the 5′ terminus of V-segments and downstream primers to IgM, kappa, or lambda constant regions (de Haard *et al.*, 1999). Primer sequences have also been refined by direct analysis of amino acid sequences from rearranged V-genes

(Welschof *et al.*, 1995). Additional exhaustive V-region primer sets were designed by *in silico* screening of primer sequences against all V-Base antibody sequences and then validated on cDNA from human PBLs or human cord blood lymphocytes (Sblattero and Bradbury, 1998), and optimization of amplification conditions has been reported to make best use of these primers (Loset *et al.*, 2005). Primers have also been reported for amplification of all six individual CDR regions of both V_H and V_L, five primer pairs for each CDR providing a broad distribution of naturally occurring CDR sequences (Soderlind *et al.*, 2000).

To increase specificity in amplification of V-gene repertoires, a first cycle of amplification uses gene-segment specific primers; amplification products are then typically purified and re-amplified with primers appended with cloning restriction sites (Marks *et al.*, 1991; Sheets *et al.*, 1998). Amplification of V-region genes using exclusively constant region primers has been demonstrated using circularized cDNA (Sassano *et al.*, 1994) or biotinylated capture primer (Heinrichs *et al.*, 1995) approaches, thereby eliminating V-region sequence "blind spots" or bias that occurs using primers specific for V-gene segments. Rapid amplification of cDNA ends (RACE) has also been used to amplify human antibody repertoires (Ruberti *et al.*, 1994; Doenecke *et al.*, 1997), as well as RACE followed by strand-directed restriction cleavage (Schoonbroodt *et al.*, 2005).

2.2.4 *Antibody sequence databases*

Compiled antibody sequence data originally enabled the prediction of hypervariable regions as antigen-binding loops and later documented the mapping of each of the immunoglobulin loci. Antibody sequence databases continue to hold tremendous value for the design, construction, and analysis of antibody fragments used in phage display libraries. The IMGT database (Lefranc *et al.*, 2009) contains both germline and rearranged gene sequences. These are provided on a publicly accessible website (http://imgt.cines.fr), and annual reviews from the LeFranc group summarize features and updates of the database and website (Lefranc *et al.*, 2009). For example, the IMGT/V-Quest search tool allows rapid assignment of query sequences to germline segments, as well as detailed annotations of changes from germline sequences (Brochet *et al.*, 2008). An extensive set of primer management tools for amplifying immunoglobulin sequences is also provided.

A second important resource for accessing germline sequence data is V-Base (http://vbase.mrc-cpe.cam.ac.uk/), developed by Tomlinson and co-workers in the Winter Lab (Tomlinson *et al.*, 1992). This site also provides primer sequences for amplifying V-region genes from genomic DNA by targeting conserved sequence regions.

The Kabat database remains an important resource (www.kabatdatabase.com) but is only available through commercial license. Finally, Andrew Martin's website (http://www.bioinf.org.uk/abs/) provides a wide range of functionality, including access to Kabat-derived sequences. An informative review was also published from the Martin Lab listing various bioinformatics tools, including additional sequence databases and analysis tools (Martin, 2007).

2.3 Antibody Display Libraries in Ff Bacteriophage

Extensive knowledge of the immunoglobulin sequence repertoire provides one set of requirements for constructing recombinant human antibody systems completely *ex vivo*. Another critical component is an appropriate display and selection system that allows individual antibody library members to be selected through antigen-binding interactions, recovered as clonal isolates, and sequenced to decode the identity of the binder. Various display formats that share these overall features have been developed, including yeast display (Boder and Wittrup, 1997; Feldhaus and Siegel, 2004), ribosome and mRNA display (Lipovsek and Pluckthun, 2004; Groves and Osbourn, 2005), and bacterial display (Lipovsek and Pluckthun, 2004; Groves and Osbourn, 2005; Mazor *et al.*, 2007; Mazor *et al.*, 2009). However, display and selection of antibody libraries in Ff phage has been the most widely used of the different formats. This can be explained by Ff's robust nature as a molecular engineering system, but it is likely also because of the fortuitous timing of its maturation as a practical library screening technology at the same time that antibody repertoire amplification and cloning techniques became available. The following sections provide a brief outline of phage biology and practical considerations for generating phage display antibody libraries.

2.3.1 *Phage biology and early use as a display system*

Ff is a filamentous single-stranded DNA (ssDNA) bacteriophage that infects bacteria harboring the F-conjugative pilus, carried on the F episome. Ff's small size, non-lytic nature, and ease of propagation have made it a versatile laboratory vector for cloning and for generating cDNA libraries (Geider, 1986). Its single-stranded form is amenable to rapid and efficient sequencing using Sanger dye terminator methods (Messing and Vieira, 1982; Vieira and Messing, 1987). Excellent reviews and laboratory manuals describing Ff bacteriophage in library construction and display are available (Lerner *et al.*, 1991; Winter and Milstein, 1991; Winter *et al.*, 1994; Kay and Paul, 1996; Barbas, 2001; Hoogenboom, 2005; Kehoe and Kay, 2005; Lerner, 2006; Hust, 2007; Thie *et al.*, 2008).

The bacteriophage species Ff (family Inoviridae, genus Inovirus) includes strains fd, f1, and M13 (Marvin and Hohn, 1969). These three strains share the same life cycle and genome structure and differ in overall genomic sequence by less than about 3% (van Wezenbeek *et al.*, 1980; Beck and Zink, 1981). The circular ssDNA genome of about 6,400 nucleotides encodes 11 viral genes, including five different capsid proteins. The major capsid protein gene, gVIII, encodes a 50-residue polypeptide that assembles in a shingled arrangement along the sides of the elongated phage particle, with about 2,700 copies per phage. The particle dimensions are approximately 900 nm × 6 nm, comparable in aspect ratio to a 1.5 cm length of human hair. In addition to the major pVIII coat protein, two pairs of proteins at five copies each form the tips of the elongated phage particle: pIII and pVI at one end, and pVII and pIX at the opposite end. Additional Ff genes encode proteins for DNA replication (gII, gX), binding of ssDNA (gV) and assembly (gIV, gI, and gXI) (Barbas, 2001).

The first stage of the virus life cycle involves infection of a pilus-expressing (F-positive) bacterial host, initiated by interaction of the N-terminal domain of the phage capsid protein pIII with the bacterial conjugative pilus (Armstrong *et al.*, 1981; Crissman and Smith, 1984). The virus is drawn toward the cytoplasmic membrane after its contact with the pilus, and ssDNA is released into the bacterial cytosol as the viral coat proteins are subsumed into the host's cytoplasmic membrane (Fig. 2.3). The newly introduced viral DNA is rapidly replicated to form double-stranded DNA (dsDNA) suitable for gene transcription/translation by the bacterial host. During an initial exponential burst phase, viral replicative enzymes pII and pX arising from the initially synthesized dsDNA first work in concert with the bacterial replicative machinery to generate double-stranded copies of the viral genome. This stage involves rolling circle amplification and is dependent on nicking of one of the duplex DNA strands by the pII protein. The resulting dsDNA duplexes, termed replicative form (RF) DNA, reach about 250 copies per cell (Marvin and Hohn, 1969) and can be isolated in good yield. This provides a convenient source of dsDNA for standard molecular cloning procedures. In a second stage, the gV ssDNA binding protein surpasses a threshold level and diverts the ssDNA from RF propagation by forming an extensive pV-ssDNA complex. As the capsid proteins pVIII, pIII, pVI, pVII, and pIX are formed, they are anchored to the bacterial cytoplasmic membrane. For pIII the N-terminal region is exposed to the oxidizing environment of the periplasmic space, while a C-terminal hydrophobic region remains embedded within the cytoplasmic membrane. All capsid proteins remain anchored to the cytoplasmic membrane through similar hydrophobic regions until the pV/ssDNA complex is extruded from the cytosol through a pore formed by the junction between pXI/pI anchored at the cytoplasmic membrane and a homo-multimeric complex of pIV assembled within the outer membrane as a cylindrical pore. During the extrusion process, the pVIII major coat protein displaces pV from the nascent viral genome, and terminating capsid proteins pIX/VII are added at the leading tip and pIII/pVI at the trailing terminus of the phage particle for final assembly and secretion through the pIV pore. The viral titer typically reaches $\sim 10^{12}$ particles per mL in laboratory culture conditions. The bacterial host cells exhibit retarded growth rates of about 50% through

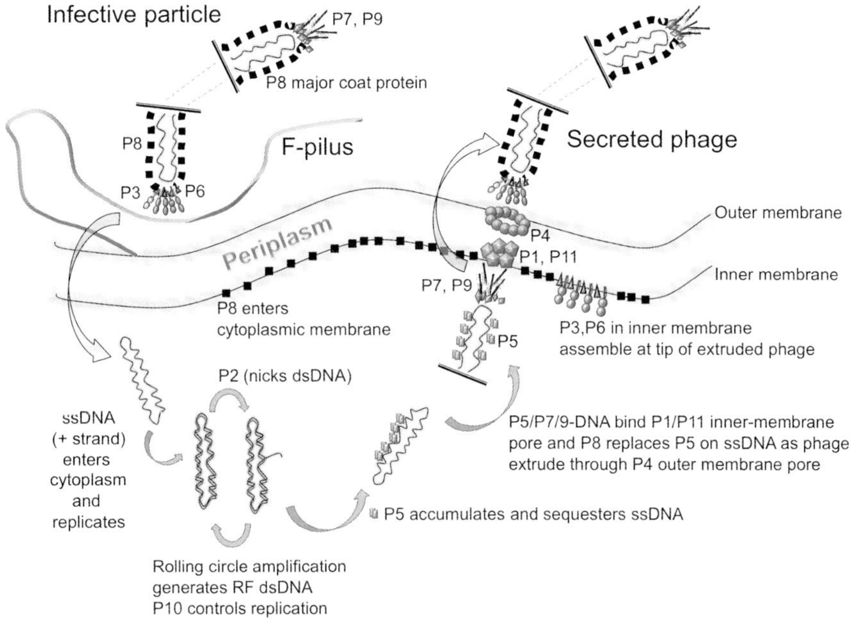

Figure 2.3 Ff bacteriophage life cycle

The infective Ff particle enters the host after binding of the P3 capsid protein to the bacterial conjugative pilus. Once inside the cytosol, host enzymes replicate the circular, single-stranded bacteriophage DNA, and an iterative DNA amplification involving rolling circle amplification results in high levels of dsDNA (replicative form). Accumulation of Ff DNA-binding protein P5 sequesters single-strand progeny DNA, and extrusion of phage through the P1/P11-derived pore results in replacement of P5 with major coat protein P8.

infection cycles, but otherwise remain functional and capable of further growth (Barbas, 2001).

A breakthrough in display technologies was reported in 1985 when Smith introduced an EcoRI epitope fragment into the Fd phage genome (Smith, 1985). The fragment was introduced between the N- and C-terminal domains of the pIII terminal capsid protein and, therefore, displayed on the surface of the phage particle, allowing affinity selection using an anti-EcoRI antibody. In a test selection an enrichment of $>10^5$ over wild-type background phage was achieved (Smith, 1985). This opened the possibility of isolating rare fragments from $>10^9$ libraries using affinity selection with proteins or haptens. Further efforts through 1990 expanded the Ff bacteriophage display technology by demonstrating its utility in affinity selection of various display fragments fused to gIII, and an optimized fusion position was found to be between the N-terminal leader sequence and the mature N-terminus. Some of these first reports in the literature included the display of beta-galctosidase (Parmley and Smith, 1988), bovine pancreatic trypsin inhibitor (BPTI) (Ladner and Gutterman, 1990), and combinatorial 6-residue (Cwirla *et al.*, 1990; Scott and Smith, 1990) and 15-residue (Devlin *et al.*, 1990) peptide libraries of $>10^7$ recombinants that were used to identify peptide sequences that bound to antibodies or proteins. These works established the suitability of expressing reasonably large fragments (e.g. 112 amino acid fragment of beta-galactosidase) (Parmley and Smith, 1988) at the N-terminus of the surface pIII viral capsid. The use of the streptavidin/biotin interaction during the affinity separation was also described first in these early studies (Parmley and Smith, 1988), and provided a versatile selection tool that since has been used extensively.

The display of antibody fragments on Ff bacteriophage followed from the discovery that both Fab and Fv fragments could be functionally expressed in *E. coli* by directing the translated polypeptides to the oxidizing environment of the periplasm using specific signal peptides (Better *et al.*, 1988; Skerra and Pluckthun, 1988). Once in the periplasm, the fragments could undergo proper refolding, including the formation of native disulfide bonds, and pairing of separately expressed V_H and V_L fragments through the Fv interface. Using a bacteriophage lambda cloning system, it was shown that V_H and V_L fragments from a mouse immune repertoire engineered with the appropriate signal peptide could be randomly combined in the periplasm of *E. coli*, providing the first published example of a random, combinatorial antibody fragment library (Huse *et al.*, 1989). Unfortunately, brute force screening would be required for a phage lambda library since there was no display/selection component in this system. However, the successful use of phage to display and select antibody fragments was reported soon thereafter (Ladner and Gutterman, 1990; McCafferty *et al.*, 1990). The approach adopted by the Winter lab (McCafferty *et al.*, 1990) involved insertion of an anti-lysozyme scFv into a tetracycline-resistant Fd clone, Fd-tet (Zacher *et al.*, 1980). It was demonstrated that the desired binder could be selected from a high background ($>10^4$-fold excess) of parental phage by two rounds of affinity purification using a lysozyme-sepharose column. As the feasibility of antibody fragment display on phage was established with these first reports, approaches rapidly developed in vector and library design to expand its use.

Theoretical considerations indicated that the potential affinity of recovered binders should increase as the library size increases (Perelson and Oster, 1979). Therefore, generation of large, highly diverse libraries was a primary objective in the construction of early libraries. Phagemid vectors were adopted for construction of phage display libraries since transformation efficiencies are much greater using phagemid DNA than for the substantially larger complete phage genome. Phagemid vectors are similar to standard cloning vectors but include a phage intergenic region (Cleary and Ray, 1980), which includes both a viral origin of replication and a packaging signal that allow the DNA to be replicated and packaged using viral processing machinery provided by a helper phage. Helper phage (e.g. R408, VCM13, and M13K07) supply the full complement of Ff genes and are typically engineered with an antibiotic resistance marker as well as with a defective packaging signal that results in selective packaging of the phagemid DNA over helper phage (Russel *et al.*, 1986). The use of phagemid vectors in phage display was first reported in the literature for proteins (Bass *et al.*, 1990; Ladner and Gutterman, 1990), and for Fab (Barbas *et al.*, 1991; Hoogenboom *et al.*, 1991) and scFv (Breitling *et al.*, 1991) antibody fragment libraries.

Since the feasibility of displaying antibody fragments in both full phage and phagemid systems was established, many features have been examined to optimize the system through control of display levels and display format. These include choice of capsid protein for optimal display, antibody fragment format (scFv *vs.* Fab), and various aspects of the expression vector.

2.3.2 *Variation of display format and control of display level*

2.3.2.1 *Coat protein display position*

The earliest reports in the literature of antibody phage display used fusion to gIII (Ladner and Gutterman, 1990; McCafferty *et al.*, 1990), as for the peptide and small protein libraries. This site offers several advantages compared to other coat proteins and has been the most widely used. The pIII protein, along with pVI, is added at the tip of the extruding phage as a final event and, therefore, minimally impedes particle assembly. Intra-host expression of pIII, however, causes immunity to hyperinfection by helper phage and therefore fusion of display fragments to full-length gIII requires growth under conditions that suppress its expression until helper infection. Alternatively, fusion to a truncated C-terminal domain of pIII precludes the problem of immunity. Fusion to coat proteins other than pIII appears more sensitive to fusion size and position. For example, systems in which all copies of pVIII include the display fusion are limited to display fragments of no more than about six residues (Iannolo *et al.*, 1995). If a hybrid system is used in which a source of wild-type pVIII is added in addition to the fusion, larger fragments including Fabs can be displayed (Kang *et al.*, 1991), but optimization is sensitive to many factors including structural effects on processing of the leader peptide (Malik *et al.*, 1996).

Display by C-terminal fusion to pVI has also been reported for medium-sized protein fragments. A Dog hookworm cDNA library displayed on Ff pVI and selected with serine proteases trypsin and factor Xa yielded Kunitz inhibitors (Jespers *et al.*, 1995). In another example, pVI display was demonstrated using the CH3 domain of human IgG and a bacterial enzyme PhoA, and this same system was then used for display of a colorectal tumor cDNA library that was screened with human antibodies (Hufton *et al.*, 1999). Coat proteins pVII and pIX, present at the opposite particle end as pIII/pVI, have also been used for Ff phage display. Initially V_H and V_L fragments were separately fused to the genes for pVII and pIX, respectively, and the antibody fragments were displayed as functionally paired Fv (Gao *et al.*, 1999). Later, a large (4×10^9) scFv library was displayed as a single fusion to pIX and yielded high affinity (including sub-nanomolar) binders against a broad range of targets (Gao *et al.*, 2002).

2.3.2.2 *Display format, scFv vs. Fab*

The scFv and Fab formats (Fig. 1) have been the most extensively used antibody fragments for phage library display. It was recognized early that even isolated V_H domains or dAbs (Ward *et al.*, 1989) can bind antigen, but the assembled V_H and V_L, i.e. the complete Fv fragment, has typically been used to obtain high affinity and specific binding. To stabilize an expressed Fv, flexible linkers such as ($[Gly]_4$-$Ser)_3$ (Huston *et al.*, 1988) or linkers chosen from a database of known polypeptide structures to meet spatial requirements (Bird *et al.*, 1988) were used to link the C- and N-termini, respectively, either from V_H to V_L (Huston *et al.*, 1988) or from V_L to V_H (Bird *et al.*, 1988). Such strategies were then incorporated in the first demonstrations of phage antibody display (McCafferty *et al.*, 1990; Breitling *et al.*, 1991; Clackson *et al.*, 1991; Marks *et al.*, 1991) and continued for the construction of large libraries. The single-chain format offers the benefit of small size and simplicity of design as a single polypeptide chain, important for ease of cloning and expression and numerous applications including imaging and fusion toxins (Huston *et al.*, 1993), and intracellular delivery (Lo *et al.*, 2008). However, the single chain presents practical caveats such as propensity to multimerize, forming either bivalent diabodies (Holliger *et al.*, 1993) or higher-level aggregates (Griffiths *et al.*, 1993), since hydrophobic residues that are otherwise buried at the variable/constant interface are exposed to solvent (Nieba *et al.*, 1997). Aggregation of scFvs is sensitive to both linker length (Holliger *et al.*, 1993) and residue composition outside of the linker regions (Ewert *et al.*, 2003). The aggregation of scFv not only reduces the fraction of properly folded and functional Fv in the library but also causes artifacts in affinity measurements (Townsend *et al.*, 2006).

The Fab format, in contrast, offers a more natural presentation of the binding domain since the variable/constant interfaces for both heavy and light chains are intact. However, its expression in bacteria requires assembly of separate V_H-C_H1 and V_L-C_L chains in the periplasm. The phage lambda library referenced above (Huse *et al.*, 1989) demonstrated the feasibility of Fab assembly for a diverse combination of heavy and light chains in the periplasm of *E. coli.* Hoogenboom and colleagues then reported on a phagemid vector pHEN1 (Hoogenboom *et al.*, 1991) in which a model Fab was functionally displayed (after rescue with VCSM13) using a single *lac* promoter and the *pelB* leader sequence for both chains with the light chain in the downstream position and fused to full-length gIII. Effective display was also demonstrated in the alternative orientation with the heavy chain fused to gIII in pHEN1 and the phagemid rescued with an fd-tet into which the light chain was inserted downstream of a gIII leader sequence (Hoogenboom *et al.*, 1991). In parallel to these reports, the Lerner and Benkovic labs reported advances with phagemid vectors in which Fab was fused to a truncated C-terminal domain of pIII that lacked the infective N-terminus (Barbas *et al.*, 1991); or to pVIII, in which wild-type pVIII was also provided from the helper phage so that only a fraction of pVIII major coat protein included the fusion (Kang *et al.*, 1991). Thus, the Fab format has been adopted in many of the large, modern antibody libraries (Griffiths *et al.*, 1994; de Haard *et al.*, 1999; Hoet *et al.*, 2005; Kim *et al.*, 2008; Rothe *et al.*, 2008).

2.3.2.3 *Signal peptide and vector design*

For targeting antibody fragments to the periplasm, most display systems use signal sequences targeting Sec, a post-translational translocation pathway. Sec signal peptides have been used from *Erwinia carotovora* pectate lyase (pelB) (Better *et al.*, 1988) or from *E. coli* OmpA/PhoA (Skerra and Pluckthun, 1988) for phagemid vectors (Huse *et al.*, 1989; Barbas *et al.*, 1991; Breitling *et al.*, 1991; Hoogenboom *et al.*, 1991; Marks *et al.*, 1991; Knappik *et al.*, 2000; Rothe *et al.*, 2008), and the viral gIII signal peptide for both phagemid (de Haard *et al.*, 1999) and phage vectors (McCafferty *et al.*, 1990; Clackson *et al.*, 1991). More recently, display has been tested using signal peptides from the signal recognition particle (SRP) translocation pathway (Steiner *et al.*, 2006), which is a co-translational pathway more amenable to fast-folding proteins that may exhibit poor translocation yields through the cytoplasmic membrane. Although Sec and SRP signal peptides showed similar results for display of scFv (Thie *et al.*, 2008), SRP has been demonstrated to provide dramatic advantages for fast-folding display proteins such as DARPins (Steiner *et al.*, 2008).

Typical display vector designs allow control of antibody fragment transcription by a promoter-operator, most often *lac* (Better *et al.*, 1988), inducible by lactose analogs such as IPTG and repressed by glucose. Translation requires the Shine-Dalgarno ribosomal binding sequence (RBS) immediately upstream of the signal-peptide/antibody fragment sequences. Fab display requires translation of heavy and light as two separate chains. This has been achieved through monocistronic transcription (separate promoter-operators upstream of both heavy and light chains); however, a bicistronic format (single promoter-operator resulting in mRNA carrying both light and heavy chain on a single message) is most commonly used in Fab display. For Fab vectors, gIII can be fused to the 3′ end of either the heavy or light chain, and the 5′ to 3′ orientation can be either heavy-light or light-heavy, giving four possibilities for chain orientation and gIII fusion position. Most commonly, the light chain is positioned most closely to the promoter-operator, upstream of the heavy-chain/gIII fusion (either full-length or C-terminal domain). For scFv, it can be noted from the tertiary structure of antibody Fv domains that the N-termini of both V_H and V_L extend from the "upper" or CDR-containing ends of the heterodimer, while the C-termini extend from the "lower" end of the Fv domain (Fig. 2.1). This allows insertion of the scFv linker either between the C-terminus (3′ end) of V_H and N-terminus (5′ end) of V_L, or *vice versa*. Although the V_L-linker-V_H orientation has been reported (Bird *et al.*, 1988; Knappik *et al.*, 2000; Sblattero and Bradbury, 2000), the V_H-linker-V_L orientation has been used nearly exclusively in library construction. For assembly of scFv libraries, overlap extension PCR was initially developed (Clackson *et al.*, 1991), which involves PCR amplification of heavy and light V-region genes which are then randomly combined to form scFv by overlap and amplification with oligonucleotides that include the linker region flanked by sequences complimentary to the termini of the amplified V-genes. Overlap annealing can involve a three-fragment strategy, in which a separate linker piece anneals via its termini to V_H and V_L segments, (Clackson *et al.*, 1991; Marks *et al.*, 1991) or a two-fragment strategy in which the variable segments each contain designed complementary sequences at 3′ and 5′ termini, respectively, that include within them the linker sequence (Vaughan *et al.*, 1996; Sheets *et al.*, 1998). scFv libraries have also been constructed by directed cloning of one chain segment into a sub-library containing the other chain (Little *et al.*, 1999).

2.3.2.4 *Control of functional display level and valency*

A portion of each antibody phage display library includes non-functional members resulting from capture of non-functional gene segments, PCR amplification errors, or sequence errors in synthetic oligonucleotides used in library construction. This can be quantified by sequencing or by immunoblotting isolates from the unselected libraries, and typically the non-functional fraction accounts for 20–40% of the library (Knappik *et al.*, 2000; Soderlind *et al.*, 2000; Desiderio *et al.*, 2001; Rauchenberger *et al.*, 2003). Various strategies have been devised to minimize the fraction of non-functional Fab or scFv at both the selection and library construction stages. These include protease-driven elution of bound phage after antigen selection and washing, which promotes selective release of functional over "sticky" phage harboring non-functional fusions (Johansen *et al.*, 1995; Ward *et al.*, 1996). In addition, the initial library can be passed through an affinity purification step to pre-select functional Fabs, e.g. using protein-L for a kappa-containing library (Loset *et al.*, 2005). Non-functional Fabs can also be removed during construction of sub-libraries by inclusion of β-lactamase or other selectable markers downstream from the display sequence, allowing only bacteria harboring fully translated Fab or scFv to propagate (Seehaus *et al.*, 1992; Zacchi *et al.*, 2003; Rothe *et al.*, 2008).

When antibody fragments are displayed on one of the Ff coat proteins, the valency of display is an important factor that affects the affinity of recovered binders. Multivalent display results in binding avidity that promotes the recovery of lower affinity binders (O'Connell *et al.*, 2002), while monovalent display yields higher affinity binders. Traditional phagemid libraries, when rescued with standard helper phage that include wild-type gIII (R408, VCSM13, M13K08), provide populations of packaged phagemid with mostly non-display, a small amount (0.1–5%) of monovalent display, and very small fractions with greater than monovalent display. In comparison to phagemid, phage antibody libraries have been engineered with display protein fused to full-length gIII (Clackson *et al.*, 1991; Griffiths *et al.*, 1994; O'Connell *et al.*, 2002); or, alternatively, fused to a truncated gIII with a second copy of full-length gIII present in the engineered phage genome (Hoet *et al.*, 2005). These two approaches led to multivalent and monovalent display, respectively. To provide multivalent display with phagemid systems, "hyperphage" or "ex-phage" systems were developed (Griffiths *et al.*, 1993; Rondot *et al.*, 2001; Baek *et al.*, 2002; Kramer *et al.*, 2003; Chasteen *et al.*, 2006; Soltes *et al.*, 2007), where gIII in the helper phage genome is either removed or truncated, and therefore all copies of gIII on the phagemid particle surface include the display fusion protein. An alternative strategy for providing bivalent display in the phagemid system is to include a dimerization domain, such as the GCN4 leucine zipper motif, or even a single unpaired cysteine between the expressed antibody chain and the gIII fusion (Lee *et al.*, 2004).

2.3.2.5 *Maximizing library size*

Since recovered affinities correlate with library size (more exactly, with the number of distinct functional antibody fragments displayed in the library) a major focus has been to generate the largest possible libraries. The most commonly used methods involve generation of desired heavy/light-chain diversity in phage or phagemid vectors followed by brute-force transformation by electroporation (Dower *et al.*, 1988). An alternative designed to achieve still larger library sizes involves infection instead of transformation and in-cell recombination of heavy and light chains by the bacteriophage P1 cre-loxP system (Waterhouse *et al.*, 1993). In this system, wild-type and mutant loxP recombination sequences flank heavy chain sequences in both a donor cloning vector (containing the source of diverse heavy-chain sequences) and an acceptor phage vector, which contains light chain diversity plus gIII fused to a "dummy" heavy chain sequence to be replaced by the donor. A sub-library of cells containing the donor vector diversity is first infected with both the acceptor phage and a P1 phage that provides the necessary Cre recombinase. Two cycles of recombination are then required to produce the rearranged acceptor light-chain sequence linked to donor heavy-chain sequence fused to gIII, plus additional recombination products. In a model study 71% of the resulting clones were found to contain the desired recombinants (Waterhouse *et al.*, 1993), and a Fab library of 6×10^{10} was constructed using this approach (Griffiths *et al.*, 1994). In a modification of this approach for scFv libraries, the V_H–gIII segment of a donor scFv vector is flanked by a wild-type and mutant loxP recombination sequences, and cells super-infected with donor phage undergo recombination, resulting in new combinations of V_H and V_L (Sblattero and Bradbury, 2000). Using this method, a 3×10^{11} scFv library was constructed (Sblattero and Bradbury, 2000).

2.3.2.6 *Assessment of diversity*

The dogma of naive antibody phage display is that, unlike antibodies generated by classical immunization routes or even from immune libraries using phage display, there should be essentially no immunodominant epitopes on any given antigen presented to the library. Therefore, theoretically, antibodies can be generated to any accessible part of the selection antigen. As a consequence of this, within a population of antigen-specific antibodies there will be a broad spectrum of antibody functionalities generated, from purely target binders (to both conformational and linear epitopes) to antibodies that exert a biological function (e.g. receptor agonism/antagonism, apoptosis, receptor internalization, etc.). An elegant demonstration of the breadth of specificities generated from naive antibody phage display was given in a report in which a naive antibody phage display library was interrogated for binders to the human TRAIL-R1 and TRAIL-R2 receptors (Dobson *et al.*, 2009). The selections generated 234 different scFvs specific for TRAIL-R1, 269 different scFvs specific for TRAIL-R2, and an additional 134 different scFvs that were cross-reactive for both receptors. A high-throughput surrogate apoptosis assay identified ten TRAIL-R1 specific scFv and six TRAIL-R2 specific scFv with agonistic activity *in vitro*. These scFv were all highly specific for either TRAIL-R1 or TRAIL-R2, potently inhibited tumor cell proliferation, and were antagonists of TRAIL binding.

Glanville and colleagues utilized the latest high-throughput long-read pyrosequencing and a novel application of profile Hidden Markov Model-based sequence analysis to give, to date, the most in-depth sequence analysis of the genetic diversity in a naive antibody phage display library pre- and post-selection on a panel of antigens (Glanville *et al.*, 2009). Analysis of a naive library by these methodologies estimated the combined library diversity of unique non-redundant paratopes to be approximately 3.5×10^{10}, compared to 3.1×10^{10} as determined by the traditional colony count of transformants. The breakdown in terms of heavy and light chain diversity based on CDR sequence was approximately 2.2×10^5 members for the heavy chain CDRs, and 1.6×10^5 for the light chain CDRs. From selections on 16 different antigens, over 17,000 leads were sequenced. The selected antibodies were derived from the majority of germlines available in the library. While some variation in germline frequency and chain pairings was observed, the overall distribution for the naive library and selected binders suggested that germline utilization was proportional to their availability. This was the first such study to truly shed light on the actual genetic diversity of a large library in a statistically significant way.

2.3.3 *Representative large non-immune libraries*

This review of key technical advances together with the most extensively described and referenced libraries will provide a historical perspective of the evolution of antibody phage display to its current status as a robust drug discovery platform. Landmark large ($>10^9$ recombinants) libraries are summarized in Table 2.1, with some brief comments describing key features of these in the section below.

Antibody fragment display on phage (McCafferty *et al.*, 1990) was rapidly followed by a report from the Winter lab describing a relatively small ($\sim 10^5$) scFv library generated by random pairing of V_H and V_L segments from the cDNA of immunized mouse spleen by overlap extension PCR (Clackson *et al.*, 1991). This "immune" library produced 10 nM binders, while a 10^6 library generated in parallel from non-immunized mice ("non-immune" library) failed to produce measurable binders, even after two selection rounds (Clackson *et al.*, 1991). These first reports raised the critical question of whether non-immune libraries could produce binders against any antigen and, if so, what size would be sufficient.

While the scFv format provided a minimal binding unit encompassing all six CDRs and readily amenable to combinatorial engineering, both the Winter and Lerner labs were simultaneously establishing display of Fab for library generation. Fab requires assembly of two separate chains in the periplasm, but more naturally represents a full-length antibody, as discussed in Section 3.2.2. Phagemid vectors for Fab display were developed that encode fusions either to the C-terminal domain of gIII (Barbas *et al.*, 1991) or to full-length gIII (Hoogenboom *et al.*, 1991), with a *lac* promoter-operator and pelB signal peptide for directing Fab expression to the periplasm. One of these systems (Hoogenboom *et al.*, 1991) also introduced an amber stop codon between Fab chain and gIII, allowing interconversion between phagemid display and soluble Fab expression by transformation into either suppressor or non-suppressor *E. coli* strains (e.g. HB2151), respectively.

Following demonstrations of successful scFv and Fab phage display, relatively small ($<10^8$) non-immune human antibody libraries were constructed that yielded binders against both non-human antigens (Marks *et al.*, 1991) and, importantly, self-antigens (Griffiths *et al.*, 1993), further reinforcing the technology as a powerful platform for discovery of human therapeutic antibody leads. With some key feasibility questions answered and basic cloning and construction strategies established, large ($>10^9$) non-immune libraries were next generated, and their results demonstrated that libraries exceeding 10^9 were sufficient for

Table 2.1 Representative non-immune human antibody phage display libraries

Reference	Laboratory	Format	Size ($\times 10^9$)	Diversity Type*	Diversity Source	VH cDNA Amplification Target	Strongest Reported Affinity
Griffiths *et al.*, 1994	Winter (MRC)	Fab/phage	65	N + S	Germline CDR1,2; synthetic CDR3 Cre-Lox recombination H3 length range 4–12	NA (amplified from pre-rearranged genomic DNA)	4 nM (N = 5)
Vaughan *et al.*, 1996	CAT	scFv/phagemid	14	N	PBL, bone marrow, tonsil	J-segment	0.3 nM (N = 7)
Sheets *et al.*, 1998	Marks (UCSF)	scFv/phagemid	6.7	N	PBL, spleen	IgM	0.2 nM (N = 8)
de Haard *et al.*, 1999	Target Quest	Fab/phagemid	43	N	PBL, spleen	IgM	2.7 nM (N = 3)
Little *et al.*, 1999	Little (Heidleberg)	scFv/phagemid	4	N	PBL	IgM, IgG	ND
Sblattero and Bradbury, 2000	Bradbury (Los Alamos)	scFv/phagemid	300	N	PBL Cre-Lox recombination PBL, spleen, lymph, tonsil	IgM	16 nM (N = 4)
Soderlind *et al.*, 2000	BioInvent	scFv/phagemid	2	(N)	Natural CDRs introduced into static VH and V-lambda frameworks	IgM, IgG, IgA	0.9 nM (N = 11)
Lee *et al.*, 2004	Genentech	Fab/phagemid	40	S	Static VH framework with synthetic diversity in H-CDR1, 2 & 3 and single sequence for VL H3 length range 7–19	NA (fully synthetic)	1.4 nM (N = 23)

(Continued)

Table 2.1 (*Continued*)

Reference	Laboratory	Format	Size ($\times10^9$)	Diversity Type*	Diversity Source	VH cDNA Amplification Target	Strongest Reported Affinity
Hoet *et al.*, 2005	Dyax	Fab/phagemid Fab/phage	30 10	N + S	Static VH framework with synthetic diversity in H-CDR1 & 2 and natural diversity (PBL) in H-CDR3 and VL	IgM	0.22 nM (N > 100)
Silacci *et al.*, 2005	Philogen	scFv/Phagemid	3	S	Static VH, V-lambda, V-kappa frameworks with synthetic diversity in fixed-length CDR3 PBL	NA	670 nM (N = 1)
Loset *et al.*, 2005	Affitech	scFv/Phagemid	6.4	N	Protein-L purification applied to VH-V-kappa sublibrary increasing functional fraction from ~ 20% to ~ 75%	IgM, IgD	ND
Rothe *et al.*, 2008	Morphosys	Fab/phagemid	3	S	Synthetic CDR diversity in 14 static frameworks (7 VH, 4 V-kappa, 3 V-lambda) H3 length range 4–22	NA	0.04 nM (N = 23)

*N = diversity source is from "natural" human immunoglobulin repertoire; S = synthetic diversity obtaining leads with nanomolar and sub-nanomolar binding constants against a wide range of antigens and haptens (Table 1).

The earliest reported library exceeding 10^9 members utilized a number of innovative design features and, at 6.5×10^{10} members, remains one of the largest libraries reported (Griffiths *et al.*, 1994). Nearly complete sets of human germline V_H and V_L segments were amplified from chromosomal (non-rearranged) DNA, encompassing germline CDR1 and CDR2 sequences for both heavy and light chain fragments. CDR3 variability was introduced using synthetic oligonucleotides, providing loop sizes of 4–12, 7–9, and 8–11 for CDR3 sequences of V_H, V-kappa, and V-lambda, respectively. The heavy and light chain sequences containing both germline and synthetic variation were then paired randomly to form a Fab library using the cre-lox recombination system, allowing the large library size. This library was tested against over 30 haptens and protein antigens (both self and non-self), yielding high affinities, down to 4 nM for many of the targets (Griffiths *et al.*, 1994).

Additional non-immune human scFv and Fab libraries exceeding 10^9 have been subsequently reported, mostly using brute-force, large-scale transformation procedures to achieve large library size. These "natural" libraries derive variable region diversity by amplification of cDNA from lymphocytes, providing fully rearranged antibody sequences, including the highly diverse H-CDR3 segment. As summarized in Table 2.1, initial PCR amplification strategies for the V_H segments have targeted different gene pools: e.g. exclusively IgM (Sheets *et al.*, 1998; de Haard *et al.*, 1999; Sblattero and Bradbury, 2000) combinations of IgM and IgD (Loset *et al.*, 2005), or combinations of IgM and IgG (Little *et al.*, 1999). Inclusion of the IgG pool should increase levels of sequences with somatic hypermutations, although this was not addressed in the publication (Little *et al.*, 1999). In the early library from CAT (Vaughan *et al.*, 1996), V_H amplification used J-segment primers. As a variant on capture of full-length V_H and V_L sequences for library construction, the BioInvent scFv library was constructed by PCR amplification of exclusively the CDR regions from the IgM, IgA, and IgG gene pools (Soderlind *et al.*, 2000) using precisely-designed primers. These CDR segments were then inserted into static V_H and V_L framework scaffolds (Soderlind *et al.*, 2000).

A series of large libraries designed with partial or fully synthetic diversity have also been constructed and validated, as also shown in Table 2.1. Philogen, building on the approaches developed by the Neri lab at ETH, introduced synthetic diversity exclusively in CDR3 regions with static frameworks for V_H, V-kappa, and V-lambda (Silacci *et al.*, 2005). Although recovered affinities appear to be low, the binders from this library appear to provide high specificity and utility in various immunoassays and can be readily affinity-matured to high affinity leads, e.g. sufficient for development as therapeutic candidates (Silacci *et al.*, 2005). At Genentech, exhaustive design approaches were described that led to the generation of a 4×10^{10} Fab library with synthetic diversity at each of the V_H CDRs introduced into a static V_H framework, together with a static V_L sequence (Lee *et al.*, 2004; Sidhu *et al.*, 2004). By focusing diversity variation at positions expected to be solvent-exposed and/or in contact with antigen while keeping structural positions constant, a highly-effective library was obtained that produced high-affinity binders, down to 1 nM against a broad range of antigens (Lee *et al.*, 2004). Applying an alternate design approach, the Morphosys libraries (Knappik *et al.*, 2000; Rothe *et al.*, 2008) include a set of 14 static frameworks (7 V_H, 4 V-kappa, 3 V-lambda) into which synthetic diversity is introduced at likely antigen contact sites in CDRs based on exhaustive sequence alignments of rearranged antibody sequences and review of structural data. The Morphosys libraries also include a novel Cys-display technology in which heavy chain fragments are tethered to gIII coat protein for display by covalent disulfide linkage rather than through direct fusion of the primary sequence (Rothe *et al.*, 2008). Finally, the Dyax Fab libraries are designed following a "3-slot" concept, in which a static V_H framework is used for introduction of synthetic diversity at selected positions in CDR1 and CDR2 (slot 1), while natural diversity is introduced for H-CDR3 (slot 2) and for full-length V_L (slot 3) segments (Hoet *et al.*, 2005).

2.4 Library Selection

Isolating antibodies from a phage display library that bind a given target involves a process known as panning or selection (Fig. 2.4) (Coomber, 2002). Together with library diversity, the stringency of the selection process is also a key consideration in consistently obtaining high affinity antibodies against a

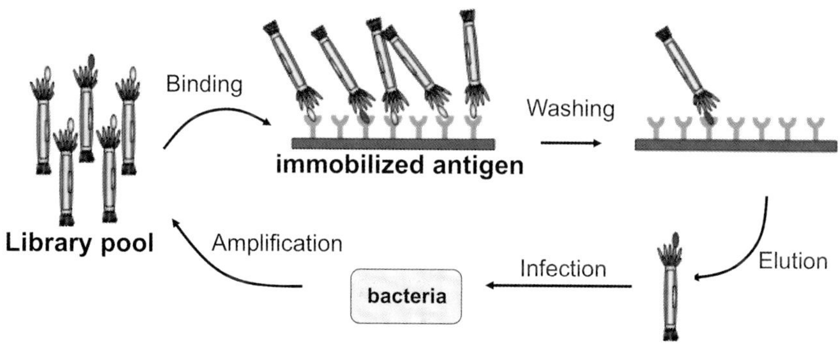

Figure 2.4 Overview of a phage display selection process

The phage library pool is incubated with immobilized antigen. Non-binding phage or weakly associating phage are washed away and bound phage are eluted. Selected phage are amplified by infection in *E. coli* and used as the input for a subsequent round of selection.

wide range of antigens and haptens (Table 1). Other key characteristics of the library include the display valency (see Section 3.24) and the expression level of the antibody fusions. Affinity maturation approaches are widely used to improve affinities to desired levels (i.e. $K_d < 10$ nM) after identification of leads with desired functionality (Groves *et al.*, 2006; Yang *et al.*, 2007; Kim *et al.*, 2008; Thie *et al.*, 2009). However, if a well-designed and highly diverse library is paired with a stringent selection process and a content-rich screening method, antibodies with subnanomolar affinities and desired function can be obtained without resorting to affinity maturation (Wassaf *et al.*, 2006). Selection methods vary widely in how the target is immobilized and in the conditions used to exert selection stringency in order to obtain antibodies with the desired affinity and functionality.

2.4.1 *Antigen immobilization*

Selection strategies require a means of isolating phage that bind the target from non-binding phage. Methods for selecting binders from phage display libraries generally involve immobilizing the target antigen on a solid support, such as the well of a microtiter plate, an immunotube, or a bead, and panning against an antibody phage display library. While passive immobilization of the target antigen is convenient, it may lead to protein denaturation and subsequent masking of native epitopes (Butler *et al.*, 1993).

The use of biotinylated antigens and streptavidin-coated magnetic beads offers a convenient means of antigen immobilization that is amenable to selection. Immobilizing antigens via biotin offers control over immobilized antigen density and may favor the presentation of the target in an active conformation. Generally, immobilization of the target protein in an active conformation is favored by either minimizing the number of biotins incorporated per molecule of target protein or by site-specific labeling.

Many biotinylation procedures use amine-reactive biotin reagents, such as *N*-hydroxysuccinimide esters of biotin, under molar ratios designed to minimize the number of biotins conjugated to the protein. There are a number of methods reported to quantify the number of biotins incorporated per mole of protein (Green, 1965; Wu *et al.*, 2005; Batchelor *et al.*, 2007). Alternatively, site-specific approaches have been applied to protein biotinylation. For example, Koide *et al.* recently showed that proteins with a His tag can be labeled with biotin using a reagent that links a biotin to a nitrilotriacetic acid moiety capable of specifically binding His-tagged antigens (Koide *et al.*, 2009). Tan *et al.* explored intein-mediated site specific biotinylation (Tan *et al.*, 2004). For use in phage selections, site-specific biotinylation at engineered sites has also been achieved enzymatically using BirA biotin ligase (Scholle *et al.*, 2004; Moutel *et al.*, 2009).

2.4.2 Overview of phage display selection

Selection, or panning against a phage display library requires a means of immobilizing the target antigen. The antigen can be either immobilized on the solid support prior to, or after the introduction of the phage display library. The amount of total phage used depends on the diversity of the library, such that the total amount of phage added should be at least 1,000-fold greater than the expected diversity (Kotlan and Glassy, 2009). Phage displayed antibodies that bind the immobilized antigen are then washed to remove non-binding phage. Bound phage particles are usually eluted from immobilized antigen with acid or base treatment. However, it is also possible to directly infect *E. coli* with phage still bound to immobilized antigens (for example, on streptavidin-coated magnetic beads) (Bradbury and Marks, 2004; Wassaf *et al.*, 2006). The amplified phage is then used as input for the next round.

Immobilizing the antigen first, followed by washing to remove free antigen, ensures that the phage library is only exposed to immobilized antigen. However, interactions with immobilized antigen may differ from those observed with soluble antigen. Immobilized antigens may have masked epitopes, and the binding to antibody-phage may be more influenced by avidity, especially for multimeric antibody-phage constructs. Koide *et al.* showed that binders identified by phage display bind target protein immobilized via a site-specific biotin on a streptavidin-coated plate better than the same protein immobilized using non-specifically incorporated biotin or passively immobilized protein (Koide *et al.*, 2009).

Rather than immobilizing the target antigen prior to incubation with the phage display library, the soluble antigen can be incubated with the phage display library first and then captured. Hawkins *et al.* initially described driving high affinity during selections by using soluble biotinylated antigen at concentrations below the binding constant (K_d) of the antibody, or by diluting the phage/biotinyl-antigen complex into a solution of non-biotinylated antigen before capture on streptavidin beads to promote recovery of binders with slowest off-rates (Hawkins *et al.*, 1992). Since the impact of target immobilization strategy on selection success is difficult to predict *a priori* for an individual target, selection strategies are often designed that include in parallel both pre-immobilized antigen and captured soluble antigen.

Multiple rounds of panning and amplification are usually performed in the selection phase. In principle, increasing the number of rounds can return a manageable number of phage clones that display antibodies with the highest affinity. However, each successive round enriches for clones based on phage propagation efficiency as well as affinity and may thereby limit the number of different antibodies obtained. Fewer antibodies recovered may reduce the likelihood of discovering antibodies with high affinity and functional activity. By limiting the number of rounds of selection, it becomes necessary to evaluate hundreds or thousands of phage clones, a task made more standard by the use of automated screening methods (Hallborn and Carlsson, 2002; Vanhercke *et al.*, 2005; Persson *et al.*, 2008; Turunen *et al.*, 2009).

2.4.3 *Advantages and successes of phage display*

Phage display of antibody fragments has developed in multiple directions and techniques since the earliest breakthroughs in 1990. During that time clear advantages over classical hybridoma technologies in certain instances have been demonstrated. For example, phage display allows for the rapid isolation of antibodies to any conceivable antigen, including toxins and antigens that have very high degrees of sequence homology across mammalian species and are difficult to obtain via conventional immunization of rodents. From large naive antibody libraries, diverse panels of specific antibodies can be generated to a plethora of different epitopes on a single protein. In combination with high-throughput methodologies for selection, screening and analysis, it is now possible to produce specific antibodies on a proteomic scale. Moreover, the concept of designer antibodies with exquisite specificity, functionality and picomolar affinities is now a reality with more and more of these types of antibodies entering the clinic for therapeutic intervention of a multitude of diseases. Such successes are exemplified below.

2.4.3.1 *Species cross-reactivity*

Historically, species cross-reactivity has been assessed post-isolation of an antibody using the classical hybridoma approach. However, antibody phage display allows the scientist to drive the panning selection towards isolating binders which are specifically cross-reactive with protein orthologs from other animal species. This can be particularly useful in the design of antibody therapeutics where the toxicological effect of treatment with a given antibody specificity can be assessed in rodents or primates, prior to administration in humans. Such species cross-reactivity is typically achieved by performing selections on a combination of human, mouse and primate orthologs either separately or in succession, and subsequently screening for the appropriate cross-species reactivity profile. Examples of selections for cross-species reactivity include: human and rodent Nogo-66 (Hofer *et al.*, 2007), human and murine BR3 (Lee *et al.*, 2006), human and murine VEGF (Liang *et al.*, 2006), human and murine VEGF receptor (Popkov *et al.*, 2004), and human and rhesus monkey Lymphocyte Function-Associated Antigen 1 (Werther *et al.*, 1996).

2.4.3.2 *Homolog binders*

Antibody binding to protein homologs can be either a desirable or an undesirable outcome depending on the particular biology of the target protein.

However, it serves as an illustration of the power of antibody phage display, in that by careful experimental planning homolog binding can be specifically selected for or selected against as required. For example, Chen and colleagues generated an immune library to select for antibodies specific to human Toll-like receptor 2 (TLR2) (Chen *et al.*, 2009). They then performed homolog mining on the hTLR2 library to generate clones specific for murine TLR2, which did not recognize the human homolog. They extended their panel of homolog binders further by isolating antibodies to TLR5 and hTLR6, which again bound specifically to their target, with no cross-reactivity to hTLR2 or other TLRs tested. Interestingly, the immune library was generated by immunization with human TLR2, and the immune sera collected from the mice that were used to generate the library exhibited no reactivity to any of the homologs to which antibodies were generated. This suggests that the antibodies identified were the products of the combinatorial re-assortment of heavy and light chain variable gene fragments.

Commonly, homolog binding is an undesirable trait in an antibody, and this can be especially true for therapeutic antibody generation programs. Parsons and colleagues performed selections for antibodies which were specific to fetal hemoglobin (HbF) (Parsons *et al.*, 1996). However, they found that the selection was being dominated by clones that bound HbF but cross-reacted with adult hemoglobin (HbA) as well. In order to direct selection towards HbF-specific epitopes, they performed "competitive de-selection" whereby the repertoire was pre-incubated with HbA in solution before each panning. The cross-reactive scFvs can form complexes with the soluble HbA and were therefore prevented from binding the immobilized HbF. Clones with preferential binding to HbF emerged under these conditions, and one in particular exhibited exquisite specificity and could distinguish between cells expressing HbF from those expressing HbA by immunocytochemistry and flow cytometry. This clone had an affinity that was 600-fold lower than the dominant cross-reactive clone and thus only emerged as a result of competitive de-selection.

2.4.3.3 *Rare epitopes*

Antibody phage display can also be used to target rare epitopes which can be presented to the library in the form of specific, designed peptides. For example, human immunodeficiency virus-1 (HIV-1) entry into cells is mediated by the envelope glycoprotein receptor-binding (gp120) and membrane fusion-promoting (gp41) subunits. The gp41 heptad repeat 1 (HR1) domain is the target of small molecule fusion inhibitors, and its sequence is highly conserved amongst HIV-1 isolates. Therefore, it is considered an attractive target for vaccine development. However, it was not known whether antibodies could access HR1 within the context of the virion. Gp41-based peptides were used to select an antibody that bound to HR1 and inhibited the assembly of fusion intermediates *in vitro* (Miller *et al.*, 2005). This antibody inhibited the replication of a diverse panel of HIV-1 clinical isolates and represents a previously unknown example of a broadly cross-neutralizing anti-HIV-1 IgG. NMR studies and functional analyses mapped the antibody binding site to a hydrophobic pocket situated in the HR1 groove. This hydrophobic pocket was previously proposed as a drug target and has been identified as a common binding site for peptide and peptidomimetic fusion inhibitors. This suggests that the hydrophobic pocket is a "hot spot" for fusion inhibition and an ideal target for a vaccine-mediated antibody response (Miller *et al.*, 2005).

2.4.3.4 *Exquisite specificity*

Nizak and colleagues exemplified how antibody phage display can be used to generate binders with exquisite specificity and use them as tools to understand protein functions within living cells (Nizak *et al.*, 2003). They generated an antibody that was able to recognize the Golgi-localized small GTPase Rab6A, which regulates the traffic between the Golgi apparatus and the endoplasmic reticulum (ER). They were able to generate this highly specific antibody by selection on a mutant Rab6A protein with only a single amino acid mutation in Rab6A in which Gln72 is replaced by Leu, which was sufficient to lock the mutant into the guanosine triphosphate (GTP)-bound conformation. This antibody was then used to used to locate Rab6-GTP both in fixed cells, and, after green fluorescent protein (GFP) tagging and intracellular expression, to follow the fate of Rab6-GTP *in vivo*.

Synthetic peptides based on solvent accessible regions of the target antigen can be used as a surrogate antigen when the pure antigen is not available or in combination with selections on the pure antigen to direct the selected antibodies towards distinct epitopes (Pavlik *et al.*, 2003). Alternatively, strategies have been designed to focus the selection on obtaining antibodies that bind a specific site on the antigen, such as the active site of an enzyme or the ligand-binding site of a receptor (Huang *et al.*, 2006; Wassaf *et al.*, 2006). For example, an antibody inhibitor of the matrix metalloproteinase MMP-14 that binds near the active site was identified using phage display by first depleting against an immobilized complex of MMP-14 bound to a natural, active site inhibitor (TIMP-2) (Devy *et al.*, 2009). This depletion step removes a majority of the antibodies that bind MMP-14 outside the active site of the enzyme. The depleted library was then panned against active MMP-14, in which newly exposed epitopes are expected to be in the vicinity of the active site. The antibody discovered was shown by enzyme inhibition kinetics to have a competitive inhibition mechanism, and to be competed with by TIMP-2; both are consistent with the antibody binding to the active site of MMP-14 (Devy *et al.*, 2009).

The concept of antibody-guided selections is another method to focus the selection towards antibodies that specifically bind an epitope of interest (Osbourn *et al.*, 2005). This method uses existing antibodies that bind the desired epitope to identify additional antibodies that bind the same epitope through a selection process involving sequential substitution of a subset of original antibody binding segments with a new diversity source. Osbourn *et al.* (2005) used antibody-guided selection for the generation of human versions of rodent antibodies that bind tumor necrosis factor-alpha and block its binding to receptor, leading to Humira®, the first phage-derived fully-human therapeutic antibody to receive marketing approval. A similar selection strategy was used to obtain alternative antibodies that recognize the same epitope on the Ki-67 protein biomarker as a parental antibody (Jarutat *et al.*, 2006).

A further example of both exquisite specificity and ortholog cross-reactivity was demonstrated by Wu *et al.* (2010), where antibody phage display was used to generate antibodies to Notch 1 and Notch 2 receptor negative regulatory region. These antibodies specifically antagonize each receptor prologue and yet cross-react with the human and mouse sequences, enabling the discrimination of Notch1 and Notch2 receptor function both in human patients and rodent models. These antibodies demonstrated that selective blocking of Notch1 inhibited tumor growth through both the inhibition of cancer cell growth and deregulation of angiogenesis, whereas inhibitors of Notch1 plus Notch 2 caused severe intestinal toxicity.

2.4.3.5 *Cell surface antigens*

Selections can be directed towards antigen expressed on cell surfaces or membrane vesicles (Labrijn *et al.*, 2002) rather than merely purified proteins (Bradbury and Marks 2004). Subtractive strategies have even been developed to allow the selection of antibodies directed against cell surface antigens. Eisenhard *et al.* described a detailed subtractive selection protocol, whereby phage are first depleted against cells that do not express antigen, then incubated with antigen-expressing cells and cell-bound phage isolated by centrifugation (Eisenhardt *et al.*, 2007). Williams and Sharon developed a method of selecting antibody binders against colorectal cells using a percoll density gradient rather than sequential washes to separate cell-bound and free phage (Williams and Sharon, 2002). Sorting flow cytometry has also been applied successfully to antibody phage display against cell surface targets (Yuan *et al.*, 2008). Heitner *et al.* describe a microfluidic system that enables the use of live cells in the selection of an antibody against native tomoregulin, a protein that is present on cell surfaces as a multimer (Heitner *et al.*, 2006). Selection strategies designed to discover phage-derived antibodies that are internalized by the cell have been recently reviewed (Poul, 2009). Such internalizing antibodies have been shown to be useful for the delivery of cytotoxic agents (An *et al.*, 2008).

2.4.3.6 *Toxins*

The generation of antibodies to biologically active toxins is a clear area where antibody phage display has an advantage over the classical hybridoma approach. Toxins are often highly potent in small doses; hence immunization of small rodents would result in death of the animal long before any significant immune response could be mounted. To this end antibody phage display libraries have been used to generate human antibodies that could be used to neutralize toxins as both passive immunoprophylatic reagents as well as for post-exposure treatment. Zhou *et al.* isolated a panel of neutralizing human monoclonal antibodies using antibody phage display against the tripartite anthrax toxin of *Bacillus anthracis*, comprising protective antigen (PA), lethal factor (LF), and edema factor (EF) (Zhou *et al.*, 2008). Neutralizing antibodies to the lethal factor binding domain and the cell receptor binding domain within the protective antigen were identified by cell cytotoxicity assays.

Anti-*Naja kaouthia* (Thai cobra) venom anti-serum is used for the post-exposure treatment of snake bite victims. However, recipients frequently develop either an allergic reaction or an anti-isotype response causing serum sickness. Kulkeaw *et al.* (2009) used the long alpha-neurotoxin purified from *N. kaouthia* holovenom to select antibodies from a naive human antibody phage display library. Antibodies that specifically bound to and neutralized the neurotoxin were able to rescue envenomized mice from neurotoxin mediated lethality.

2.4.3.7 *High-throughput generation of reagents for proteomics*

Antibody phage display libraries are commonly used to identify binders to a single protein or small panel of proteins in relation to one specific area of research or, in the case of biotherapeutics, to answer an unmet clinical need. Antibody phage display is, however, also highly amenable to high-throughput and automated production processes. As such, it is now beginning to be employed as a tool to address the insatiable demand of researchers for high quality antibodies for use as tools to address questions related to proteins and protein: protein interactions on a proteomic scale (Mersmann *et al.*, 2010).

The largest high-throughput approach to date, the Atlas of Protein Expression at the Wellcome Trust Sanger Institute, aimed to produce antibodies as probes for immunohistochemistry to map the location of protein expression in normal tissue (Schofield *et al.*, 2007). Selections were performed on up to 94 different antigens simultaneously in a microtiter well format, screening for specific antibody clones by ELISA used automated liquid handling robots, and automated solutions to scFv purification and immunohistochemistry were also employed. This study selected, screened, and sequenced over 38,000 recombinant antibodies to 292 antigens, yielding over 7,200 unique scFvs. Of these, 4,400 antibodies were characterized by specificity testing and detailed sequence analysis. In pilot immunohistochemical studies, positive staining on tissue microarrays was found for 37% (143/381) of antibodies generated. The work highlighted the feasibility of antibody generation using phage display in a high-throughput manner. The process bottlenecks identified were not antibody generation by phage display but rather the cloning, verification and generation of high quality antigen and the preparation and use of the antibodies in downstream applications such as immunohistochemistry.

2.4.4 *Advances in selection technology*

The basic steps of phage display selections have been repeatedly demonstrated to provide a robust means of obtaining antigen-specific antibodies. However, various investigators continue to explore the limits of the technique with the aim of further optimizing the method and expanding its range of applications. Hogan *et al.* (2005) described a variation called URSA (ultra rapid selection of antibodies) that accelerates the process by omitting prior phage amplification. A selection method called Pathfinder selections was also developed and shown to be especially useful for selections on cell surface antigens, where the antigen may not be available or functional in a purified form (Osbourn *et al.*, 1998). Briefly, a "pathfinder" molecule, such as a commercially available antibody against the target antigen, is conjugated to horseradish peroxidase. After incubating the cells with the phage library and the pathfinder molecule, the HRP substrate biotin tyramine is added and results in the biotinylation of bound phage. Biotinylated phage are then specifically recovered using streptavidin-coated beads.

An interesting technology has recently been applied to phage display selections that does not involve target immobilization on beads or surfaces. Instead of target immobilization, the biotinylated target is bound to avidin coupled to a temperature sensitive reagent (poly-(N-isopropylacrylamide)) referred to as a stimuli-responsive bioconjugate (Stocker *et al.*, 2008). When the temperature is increased above 30°C, the stimuli-responsive bioconjugate precipitates along with phage that specifically bound the biotinylated antigen.

2.5 Library Screening

Outputs from the selected library are plated to pick individual phage isolates that are screened to identify antibody binders. Phage display screening methods can be differentiated as either primary or secondary screening assays. Primary assays are those that directly screen amplified outputs from selections and are amenable to being run in high throughput with automation equipment (Hust and Dubel, 2004). Increasing the number of antibodies screened using automated primary screening assays can increase the probability of obtaining high affinity antibodies

standard in terms of generating a single antibody, or panel of antibodies having the desired characteristics of specificity and affinity for any given application (e.g. immunoprecipitation, ELISA, Western blot, immunohistochemistry). Some degree of affinity maturation would be desirable, but would need to be compatible with a high-throughput process. McCafferty *et al.* (unpublished) offer a simple and effective method for affinity maturation of the selected binder population which can meet the criteria above. The VH genes from selected antibody populations are sub-cloned into naive VL gene libraries using standard cloning techniques. These new chain shuffled libraries are then used for fresh selections on the target of interest. In inhibition assays, the IC_{50} values for the chain-shuffled population post-selection are improved on average by around one $\log_{10}$ in value compared to the original selected population. This $\log_{10}$ improvement in potency may be sufficient to generate antibodies with high enough specificity and affinity for a wide range of applications in proteomics.

2.6.4 *Antibody format and production*

Most non-therapeutic uses of antibodies are as tools for the specific binding of a given target, e.g. for detection, or for intervention in an *in vitro* or *in vivo* biological system. Antibody fragments are often not suitable for such applications, due to biophysical limitations such as stability or unfavorable pharmacokinetics, or to incompatability with secondary detection reagents. In order to overcome these limitations, fragments such as scFvs or Fabs must be presented in a more "antibody-like" format, such as scFv-Fc or whole IgG molecule. This can lead to an improvement in an antibody's apparent affinity due to increased avidity. However, in terms of production, this can present a bottleneck to throughput. Typically, IgGs are generated by mammalian cell line expression, and for a reproducible and replenishable resource a stably transfected cell line would be the route of choice. The production of whole aglycosylated IgG in bacterial cells is technically feasible (Simmons *et al.*, 2002) but has not been routinely employed, suggesting that *E. coli* production is problematic. However, libraries of full-length IgGs tethered to *E. coli* spheroplasts (Mazor *et al.*, 2007) and even filamentous phage particles (Mazor *et al.*, 2010) recently have been generated, antigen-specific clones selected, and soluble IgGs produced. This will undoubtedly rekindle interest in using bacteria to express full-length IgGs, which would facilitate the large-scale generation of antibodies from phage display libraries and broad use by researchers as probes for a plethora of proteomic studies.

2.7 Conclusions

Antibody phage display has matured to a point where high affinity lead candidates are consistently obtained, enabling rapid development of human antibody therapeutics. Since the first report in the literature of antibody phage display (McCafferty *et al.*, 1990), numerous advances in library design, selection strategies and screening methods have been reported. Investigators have applied antibody phage display to a nearly limitless array of protein targets. Efforts continue to develop high-throughput functional screening assays or, even better, to optimize selection strategies such that antibodies with a desired functionality are obtained from a first pass.

References

Abdiche Y.N., Malashock D.S., Pinkerton A., Pons J. (2009). Exploring blocking assays using Octet, ProteOn, and Biacore biosensors. *Anal Biochem* **386**: 172–80.

Abergel C., Claverie J.M. (1991). A strong propensity toward loop formation characterizes the expressed reading frames of the D segments at the Ig H and T cell receptor loci. *Eur J Immunol* **21**: 3021–5.

Al-Lazikani B., Lesk A.M., Chothia C. (1997). Standard conformations for the canonical structures of immunoglobulins. *J Mol Biol* **273**: 927–48.

Alt F.W., Baltimore D. (1982). Joining of immunoglobulin heavy chain gene segments: implications from a chromosome with evidence of three D-JH fusions. *Proc Natl Acad Sci USA* **79**: 4118–22.

Amit A.G., Mariuzza R.A., Phillips S.E., Poljak R.J. (1986). Three-dimensional structure of an antigen-antibody complex at 2.8 A resolution. *Science* **233**: 747–53.

An F., Drummond D.C., Wilson S., Kirpotin D.B., Nishimura S.L., Broaddus V.C., Liu B. (2008). Targeted drug delivery to mesothelioma cells using functionally selected internalizing human single-chain antibodies. *Mol Cancer Ther* **7**: 569–78.

Anderson G.P., Matney R., Liu J.L., Hayhurst A., Goldman E.R. (2007). Multiplexed fluid array screening of phage displayed anti-ricin single domain antibodies for rapid assessment of specificity. *Biotechniques* **43**: 806–11.

Angenendt P., Wilde J., Kijanka G., Baars S., Cahill D.J., Kreutzberger J., Lehrach H., Konthur Z., Glokler J. (2004). Seeing better through a MIST: evaluation of monoclonal recombinant antibody fragments on microarrays. *Anal Chem* **76**: 2916–21.

Armstrong J., Perham R.N., Walker J.E. (1981). Domain structure of bacteriophage fd adsorption protein. *FEBS Lett* **135**: 167–72.

Baek H., Suk K.H., Kim Y.H., Cha S. (2002). An improved helper phage system for efficient isolation of specific antibody molecules in phage display. *Nucleic Acids Res* **30**: e18.

Barbas C.F. (2001). *Phage Display: A Laboratory Manual.* Cold Spring Harbor Laboratory Press, Cold Spring Harbor, NY.

Barbas C.F., 3rd, Kang A.S., Lerner R.A., Benkovic S.J. (1991). Assembly of combinatorial antibody libraries on phage surfaces: the gene III site. *Proc Natl Acad Sci USA* **88**: 7978–82.

Barbie V., Lefranc M.P. (1998). The human immunoglobulin kappa variable (IGKV) genes and joining (IGKJ) segments. *Exp Clin Immunogenet* **15**: 171–83.

Bass S., Greene R., Wells J.A. (1990). Hormone phage: an enrichment method for variant proteins with altered binding properties. *Proteins* **8**: 309–14.

Batchelor R.H., Sarkez A., Cox W.G., Johnson I. (2007). Fluorometric assay for quantitation of biotin covalently attached to proteins and nucleic acids. *Biotechniques* **43**: 503–7.

Beck E., Zink B. (1981). Nucleotide sequence and genome organisation of filamentous bacteriophages fl and fd. *Gene* **16**: 35–58.

Berek C., Milstein C. (1987). Mutation drift and repertoire shift in the maturation of the immune response. *Immunol Rev* **96**: 23–41.

Berek C., Milstein C. (1988). The dynamic nature of the antibody repertoire. *Immunol Rev* **105**: 5–26.

Bergander T., Nilsson-Valimaa K., Oberg K., Lacki K.M. (2008). High-throughput process development: determination of dynamic binding capacity using microtiter filter plates filled with chromatography resin. *Biotechnol Prog* **24**: 632–9.

Better M., Chang C.P., Robinson R.R., Horwitz A.H. (1988). Escherichia coli secretion of an active chimeric antibody fragment. *Science* **240**: 1041–3.

Bird R. E., Hardman K.D., Jacobson J.W., Johnson S., Kaufman B.M., Lee S.M., Lee T., Pope S.H., Riordan G.S., Whitlow M. (1988). Single-chain antigen-binding proteins. *Science* **242**: 423–6.

Blazek D., Celer V., Navratilova I., Skladal P. (2004). Generation and characterization of single-chain antibody fragments specific against transmembrane envelope glycoprotein gp46 of maedi-visna virus. *J Virol Methods* **115**: 83–92.

Boder E.T., Wittrup K.D. (1997). Yeast surface display for screening combinatorial polypeptide libraries. *Nat Biotechnol* **15**: 553–7.

Boozer C., Kim G., Cong S., Guan H., Londergan T. (2006). Looking towards label-free biomolecular interaction analysis in a high-throughput format: a review of new surface plasmon resonance technologies. *Curr Opin Biotechnol* **17**: 400–5.

Bornhop D.J., Latham J.C., Kussrow A., Markov D.A., Jones R.D., Sorensen H.S. (2007). Free-solution, label-free molecular interactions studied by back-scattering interferometry. *Science* **317**: 1732–6.

Bradbury A.R., Marks J.D. (2004). Antibodies from phage antibody libraries. *J Immunol Methods* **290**: 29–49.

Breitling F., Dubel S., Seehaus T., Klewinghaus I., Little M. (1991). A surface expression vector for antibody screening. *Gene* **104**: 147–53.

Brochet X., Lefranc M.P., Giudicelli V. (2008). IMGT/V-QUEST: the highly customized and integrated system for IG and TR standardized V-J and V-D-J sequence analysis. *Nucleic Acids Res* **36**: W503–8.

Buckler D.R., Park A., Viswanathan M., Hoet R.M., Ladner R.C. (2008). Screening isolates from antibody phage-display libraries. *Drug Discov Today* **13**: 318–24.

Burnet F.M. (1976). A modification of Jerne's theory of antibody production using the concept of clonal selection. *CA Cancer J Clin* **26**: 119–21.

Butler J.E., Ni L., Brown W.R., Joshi K.S., Chang J., Rosenberg B., Voss E.W., Jr. (1993). The immunochemistry of sandwich ELISAs — VI. Greater than 90% of monoclonal and

75% of polyclonal anti-fluorescyl capture antibodies (CAbs) are denatured by passive adsorption. *Mol Immunol* **30**: 1165–75.

Canziani G.A., Klakamp S., Myszka D.G. (2004). Kinetic screening of antibodies from crude hybridoma samples using Biacore. *Anal Biochem* **325**: 301–7.

Chasteen L., Ayriss J., Pavlik P., Bradbury A.R. (2006). Eliminating helper phage from phage display. *Nucleic Acids Res* **34**: e145.

Chen T., Hogan S., Conley G., Pazmany C., Wu Q.L., McNeil G.L., Bassill G., Macintyre A. N., Martik D., Nixon A.E., Sexton D.J. (2007). Discovery and characterization of human antibody inhibitors of pregnancy-associated plasma protein-A. *Biol Chem* **388**: 507–12.

Chen Y.H., Lipes B.D., Kenan D.J., Staats H.F., Gunn M.D. (2009). Identification of recombinant antibodies against multiple distinct toll-like receptors by homolog mining a single immune scFv phage library. *J Immunol Methods* **340**: 144–53.

Chothia C., Lesk A.M. (1987). Canonical structures for the hypervariable regions of immunoglobulins. *J Mol Biol* **196**: 901–17.

Chothia C., Lesk A.M., Gherardi E., Tomlinson I.M., Walter G., Marks J.D., Llewelyn M.B., Winter G. (1992). Structural repertoire of the human VH segments. *J Mol Biol* **227**: 799–817.

Clackson T., Hoogenboom H.R., Griffiths A.D., Winter G. (1991). Making antibody fragments using phage display libraries. *Nature* **352**: 624–8.

Clark L.A., Ganesan S., Papp S., van Vlijmen H.W. (2006). Trends in antibody sequence changes during the somatic hypermutation process. *J Immunol* **177**: 333–40.

Cleary J.M., Ray D.S. (1980). Replication of the plasmid pBR322 under the control of a cloned replication origin from the single-stranded DNA phage M13. *Proc Natl Acad Sci USA* **77**: 4638–42.

Concepcion J., Witte K., Wartchow C., Choo S., Yao D., Persson H., Wei J., Li P., Heidecker B., Ma W., Varma R., Zhao L.S., Perillat D., Carricato G., Recknor M., Du K., Ho H., Ellis T., Gamez J., Howes M., Phi-Wilson J., Lockard S., Zuk R., Tan H. (2009). Label-free detection of biomolecular interactions using BioLayer interferometry for kinetic characterization. *Comb Chem High Throughput Screen* **12**: 791–800.

Cook G.P., Tomlinson I. M. (1995). The human immunoglobulin VH repertoire. *Immunol Today* **16**: 237–42.

Coomber D.W. (2002). Panning of antibody phage-display libraries. Standard protocols. *Methods Mol Biol* **178**: 133–45.

Cox J.P., Tomlinson I.M., Winter G. (1994). A directory of human germ-line V kappa segments reveals a strong bias in their usage. *Eur J Immunol* **24**: 827–36.

Crissman J.W., Smith G.P. (1984). Gene-III protein of filamentous phages: evidence for a carboxyl-terminal domain with a role in morphogenesis. *Virology* **132**: 445–55.

Cummins E., Luxenberg D.P., McAleese F., Widom A., Fennell B.J., Darmanin-Sheehan A., Whitters M.J., Bloom L., Gill D., Cunningham O. (2008). A simple high-throughput purification method for hit identification in protein screening. *J Immunol Methods* **339**: 38–46.

Cunningham B.T., Laing L. (2006). Microplate-based, label-free detection of biomolecular interactions: applications in proteomics. *Expert Rev Proteomics* **3**: 271–81.

Cwirla S.E., Peters E.A., Barrett R.W., Dower W.J. (1990). Peptides on phage: a vast library of peptides for identifying ligands. *Proc Natl Acad Sci USA* **87**: 6378–82.

de Haard H.J., van Neer N., Reurs A., Hufton S.E., Roovers R.C., Henderikx P., de Bruine A.P., Arends J.W., Hoogenboom H.R. (1999). A large non-immunized human Fab fragment phage library that permits rapid isolation and kinetic analysis of high affinity antibodies. *J Biol Chem* **274**: 18218–30.

de Wildt R.M., Hoet R.M., van Venrooij W.J., Tomlinson I.M., Winter G. (1999). Analysis of heavy and light chain pairings indicates that receptor editing shapes the human antibody repertoire. *J Mol Biol* **285**: 895–901.

de Wildt R.M., Mundy C.R., Gorick B.D., Tomlinson I.M. (2000). Antibody arrays for high-throughput screening of antibody-antigen interactions. *Nat Biotechnol* **18**: 989–94.

Desiderio A., Franconi R., Lopez M., Villani M.E., Viti F., Chiaraluce R., Consalvi V., Neri D., Benvenuto E. (2001). A semi-synthetic repertoire of intrinsically stable antibody fragments derived from a single-framework scaffold. *J Mol Biol* **310**: 603–15.

Devlin J.J., Panganiban L.C., Devlin P.E. (1990). Random peptide libraries: a source of specific protein binding molecules. *Science* **249**: 404–6.

Devy L., Huang L., Naa L., Yanamandra N., Pieters H., Frans N., Chang E., Tao Q., Vanhove M., Lejeune A., van Gool R., Sexton D.J., Kuang G., Rank D., Hogan S., Pazmany C., Ma Y.L., Schoonbroodt S., Nixon A.E., Ladner R.C., Hoet R., Henderikx P., Tenhoor C., Rabbani S.A., Valentino M.L., Wood C.R., Dransfield D.T. (2009). Selective inhibition of matrix metalloproteinase-14 blocks tumor growth, invasion, and angiogenesis. *Cancer Res* **69**: 1517–26.

Dobson C.L., Main S., Newton P., Chodorge M., Cadwallader K., Humphreys R., Albert V., Vaughan T.J., Minter R.R., Edwards B.M. (2009). Human monomeric antibody fragments to TRAIL-R1 and TRAIL-R2 that display potent in vitro agonism. *MAbs* **1**: 552–62.

Doenecke A., Winnacker E.L., Hallek M. (1997). Rapid amplification of cDNA ends (RACE) improves the PCR-based isolation of immunoglobulin variable region genes from murine and human lymphoma cells and cell lines. *Leukemia* **11**: 1787–92.

Dower W.J., Miller J.F., Ragsdale C.W. (1988). High efficiency transformation of *E. coli* by high voltage electroporation. *Nucleic Acids Res* **16**: 6127–45.

Dyson M.R., Perera R.L., Shadbolt S.P., Biderman L., Bromek K., Murzina N.V., McCafferty J. (2008). Identification of soluble protein fragments by gene fragmentation and genetic selection. *Nucleic Acids Res* **36**: e51.

Eddings M.A., Miles A.R., Eckman J.W., Kim J., Rich R.L., Gale B.K., Myszka D.G. (2008). Improved continuous-flow print head for micro-array deposition. *Anal Biochem* **382**: 55–9.

Edwards B.M., Barash S.C., Main S.H., Choi G.H., Minter R., Ullrich S., Williams E., Du Fou L., Wilton J., Albert V.R., Ruben S.M., Vaughan T.J. (2003). The remarkable flexibility of the human antibody repertoire; isolation of over one thousand different antibodies to a single protein, BLyS.*J Mol Biol* **334**: 103–18.

Eisenhardt S.U., Schwarz M., Bassler N., Peter K. (2007). Subtractive single-chain antibody (scFv) phage-display: tailoring phage-display for high specificity against function-specific conformations of cell membrane molecules. *Nat Protoc* **2**: 3063–73.

Ewert S., Huber T., Honegger A., Pluckthun A. (2003). Biophysical properties of human antibody variable domains. *J Mol Biol* **325**: 531–53.

Farady C.J., Egea P.F., Schneider E.L., Darragh M.R., Craik C.S. (2008). Structure of an Fab-protease complex reveals a highly specific non-canonical mechanism of inhibition. *J Mol Biol* **380**: 351–60.

Farady C.J., Sun J., Darragh M.R., Miller S.M., Craik C.S. (2007). The mechanism of inhibition of antibody-based inhibitors of membrane-type serine protease 1 (MT-SP1). *J Mol Biol* **369**: 1041–51.

Feldhaus M.J., Siegel R.W. (2004). Yeast display of antibody fragments: a discovery and characterization platform. *J Immunol Methods* **290**: 69–80.

Fredericks Z.L., Forte C., Capuano I.V., Zhou H., Vanden Bos T., Carter P. (2004). Identification of potent human anti-IL-1RI antagonist antibodies. *Protein Eng Des Sel* **17**: 95–106.

Frippiat J.P., Williams S.C., Tomlinson I.M., Cook G.P., Cherif D., Le Paslier D., Collins J.E., Dunham I., Winter G., Lefranc M. P. (1995). Organization of the human immunoglobulin lambda light-chain locus on chromosome 22q11.2. *Hum Mol Genet* **4**: 983–91.

Gao C., Mao S., Kaufmann G., Wirsching P., Lerner R.A., Janda K.D. (2002). A method for the generation of combinatorial antibody libraries using pIX phage display. *Proc Natl Acad Sci USA* **99**: 12612–6.

Gao C., Mao S., Lo C.H., Wirsching P., Lerner R.A., Janda K.D. (1999). Making artificial antibodies: a format for phage display of combinatorial heterodimeric arrays. *Proc Natl Acad Sci USA* **96**: 6025–30.

Geider K. (1986). DNA cloning vectors utilizing replication functions of the filamentous phages of Escherichia coli. *J Gen Virol* **67** (Pt 11): 2287–303.

Geuijen C.A., Clijsters-van der Horst M., Cox F., Rood P.M., Throsby M., Jongeneelen M. A., Backus H.H., van Deventer E., Kruisbeek A.M., Goudsmit J., de Kruif J. (2005). Affinity ranking of antibodies using flow cytometry: application in antibody phage display-based target discovery. *J Immunol Methods* **302**: 68–77.

Glanville J., Zhai W., Berka J., Telman D., Huerta G., Mehta G.R., Ni I., Mei L., Sundar P.D., Day G.M., Cox D., Rajpal A., Pons J. (2009). Precise determination of the diversity of a combinatorial antibody library gives insight into the human immunoglobulin repertoire. *Proc Natl Acad Sci USA* **106**: 20216–21.

Green N.M. (1965). A Spectrophotometric Assay for Avidin and Biotin Based on Binding of Dyes by Avidin. *Biochem J* **94**: 23C–24C.

Griffiths A.D., Malmqvist M., Marks J.D., Bye J.M., Embleton M.J., McCafferty J., Baier M., Holliger K.P., Gorick B.D., Hughes-Jones N.C., *et al.* (1993). Human anti-self antibodies with high specificity from phage display libraries. *Embo J* **12**: 725–34.

Hust M., Toleikis, L., Dubel, S. (2007). *Selection Strategies II: Antibody Phage Display*. 45–62, Wiley-VCH, Weinheim.

Huston J.S., Levinson D., Mudgett-Hunter M., Tai M.S., Novotny J., Margolies M.N., Ridge R.J., Bruccoleri R. E., Haber E., Crea R., *et al.* (1988). Protein engineering of antibody binding sites: recovery of specific activity in an anti-digoxin single-chain Fv analogue produced in Escherichia coli. *Proc Natl Acad Sci USA* **85**: 5879–83.

Huston J.S., McCartney J., Tai M.S., Mottola-Hartshorn C., Jin D., Warren F., Keck P., Oppermann H. (1993). Medical applications of single-chain antibodies. *Int Rev Immunol* **10**: 195–217.

Iannolo G., Minenkova O., Petruzzelli R., Cesareni G. (1995). Modifying filamentous phage capsid: limits in the size of the major capsid protein. *J Mol Biol* **248**: 835–44.

Ignatovich O., Tomlinson I.M., Jones P.T., Winter G. (1997). The creation of diversity in the human immunoglobulin V(lambda) repertoire. *J Mol Biol* **268**: 69–77.

Jarutat T., Frisch C., Nickels C., Merz H., Knappik A. (2006). Isolation and comparative characterization of Ki-67 equivalent antibodies from the HuCAL phage display library. *Biol Chem* **387**: 995–1003.

Jespers L.S., Messens J.H., De Keyser A., Eeckhout D., Van den Brande I., Gansemans Y.G., Lauwereys M.J., Vlasuk G.P., Stanssens P.E. (1995). Surface expression and ligand-based selection of cDNAs fused to filamentous phage gene VI. *Biotechnology (N Y)* **13**: 378–82.

Jia X.C., Raya R., Zhang L., Foord O., Walker W.L., Gallo M.L., Haak-Frendscho M., Green L. L., Davis C. G. (2004). A novel method of Multiplexed Competitive Antibody Binning for the characterization of monoclonal antibodies. *J Immunol Methods* **288**: 91–8.

Jiao Y., Zhao P., Zhu J., Grabinski T., Feng Z., Guan X., Skinner R.S., Gross M.D., Hay R.V., Tachibana H., Cao B. (2005). Construction of human naive Fab library and characterization of anti-met Fab fragment generated from the library. *Mol Biotechnol* **31**: 41–54.

Johansen L.K., Albrechtsen B., Andersen H.W., Engberg J. (1995). pFab60: a new, efficient vector for expression of antibody Fab fragments displayed on phage. *Protein Eng* **8**: 1063–7.

Kabat E.A., National Institutes of Health (U.S.), Columbia University. (1991). *Sequences of proteins of immunological interest.* U.S. Dept. of Health and Human Services, Public Health Service, National Institutes of Health, Bethesda, MD.

Kabat E.A., Wu T.T. (1971). Attempts to locate complementarity-determining residues in the variable positions of light and heavy chains. *Ann N Y Acad Sci* **190**: 382–93.

Kanda V., Kariuki J.K., Harrison D.J., McDermott M.T. (2004). Label-free reading of microarray-based immunoassays with surface plasmon resonance imaging. *Anal Chem* **76**: 7257–62.

Kang A.S., Barbas C.F., Janda K.D., Benkovic S.J., Lerner R.A. (1991). Linkage of recognition and replication functions by assembling combinatorial antibody Fab libraries along phage surfaces. *Proc Natl Acad Sci USA* **88**: 4363–6.

Karlsson R., Larsson A. (2004). Affinity measurement using surface plasmon resonance. *Methods Mol Biol* **248**: 389–415.

Kawasaki K., Minoshima S., Nakato E., Shibuya K., Shintani A., Schmeits J.L., Wang J., Shimizu N. (1997). One-megabase sequence analysis of the human immunoglobulin lambda gene locus. *Genome Res* **7**: 250–61.

Kay B.K., Paul J.I. (1996). High-throughput screening strategies to identify inhibitors of protein-protein interactions. *Mol Divers* **1**: 139–40.

Kehoe J.W., Kay B.K. (2005). Filamentous phage display in the new millennium. *Chem Rev* **105**: 4056–72.

Kim S.J., Jang M.H., Ahn H.J., Kim J.H., Lim J.H., Ryu C.J., Lim N.K., Kim K.S., Park M.J., Park I., Hong H.J. (2008). Selection of an affinity-matured antibody against a defined epitope by phage display of an immune antibody library. *J Immunol Methods* **329**: 176–83.

Kipriyanov S.M. (2009). High-Level Periplasmic Expression and Purificationof scFvs. *Methods Mol Biol* **562**: 205–14.

Knappik A., Ge L., Honegger A., Pack P., Fischer M., Wellnhofer G., Hoess A., Wolle J., Pluckthun A., Virnekas B. (2000). Fully synthetic human combinatorial antibody libraries (HuCAL) based on modular consensus frameworks and CDRs randomized with trinucleotides. *J Mol Biol* **296**: 57–86.

Kohler G., Milstein C. (1975). Continuous cultures of fused cells secreting antibody of pre-defined specificity. *Nature* **256**: 495–7.

Koide A., Wojcik J., Gilbreth R.N., Reichel A., Piehler J., Koide S. (2009). Accelerating phage-display library selection by reversible and site-specific biotinylation. *Protein Eng Des Sel* **22**: 685–90.

Kotlan B., Glassy M.C. (2009). Antibody phage display: Overview of a powerful technology that has quickly translated to the clinic. *Methods Mol Biol* **562**: 1–15.

Kramer R.A., Cox F., van der Horst M., van der Oudenrijn S., Res P.C., Bia J., Logtenberg T., de Kruif J. (2003). A novel helper phage that improves phage display selection efficiency by preventing the amplification of phages without recombinant protein. *Nucleic Acids Res* **31**: e59.

Krishnamoorthy G., Carlen E.T., Kohlheyer D., Schasfoort R.B., van den Berg A. (2009). Integrated Electrokinetic Sample Focusing and Surface Plasmon Resonance Imaging System for Measuring Biomolecular Interactions. *Anal Chem* **81**: 1957–63.

Kulkeaw K., Sakolvaree Y., Srimanote P., Tongtawe P., Maneewatch S., Sookrung N., Tungtrongchitr A., Tapchaisri P., Kurazono H., Chaicumpa W. (2009). Human monoclonal ScFv neutralize lethal Thai cobra, Naja kaouthia, neurotoxin. *J Proteomics* **72**: 270–82.

Kwong K.Y., Baskar S., Zhang H., Mackall C.L., Rader C. (2008). Generation, affinity maturation, and characterization of a human anti-human NKG2D monoclonal antibody with dual antagonistic and agonistic activity. *J Mol Biol* **384**: 1143–56.

Labrijn A.F., Koppelman M.H., Verhagen J., Brouwer M.C., Schuitemaker H., Hack C.E., Huisman H.G. (2002). Novel strategy for the selection of human recombinant Fab

Popkov M., Jendreyko N., Gonzalez-Sapienza G., Mage R.G., Rader C., Barbas C.F., 3rd (2004). Human/mouse cross-reactive anti-VEGF receptor 2 recombinant antibodies selected from an immune b9 allotype rabbit antibody library. *J Immunol Methods* **288**: 149–64.

Poul M.A. (2009). Selection of antibodies able to rapidly enter Mammalian cells. *Methods Mol Biol* **562**: 155–63.

Presta L.G. (2008). Molecular engineering and design of therapeutic antibodies. *Curr Opin Immunol* **20**: 460–70.

Rahim A.A. (2006). Pyrosequencing(R) of phage display libraries for the identification of cell-specific targeting ligands. *Methods Mol Biol* **373**: 135–46.

Rajewsky K. (1996). Clonal selection and learning in the antibody system. *Nature* **381**: 751–8.

Rauchenberger R., Borges E., Thomassen-Wolf E., Rom E., Adar R., Yaniv Y., Malka M., Chumakov I., Kotzer S., Resnitzky D., Knappik A., Reiffert S., Prassler J., Jury K., Waldherr D., Bauer S., Kretzschmar T., Yayon A., Rothe C. (2003). Human combinatorial Fab library yielding specific and functional antibodies against the human fibroblast growth factor receptor 3. *J Biol Chem* **278**: 38194–205.

Rhyner C., Weichel M., Fluckiger S., Hemmann S., Kleber-Janke T., Crameri R. (2004). Cloning allergens via phage display. *Methods* **32**: 212–8.

Rich R.L., Cannon M.J., Jenkins J., Pandian P., Sundaram S., Magyar R., Brockman J., Lambert J., Myszka D. G. (2008). Extracting kinetic rate constants from surface plasmon resonance array systems. *Anal Biochem* **373**: 112–20.

Rondot S., Koch J., Breitling F., Dubel S. (2001). A helper phage to improve single-chain antibody presentation in phage display. *Nat Biotechnol* **19**: 75–8.

Rothe C., Urlinger S., Lohning C., Prassler J., Stark Y., Jager U., Hubner B., Bardroff M., Pradel I., Boss M., Bittlingmaier R., Bataa T., Frisch C., Brocks B., Honegger A., Urban M. (2008). The human combinatorial antibody library HuCAL GOLD combines diversification of all six CDRs according to the natural immune system with a novel display method for efficient selection of high-affinity antibodies. *J Mol Biol* **376**: 1182–200.

Ruberti F., Cattaneo A., Bradbury A. (1994). The use of the RACE method to clone hybridoma cDNA when V region primers fail. *J Immunol Methods* **173**: 33–9.

Russel M., Kidd S., Kelley M.R. (1986). An improved filamentous helper phage for generating single-stranded plasmid DNA. *Gene* **45**: 333–8.

Safsten P., Klakamp S.L., Drake A.W., Karlsson R., Myszka D.G. (2006). Screening antibody-antigen interactions in parallel using Biacore A100. *Anal Biochem* **353**: 181–90.

Sassano M., Repetto M., Cassani G., Corti A. (1994). PCR amplification of antibody variable regions using primers that anneal to constant regions. *Nucleic Acids Res* **22**: 1768–9.

Sastry L., Alting-Mees M., Huse W.D., Short J.M., Sorge J.A., Hay B.N., Janda K.D., Benkovic S.J., Lerner R.A. (1989). Cloning of the immunological repertoire in Escherichia coli for generation of monoclonal catalytic antibodies: construction of a heavy chain variable region-specific cDNA library. *Proc Natl Acad Sci USA* **86**: 5728–32.

Savitsky P., Bray J., Cooper C.D., Marsden B.D., Mahajan P., Burgess-Brown N.A., Gileadi O. (2010). High-throughput production of human proteins for crystallization: The SGC experience. *J Struct Biol* **172**: 3–13.

Sblattero D., Bradbury A. (1998). A definitive set of oligonucleotide primers for amplifying human V regions. *Immunotechnology* **3**: 271–8.

Sblattero D., Bradbury A. (2000). Exploiting recombination in single bacteria to make large phage antibody libraries. *Nat Biotechnol* **18**: 75–80.

Scaviner D., Barbie V., Ruiz M., Lefranc M.P. (1999). Protein displays of the human immunoglobulin heavy, kappa and lambda variable and joining regions. *Exp Clin Immunogenet* **16**: 234–40.

Schable K.F., Zachau H.G. (1993). The variable genes of the human immunoglobulin kappa locus. *Biol Chem Hoppe Seyler* **374**: 1001–22.

Schaible U., Liss M., Prohaska E., Decker J., Stadtherr K., Wolf H. (2004). Affinity measurements of biological molecules by a quartz crystal microbalance (QCM) biosensor. *Methods Mol Med* **94**: 321–30.

Schofield D.J., Pope A.R., Clementel V., Buckell J., Chapple S., Clarke K.F., Conquer J.S., Crofts A.M., Crowther S.R., Dyson M.R., Flack G., Griffin G.J., Hooks Y., Howat W.J., Kolb-Kokocinski A., Kunze S., Martin C.D., Maslen G.L., Mitchell J.N., O'Sullivan M., Perera R.L., Roake W., Shadbolt S.P., Vincent K.J., Warford A., Wilson W.E., Xie J., Young J.L., McCafferty J. (2007). Application of phage display to high throughput antibody generation and characterization. *Genome Biol* **8**: R254.

Scholle M.D., Collart F.R., Kay B.K. (2004). *In vivo* biotinylated proteins as targets for phage-display selection experiments. *Protein Expr Purif* **37**: 243–52.

Schoonbroodt S., Frans N., DeSouza M., Eren R., Priel S., Brosh N., Ben-Porath J., Zauberman A., Ilan E., Dagan S., Cohen E.H., Hoogenboom H.R., Ladner R.C., Hoet R.M. (2005). Oligonucleotide-assisted cleavage and ligation: a novel directional DNA cloning technology to capture cDNAs. Application in the construction of a human immune antibody phage-display library. *Nucleic Acids Res* **33**: e81.

Scott J.K., Smith G.P. (1990). Searching for peptide ligands with an epitope library. *Science* **249**: 386–90.

Secchiero P., Sblattero D., Chiaruttini C., Melloni E., Macor P., Zorzet S., Tripodo C., Tedesco F., Marzari R., Zauli G. (2009). Selection and characterization of a novel agonistic human recombinant anti-TRAIL-R2 minibody with anti-leukemic activity. *Int J Immunopathol Pharmacol* **22**: 73–83.

Seehaus T., Breitling F., Dubel S., Klewinghaus I., Little M. (1992). A vector for the removal of deletion mutants from antibody libraries. *Gene* **114**: 235–7.

Sexton D.J., Chen T., Martik D., Kuzmic P., Kuang G., Chen J., Nixon A.E., Zuraw B.L., Forteza R.M., Abraham W. M., Wood C.R. (2009). Specific inhibition of tissue kallikrein 1 with a human monoclonal antibody reveals potential role in airway diseases. *Biochem J* **422**: 383–92.

Sheets M.D., Amersdorfer P., Finnern R., Sargent P., Lindquist E., Schier R., Hemingsen G., Wong C., Gerhart J.C., Marks J.D. (1998). Efficient construction of a large nonimmune phage antibody library: the production of high-affinity human single-chain antibodies to protein antigens. *Proc Natl Acad Sci USA* **95**: 6157–62.

Sidhu S.S., Li B., Chen Y., Fellouse F.A., Eigenbrot C., Fuh G. (2004). Phage-displayed antibody libraries of synthetic heavy chain complementarity determining regions. *J Mol Biol* **338**: 299–310.

Silacci M., Brack S., Schirru G., Marlind J., Ettorre A., Merlo A., Viti F., Neri D. (2005). Design, construction, and characterization of a large synthetic human antibody phage display library. *Proteomics* **5**: 2340–50.

Simmons L.C., Reilly D., Klimowski L., Raju T.S., Meng G., Sims P., Hong K., Shields R.L., Damico L.A., Rancatore P., Yansura D.G. (2002). Expression of full-length immunoglobulins in Escherichia coli: rapid and efficient production of aglycosylated antibodies. *J Immunol Methods* **263**: 133–47.

Skerra A., Pluckthun A. (1988). Assembly of a functional immunoglobulin Fv fragment in Escherichia coli. *Science* **240**: 1038–41.

Smith G.P. (1985). Filamentous fusion phage: novel expression vectors that display cloned antigens on the virion surface. *Science* **228**: 1315–17.

Soderlind E., Strandberg L., Jirholt P., Kobayashi N., Alexeiva V., Aberg A.M., Nilsson A., Jansson B., Ohlin M., Wingren C., Danielsson L., Carlsson R., Borrebaeck C.A. (2000). Recombining germline-derived CDR sequences for creating diverse single-framework antibody libraries. *Nat Biotechnol* **18**: 852–6.

Soltes G., Hust M., Ng K.K., Bansal A., Field J., Stewart D.I., Dubel S., Cha S., Wiersma E.J. (2007). On the influence of vector design on antibody phage display. *J Biotechnol* **127**: 626–37.

Steiner D., Forrer P., Pluckthun A. (2008). Efficient selection of DARPins with subnanomolar affinities using SRP phage display. *J Mol Biol* **382**: 1211–27.

Steiner D., Forrer P., Stumpp M.T., Pluckthun A. (2006). Signal sequences directing cotranslational translocation expand the range of proteins amenable to phage display. *Nat Biotechnol* **24**: 823–31.

Steukers M., Schaus J.M., van Gool R., Hoyoux A., Richalet P., Sexton D.J., Nixon A.E., Vanhove M. (2006). Rapid kinetic-based screening of human Fab fragments. *J Immunol Methods* **310**: 126–35.

Stocker G., Dumoulin D., Vandevyver C., Hilbrig F., Freitag R. (2008). Screening of a human antibody phage display library against a peptide antigen using stimuli-responsive bioconjugates. *Biotechnol Prog* **24**: 1314–24.

Su B., Hrin R., Harvey B.R., Wang Y.J., Ernst R.E., Hampton R.A., Miller M.D., Strohl W. R., An Z., Montgomery D.L. (2007). Automated high-throughput purification of antibody fragments to facilitate evaluation in functional and kinetic based assays. *J Immunol Methods* **322**: 94–103.

Tan L.P., Lue R.Y., Chen G.Y., Yao S.Q. (2004). Improving the intein-mediated, site-specific protein biotinylation strategies both in vitro and in vivo. *Bioorg Med Chem Lett* **14**: 6067–70.

Thie H., Meyer T., Schirrmann T., Hust M., Dubel S. (2008). Phage display derived therapeutic antibodies. *Curr Pharm Biotechnol* **9**: 439–46.

Thie H., Voedisch B., Dubel S., Hust M., Schirrmann T. (2009). Affinity maturation by phage display. *Methods Mol Biol* **525**: 309–22, xv.

Tomlinson I.M., Cox J.P., Gherardi E., Lesk A.M., Chothia C. (1995). The structural repertoire of the human V kappa domain. *Embo J* **14**: 4628–38.

Tomlinson I.M., Walter G., Jones P.T., Dear P.H., Sonnhammer E.L., Winter G. (1996). The imprint of somatic hypermutation on the repertoire of human germline V genes. *J Mol Biol* **256**: 813–7.

Tomlinson I.M., Walter G., Marks J.D., Llewelyn M.B., Winter G. (1992). The repertoire of human germline VH sequences reveals about fifty groups of VH segments with different hypervariable loops. *J Mol Biol* **227**: 776–98.

Townsend S., Finlay W.J., Hearty S., O'Kennedy R. (2006). Optimizing recombinant antibody function in SPR immunosensing. The influence of antibody structural format and chip surface chemistry on assay sensitivity. *Biosens Bioelectron* **22**: 268–74.

Turunen L., Takkinen K., Soderlund H., Pulli T. (2009). Automated panning and screening procedure on microplates for antibody generation from phage display libraries. *J Biomol Screen* **14**: 282–93.

Valentine R.C., Green N.M. (1967). Electron microscopy of an antibody-hapten complex. *J Mol Biol* **27**: 615–7.

van Dijk K.W., Mortari F., Kirkham P.M., Schroeder H.W., Jr., Milner E.C. (1993). The human immunoglobulin VH7 gene family consists of a small, polymorphic group of six to eight gene segments dispersed throughout the VH locus. *Eur J Immunol* **23**: 832–9.

van Wezenbeek P. M., Hulsebos T.J., Schoenmakers J.G. (1980). Nucleotide sequence of the filamentous bacteriophage M13 DNA genome: comparison with phage fd. *Gene* **11**: 129–48.

Vanhercke T., Ampe C., Tirry L., Denolf P. (2005). Rescue and in situ selection and evaluation (RISE): a method for high-throughput panning of phage display libraries. *J Biomol Screen* **10**: 108–17.

Vaughan T.J., Williams A.J., Pritchard K., Osbourn J.K., Pope A.R., Earnshaw J.C., McCafferty J., Hodits R.A., Wilton J., Johnson K.S. (1996). Human antibodies with sub-nanomolar affinities isolated from a large non-immunized phage display library. *Nat Biotechnol* **14**: 309–14.

Vieira J., Messing J. (1987). Production of single-stranded plasmid DNA. *Methods Enzymol* **153**: 3–11.

Wagner S.D., Neuberger M.S. (1996). Somatic hypermutation of immunoglobulin genes. *Annu Rev Immunol* **14**: 441–57.

Ward E.S., Gussow D., Griffiths A.D., Jones P.T., Winter G. (1989). Binding activities of a repertoire of single immunoglobulin variable domains secreted from Escherichia coli. *Nature* **341**: 544–6.

Ward R.L., Clark M.A., Lees J., Hawkins N.J. (1996). Retrieval of human antibodies from phage-display libraries using enzymatic cleavage. *J Immunol Methods* **189**: 73–82.

Wassaf D., Kuang G., Kopacz K., Wu Q.L., Nguyen Q., Toews M., Cosic J., Jacques J., Wiltshire S., Lambert J., Pazmany C.C., Hogan S., Ladner R.C., Nixon A.E., Sexton D.J. (2006). High-throughput affinity ranking of antibodies using surface plasmon resonance microarrays. *Anal Biochem* **351**: 241–53.

Waterhouse P., Griffiths A.D., Johnson K.S., Winter G. (1993). Combinatorial infection and in vivo recombination: a strategy for making large phage antibody repertoires. *Nucleic Acids Res* **21**: 2265–6.

Welschof M., Terness P., Kolbinger F., Zewe M., Dubel S., Dorsam H., Hain C., Finger M., Jung M., Moldenhauer G., *et al.* (1995). Amino acid sequence based PCR primers for amplification of rearranged human heavy and light chain immunoglobulin variable region genes. *J Immunol Methods* **179**: 203–14.

Werther W.A., Gonzalez T.N., O'Connor S.J., McCabe S., Chan B., Hotaling T., Champe M., Fox J.A., Jardieu P.M., Berman P.W., Presta L.G. (1996). Humanization of an anti-lymphocyte function-associated antigen (LFA)-1 monoclonal antibody and reengineering of the humanized antibody for binding to rhesus LFA-1. *J Immunol* **157**: 4986–95.

Williams B.R., Sharon J. (2002). Polyclonal anti-colorectal cancer Fab phage display library selected in one round using density gradient centrifugation to separate antigen-bound and free phage. *Immunol Lett* **81**: 141–8.

Williams S.C., Frippiat J.P., Tomlinson I.M., Ignatovich O., Lefranc M.P., Winter G. (1996). Sequence and evolution of the human germline V lambda repertoire. *J Mol Biol* **264**: 220–32.

Winter G., Griffiths A.D., Hawkins R.E., Hoogenboom H.R. (1994). Making antibodies by phage display technology. *Annu Rev Immunol* **12**: 433–55.

Winter G., Milstein C. (1991). Man-made antibodies. *Nature* **349**: 293–9.

Wu S., Cygler M. (1993). Conformation of complementarity determining region L1 loop in murine IgG lambda light chain extends the repertoire of canonical forms. *J Mol Biol* **229**: 597–601.

Wu T.T., Johnson G., Kabat E.A. (1993). Length distribution of CDRH3 in antibodies. *Proteins* **16**: 1–7.

Wu T.T., Kabat E.A. (1970). An analysis of the sequences of the variable regions of Bence Jones proteins and myeloma light chains and their implications for antibody complementarity. *J Exp Med* **132**: 211–50.

Wu Y., Cain-Hom C., Choy L., Hagenbeek T.J., de Leon G.P., Chen Y., Finkle D., Venook R., Wu X., Ridgway J., Schahin-Reed D., Dow G.J., Shelton A., Stawicki S., Watts R.J., Zhang J., Choy R., Howard P., Kadyk L., Yan M., Zha J., Callahan C.A., Hymowitz S.G., Siebel C.W. (2010). Therapeutic antibody targeting of individual Notch receptors. *Nature* **464**: 1052–7.

Wu Y., Simons P.C., Lopez G.P., Sklar L.A., Buranda T. (2005). Dynamics of fluorescence dequenching of ostrich-quenched fluorescein biotin: a multifunctional quantitative assay for biotin. *Anal Biochem* **342**: 221–8.

Yang G.H., Yoon S.O., Jang M.H., Hong H.J. (2007). Affinity maturation of an anti-hepatitis B virus PreS1 humanized antibody by phage display. *J Microbiol* **45**: 528–33.

Yuan Q.A., Robinson M.K., Simmons H.H., Russeva M., Adams G.P. (2008). Isolation of anti-MISIIR scFv molecules from a phage display library by cell sorter biopanning. *Cancer Immunol Immunother* **57**: 367–78.

Zacchi P., Sblattero D., Florian F., Marzari R., Bradbury A.R. (2003). Selecting open reading frames from DNA. *Genome Res* **13**: 980–90.

Zachau H.G. (1993). The immunoglobulin kappa locus-or-what has been learned from looking closely at one-tenth of a percent of the human genome. *Gene* **135**: 167–73.

Zacher A.N., 3rd, Stock C.A., Golden J.W., 2nd, Smith G.P. (1980). A new filamentous phage cloning vector: fd-tet. *Gene* **9**: 127–40.

Zemlin M., Klinger M., Link J., Zemlin C., Bauer K., Engler J.A., Schroeder H.W., Jr., Kirkham P.M. (2003). Expressed murine and human CDR-H3 intervals of equal length exhibit distinct repertoires that differ in their amino acid composition and predicted range of structures. *J Mol Biol* **334**: 733–49.

Zhao Q., Chan Y.W., Lee S.S., Cheung W.T. (2009). One-step expression and purification of single-chain variable antibody fragment using an improved hexahistidine tag phagemid vector. *Protein Expr Purif* **68**: 190–5.

Zhou B., Carney C., Janda K.D. (2008). Selection and characterization of human antibodies neutralizing Bacillus anthracis toxin. *Bioorg Med Chem* **16**: 1903–13.

3

Affinity Maturation Approaches for Antibody Lead Optimization

David Lowe, Trevor Wilkinson* and Tristan J. Vaughan**

3.1 Introduction

Monoclonal antibodies (mAbs) and their derivatives (e.g. antibody fragments) represent the largest group of therapeutic proteins in development, with approximately two dozen mAbs currently approved across a range of therapeutic areas (Parren and van de Winkel, 2008; Dimitrov and Marks, 2009). The design of a therapeutic antibody now involves multiple considerations once the target has been selected. The antibody should bind to the target antigen epitope with high affinity and specificity compared to closely related human homologs. It is often also desirable to show binding to a relevant species ortholog to facilitate preclinical studies required for translational strategies to the clinic. For some therapeutic indications, modulation of the effector functions (CDC, ADCC) of the antibody may be desirable. These modulations are targeted to the Fc region of the antibody and can include effector function enhancement, effector function abrogation, or modulation of the half life of the antibody by engineering changes in FcRn binding (Wang *et al.*, 2008; Tabrizi and Roskos, 2007). Optimization of a lead antibody can therefore be focused on both antigen recognition and binding and/or effector function. For the purpose of this review, we will concentrate on how antigen recognition is improved through various affinity maturation platforms and methodologies. The topic of Fc engineering has been covered in a number of recent reviews and key articles (Presta, 2008; Labrijn *et al.*, 2008; Lazar *et al.*, 2006). In this chapter we will address considerations for the optimization of

* Medimmune Ltd., UK.

affinity, technologies available for optimization and the impact of structural knowledge on affinity maturation.

3.2 Measuring the Affinity of Antibody–Antigen Complexes

High-affinity antibody binding interactions are important for pharmaceutical applications. Antibody affinity is a quantitative measure of the strength of association between a single antibody-binding site and a single antigen-binding site. The affinity for a defined antibody–antigen interaction determines the number of antigen molecules that will be bound by antibody at any given concentration of antibody and antigen. Affinity is, therefore, a critical parameter affecting the potency of an antibody therapeutic and will affect the dose and dosing interval for the antibody (Roskos *et al.*, 2007).

Affinity describes the strength of the reversible association between an antibody and antigen as described in the following equation:

$$\mathrm{Ab} + \mathrm{Ag} \underset{k_{off}\,(\text{off rate, } s^{-1})}{\overset{k_{on}\,(\text{on rate, } M^{-1}s^{-1})}{\rightleftarrows}} \mathrm{Ab}\cdot\mathrm{Ag}.$$

The affinity is commonly described as the equilibrium dissociation constant (K_D), which may be calculated by dividing k_{off} by k_{on} and describes the microscopic or site-binding constant (Tabrizi *et al.*, 2008). This definition assumes that the affinity is measured between an antibody-binding site and the antigen in a monovalent interaction. If multivalent interactions occur, giving rise to cooperative interactions, this can produce avid interactions, and the strength of such interactions may be several orders of magnitude stronger than the single antibody-binding site–antigen interaction. The potential for multivalent interaction must be taken into account when undertaking an antibody optimization process. Given the relationship between dissociation constant and the on and off rate, affinity maturation is typically monitored by techniques that measure these parameters, such as surface plasmon resonance. Improvements in affinity typically translate into *in vitro* potency improvements and improved clinical efficacy. These techniques can also be used during affinity maturation to provide information on how affinity changes correlate with changes in antibody sequence, and to monitor and select antibody drug candidates with suitable properties with respect to binding to orthologous antigens (for selection of relevant pharmacology and/or toxicology species) and to homologous antigens (to address potential safety considerations).

3.2.1 *Improving antibody affinity and potency*

Therapeutic applications typically require antibodies with affinities in the sub-nanomolar range; this is true for the large majority of mAbs in clinical development. There are certain scenarios, such as neutralizing a cytokine that is present at extremely low concentrations, that require sub-picomolar affinities (Rathanaswami *et al.*, 2005). In such cases, antibodies derived from either *in vitro* display platforms or from *in vivo* immunizations will likely require some additional affinity optimization. Improvements in antibody affinity often correlate with improvements in potency, but more importantly, *in vivo* efficacy (as a surrogate for clinical efficacy) is the driving force behind engineering to increase affinity (Wu *et al.*, 1998; Viti *et al.*, 1999; Johnson *et al.*, 1999; Maynard *et al.*, 2002). It is important during the optimization process to monitor both affinity and *in vitro* potency, as affinity improvements do not always necessarily correlate with improved potency. This is often true for an antibody with an agonistic function, such as one that acts as an erythropoietin mimetic. This antibody, Ab 12.6 ($K_D = 32$ nM), was derived from a parental antibody, Ab 12 ($K_D = 9$ nM), and was demonstrated to have enhanced biological efficacy when compared to the parental antibody (Lacy *et al.*, 2008). In this case the affinity and potency were inversely correlated. In general, however, affinity, potency, and *in vivo* efficacy have a close relationship with each other.

A recent example illustrating the link between increased affinity for a target antigen and improved potency is the development of motavizumab, an optimized variant of the antibody palivizumab. Palivizumab is directed against the A antigenic site of the respiratory syncytial virus (RSV) F protein and has the ability to broadly neutralize clinical isolates of RSV A and B subtypes (Wu *et al.*, 2005). Motavizumab was developed by affinity maturation of palivizumab. Initial maturation studies led to the isolation of an antibody, A4b4, which demonstrated a 27 pM affinity for RSV F protein (127-fold enhancement over parent) and a 44-fold enhanced RSV neutralization potency *in vitro* (Wu *et al.*, 2007). *In vivo* studies demonstrated that this enhanced *in vitro* potency did not translate well to efficacy, and this was attributed to broad tissue binding. Further directed engineering of a few specific residues, by reversion back to the original residues, diminished tissue binding and produced the final affinity-matured antibody, motavizumab. Motavizumab demonstrates a 70-fold improvement in binding affinity to the RSV F protein ($K_D = 34.6$ pM compared to 2634 pM) (Wu *et al.*, 2007).

3.3 *In Vivo* Affinity Maturation

When considering strategies for the optimization of antibodies *in vitro*, it is instructive to understand the *in vivo* processes that occur during an immune response. Generation of high-affinity specific antibodies is critical for the resistance of vertebrate species to pathogenic organisms. During the humoral response, low-affinity immunoglobulin M (IgM) antibodies to a given target, such as a toxin or virus, are initially generated; these are then superseded by higher-affinity IgG, IgE or IgA molecules to the same epitope. Maturation of the immune response is critical for survival. There are a number of hyper-IgM syndromes (HIGM) that exist in humans, all characterized by an inability to switch to these higher-affinity antibodies and often fatal early in life (Revy *et al.*, 2000). We focus here on ways in which these affinity maturation mechanisms can be mimicked *in vitro*. While an overview is provided below, more detailed reviews of the mechanisms involved in the natural immune response include Maizels (2005); Di Noia and Neuberger (2007); and Shivarov *et al.* (2009).

At a cellular level, affinity maturation of an immune response is characterized by recognition of the cognate antigen by a B lymphocyte and co-stimulation with helper T lymphocytes, which lead to the entry of the B cells into the germinal center of the peripheral lymphoid organs and rapid proliferation before finally differentiating into either plasma or memory cells. This entire process, from the initial encounter with the antigen to the transformation of the low-affinity IgM response to a high-affinity IgG, IgA or IgE response, is rapid and can occur within a few days.

3.3.1 *Mechanisms of diversification*

Somatic hypermutation (SHM) is characterized by the introduction of point mutations into the variable regions of immunoglobulin genes. This occurs at a rate of approximately one mutation per 1,000 bp per cell generation, a rate that is 10^6 times higher than that for spontaneous mutagenesis in somatic cells (Liu and Schatz, 2009). Within the V genes, SHM hotspots have been shown to correlate with the regions of the antibody that form the main antigen-binding sites, known as the complementarity determining regions (CDRs) (Berek and Milstein, 1987). SHM consists of two steps: the generation of a DNA lesion through base deamination, followed by error-prone repair of the lesion. The key enzyme that is central to the process of SHM is activation-induced deaminase (AID) (Muramatsu *et al.*, 1999). AID expression is specifically restricted to activated B cells within the germinal centers and is thought to be induced by mechanisms that include interleukin (IL) 4 signaling and CD40 ligation (Zhou *et al.*, 2003; Dedeoglu *et al.*, 2004). AID targets single-stranded DNA and deaminates deoxycytidine residues, converting them to uracils. This creates a mismatch between U:G that can be resolved by a number of enzymes, such as uracil nucleoside glycosylase (UNG), which removes the uracil base, and can then lead to transversions and transitions, via the effects of other enzymes. Alternatively, the U:G mispairing can be recognized by the MSH2/MSH6 mismatch recognition heterodimer, leading to a DNA synthesis process whereby further mutations can be introduced. AID preferentially deaminates deoxycytidines within the sequence context WRC, where W = A/T and R = purine. There is a good correlation between the sites of hypermutation hotspots within the mammalian V genes in the CDRs and the presence of this WRC consensus motif (Rogozin and Kolchanov, 1992).

The process by which an IgM molecule is converted to an IgG, IgA, or IgE molecule during affinity maturation is similarly triggered by AID-mediated DNA deamination (Chaudhuri and Alt, 2004). The heavy chain variable (VDJ) region is detached from the constant region gene of IgM (Cμ) and joined with an alternative constant region gene such as IgG (Cγ), deleting the DNA between the two gene segments. Recombination junctions occur within so-called switch (S) regions, which are guanine-rich regions within the introns within the constant region genes. AID-mediated deamination in the vicinity of the switch regions of both the donor and acceptor constant genes drives this recombination at a molecular level.

Following the generation of a repertoire of somatically hypermutated, class-switched B cell receptor variants within the germinal centers, the small proportion of cells expressing receptors with a higher affinity for the antigen is selectively expanded, due to preferential engagement with the helper T cells compared to those with a lower affinity for the antigen. This allows the formation of long-lived plasma cells and memory B cells, via a helper T cell-dependent process (Allen *et al.*, 2007).

3.3.2 *Kinetic limitations of the in vivo immune response*

Antibodies generated *in vivo* generally fall within a broad affinity (K_D) range of 50 pM to 1 μM (Foote and Eisen, 2000). The upper limit of affinity selection *in vivo* is composed of a combination of k_{on} and k_{off} limiting factors. Protein–protein interaction intrinsic k_{on} rates are limited by the diffusion coefficients of the two molecules in a 1:1 interaction (Foote and Eisen, 1995). For antibodies, this corresponds to an upper k_{on} limit of approximately 10^6 $M^{-1}s^{-1}$ and should be identical for antibodies derived from both *in vivo* and *in vitro* methodologies. For k_{off}, the limitation *in vivo* is affected by the residence time of an antigen complexed to a B cell surface receptor. Once the k_{off} of an antigen complexed to a B cell receptor is above the threshold necessary for uptake within the B cell (leading to the internalization and processing of the antigen), any further improvements in these kinetics will not manifest as a selective advantage. These limitations result in an approximate K_D ceiling of 100 pM for the positive selection of antibody affinity-matured antibodies *in vivo*. It should be noted, however, that antibodies with higher affinities than this can be derived, but are simply rarer in the population. The isolation of a mAb with a reported sub-picomolar affinity for the chemokine IL-8 has, for example, been reported following immunization of a humanized transgenic mouse (Rathanaswami *et al.*, 2005).

3.4 *In Vitro* Affinity Maturation

With the invention of *in vitro* antibody engineering techniques in the early 1990s, the ability to generate antibodies with affinities beyond those that are readily

multiple laboratories around the world. Some of the earliest uses of phage display were to develop methods for improving the affinity of antibodies, whether derived from naïve immune repertoires or from *in vivo* immunization. In order to achieve this, two key steps are required: the diversification of the antibody sequence, and the enrichment and isolation of the clones with improved binding kinetics.

Although phage display was largely developed as the platform of choice, other display systems can also be employed (Fig. 3.2). The wide use of phage display is based on its advantages over other systems, such as its simplicity and the stability of the phage particles, which lend an overall robustness to the methodology. This, in turn, facilitates working with more complex antigen classes, such as cell or tissue surfaces (Liu *et al.*, 2004; Tordsson *et al.*, 1997).

A large number of mutagenesis strategies have been employed to introduce diversity into the parental antibody sequence prior to selection of improved variants. The earliest approach utilized random mutagenesis using error-prone PCR amplification of the antibody V genes, followed by iterative rounds of phage selection using decreasing concentrations of antigen in solution, in order to enrich for variants with lower K_D (Hawkins *et al.*, 1992). This method relies on having an antigen that is labeled (e.g. by biotinylation), as well as a way of capturing the labeled antigen-phage complexes (e.g. with use of a streptavidin-coated surface, such as a bead). Keeping the antigen in solution during the selection allows the concentration to be controlled below the K_D of the parental clone and also reduces the effects of avidity due to selection of antibody fragments with a greater propensity to dimerize (Schier *et al.*, 1996b). Alternatively, specific selection for antibody variants with lower off rates can be performed by carrying out solution phase selections in the presence of excess levels of a competitor such as unlabeled antigen or antibody (Hawkins *et al.*, 1992).

3.4.1.1 *Diversification of antibody sequences*

Generation of diverse variants of a given antibody can be achieved in three general ways that may be carried out individually, or in combination: (1) antibody chain shuffling; (2) targeting the mutagenesis to particular areas of the sequence, such as CDRs; (3) random mutagenesis.

3.4.1.2 *Antibody chain shuffling*

One of the processes by which antibodies can undergo affinity maturation *in vivo* is the exchange of heavy and light chains following the initial immune response, and the selection of new pairings that confer high binding affinity, a process known as “repertoire shift” (Bereck and Milstein, 1987). This process can be readily mimicked *in vitro*, thanks to the modular nature of antibody structures coupled with the technical simplicity of manipulating antibody genes using standard molecular biological methods. An early successful use of phage display for antibody affinity maturation employed a procedure whereby the light chain of an antibody against the hapten 2-phenyloxazol-one (phOx) was replaced by a random repertoire of light chains obtained from unimmunized donors, followed by selection for higher affinity, leading to a 20-fold improvement in affinity (Marks *et al.*, 1992). This experiment also demonstrated that the heavy chain CDR1 and 2 loops could also be replaced by repertoires derived from random donors, leading to further improvements. Other laboratories have subsequently demonstrated the utility of this approach for the maturation of numerous antibodies to a wide range of targets, such as HIV-1 gp120 (Barbas *et al.*, 1993; Barbas *et al.*, 1994), c-erbB-2 (Schier *et al.*, 1996b), and the chemokine MCP-1 (Yoshinga *et al.*, 2008). Indeed, experience in our own laboratory indicates that approximately ten-fold improvements in affinity may be readily achieved with this approach and considerably more when combined with mutations in heavy chain CDRs (Main *et al.*, 2006).

3.4.1.3 *Targeted mutagenesis*

The high degree of sequence similarity that characterizes antibodies, as well as the extensive knowledge of antibody protein structure gained from solved crystal structures, make antibodies ideal candidates for targeted diversification. There are many examples of *in vitro* affinity maturation approaches that have targeted the CDR loops that make up the antigen-binding pocket. The most common strategy for targeted randomization of antibodies is to employ degenerate oligonucleotides containing codon equivalents NNS or NNK. In these, N represents an equal mixture of the four possible nucleotides A, C, G, and T, whereas S is a mixture of only G and C, and K is a mixture of just G and T. These designs allow the incorporation of any of the 20 common amino acids, while minimizing the number of potential stop codons. It should be noted that each of the possible amino acids will not necessarily be represented equally in the oligonucleotide mix. For example, NNS mutagenesis results in leucine, serine, and arginine being three times more likely to be encoded than histidine. The choice of how many and which amino acids within the CDRs to mutate is complex. The six CDRs of an antibody can contain 60–80 amino acids in total. If each of these positions was theoretically mutated to any one of the twenty common amino acids via NNS mutagenesis, a total diversity of $1 \times 32^{60} = 2.0 \times 10^{90}$ individual recombinants would be required to cover the entire sequence space. Clearly, this is far too large to be practical! Perhaps more importantly, mutation to all 20 amino acids at multiple positions in the CDRs simultaneously is likely to drastically change the architecture of the binding

Table 3.1 Library sizes required to ensure coverage of theoretical diversity encoded by NNS or NNK codon mutagenesis in up to 7 positions[a]

No. Residues	Total Diversity	Library Size (90% Confidence[a])	Library Size (99% Confidence[a])
3	3.3×10^4	7.6×10^4	1.5×10^5
4	1.1×10^6	2.4×10^6	4.9×10^6
5	3.4×10^7	7.7×10^7	1.6×10^8
6	1.1×10^9	2.5×10^9	5.0×10^9
7	3.4×10^{10}	7.9×10^{10}	1.6×10^{11}

[a] calculated based on Poisson distribution.

pocket, leading to changes in specificity and thus reducing the functional library size. Practically, the maximum library sizes obtainable for phage display are between 10^9 and 10^{11}, due to the limitations of bacterial transformation efficiency (Vaughan *et al.*, 1996). These limitations have resulted in affinity maturation strategies whereby smaller blocks of residues are mutated (Glaser *et al.*, 1992; Schier *et al.*, 1996b). Mutating six positions of an antibody to any of the 20 common amino acids, via NNS or NNK codon mutagenesis, generates a diversity of $1 \times 32^6 = 1.1 \times 10^9$, which should therefore be covered efficiently in a phage library with a size of 5×10^9 transformants, based upon a 99% confidence value in the Poisson distribution (Table 3.1). This is a highly effective method as it explores all the available sequence space at the chosen positions. Given this practical limitation of six positions for mutagenesis per phage library, the choice of which CDR loops to target in such a way is central to this strategy. Heavy and light chain CDR3 loops are typically found at the center of the antigen-binding pocket, with heavy chain CDR3 in particular forming multiple interactions with the antigen (Amit *et al.*, 1986). In addition to its importance in conferring the specificity of an antibody, V_H CDR3 is an area that can be readily mutated to increase affinity (Chen *et al.*, 1999; Furukuwa *et al.*, 2001; Main *et al.*, 2006; Thom *et al.*, 2006). In our laboratory, numerous antibodies have been optimized with up to >1000-fold improvements in affinity by targeted mutagenesis of V_H and V_L CDR3 using phage display (unpublished results).

Alternatives to this saturation approach include methods to bias the mutagenesis towards the parent sequence or towards sterically or chemically conservative changes, a strategy that has been termed parsimonious mutagenesis (Yelton *et al.*, 1995; Schier *et al.*, 1996c; Thompson *et al.*, 1996). Another approach is to generate smaller libraries of antibody variants, each with a single amino acid mutation at every single position of the six CDRs (Wu *et al.*, 1998). Targeting the naturally occurring hotspots seen in the *in vivo* somatic hypermutation has also proved to be a successful approach (Chowdhury and Pastan, 1999), where an anti-CD22 antibody is mutated at specific positions in its CDRs, leading to ten-fold improvements in potency and affinity (Ho *et al.*, 2005).

Whichever methodology is employed, one of the most powerful advantages of such *in vitro* systems is the ability to recombine beneficial mutations within a molecule to achieve exponential increases in affinity due to synergistic kinetic or thermodynamic improvements. Examples from multiple laboratories include the affinity maturation of an anti-Lewis Y antibody by recombining mutations in the three heavy chain CDRs, leading to 65-fold improvement in binding to carcinoma cells (Yelton *et al.*, 1995); an anti-VEGF antibody, whereby mutations similarly in the heavy chain CDRs were recombined resulting in a 100-fold improvement in affinity (Chen *et al.*, 1999); and an anti-human GM-CSF antibody, whereby mutations in the light chain CDR3 were recombined with changes in the heavy chain CDR2, leading to a 5000-fold affinity improvement (Steidl *et al.*, 2008).

An alternative strategy for recombining advantageous mutations, termed "CDR-walking" (Yang *et al.*, 1995), has been used successfully in a number of cases. This method involves the targeting of a CDR, or several CDRs, in independent libraries by saturation mutagenesis, followed by screening for affinity improvements. The mutated variant with the highest affinity gain is then used as the starting point for further mutagenesis of a different CDR, leading to the isolation of variants with even greater affinity improvements. Using this strategy, an anti-HIV-1 gp120 antibody was optimized from a starting affinity of 6 nM to a final affinity of 15 pM. The disadvantage of this method, aside from its sequential nature, is that only one set of CDR changes is recombined with new mutations at any one time, risking the potential loss of synergistic improvements from combinations of mutations that alone may show only modest effects on affinity, but that combined may result in the most optimal gains. An alternative approach is to recombine whole populations of independent CDR-mutated phage library selection outputs and select for further affinity gains. This enables the pairing of potentially thousands of improved variants with mutations in particular CDRs with each other, providing a much greater chance of generating large synergistic improvements (Dufner *et al.*, 2006).

3.4.1.4 *Random mutagenesis*

Random mutagenesis refers to the introduction of mutations anywhere within the antibody sequence, and so targets both the CDR loops and the underlying framework regions within the Fv. The earliest examples of *in vitro* affinity maturation using phage display employed an error-prone PCR methodology (Leung *et al.*, 1989) to introduce variation into the antibody sequence (Hawkins *et al.*, 1992), and this remains a simple and successful way to generate a randomly mutated library for phage (Crameri *et al.*, 1996) as well as other protein display approaches such as ribosome (Jermutus *et al.*, 2001), bacterial (Daugherty *et al.*, 2000), and yeast surface display (Boder *et al.*, 2000). The DNA polymerase used in PCR will naturally introduce a low level of error during replication, which can be exploited to introduce low levels of diversity into a protein sequence. Error-prone PCR extends this property further and relies on manipulation of the Mg^{2+} concentrations in the PCR mixture to provoke imperfect copying and hence mutagenesis. By adjusting the Mg^{2+} concentrations, the frequency of mutation can be broadly controlled. An alternative but less widely used method is the use of *E. coli* mutator strains, such as MutD5 (Irving *et al.*, 1996; Low *et al.*, 1996), between rounds of phage display selection, which have been used to successfully improve antibody fragment affinities.

DNA shuffling (Stemmer, 1994) is a method whereby random segments of genes are homologously recombined and screened for improved properties such as affinity. This technique has been applied to antibody variable domain genes in both phage (Crameri *et al.*, 1996; Fermer *et al.*, 2004) and ribosome display (Jermutus *et al.*, 2001; Chodorge *et al.*, 2008), and when coupled with stringency selections with decreasing concentrations has led to affinity improvements of up to three orders of magnitude (Fermer *et al.*, 2004). Both shuffling and random point mutagenesis have been shown to be particularly useful when coupled with targeted (CDR) mutagenesis approaches. Simultaneous evolution of both targeted and non-targeted regions, which could not necessarily be predicted without detailed structural information, have the potential to generate variants with significant synergistic improvements in affinity and potency. Additionally, random mutagenesis, which naturally interrogates the entire length of the antibody V-region, is capable of identifying antibody residues that impart affinity improvements that may be non-obvious, such as within the more outlying CDRs or within the framework regions. Our laboratory used such an approach to identify individual hotspots within multiple CDRs within an anti-IL-13 antibody, which were combined with mutations derived from targeted CDR mutagenesis to generate a highly potent therapeutic lead molecule over 300-fold more potent than the parent antibody (Thom *et al.*, 2006).

Phage display has been the leading methodology for *in vitro* antibody evolution for many years thanks to its simplicity, robustness, and high frequency of success. The major limitation of phage display, common to all protein evolution systems involving cellular expression of the variant library, is the transformation efficiency of the host strain. For phage display, the transformation efficiency of bacteria such as *Escherichia coli* limits the library size to around 10^{10}. The potential sequence space that can thus be interrogated is limited to a handful of positions per individual variant library. As detailed above, numerous strategies have been developed to surmount this limitation, often exploiting combinatorial approaches to achieve synergistic affinity gains of several orders of magnitude. Cell-free display systems, described later in this chapter, represent an alternative strategy to cover a greater representation of the antibody sequence space during maturation.

3.4.1.5 *Yeast surface display*

Yeast surface display has been in use for over a decade as a technology platform for affinity maturation and protein engineering but is considerably less widespread in its use. Its application was first described by Boder and Wittrup (1997) and the most significant feature of this approach is the eukaryotic host which possesses the secretory biosynthetic apparatus to facilitate oxidative protein folding and glycosylation. It is therefore well suited to the display of mammalian-derived cell surface or secreted proteins (Levin and Weiss, 2006; Pepper *et al.*, 2008). A further advantage of this technology is the ability to perform rapid and quantitative cell library screening by fluorescence activated cell sorting. In addition to antibodies, yeast surface display has been used to optimize a variety of other proteins including IL-2, T-cell receptors, integrins and epidermal growth factor receptors (Rao *et al.*, 2004., Holler *et al.*, 2000; Jin *et al.*, 2006; Chao *et al.*, 2004).

In a seminal example of the use of yeast display for affinity maturation of antibodies, Boder *et al.* (2000) performed kinetic screening of a randomly mutagenized scFv library to obtain improvement in binding parameters of a scFv recognizing a fluorescein-biotin hapten. This resulted in a >10,000-fold decrease in the rate of dissociation following four rounds of affinity mutagenesis. The improved scFv mutant had an antigen-binding equilibrium dissociation constant of 48 fM measured in a low-salt buffer, representing one of the highest affinities for an antibody reported to date (Boder *et al.*, 2000). Further examples of affinity improvements are reported for carcinoembryonic antigen (Graff *et al.*, 2004), TNF-α (Rajpal *et al.*, 2005), botulinum neurotoxins (Razai *et al.*, 2005), lysozyme (VanAntwerp *et al.*, 1998), and CD3 diphtheria toxin (Wang *et al.*, 2007). In the case of TNF-α, a starting mAb with sub-nM K_D was subjected to a mutagenesis approach described as "look-through" mutagenesis, in which nine amino acid residues representing the major chemical functionalities of amino acids (i.e. basic, aromatic, acidic, etc.) were introduced at every position in each CDR. The corresponding antibody libraries were displayed on yeast, and variants were identified which exhibited 500–870-fold higher affinities than the parent antibody. In the best case, this represented an improvement from a reported K_D of 1 nM for D2E7 to 1 pM for the best variant.

A potential disadvantage of yeast and other cellular display systems could result from the multiple copies of the expressed protein on the cellular surface.

only Lpp-OmpA to date being reported to display scFv antibody fragments (Daugherty, 2007).

Similar *in vitro* methods for ribosome display have also been developed and exemplified, although they are largely unproven in regards to affinity maturation. Covalent display (Reiersen *et al.*, 2005) exploits the phage endonuclease P2A, which forms covalent bonds with the phosphate backbone. By constructing genetic fusions of scFv genes with P2A and performing *in vitro* transcription and translation, complexes of scFv-P2A fusion proteins covalently linked to their own DNA are generated, allowing selection and recovery of specific binding clones. Using a similar principle, a technique called CIS Display has been developed (Odegrip *et al.*, 2004) that utilizes genetic fusion of a target gene, such as an antibody fragment, to the DNA replication initiator protein RepA, which binds specifically to template DNA from which it has been expressed, again maintaining a link between genotype and phenotype. Both of these systems have potential for use in evolving antibodies and other proteins, although no examples of their success for antibody affinity maturation have yet been reported. By contrast, a technique called mRNA display has been demonstrated to have utility in selecting for antibody fragments with improved affinities (Fukuda *et al.*, 2006). mRNA display relies on covalent linkage of the gene of interest to its expressed polypeptide through the action of puromycin, which is chemically annealed to the end of mRNA molecule prior to *in vitro* transcription and translation. Using this system, an anti-fluorescein scFv fragment was mutated by error-prone PCR and DNA shuffling, and higher affinity variants were selected using an off-rate selection procedure. Antibody fragments were isolated that were improved in affinity by approximately 30-fold.

An alternative potential *in vitro* approach to protein engineering is a method of linking a gene with its encoded protein and substrate via compartmentalization due to the formation of water-in-oil emulsions, which was developed by Griffiths and co-workers (Tawfik and Griffiths, 1998). This has been used to successfully improve the catalytic properties of many enzymes, such as DNA methyltransferases, DNA polymerase and phosphotriesterase. Aqueous microdroplets are produced, each containing on average one gene that can be transcribed and translated *in situ.* Linkage of the gene with its encoded enzyme can be achieved by covalent binding of the enzyme gene to an avidin-coated bead, followed by addition of a substrate which is itself linked to a caged biotin molecule. UV light-mediated release of the biotin leads to attachment of the enzymatic product to the gene via the avidin-coated bead. The recovered gene-product complex can then be enriched and recovered using an anti-product antibody. Alternatively, this approach has also been adapted by the creation of microbead-display libraries, with each bead displaying both the enzyme gene and protein via HA tagging of the enzyme and captured using anti-HA antibodies to avidin-coated microbeads (Griffiths and Tawfik, 2003). Interestingly, although this approach has been reported for the optimization of activity of several enzymes, there have been no reports of its use for antibody engineering.

3.4.1.8 *Comparison of approaches*

Given the wide range of potential protein display systems and mutagenesis strategies available to the researcher, it is tempting to try to compare the different systems to objectively determine which has the most utility. Direct comparisons are challenging, however, as many protein display platforms are proprietary to commercial organizations and are patent-protected. Furthermore, rarely is a single antibody optimized using a variety of different platforms and mutagenesis strategies, to enable a direct comparison. Nevertheless, analysis of the literature can provide information regarding the relative advantages and disadvantages of the different technologies (Table 3.2), as well as the scale of the affinity gain and affinity of the best-optimized variant (Table 3.3).

In vitro optimization strategies can now achieve K_D values in the low picomolar to femtomolar range (Table 3.3) from antibodies originating from displayed repertoires (Steidl *et al.*, 2008) or from immunization (Boder *et al.*, 2000). Additionally, the degree of improvement in affinity from the parent antibody can often be more than three orders of magnitude. There are significant numbers of successful examples of both targeted and random mutagenesis strategies, but only a handful of examples of the use of chain shuffling, making comparisons difficult (Table 3.3). Nevertheless, chain shuffling does appear to give more modest affinity improvements compared to the other strategies, likely due to difficulties in obtaining a "best fit" for the antigen within a limited structural repertoire. By contrast, CDR targeting strategies and random approaches both appear to give more substantial improvements in affinity. The DNA shuffling methodology in particular has been successfully shown to generate low picomolar affinity antibodies when used in yeast, ribosome, or mRNA display formats. Numerous mutagenesis strategies have also been successfully pursued to target CDR loops, ranging from simply targeting V_H CDR3, to targeting of mutational hotspots within the CDRs, to combinatorial methods targeting all six CDRs. As a general guide, the most successful approaches have been those that have involved a combinatorial selection of mutations in different CDRs, exploiting synergistic improvements. For example, the highest affinity antibody derived from phage display described to date is an anti-GM-CSF antibody with a K_D of 400 fM, which was derived by combining mutations independently selected for within the V_H CDR2 and V_L CDR3. For many years, predicting beforehand which approaches will lead to optimal affinity gains has been extremely challenging, resulting in empirical and oftentimes consuming strategies for lead optimization. Recent developments based on the constantly growing bank of solved antibody structures, as

Table 3.2 Comparison of different display technologies used for antibody affinity optimization

Display Technology	Advantages	Disadvantages
Phage display	Robust, well-established technology with proven record Simple selection protocols that do not require complex equipment Can be used easily with complex antigens such as cell surfaces	Library sizes limited by bacterial transformation efficiency Possibility to select for expression, dimerization or optimal protein folding, rather than affinity
Ribosome display	*In vitro* → 10^{11} library sizes possible Mutagenesis can be introduced between rounds of selection by PCR	Relative instability of mRNA Not as suitable for heterogeneous antigens such as cell surfaces Possibility to select for clones based upon favored PCR amplification
Yeast display	Eukaryotic system — post-translation modifications, e.g. glycosylation Flow cytometry/cell sorting allows selection of properly folded antibody variants Whole population can be sorted — rapid selection of most improved affinity variants	Library sizes limited by yeast transformation efficiency/ cloning protocols Not as suitable for heterogeneous antigens such as cell surfaces Potential for selection of avid interactions with multivalent antigens
Bacterial display	Flow cytometry/cell sorting allows selection of properly folded antibody variants Whole population can be sorted — rapid selection of most improved affinity variants	Library sizes limited by bacterial transformation efficiency Not as suitable for heterogeneous antigens such as cell surfaces Few host proteins shown to be capable of efficient display of antibody fragments Potential for selection of avid interactions with multivalent antigens

(*Continued*)

Table 3.2 (*Continued*)

Display Technology	Advantages	Disadvantages
mRNA display	*In vitro* → 10^{11} library sizes possible Mutagenesis can be introduced between rounds of selection by PCR	Not as widely validated as other technologies
CIS display	*In vitro* → 10^{11} library sizes possible Mutagenesis can be introduced between rounds of selection by PCR	Not as widely validated as other technologies
In vitro Compartmentalization	*In vitro* → 10^{11} library sizes possible	Has not been demonstrated for display of antibodies

well as advances in *in silico* protein modeling, are now beginning to yield results that will bring the field closer towards more rationally designed methods of antibody affinity maturation.

3.5 Antibody–Antigen Structural Studies

During the last decade, a significant number of antibody (Fab, scFv)-antigen complex structures have been solved by X-ray crystallography, resulting in nearly 300 structures being deposited in the Protein Data Bank (Argiriadi *et al.*, 2009). These structures cover antibody–hapten complexes, antibody–peptide complexes and antibody–protein complexes. The increased interest in antibodies as therapeutics has also resulted in a significant number of antibody–antigen structures being solved in which the antigen is a potential therapeutic target (see Table 3.4 and references therein). A central paradigm that emerges from these structures is that the three-dimensional structure formed by the six CDRs recognizes and binds to a complementary surface on the antigen. The extent of the binding surface is clearly different for hapten and protein recognition. In the case of haptens, the binding is in a cleft between the heavy and light chain CDR3s with a burial of ~400 $Å^2$ surface. Proteins, by contrast, occupy the whole antibody binding site contacting all six CDRs with the burial of ~1,400 to 2,300 $Å^2$ surface (Li *et al.*, 2003). These structures have also revealed that antibody–antigen complexes exhibit a high degree of both shape and chemical complementarity at the interacting surface (Conte *et al.*, 1999; Davies and Cohen, 1996).

Table 3.3 Examples of successful *in vitro* antibody affinity optimization

Antigen	Technology	Mutagenesis Strategy	No. CDRs Targeted	Lead Affinity (pM)	Fold Improvement	Reference
Chain Shuffling						
phOx	PD	Chain shuffling	—	1000	20	Marks *et al.*, (1992)
MCP-1	PD	Chain shuffling	—	1400	15	Yoshinaga *et al.*, (2008)
c-erbB-2	PD	Chain shuffling	—	2500	6	Schier *et al.*, (1996a)
Targeted Mutagenesis						
mVEGF	PD	CDR target/recombination	6	16	6250	Lee *et al.*, (2004)
GM-CSF	PD	CDR target/recombination	2	0.4	5000	Steidl *et al.*, (2008)
TNF-α	YD	CDR target/recombination	6	1	870	Rajpal *et al.*, (2005)
Gastrin	PD	CDR target + rational design	4	13200	454	Barderas *et al.*, (2008)
HIV gp120	PD	CDR target/recombination	5	15	420	Yang *et al.*, (1995)
IL-13	PD/RD	CDR target EP-PCR	3	81	167	Thom *et al.*, (2006)
RAGE	PD/RD	CDR target EP-PCR	2	30	148	Finlay *et al.*, (2009)
Lysozyme	YD	Rational design	4	30	140	Lippow *et al.*, (2007)
VEGF	PD	CDR target/recombination	4	140	140	Chen *et al.*, (1999)
Botulinum toxin	YD	EP-PCR + CDR target	1	0.5	81	Razai *et al.*, (2005)
RSV F Protein	Colony screening	CDR target/recombination	6	35	76	Wu *et al.*, (2005)

(*Continued*)

Table 3.3 (*Continued*)

Antigen	Technology	Mutagenesis Strategy	No. CDRs Targeted	Lead Affinity (pM)	Fold Improvement	Reference
Lewis Y	PD	CDR target	2	3200*	66	Yelton *et al.*, (1995)
Mesothilin	PD	CDR hotspot	1	200	55	Chowdhury *et al.*, (1999)
CD22	PD	CDR hotspot	3	800	7	Ho *et al.*, (2005)
c-erbB-2	PD	Parsimonious mutagenesis	3	2400	6	Schier *et al.*, (1996b)
Random Mutagenesis						
Fluorescein	YD	DNA shuffling + EP-PCR	—	0.05	1400	Boder *et al.*, (2000)
CEA	YD	EP-PCR/DNA shuffling	—	30	283	Graff *et al.*, (2004)
phOx	PD	Mutator Strain	—	3200	100	Low *et al.*, (1996)
Insulin	RD	PCR	—	40	82	Hanes *et al.*, (2000)
Anthrax toxin	PD	EP-PCR	—	250	48	Maynard *et al.*, (2002)
Botulinum toxin	YD	EP-PCR	—	1.7	35	Razai *et al.*, (2005)
Fluorescein	mRD	DNA shuffling + EP-PCR	—	880	30	Fukuda *et al.*, (2006)
Fas	RD	DNA shuffling	—	600	22	Chodorge *et al.*, (2008)
Digoxigenin	BD	EP-PCR	—	47	19	Daugherty *et al.*, (2000)
Prion peptide	RD	DNA shuffling + EP-PCR	—	1	13	Luginbuhl *et al.*, (2006)
NIP	PD	EP-PCR	—	9000	4	Hawkins *et al.*, (1992)

PD = Phage Display, RD = Ribosome Display, YD = Yeast Display, mRD = mRNA Display, BD = Bacterial cell surface display, EP-PCR = Error-prone PCR.

Table 3.4 Crystal structures of antibody–antigen complexes of therapeutic or diagnostic interest

Antigen	Antibody	Reference	Pdb ID
CD20 peptide	Rituximab Fab	Du *et al.*, 2007	2OSL
CD20 peptide	C2H7 Fab	Du *et al.*, 2008a	3BKY
Osteopontin	23C3 Fab	Du *et al.*, 2008b	3CXD
EGF-R	IMC-11F8 Fab	Li *et al.*, 2008	3B2V
EGF-R	Cetuximab Fab	Li *et al.*, 2005	1YY9
ErbB2	Pertuzumab Fab	Franklin *et al.*, 2004	1S78
VEGF-B	2H10 Fab	Leonard *et al.*, 2008	2VWE
Interleukin-23	7G10 Fab	Beyer *et al.*, 2008	3D85
CD8	YTS156.7 Fab	Shore *et al.*, 2008	2ARJ
HER-2	Trastuzumab Fab	Cho *et al.*, 2003	1N8Z
EGF-R	Matuzumab Fab	Schmeidel *et al.*, 2008	3C09
BNP	mAb 106.3 Fab	Longenecker *et al.*, 2009	3E8U
IL-13	CNTO607 Fab	Teplyakov *et al.*, 2009	3G6A
Rotavirus VP7	mAb 4F8 Fab	Aoki *et al.*, 2009	3FMG
Mycobacterial polysaccharide	CS-35 Fab	Murase *et al.*, 2009	3HNV
EGF-R	Nimotuzumab Fab	Talavera *et al.*, 2009	3GKW
Interleukin-18	125-2H Fab	Argiriadi *et al.*, 2009	2VXT
Anthrax antigen	M18 scFv	Leysath *et al.*, 2009	3ETB
Prion protein (PrP)	ICSM18 Fab	Antonyuk *et al.*, 2009	2W9E
IL-17A	CAT-2200 Fab	Gerhardt *et al.*, 2009	2VXS
TGF-β3	GC-1008 Fab	Grütter *et al.*, 2008	3EO1

Improvements in affinity to an antigen can be produced by a variety of structural changes in the antibody, which improve binding site complementarity to the antigen. Theoretically this could involve several mechanisms, such as improved water-mediated contacts between antigen and antibody, CDR loop conformation changes and rotation of the variable domain of the heavy chain–light chain interface, among others (Acierno *et al.*, 2007; Bhat *et al.*, 1994, Rini *et al.*, 1992; Stanfield *et al.*, 1993). Analysis of structural information is leading to insights as to how antibody affinity maturation can be explained at the structural level. Early studies in this area focussed on antibody maturation in relation to haptens, but more recent data have given some insights into the process for antibody–protein interactions.

A seminal article by Wedemayer *et al.* (1997) described the analysis of antibody structures for the affinity-matured anti-nitrophenyl phosphonate antibody (Fab48G7) and for its germline counterpart (Fab48G7g). The affinity-matured antibody has a 30,000-fold greater affinity for the hapten compared to the germline variant as a result of nine somatic replacement mutations (K_D values: 48G7, 50 μM; 48G7 g, 16 nM). A crystal structure for the germline antibody Fab and its complex with hapten were solved at 2.1 Å[2] and compared to the affinity-matured antibody. The key observations from this study were that significant changes in configuration of the binding site occurred on binding of the hapten to the germline antibody (induced fit), while the hapten binds to the matured antibody by a lock and key mechanism. This suggests that the affinity maturation is driven by a mechanism in which the binding site is pre-organized into a conformation that favors hapten binding. It is of significant interest that structural analysis of the nine mutations (three light chain, two in CDR1 and one in CDR2; six heavy chain, three in CDR2, three in framework regions) revealed that none of the mutated residues directly contacts the hapten. This suggests that structural changes distant from the hapten binding site somehow favorably influence the binding site. Mutational and energetic analysis of the affinity maturation of 48G7 (Yang and Schultz, 1999) shows that the conformation of the hapten is different in the low and high affinity forms of the antibody. As an example, mutation of S34L and D55L in the light chain is associated with a change of 1.6 Å in the position of R46. This has the overall effect of allowing the nitro group of the hapten to adopt the high affinity conformation. This example illustrates how apparently distant mutations can give rise to favorable effects on binding.

In a further example of affinity maturation for a hapten-binding antibody, studies have been carried out on an anti-fluorescein antibody (4-4-20), which binds with high affinity to the chromophore fluorescein (Zimmermann *et al.*, 2006). During the maturation of 4-4-20, the affinity of the germline antibody proceeds through two intermediates before the highest affinity 4-4-20 antibody is attained. Thorpe and Brooks (2007) have provided evidence from molecular dynamics that show the increased affinity is associated with a decreased entropic cost to binding. The results from this study support the concept that the high affinity interaction correlates with decreased flexibility of the antibody paratope and the formation of a pre-organized ligand binding site, consistent with a lock and key model of binding for high affinity recognition.

More recently, Li *et al.* (2003) solved the crystal structures of four closely related anti-hen egg-white lysozyme (HEL) antibodies (HyHEL8, HyHEL10, HyHEL26 and HyHEL63), which represented different stages of affinity maturation and bound to the same epitope on HEL. The affinity of the interactions (K_D) were in the range 7 nM (HyHEL26) to 0.2 nM (HyHEL8). In this study binding was enhanced, not by additional hydrogen bonds or van der Waals contacts, but by a burial of increased amounts of apolar surface at the expense of polar surface in conjunction with improved shape complementarity.

Structural analysis of antibody–antigen and antibody–hapten interactions has revealed information about the key changes that occur on affinity maturation of antibodies. It is difficult, however, to draw generic conclusions, as multiple factors influence antibody–antigen interactions, and the magnitude of affinity increases on maturation can vary over orders of magnitude. For antibody–hapten interactions, a general principle is emerging that antibodies adopt a lock and key mechanism upon affinity maturation. In the case of antibody–protein interactions, key hot spot residues drive the energetics of antibody binding, and modulation of peripheral residues can influence binding affinity. Structural analysis has also revealed that in addition to changes occurring in CDR regions, however, antibody interactions can also involve framework residues (Du *et al.*, 2008b) and water-mediated interactions (Argiriadi *et al.*, 2009). This observation suggests that affinity maturation could involve multiple mechanisms and that exploration of CDR mutations alone may not produce the highest affinity antibodies.

3.5.1 *Can we design a predictive approach to affinity maturation?*

Computational design technology for small molecule drug optimization is an established approach within the industry. This structure-based design process has led to molecules that bind to their targets with high affinity (Jorgensen, 2004). However, the application of computational design methods to improving protein–protein affinity is not well established and may be described as an emerging area of research. This reflects the computational difficulty of addressing whole proteins, the availability of structural information, and the effectiveness of stochastic and non-stochastic directed evolution approaches. Specific challenges that exist in the application of rational design to protein–protein affinity improvements include conformational changes that may occur on binding, the presence of interfacial water molecules, polar and charged side chains and the trade off of protein–solvent with protein–protein interactions from the unbound to the bound state (Lippow *et al.*, 2007). The accurate computation of free energies of binding is a central feature of computer-aided design and is particularly challenging for protein–protein interactions.

Given the challenges in this area, there are relatively few examples of computational design being applied to antibody–antigen affinity improvement where the antigen is itself a protein. Lippow *et al.* (2007) demonstrated how an iterative computational design procedure can be used to improve the affinity of antibody interactions. In their study mutations were made in each of the CDRs of an anti-lysozyme antibody D1.3, and the free energy changes on binding were calculated using a physics-based energy function. Improvements in binding were assessed by experimental determination of the binding of mutants to lysozyme. The result of this study indicated that the electrostatic contribution to free energy alone was a better predictor of the binding affinity than the total free energy. The study was continued with the anti-epidermal growth factor receptor drug cetuximab (Erbitux). In this example, several positions were identified for affinity-enhancing mutations based on improved electrostatics. Single mutations were introduced into the molecule for experimental verification at five positions based on the largest magnitude of predicted affinity improvement. These single mutations produced minimal improvements in binding, but on combination of three such mutations the binding affinity was synergistically improved ten-fold to give a final K_D of 50 pM. This computational approach was further validated by examining bevacizumab (Avastin), which binds to VEGF. In this instance, five positions were determined to be suitable for electrostatics-based improvement, and three of these mutations were found in a published high affinity variant of bevacizumab (Chen *et al.*, 1999). This demonstrates some promise for such computational approaches.

A further example of computational design being used to improve the affinity of an antibody was described by Clark *et al.* (2006). In this study, the authors describe the use of structure-based computational methods to optimize the binding affinity of an antibody fragment directed against the I-domain of the integrin VLA1. This antibody had a dissociation constant (K_D) of 7 nM, and a 2.8 Å crystal structure of the Fab-antigen complex was solved. Structure-based computational design was used to determine potentially beneficial mutations, and approximately 80 designed variants were prepared by mutagenesis of the parent antibody followed by experimental affinity determination for the antibody–antigen interaction. Combination of four of the higher affinity mutations (LC S28Q/N52E and HC T50V/K64E) led to an antibody with a K_D of 850 pM. Crystallographic data were collected for a related quadruple mutant (LC S28Q/N52Y and HC T50V/K64E) and showed that the designed protein made the predicted contacts. This study illustrates the promise of computational design for affinity maturation. However, it is worth noting that the authors acknowledged that many of the predicted amino acid mutations were not successful in improving affinity. Key factors here may have been the low resolution crystal structure used as a starting point, combined with a relatively small predicted effect on binding energy (<1 kcal/mol).

Perhaps the greatest challenge to computational design exists when the structure of the antibody is not known. Barderas *et al.* (2008) have demonstrated that computational methods may be applied. In this example, rational engineering methods were applied to an anti-gastrin scFv. The approach here was to make homology models of an anti-gastrin scFv (TA4) followed by docking of the experimentally-determined epitope fragment of gastrin into the binding sites. This information was used to identify areas of suboptimal packing, and semi-rational mutagenesis was carried out resulting in a 454-fold improvement in affinity. This translated into improved potency for inhibition of gastrin-induced proliferation of tumoral cells.

A final example of the application of computational methods addresses the issue of designing antibodies that recognize both the human target and a species ortholog. Antibodies raised against a specific human antigen often do not cross-react with non-human versions of the antigen. This can affect choice of pre-clinical animal models and impacts on the efficacy of the antibody in such models. The ideal scenario is to design an antibody that has identical or very close affinities between the human and species ortholog. Farady *et al.*, (2009) made use of computational methods to guide the selection of antibody CDR residues, which could be mutated to improve the species cross-reactivity of an antibody that inhibits the cancer-associated protease MT-SP1. The antibody, E2, is a potent inhibitor (K_I = 12 pM) of MT-SP1. E2 is a 300-fold less effective inhibitor of the mouse orthologue (epithin), despite the mouse enzyme showing 87% identity. The basis for this difference is not apparent from biochemical and structural analyses of the E2/MT-SP1 complex, as only three of the residues on the protease surface that make contact with the antibody differ between human and mouse. Computational methods were used to select residues in E2 that were structurally close to the three residues differing between the mouse and human enzymes, and free energies were calculated based on mutations. Eight mutations were identified and experimentally tested by making the Fab constructs with the individual mutations. One mutation, T98R, in the H3 loop improved the binding to the mouse ortholog 14-fold, resulting in a K_I of 340 pM. It is worth noting that the other seven mutations were either neutral or resulted in a dramatic loss of potency against the human enzyme.

In silico methods have, therefore, been applied successfully to guide the optimization of antibody affinity and species cross-reactivity. There are relatively few examples available in the public domain, but the success of such approaches suggests that *in silico* methods could be developed to aid affinity maturation strategies. It is worth noting, however, that not all mutations suggested by *in silico* methods have the desired properties, and experimental confirmation is required. The routine design of affinity maturation strategies would, therefore, appear to require significantly more research.

3.6 Concluding Remarks

A broad array of display technologies now exists to aid the antibody engineer in affinity optimization. All utilize the approach of diversification of the DNA sequence, either randomly or through a semi-rational design strategy, followed by iterative selection of variants exhibiting improved binding kinetics. The affinity ceiling for *in vitro* maturation has been shown to be an order of magnitude beyond the limits of the B cell *in vivo*, with several reports of antibody K_D values in the femtomolar range (Boder *et al.*, 2000; Steidl *et al.*, 2008). Indeed, the major practical limitation to further increases in antibody affinities would be the incubation times for the selection experiments, which begin to stretch to weeks, rather than days or hours. For antibodies with therapeutic potential, affinities that are significantly improved with affinities of tens of picomolar K_D or higher have potential advantages in terms of effective *in vivo* therapeutic dose and cost of goods. However, it is important to monitor any potential unexpected safety disadvantages due to unforeseen toxicological consequences associated with effectively irreversible binding to a target, immunogenicity changes made to the molecule, or increased binding to a homologous protein.

Future trends in the field of antibody affinity optimization will undoubtedly incorporate advances in the fields of protein modeling and *in silico* design, and will also look to combine gains in affinity gains with other requirements for a successful protein therapeutic, such as expression level, protein stability, species cross-reactivity, and minimal immunogenicity. Evolution of established protein display technologies, as well as the invention and development of new systems, will likely play a central role in meeting these new challenges.

References

Acierno J.P., Braden B.C., Klinke S., Goldbaum F.A., Cauerhoff A. (2007). Affinity maturation increases the stability and plasticity of the Fv domain of anti-protein antibodies. *J Mol Biol* **374**: 130–46.

Allen C.D.C., Okada T., Cyster J.G. (2007). Germinal center organization and cellular dynamics. *Immunity* **27**: 190–202.

Amit A.G., Mariuzza R.A., Phillips S.E., Poljak R.J. (1986). Three dimensional structure of an antigen-antibody complex at 2.8 A resolution. *Science* **233**: 747–53.

Aoki S.T., Settembre E.C., Trask S.D., Greenberg H.B., Harrison S.C., Dormitzer P.R. (2009). Structure of rotavirus outer-layer protein VP7 bound with a neutralizing Fab. *Science* **324**: 1444–7.

Argiriadi M.A., Xiang T., Wu C., Ghayur T., Borhani D.W. (2009). Unusual water-mediated antigenic recognition of the pro-inflammatory cytokine interleukin-18. *J Biol Chem* **284**: 24478–89.

Barbas C.F. 3rd, Collet T.A., Amberg W., Roben P., Binley J.M., Hoekstra D., Cababa D., Jones T.R., Williamson R.A., Pilkington G.R., Haigwood N.L., Cabezas E., Satterthwait A.C., Sanz I., Burton D.R. (1993). Molecular profile of an antibody response to HIV-1 as probed by combinatorial libraries. *J Mol Biol* **230**: 812–23.

Barbas C.F. 3rd, Hu D., Dunlop N., Sawyer L., Cababa D., Hendry R.M., Nara P.L., Burton D.R. (1994). *In vitro* evolution of a neutralizing human antibody to human immunodeficiency virus type 1 to enhance affinity and broaden strain cross-reactivity. *Proc Natl Acad Sci USA* **91**: 3809–13.

Barderas R., Desmet J., Timmerman P., Meloen R., Casal J.I. (2008). Affinity maturation of antibodies assisted by in silico modelling. *Proc Natl Acad Sci USA* **105**: 9029–34.

Berek C., Milstein C. (1987). Mutation drift and repertoire shift in the maturation of the immune response. *Immunol Rev* **96**: 23–41.

Beyer B.M., Ingram R., Ramanthan L., Reichert P., Le H.V., Madison V., Orth P. (2008). Crystal structures of the pro-inflammatory cytokine interleukin-23 and its complex with a high-affinity neutralizing antibody. *J Mol Biol* **382**: 942–55.

Bhat T.N., Bentley G.A., Boulot G., Greene M.I., Tello D., Dall'Acqua W., Souchon H., Schwarz F.P., Mariuzza R.A., Poljak R.J. (1994). Bound water molecules and conformational stabilization help mediate an antigen-antibody association. *Proc Natl Acad Sci USA* **91**: 1089–93.

Boder E.T., Midelfort K.S., Wittrup K.D. (2000). Directed evolution of antibody fragments with monovalent femtomolar antigen-binding affinity. *Proc Natl Acad Sci USA* **97**: 10701–5.

Boder E.T., Wittrup K.D. (1997). Yeast surface display for screening combinatorial polypeptide libraries. *Nat Biotechnol* **15**: 553–7.

Chao G., Cochran J.R., Wittrup K.D. (2004). Fine epitope mapping of anti-epidermal growth factor receptor antibodies through random mutagenesis and yeast surface display. *J Mol Biol* **342**: 539–50.

Chaudhuri J., Alt F.W. (2004). Class-switch recombination: interplay of transcription, DNA deamination and DNA repair. *Nat Rev Immunol* **4**: 541–52.

Chen H.Z., Zubay G. (1983). Prokaryotic coupled transcription-translation. *Methods Enzymol* **101**: 674–90.

Chen Y., Wiesmann C., Fuh G., Li B., Christinger H.W., McKay P., de Vos A.M., Lowman H.B. (1999). Selection and analysis of an optimized anti-VEGF antibody: crystal structure of an affinity-matured Fab in complex with antigen. *J Mol Biol* **293**: 865–81.

Chodorge M., Fourage L., Ravot G., Jermutus L., Minter R. (2008). *In vitro* DNA recombination by L-Shuffling during ribosome display affinity maturation of an anti-Fas antibody increases the population of improved variants. *Protein Eng Des Sel* **21**: 343–51.

Cho H.-S., Mason K., Ramyar K.X., Stanley A.M., Gabelli S.B., Denney D.W., Leahy D.J. (2003). Structure of the extracellular region of HER2 alone and in complex with the Herceptin Fab. *Nature* **421**: 756–60.

Chowdhury P.S., Pastan I. (1999). Improving antibody affinity by mimicking somatic hypermutation in vitro. *Nat Biotechnol* **17**: 568–72.

Clark L.A., Boriack-Sodin P.A., Eldredge J., Fotch C., Friedman B., Hanf K.J.M., Jarpe M., Liparato S.F., Li Y., Lugovskoy A., Miller S., Rushe M., Sherman W., Simon K., Van Vlijmen, H. (2006). Affinity enhancement of an *in vivo* matured therapeutic antibody using structure-based computational design. *Prot Sci* **15**: 949–60.

Conte L.L., Chothia C., Janin J. (1999). The atomic structure of protein-protein recognition sites. *J Mol Biol* **285**: 2177–98.

Crameri A., Cwirla S., Stemmer W.P. (1996). Construction and evolution of antibody-phage libraries by DNA shuffling. *Nat Med* **2**: 100–2.

Daugherty P.S. (2007). Protein engineering with bacterial display. *Curr Opin Struct Biol* **17**: 474–80.

Daugherty P.S., Chen G., Iverson B.L., Georgiou G. (2000). Quantitative analysis of the effect of the mutation frequency on the affinity maturation of single chain Fv antibodies. *Proc Natl Acad Sci USA* **97**: 2029–34.

Daugherty P.S., Chen G., Olsen M.J., Iverson B.L., Georgiou G. (1998). Antibody affinity maturation using bacterial surface display. *Protein Eng* **11**: 825–32.

Davies D.R., Cohen G.H. (1996). Interactions of protein antigens with antibodies. *Proc Natl Acad Sci USA* **93**: 7–12.

Dedeoglu F., Horwitz B., Chaudhuri J., Alt F.W., Geha R.S. (2004). Induction of activation-induced cytidine deaminase gene expression by IL-4 and CD40 ligation is dependent on STAT6 and NFkappaB. *Int Immunol* **16**: 395–404.

Dimitrov D.S., Marks J.D. (2009). Therapeutic antibodies: current state and future trends — Is a paradigm change coming soon? In Dimitrov A.S. (Ed.), *Therapeutic Antibodies: Methods and Protocols*, vol. 525, pp 1–27. Humana Press.

Di Noia J.M., Neuberger M.S. (2007). Molecular mechanisms of antibody somatic hypermutation. *Annu Rev Biochem* **76**: 1–22.

Du J., Wang H., Zhong C., Peng M., Li B., Huo S., Guo Y., Ding J. (2007). Structural basis for recognition of CD20 by therapeutic antibody Rituximab. *J Biol Chem* **282**: 15073–80.

Du J., Wang H., Zhong C., Peng B., Zhang M., Li B., Hou S., Guo Y., Ding J. (2008a). Crystal structure of chimeric antibody C2H7 Fab in complex with a CD20 peptide. *Mol Immunol* **45**: 2861–8.

Du J., Hou S., Zhong C., Lai Z., Yang H., Dai J., Zhang D., Wang H., Guo Y., Ding J. (2008b). Molecular basis of recognition of human osteopontin by 23C3, a potential therapeutic antibody for treatment of rheumatoid arthritis. *J Mol Biol* **382**: 835–42.

Dufner P., Jermutus L., Minter R.R. (2006). Harnessing phage and ribosome display for antibody optimisation. *Trends Biotechnol* **24**: 523–9.

Farady C.J., Sellers B.J., Jacobsen M.P., Craik C.S. (2009). Improving the species cross reactivity of an antibody using computational design. *Bioorg Med Chem Lett* **19**: 3744–7.

Fermer C., Andersson I., Nilsson K., Nilsson O. (2004). Specificity rescue and affinity maturation of a low-affinity IgM antibody against pro-gastrin-releasing peptide using phage display and DNA shuffling. *Tumour Biol* **25**: 7–13.

Finlay W.J., Cunningham O., Lambert M.A., Darmanin-Sheehan A., Liu X., Fennell B.J., Mahon C.M., Cummins E., Wade J.M., O'Sullivan C.M., Tan X.Y., Piche N., Pittman D.D., Paulsen J., Tchistiakova L., Kodangattil S., Gill D., Hufton S.E. (2009). Affinity maturation of a humanized rat antibody for anti-RAGE therapy: comprehensive mutagenesis reveals a high level of mutational plasticity both inside and outside the complementarity-determining regions. *J Mol Biol* **388**: 541–58.

Foote J., Eisen H.N. (1995). Kinetic and affinity limits on antibodies produced during immune responses. *Proc Natl Acad Sci USA* **92**: 1254–6.

Foote J., Eisen H.N. (2000) Breaking the affinity ceiling for antibodies and T cell receptors. *Proc Natl Acad Sci USA* **97**: 10679–81.

Franklin M.C., Carey K.D., Vajdos F.F., Leahy D.J., de Vos A.M., Sliwkowski M.X. (2004). Insights into ErbB signalling from the structure of the ErbB2-pertuzumab complex. *Canc Cell* 5: 317–28.

Fukuda I., Kojoh K., Tabata N., Doi N., Takashima H., Miyamoto-Sato E., Yanagawa H. (2006). *In vitro* evolution of single-chain antibodies using mRNA display. *Nucleic Acids Res* **34**: 127.

Furukawa K., Shirai H., Azuma T., Nakamura H. (2001). A role of the third complementarity-determining region in the affinity maturation of an antibody. *J Biol Chem* **276**: 27622–8.

Gerhardt S., Abbott W.M., Hargreaves D., Pauptit R.A., Davies R.A., Needham M.R.C., Langham C., Barker W., Aziz A., Smith M.J., Dawson S., Welsh F., Wilkinson T., Vaughan T., Beste G., Bishop S., Popovic B., Sleeman M., Tuske S.J., Coales S.J., Hamuro Y., Russell C. (2009). Structure of IL-17A in complex with a potent, fully human neutralising antibody. *J Mol Biol* **394**: 905–21.

Glaser S.M., Yelton D.E., Huse W.D. (1992). Antibody engineering by codon-based mutagenesis in a filamentous phage vector system. *J Immunol* **149**: 3903–13.

Graff C.P., Chester K., Begent R., Wittrup K.D. (2004). Directed evolution of an anti-carcinoembryonic antigen scFv with a 4-day monovalent dissociation half-time at 37 degrees C. *Protein Eng Des Sel* **17**: 293–304.

Griffiths A.D., Tawfik D.S. (2003). Directed evolution of an extremely fast phosphotriesterase by *in vitro* compartmentalization. *EMBO J* **22**: 24–35.

Grütter C., Wilkinson T., Turner R., Podichetty S., Finch D., McCourt M., Loning S., Jermutus L., Grütter M.G. (2008). A cytokine-neutralizing antibody as a structural mimetic of 2 receptor interactions. *Proc Natl Acad Sci USA* **105**: 20251–6.

Hanes J., Plückthun A. (1997). *In vitro* selection and evolution of functional proteins by using ribosome display. *Proc Natl Acad Sci USA* **94**: 4937–42.

Hanes J., Jermutus L., Plückthun A. (2000). Selecting and evolving functional proteins *in vitro* by ribosome display. *Methods Enzymol* **328**: 404–30.

Hawkins R.E., Russell S.J., Winter G. (1992). Selection of phage antibodies by binding affinity. Mimicking affinity maturation. *J Mol Biol* **226**: 889–96.

Ho M., Kreitman R.J., Onda M., Pastan I. (2005). *In vitro* antibody evolution targeting germline hot spots to increase activity of an anti-CD22 immunotoxin. *J Biol Chem* **280**: 607–17.

Holler P.D., Holman P.O., Shusta E.V., O'Herrin S., Wittrup K.D., Kranz D.M. (2000). *In vitro* evolution of a T-cell receptor with high affinity for peptide/MHC. *Proc Natl Acad Sci USA* **97**: 5387–92.

Irving R.A., Kortt A.A., Hudson P.J. (1996). Affinity maturation of recombinant antibodies using E. coli mutator cells. *Immunotechnology* **2**: 127–43.

Jermutus L., Honegger A., Schwesinger F., Hanes J., Plückthun A. (2001). Tailoring *in vitro* evolution for protein affinity or stability. *Proc Natl Acad Sci USA* **98**: 75–80.

Jin M., Song G., Carman C.V., Kim Y.-S, Astrof N.S., Shimaoka M., Wittrup K.D., Springer T.A. (2006). Directed evolution to probe protein allostery and integrin I domains of 200,000-fold higher affinity. *Proc Natl Acad Sci USA* **103**: 5758–63.

Johnson S., Griego S.D., Pfarr D.S., Doyle M.L., Woods R., Carlin D., Prince G.A., Koenig S., Young J.F., Dillon S.B. (1999). A direct comparison of the activities of two humanized respiratory syncytial virus monoclonal antibodies: MEDI-493 and RSHZ19. *J Infect Dis* **180**: 35–40.

Jorgensen W.L. (2004). The many roles of computation in drug discovery. *Science* **303**: 1813–7.

Labrijn A.F., Aalberse R.C., Schuurman J. (2008). When binding is enough: nonactivating antibody formats. *Curr Opin Immunol* **20**: 479–85.

Lacy S.E., DeVries P.J., Xie N., Fung E., Lesniewski R.R., Reilly E.B. (2008) The potency of erythropoietin-mimic antibodies correlates inversely with affinity. *J Immunol* **181**: 1282–7.

Lazar G.A., Dang W., Kark S., Vafa O., Peng J.S., Hyun L., Chan C., Chung H.S., Eivazi A., Yoder S.C., Vielmetter J., Carmichael D.F., Hayes R.J., Dahiyat B.I. (2006). Engineered antibody Fc variants with enhanced effector function. *Proc Natl Acad Sci USA* **103**: 4005–10.

Lee C.V., Liang W.C., Dennis M.S., Eigenbrot C., Sidhu S.S., Fuh G. (2004). High-affinity human antibodies from phage-displayed synthetic Fab libraries with a single framework scaffold. *J Mol Biol* **340**: 1073–93.

Lee S.Y., Choi J.H., Xu Z. (2003). Microbial cell-surface display. *Trends Biotechnol* **21**: 45–52.

Leonard P., Scotney P.D., Jabeen T., Iyer S., Fabri L.J., Nash A.D., Acharya K.R. (2008). Crystal structure of vascular endothelial growth factor-B in complex with a neutralising antibody Fab fragment. *J Mol Biol* **384**: 1203–17.

Leung D., Chen E., Goeddel D. (1989). A method for random mutagenesis of a defined DNA segment using a modified polymerase chain reaction. *Technique* **1**: 11–15.

Levin A.M., Weiss G.A. (2006). Optimizing the affinity and specificity of proteins with molecular display. *Mol Biosyst* **2**: 49–57.

Leysath C.E., Monzingo A.F., Maynard J.A., Barnett J., Georgiou G., Iverson B.L., Robertus J.D. (2009). Crystal structure of the engineered neutralizing antibody M18 complexed to domain 4 of the Anthrax protective antigen. *J Mol Biol* **387**: 680–93.

Li S., Schmitz K.R., Jeffrey P.D., Wiltzius J.J.W., Kussie P., Ferguson K.M. (2005). Structural basis for inhibition of the epidermal growth factor receptor by cetuximab. *Canc Cell* **7**: 301–11.

Li S., Kussie P., Ferguson K.M. (2008). Structural basis for EGF receptor inhibition by the therapeutic antibody IMC-11F8. *Structure* **16**: 216–27.

Li Y., Li H., Yang F., Smith-Gill S.J., Mariuzza R.A. (2003). X-ray snapshots of the maturation of an antibody response to a protein antigen. *Nat Struct Biol* **10**: 482–8.

Lippow S.M., Wittrup K.D., Tidor B. (2007). Computational design of antibody-affinity improvement beyond in vivo maturation. *Nat Biotechnol* **25**: 1171–6.

Liu B., Conrad F., Cooperberg M.R., Kirpotin D.B., Marks J.D. (2004). Mapping tumor epitope space by direct selection of single-chain Fv antibody libraries on prostate cancer cells. *Cancer Res* **64**: 704–10.

Liu M., Schatz D.G. (2009). Balancing AID and DNA repair during somatic hypermutation. *Trends Immunol* **30**: 173–81.

Longenecker K.L., Ruan Q., Fry E.H., Saldana S.C., Brophy S.E., Richardson P.L., Tetin S.Y. (2009). Crystal structure and thermodynamic analysis of diagnostic mAb 106.3 complexed with BNP 5–13 (C10A). *Proteins* **76**: 536–47.

Low N.M., Holliger P.H., Winter G. (1996). Mimicking somatic hypermutation: affinity maturation of antibodies displayed on bacteriophage using a bacterial mutator strain. *J Mol Biol* **260**: 359–68.

Luginbühl B., Kanyo Z., Jones R.M., Fletterick R.J., Prusiner S.B., Cohen F.E., Williamson R.A., Burton D.R., Plückthun A. (2006). Directed evolution of an anti-prion protein scFv fragment to an affinity of 1 pM and its structural interpretation. *J Mol Biol* **363**: 75–97.

Main S., Handy R., Wilton J., Smith S., Williams L., Du Fou L., Andrews J., Conroy L.A., May R., Anderson I., Vaughan T.J. (2006). A potent anti-Eotaxin1 antibody, CAT-213: isolation by phage display and *in vitro* and *in vivo* efficacy. *J Pharm Exp Ther* **319**: 1395–404.

Maizels N. (2005). Immunoglobulin gene diversification. *Annu Rev Genet* **39**: 23–46.

Marks J.D., Griffiths A.D., Malmqvist M., Clackson T.P., Bye J.M., Winter G. (1992). Bypassing immunization: building high affinity human antibodies by chain shuffling. *Biotechnology (NY)* **10**: 779–83.

Maynard J.A., Maassen C.B., Leppla S.H., Brasky K., Patterson J.L., Iverson B.L., Georgiou G. (2002). Protection against anthrax toxin by recombinant antibody fragments correlates with antigen affinity. *Nat Biotechnol* **20**: 597–601.

McCafferty J., Griffiths A.D., Winter G., Chiswell D.J. (1990). Phage antibodies: filamentous phage displaying antibody variable domains. *Nature* **348**: 552–4.

Muramatsu M., Sankaranand V.S., Anant S., Sugai M., Kinoshita K., Davidson N.O., Honjo T. (1999). Specific expression of activation-induced cytidine deaminase (AID), a novel member of the RNA-editing deaminase family in germinal center B cells. *J Biol Chem* **274**: 18470–6.

Murase T., Zheng R.B., Joe M., Bai Y., Marcus S.L., Lowary T.L., Ng K.K. (2009). Structural insights into antibody recognition of mycobacterial polysaccharides. *J Mol Biol* **392**: 381–92.

Odegrip R., Coomber D., Eldridge B., Hederer R., Kuhlman P.A., Ullman C., FitzGerald K., McGregor D. (2004). CIS display: *In vitro* selection of peptides from libraries of protein-DNA complexes. *Proc Natl Acad Sci USA* **101**: 2806–10.

Parren P.W., Van de Winkel J.G. (2008). An integrated science based approach to drug development. *Curr Opin Immunol* **20**: 426–30.

Pepper L.R., Cho Y.K., Boder E.T., Shusta, E.V. (2008). A decade of yeast surface display technology: Where are we now? *Comb Chem High Through Screen* **11**: 127–34.

Presta L.G. (2008). Molecular engineering and design of therapeutic antibodies. *Curr Opin Immunol* **20**: 460–70.

Rajpal A., Beyaz N., Haber L., Cappuccilli G., Yee H., Bhatt R.R., Takeuchi T., Lerner R.A., Crea R. (2005). A general method for greatly improving the affinity of antibodies by using combinatorial libraries. *Proc Natl Acad Sci USA* **102**: 8466–71.

Rao B.M., Driver I., Lauffenburger D.A., Wittrup K.D. (2004). Interleukin 2 (IL-2) variants engineered for increased IL-2 receptor alpha-subunit affinity exhibit increased potency arising from a cell surface ligand reservoir effect. *Mol Pharmacol* **66**: 864–9.

Rathanaswami P., Roalstad S., Roskos L., Su Q.J., Lackie S., Babcock J. (2005). Demonstration of an *in vivo* generated sub-picomolar affinity fully human monoclonal antibody to interleukin-8. *Biochem Biophys Res Comm* **334**: 1004–13.

Razai A., Garcia-Rodriguez C., Lou J., Geren I.N., Forsyth C.M., Robles Y., Tsai R., Smith T.J., Smith L.A., Siegel R.W., Feldhaus M., Marks J.D. (2005). Molecular evolution of antibody affinity for sensitive detection of botulinum neurotoxin type A. *J Mol Biol* **351**: 158–69.

Reiersen H., Løbersli I., Løset G.A., Hvattum E., Simonsen B., Stacy J.E., McGregor D., Fitzgerald K., Welschof M., Brekke O.H., Marvik O.J. (2005). Covalent antibody display — an *in vitro* antibody-DNA library selection system. *Nucleic Acids Res* **33**: e10.

Revy P., Muto T., Levy Y., Geissmann F., Plebani A., Sanal O., Catalan N., Forveille M., Dufourcq-Labelouse R., Gennery A., Tezcan I., Ersoy F., Kayserili H., Ugazio A.G., Brousse N., Muramatsu M., Notarangelo L.D., Kinoshita K., Honjo T., Fischer A., Durandy A. (2000). Activation-induced cytidine deaminase (AID) deficiency causes the autosomal recessive form of the Hyper-IgM syndrome (HIGM2). *Cell* **102**: 565–75.

Rini J.M., Schulze-Gahmen U., Wilson I.A. (1992). Structural evidence for induced fit as a mechanism for antibody-antigen recognition. *Science* **255**: 959–65.

Roskos L., Klakamp S., Liang M., Arends R., Green L. (2007). Molecular engineering II: Antibody affinity. In Dübel, S. (Ed.), *Handbook of Therapeutic Antibodies*, pp145–69. Wiley-VCH, Weinheim.

Rogozin I.B., Kolchanov N.A. (1992). Somatic hypermutagenesis in immunoglobulin genes II. Influence of neighbouring base sequences on mutagenesis. *Biochim Biophys Acta* **1171**: 11–18.

Schier R., Balint R.F., McCall A., Apell G., Larrick J.W., Marks J.D. (1996a). Identification of functional and structural amino-acid residues by parsimonious mutagenesis. *Gene* **169**: 147–55.

Schier R., Bye J., Apell G., McCall A., Adams G.P., Malmqvist M., Weiner L.M., Marks J.D. (1996b). Isolation of high-affinity monomeric human anti-c-erbB-2 single chain Fv using affinity-driven selection. *J Mol Biol* **255**: 28–43.

Schier R., McCall A., Adams G.P., Marshall K.W., Merritt H., Yim M., Crawford R.S., Weiner L.M., Marks C., Marks J.D. (1996c). Isolation of picomolar affinity anti-c-erbB-2 single-chain Fv by molecular evolution of the complementarity determining regions in the center of the antibody binding site. *J Mol Biol* **263**: 551–67.

Schmiedel J., Blaukat A., Li S., Knöchel T., Ferguson K.M. (2008). Matuzumab binding to EGFR prevents the conformational change rearrangement required for dimerization. *Canc Cell* **13**: 365–73.

Shivarov V., Shinkura R., Doi T., Begum N.A., Nagaoka H., Okazaki I.M., Ito S., Nonaka T., Kinoshita K., Honjo T. (2009). Molecular mechanism for generation of antibody memory. *Philos Trans R Soc Lond B Biol Sci* **364**: 569–75.

Shore D.A., Issafras H., Landais E., Teyton L., Wilson I.A. (2008). The crystal structure of CD8 in complex with YTS156.7.7 Fab and interaction with other CD8 antibodies define the binding mode of CD8 αβ to MHC class I. *J Mol Biol* **384**: 1190–202.

Stanfield R.L., Takimoto-Kamimura M., Rini J.M., Profy A.T., Wilson I.A. (1993). Major antigen-induced rearrangements in an antibody. *Structure* **1**: 83–93.

Steidl S., Ratsch O., Brocks B., Dürr M., Thomassen-Wolf E. (2008). *In vitro* affinity maturation of human GM-CSF antibodies by targeted CDR-diversification. *Mol Immunol* **46**: 135–44.

Stemmer W.P.C. (1994). Rapid evolution of a protein *in vitro* by DNA shuffling. *Nature* **370**: 389–91.

Tabrizi M.A., Roskos L.K. (2007). Preclinical and clinical safety of monoclonal antibodies. *Drug Discov Today* **12**: 540–7.

Tabrizi M.A., Bornstein G.G., Klakamp S.L., Drake A., Knight R., Roskos L. (2008). Translational strategies for development of monoclonal antibodies from discovery to the clinic. *Drug Discov Today* **14**: 298–305.

Talavera A., Friemann R., Gõmez-Puerta S., Martinez-Fleites C., Garrido G., Rabasa A., López-Requena A., Pupo A., Johansen R.F., Sánchez O., Krengel U., Moreno E. (2009). Nimotuzumab, an antitumor antibody that targets the epidermal growth factor receptor, blocks ligand binding while permitting the active receptor conformation. *Cancer Res* **69**: 5851–9.

Tawfik D.S., Griffiths A.D. (1998). Man-made cell-like compartments for molecular evolution. *Nat Biotechnol* **16**: 652–6.

Teplyakov A., Obmolova G., Wu S., Luo J., Kang J., O'Neil K., Gilliland G.L. (2009). Epitope mapping of anti-interleukin 13 neutralizing antibody CNTO607. *J Mol Biol* **389**: 115–23.

Thom G., Cockroft A.C., Buchanan A.G., Candotti C.J., Cohen E.S., Lowne D., Monk P., Shorrock-Hart C.P., Jermutus L., Minter R.R. (2006). Probing a protein-protein interaction by *in vitro* evolution. *Proc Natl Acad Sci USA* **103**: 7619–24.

Thompson J., Pope T., Tung J.S., Chan C., Hollis G., Mark G, Johnson K.S. (1996). Affinity maturation of a high-affinity human monoclonal antibody against the third hypervariable loop of human immunodeficiency virus: use of phage display to improve affinity and broaden strain reactivity. *J Mol Biol* **256**: 77–88.

Thorpe I.F., Brooks C.L. (2007). Molecular evolution of affinity and flexibility in the immune system. *Proc Natl Acad Sci* **104**: 8821–6.

Tordsson J., Abrahmsén L., Kalland T., Ljung C., Ingvar C., Brodin T. (1997). Efficient selection of scFv antibody phage by adsorption to in situ expressed antigens in tissue sections. *J Immunol Methods* **210**: 11–23.

VanAntwerp J.J., Wittrup K.D. (1998). Thermodynamic characterization of affinity maturation: the D1.3 antibody and a higher-affinity mutant. *J Mol Recognit* **11**: 10–13.

Vaughan T.J., Williams A.J., Pritchard K., Osbourn J.K., Pope A.R., Earnshaw J.C., McCafferty J., Hodits R.A., Wilton J., Johnson K.S. (1996). Human antibodies with sub-nanomolar affinities isolated from a large non-immunized phage display library. *Nat Biotechnol* **14**: 309–14.

Viti F., Tarli L., Giovannoni L., Zardi L., Neri, D. (1999). Increased binding affinity and valence of recombinant antibody fragments lead to improved targeting of tumoral angiogenesis. *Cancer Res* **59**: 347–52.

Villemagne D., Jackson R., Douthwaite J.A. (2006). Highly efficient ribosome display selection by use of purified components for *in vitro* translation. *J Immunol Methods* **313**: 140–8.

Wang Z., Kim G., Woo J., Liu Y.Y., Mathias A., Stavrou S., Neville D.M. (2007). Improvement of a recombinant anti-monkey anti-CD3 diptheria toxin based immunotoxin by yeast display affinity maturation of the scFv. *Bioconj Chem* **18**: 947–55.

Wang W., Wang E.Q., Balthasar J.P. (2008). Monoclonal antibody pharmacokinetics and pharmacodynamics. *Clin Pharm Ther* **84**: 548–58.

Wu H., Beuerlein G., Nie Y., Smith H., Lee B.A., Hensler M., Huse W.D., Watkins J.D. (1998). Stepwise *in vitro* affinity maturation of Vitaxin, an alphav beta3-specific humanized mAb. *Proc Natl Acad Sci USA* **95**: 6037–42.

Wu H., Pfarr D.S., Johnson S., Brewah Y.A., Woods R.M., Patel N.K., White W.I., Young J.F., Kiener P.A. (2007). Development of motavizumab, an ultra-potent antibody for the prevention of respiratory syncytial virus infection in the upper and lower respiratory tract. *J Mol Biol* **368**: 652–65.

Wu H., Pfarr D.S., Tang Y., An L.L., Patel N.K., Watkins J.D., Huse W.D., Kiener P.A.,Young J.F. (2005). Ultra-potent antibodies against respiratory syncytial virus: effects of binding kinetics and binding valence on viral neutralization. *J Mol Biol* **350**: 126–44.

Yang W.P., Green K., Pinz-Sweeney S., Briones A.T., Burton D.R., Barbas C.F., 3rd. (1995). CDR walking mutagenesis for the affinity maturation of a potent human anti-HIV-1 antibody into the picomolar range. *J Mol Biol* **254**: 392–403.

Yang P.L., Schultz P.G. (1999). Mutational analysis of the affinity maturation of antibody 48G7. *J Mol Biol* **294**: 1191–201.

Yelton D.E., Rosok M.J., Cruz G., Cosand W.L., Bajorath J., Hellström I., Hellström K.E., Huse W.D., Glaser S.M. (1995). Affinity maturation of the BR96 anti-carcinoma antibody by codon-based mutagenesis. *J Immunol* **155**: 1994–2004.

Yoshinaga K., Matsumoto M., Torikai M., Sugyo K., Kuroki S., Nogami K., Matsumoto R., Hashiguchi S., Ito Y., Nakashima T., Sugimura K. (2008). Ig L-chain shuffling for affinity maturation of phage library-derived human anti-human MCP-1 antibody blocking its chemotactic activity. *J Biochem* **143**: 593–601.

Zahnd C., Spinelli S., Luginbühl B., Amstutz P., Cambillau C., Plückthun A. (2004). Directed *in vitro* evolution and crystallographic analysis of a peptide-binding single chain antibody fragment (scFv) with low picomolar affinity. *J Biol Chem* **279**: 18870–7.

Zimmermann J., Oakman E.L., Thorpe I.F., Shi X., Abbyad P., Brooks C.L., 3rd, Boxer S.G., Romesberg F.E. (2006). Antibody evolution constrains conformational heterogeneity by tailoring protein dynamics. *Proc Natl Acad Sci USA* **103**: 13722–7.

Zhou C., Saxon A., Zhang K. (2003). Human activation-induced cytidine deaminase is induced by IL-4 and negatively regulated by CD45: implication of CD45 as a Janus kinase phosphatase in antibody diversification. *J Immunol* **170**: 1887–93.

4

Transgenic Mice Rearranging Human Antibodies

*Sean Stevens**

4.1 Introduction

The value of treating human disease with immune sera was recognized well before antibodies themselves were fully understood. The development of "therapeutic serum" by von Behring and Kitasato in the 1890s showed the potential of harnessing the immune system to treat infectious disease and earned the pair a Nobel Prize in 1901. However, it was not until the development of monoclonal antibody (mAb) technology in the 1970s by Kohler and Milstein that it was possible to generate unlimited amounts of a particular antibody of a desired specificity for use as a therapeutic (Kohler and Milstein, 1975). Immunization of mice with any number of different antigens resulted in collections of antibodies of high specificity and affinity. Unfortunately, the ability of the immune system to efficiently recognize and eliminate foreign antigens also severely limited the use of mouse mAbs as therapeutics in humans. Patients can rapidly produce human anti-mouse antibodies, leading to clearance and reduced efficacy of mouse mAbs and preventing their use in chronic diseases which require repeated therapeutic administration.

The immunogenicity of mouse mAbs led to a number of efforts to create antibodies encoded by human-derived amino acid sequences. Chimeric antibodies, created by the transplant of mouse variable domains onto human constant regions, and humanized antibodies, in which non-antigen binding portions of the mouse variable region are also replaced by human sequences, have greatly reduced immunogenicity but require significant engineering to maintain antigen specificity and affinity (Almagro and Fransson, 2008). The use of large libraries of

* Ablexis LLC, USA.

human heavy and light chain variable regions in different types of display approaches can help identify specific antibodies with reasonable affinity and reduced immunogenicity. However, further *in vitro* manipulation is often required to improve binding affinity and expression of the desired antibody (Hoogenboom, 2005). Attempts to apply monoclonal technology to human cells have been useful for some infectious disease targets to which patients have been vaccinated or exposed naturally, but are otherwise limited due to the infrequency of B cells expressing high affinity antibodies to desired therapeutic targets and the ethical prohibitions against immunizing humans with specific antigens (Lanzavecchia *et al.*, 2007). Although effective, the majority of these technologies were developed to circumvent the immunogenicity of mouse mAbs. The use of the genuine humoral response still has significant benefits. An ideal platform would therefore take advantage of the benefits of an *in vivo* immune response in mice and the convenience of monoclonal antibody technology, while utilizing human antibody sequences to minimize immunogenicity of the therapeutic antibodies produced.

Creation of a platform to produce human antibodies in mice requires introduction of human antibody genes into the mouse genome in their germline configuration. Human Ig segments inserted into the mouse genome must be able to productively rearrange to form variable regions representative of the naïve human antibody repertoire, and these antibodies must be able to undergo the processes of diversification and affinity maturation using the endogenous murine cellular machinery. The human antibodies generated by these mice should not interfere with the normal functioning of the murine immune system, and it is desirable to use standard mAb (and other) technology to isolate cells producing antigen-specific human antibodies.

4.2 Rearranged Immunoglobulin Transgenes

Even as the technologies necessary for the genetic manipulation of the mouse genome were initially being developed, many different approaches were used to introduce and express human antibody genes in mice. In fact, some of the very first transgenic mice were made using rearranged immunoglobulin kappa light chain transgenes derived from a mouse myeloma (Brinster *et al.*, 1983). Soon thereafter, several labs described production of transgenic mice with rearranged heavy and light chain immunoglobulin genes, either alone or in combination (Grosschedl *et al.*, 1984; Rusconi and Kohler, 1985; Yamamura *et al.*, 1984; Storb *et al.*, 1986).

Early studies revealed that transgenic mice bearing kappa light chain transgenes from plasmacytomas or hybridomas expressed the transcripts only in B cells, albeit at variable levels. Expression of the kappa light chain transgenes may require the enhancer found downstream of the kappa constant region (Betz *et al.*,

1994). Similarly, rearranged lambda light chain transgenes also exhibited B cell-specific expression (Eccles *et al.*, 1990). By contrast, rearranged mu heavy chain transgenes were expressed in both B and T cells and do not require sequences downstream of the constant region (Storb *et al.*, 1986). However, deletion of the intronic enhancer regions from heavy chain transgenes greatly reduced expression (Yamamura *et al.*, 1986). Both heavy and light chain transgenes were expressed regardless of insertion site (Storb *et al.*, 1985; Grosschedl *et al.*, 1984).

In one study, a fully-human IgG1 rearranged transgene derived from a human leukemia cell line was inserted into the mouse genome and shown to be expressed in B cells. Approximately 15% of mouse B cells expressed human IgG1 albeit not exclusively, demonstrating both the inefficient usage of the human transgene relative to the endogenous mouse IgH loci and the lack of allelic exclusion. Expression levels increased upon LPS stimulation, resulting in ten-fold more human IgG1-expressing B cells. Serum levels of the human IgG1 were nearly undetectable in unstimulated mice, but substantially increased with LPS but not ConA induction. Although the amount of human Ig was greatly increased in serum, it was still over 1,000-fold less abundant than mouse IgG1, indicating the inefficiency of expression relative to the endogenous mouse IgH loci (Yamamura *et al.*, 1986). Furthermore, mice possessing paired heavy and light chain transgenes specific for a given antigen expressed detectable serum antibody levels without stimulation, well before the mice would normally be immunocompetent. Immunization of transgenic mice with the specific antigen further increased serum antibody levels, due to a significant increase in the number of antigen-specific B cells (Pinkert *et al.*, 1989).

Productive rearrangement of one Ig locus prevents rearrangement and expression of other Ig loci in the B cell in a process called allelic exclusion (Cedar and Bergman, 2008). Rearranged mu heavy chain transgenes significantly reduce V–D–J recombination at the endogenous heavy chain loci (Rusconi and Kolher, 1985). Transgenic mice with mu transgenes lacking the membrane exons were unable to prevent rearrangement of the endogenous loci (Nussenzweig *et al.*, 1988), whereas transgenic mice expressing only the membrane-bound form suppressed endogenous loci, demonstrating that signaling from the B cell receptor is required for inhibition (Nussenzweig *et al.*, 1987). Mice bearing only rearranged kappa light chain transgenes or kappa light chain transgenes in combination with heavy chain transgenes had variable effects upon endogenous light chain rearrangement (Gollahon *et al.*, 1988; Ritchie *et al.*, 1984).

When exposed to antigen and appropriate T cell help, antigen-specific B cells are stimulated to undergo rapid proliferation and diversification of their variable regions through AID-dependent mutagenesis, a process called somatic hypermutation (Peled *et al.*, 2008). B cells possessing antibodies with mutations that create higher affinity for the target out-compete other B cells in the germinal center. Depending upon the combination of cytokines and co-receptor interactions, the

B cells will undergo class switching, a recombination event that joins the heavy chain variable region to a heavy chain constant region with different functional characteristics (IgG, IgE, IgA, etc.). The B cells will then expand and differentiate into antibody-secreting plasma cells or long-lived memory B cells (Tarlinton, 2008). Hyperimmunization of mice bearing rearranged heavy and light chain transgenes specific for the antigen have been shown to produce significant levels of antigen-specific serum antibodies and demonstrated somatic hypermutation of the transgenic variable regions, but not of the constant or non-coding regions (O'Brien *et al.*, 1987; Durdik *et al.*, 1989). Somatic hypermutation of heavy chain transgenes can also be observed in hybridomas derived from immunized mice (Giusti *et al.*, 1992; Sohn *et al.*, 1993).

Expression of rearranged transgenes in mice addresses several critical issues in the production of antibodies in mice. Inclusion of genuine Ig promoters and enhancers are necessary and sufficient for B cell expression of Ig transgenes, although the cell specificity of the heavy chain transgenes and overall levels of expression are variable. Expression of heavy chain transgenes capable of producing membrane-bound forms of Ig can induce allelic exclusion, as can some light chain transgenes. Transgenic mice with antigen-specific heavy and light chain transgenes generate an immune response to antigen by exhibiting elevated levels of serum antibody, demonstrating that rearranged transgenes are capable of producing functional antibody and are able to respond to antigenic stimulation.

4.3 Unrearranged Immunoglobulin Transgenes

Another critical factor in production of antibodies to a variety of different antigens is the ability to develop a primary repertoire through V–D–J and V–J recombination of the germline Ig loci (Tonegawa, 1983). Variable segment rearrangement requires the RAG genes, which recognize the recombination signal sequences (RSS) present in the variable gene segments and mediate the recombination reaction. Data from cell lines had indicated that the RSS and recombination machinery are conserved in mice and humans (Yancopoulos and Alt, 1986). However, it was important to test whether recombination functions were maintained when substrates were integrated into the mouse genome.

When an unrearranged chicken lambda transgene comprising VL, JL, and CL segments and flanking sequences was introduced into the mouse genome and expressed specifically in spleen but not thymus, rearrangement of the chicken lambda transgene was observed in transgenic mouse splenocytes (Bucchini *et al.*, 1987). However, mice bearing a germline rabbit kappa transgene exhibited rearrangement events in both B and T cells but were poorly transcribed, suggesting some incompatibilities exist between mouse and rabbit gene expression signals (Goodhardt *et al.*, 1987). Subsequent experiments demonstrated V–D–J and V–J recombination with a variety of different transgenic substrates

(Ferrier *et al.*, 1990), showing that the RAG-mediated recombination mechanism worked with germline variable segments from many different species and could be affected by locus accessibility, sterile transcripts of variable segments and transcription control elements (Yancopoulos and Alt, 1986).

During the immune response, the heavy chain constant region loci can undergo a deletional recombination event that removes the IgM exons and places the variable region in close proximity to a downstream constant region gene. This event creates antibodies with identical variable region sequences but different constant regions that display different functional properties or effector functions. Transgenic substrates have also been designed to test class switching between different isotypes and to study sequence requirements. Results suggest a role for various expression control sequences as well as sterile transcripts in the initiation of class switching (Stavnezer *et al.*, 2008).

Many different types of unrearranged Ig transgene constructs are able to undergo the processes of VDJ/VJ rearrangement and class switching in mice, demonstrating that the basic mechanisms of each can be used with heterologous sequences found outside of their normal genomic context. The specific control elements required can be complex, and several labs have observed recombination between transgenes and endogenous loci despite their location in disparate places within the mouse genome.

4.4 Human Antibodies in Mice

4.4.1 *Transloci mice*

The results from studies of rearranged and germline transgenes showed that many of the key processes for the production of antibodies, such as VDJ and VJ recombination, allelic exclusion, somatic hypermutation, and class switching can be reproduced with synthetic Ig constructs. Antibodies produced from transgenes are able to signal through the murine cellular machinery and respond to immunization with antigen. Therefore it should be possible to create heterologous Ig loci that can rearrange to form a primary repertoire, undergo antigen selection and affinity maturation, and produce high affinity antibodies in mice. While transgenic experiments have been able to reproduce many aspects of antibody rearrangement, maturation, and expression, it is not clear if any of the transgene functions demonstrated to date are as efficient as the genuine loci *in vivo*.

4.4.2 *Synthetic human transloci*

Initial efforts to produce human antibodies in mice used a synthetic minilocus approach. This was due in part to technical restrictions on the size of vectors that could be easily manipulated and introduced into the mouse genome. One of the earliest studies used a chimeric IgH minilocus consisting of two germline VH segments, four DH segments, six JH segments, the IgM constant region gene, and the IgH intronic enhancer individually cloned into a plasmid vector and inserted into mouse eggs via pronuclear injection. One VH and one DH segment were derived from mouse sequences, and the remaining segments were human. Mice carrying the transgenic minilocus exhibited a significant number of D–J rearrangement events, but fewer V–DJ rearrangements, and were detectable in both spleen and thymus. Human IgM was produced by 3% of hybridomas and found at concentrations of 1–100 μg/ml in serum. Most hybridomas expressing human IgM also expressed mouse heavy chains. Sequence analysis demonstrated that both human and mouse VH segments were utilized (Brüggemann *et al.*, 1989).

The kappa light chain locus was less complicated than the heavy chain, lacking diversity segments and possessing a single constant region, which facilitated creation of synthetic transgene constructs. A human kappa light chain synthetic minilocus consisting of five VK and all JK and CK segments as well as known regulatory regions was placed in a highly condensed 15 kb arrangement and introduced into the mouse genome. This human minilocus was shown to be rearranged and expressed in mice, but with low efficiency (Xian *et al.*, 1998).

These first attempts to generate human Ig chains in mice showed that it was possible to create loci through synthetic approaches that rearrange and produce Ig protein. However, the efficiency and cell specificity of these loci was reduced relative to wild-type loci, and critical functions such as allelic exclusion were not necessarily maintained. A separate approach to producing human antibodies in mice expanded upon the minilocus concept by joining distinct functional Ig gene segments into more complex constructs for pronuclear injection into mouse oocytes. Human Ig loci are very large, spreading over more than a megabase each for IgH and IgK, with functional Ig segments encoded throughout the genomic sequences. Therefore, this type of approach allowed the incorporation of only the most critical segments and reduced the overall size of the construct.

An initial study combined a human kappa light chain transgene bearing one VK, five JK, and the CK segment with human heavy chain transgenes containing one functional VH, ten DH, and six JH segments with either the IgM constant region gene, or both IgM and IgG constant region genes to allow for testing class switching *in vivo*. The transgene constructs were introduced via pronuclear injection into wild-type mouse embryos. The resulting mice expressed variable but low levels of human Ig in serum. Flow cytometry data showed that B cells expressing the human heavy chain transgene ranged from 1–20% of the population. In mice possessing both human heavy chain and human light chain transgenes, 0.5% expressed both human heavy chain and human light chain, which was proportional to the amount of cells expressing human heavy chain alone 5%) and human kappa light chain alone (10%). Comparison with endogenous mouse heavy chain suggested that the heavy chain transgene was capable of allelic exclusion. Sequence analysis showed a diverse set of rearrangements, with 20 out of 29 unique light chains and 36 out of 49 unique heavy chains, with nearly all of the variable segments utilized. No somatic hypermutation was observed; however, the mice were not immunized (Taylor *et al.*, 1992).

In subsequent work, an IgH transgene construct with greater potential diversity was generated possessing three VH, 16 DH, and all of the JH segments linked to human IgM and IgG1 constant regions. A more extended IgK transgene was created through the co-integration of two separate overlapping constructs, resulting in a human IgK minilocus consisting of four VK and five JK segments combined with the human IgK constant region. These transgenes exhibited normal variable rearrangement as well as somatic hypermutation and class switching, leading to the creation of a genuine human Ig repertoire in mice (Lonberg *et al.*, 1994).

Results from previous transgenic approaches showed that the endogenous murine Ig loci could effectively out-compete Ig transgenes, reducing the frequency of B cells expressing human Ig chains. In addition, endogenous loci may not be effectively silenced through allelic exclusion by transgenes. One mechanism to overcome these limitations is to inactivate endogenous mouse loci so that the only Ig that is produced must be derived from the human transgenes. Therefore, transgenes were introduced into mice with deletions in the endogenous murine IgH and IgK (but not IgL) loci in order to force the mice to produce predominantly human antibodies.

The B cell populations of mice possessing human IgH and IgK transgenes with inactivated murine IgH and IgK were compared to wild-type mice. The transgenic mice had smaller B cell populations, suggesting that the Ig transgenes were unable to fully reconstitute B cell development and/or expansion in the periphery. The transgenic mice also exhibited significantly lower serum Ig levels of both IgM and IgG compared with wild-type mice, which may have been a reflection of reduced B cell numbers, restricted Ig rearrangement, or expression from the transgenic loci. Although the endogenous mouse IgH and IgK loci were mutated to prevent their function, the mouse IgL locus was intact. The lambda light chain in mouse normally represents approximately 5–7% of light chain expression in wild-type mice, but significantly greater levels were observed in mice with transgenic miniloci. Interestingly, significant amounts of mouse heavy chain were also observed in serum despite having a deletion to prevent use of the endogenous mouse heavy chain loci.

Polymerase chain reaction (PCR) and sequence analysis demonstrated that nearly all of the Ig segments introduced were utilized in rearrangements. Despite possessing a limited variable repertoire, the human Ig miniloci were able to produce a response to a number of immunized antigens, with somatic hypermutation and class switching indicative of the affinity maturation process. Hybridomas generated from these mice further established the production of fully-human antibodies *in vivo* (Lonberg *et al.*, 1994).

4.4.3 *Natural human transloci*

The inefficiencies observed with synthetic minilocus constructs could be due in part to a lack of critical elements excluded from the transgenes. Therefore, several different approaches were used to reconstruct Ig loci as transgenes using portions of the genuine human germline Ig genomic sequences derived from cosmids, bacteriophage P1 clones or yeast artificial chromosomes (YACs).

4.4.3.1 *Cosmid-based transloci*

Cosmid vectors can contain heterologous sequences from 30–60 kb in length and have been used to create libraries of genomic sequences that can be propagated in prokaryotes (Hohn *et al.*, 1988). The insert size makes it more feasible and convenient to reconstruct germline Ig loci using unrearranged genomic DNA as transgenes. In one study, two cosmids containing approximately 100 kb of the human IgH sequence were isolated, co-injected into fertilized mouse pronuclei, and studied for productive human IgH chain rearrangements. Two founders expressed human Ig chain in serum at 10–100 μg/ml. Both of these founders exhibited extensive rearrangements of the human IgH loci with many extending across the two cosmids, suggesting co-integration of the vectors. The rearrangements were only detected in spleen and involved the majority of the VH, DH and JH segments inserted (Brüggemann *et al.*, 1991).

In order to remove the endogenous background, the cosmid-derived human IgH transgenic mice were crossed into the uMT mouse strain. The uMT mouse has a deletion in IgM membrane exons, preventing expression of membrane-bound IgM and thus blocking proper B cell development. The restriction of the endogenous loci led to much greater expression of the human transgenes, with human serum IgM levels of 180–800 μg/ml compared with 0.2–55 μg/ml in control animals possessing wild-type endogenous IgM alleles. Immunization of transgenic animals in the uMT background resulted in low but detectable antigen-specific antibody responses (Wagner *et al.*, 1994a).

Further studies with a variety of antigens showed that cosmid-derived transgenes in the uMT background were capable of creating antigen-specific antibody responses to haptens (NIP, phOx), protein antigens (lysozyme, phycoerythrin), and human lymphoma cells. In all cases, the primary responses were substantially less than those derived from wild-type mice. Sequence analysis demonstrated significant diversity in human transgene rearrangements and evidence of somatic hypermutation in sequences of hybridomas specific for phOx (Wagner *et al.*, 1994b).

4.4.3.2 *Bacteriophage P1-based transloci*

Bacteriophage P1 are prokaryotic vectors that can carry up to 100 kb of heterologous DNA, facilitating creation of larger transgenic loci (Sternberg, 1990). An approximately 200 kb continuous section of the human germline IgH locus was reconstructed through the co-injection of three overlapping P1 clones, containing five VH segments, all DH and JH segments, and the IgM and IgD constant region genes. The three linearized clones presumably recombined prior to integration into the genome, recreating the continuous IgH sequence. Transgenic mice had low titers of human IgM, with only 1–2% of B cells expressing human IgM. PCR showed that all five of the human VH segments were involved in rearrangement events. Two out of three hybridomas derived from the transgenic mice expressing human IgM also expressed mouse Ig, indicating a lack of allelic exclusion.

The P1-derived human IgH transgenic mice were bred with uMT mice for further study. The human transgene overcame the block in B cell development caused by the uMT defect, restoring the B cell population to about 30–50% of wild-type levels. Nearly all the B cells expressed human IgM on their surface and produced approximately 200 μg/ml serum IgM. Immunization with phOx-coupled albumin generated a substantial antigen-specific response, which was slightly reduced compared to wild-type mice. The variable rearrangements exhibited significant diversity, with substantial N and P nucleotide insertions and somatic hypermutation. While the serum Ig titers were similar, the levels of somatic hypermutation were significantly greater than those observed in the plasmid- and cosmid-derived transgenic mice (Wagner *et al.*, 1994a). Taken together, the results from the P1-derived human IgH loci suggested that larger transgene size and complexity yielded greater sequence diversity, improved rates of somatic hypermutation and improved antigen responses (Wagner *et al.*, 1996).

4.4.3.3 *Yeast artificial chromosome-based transloci*

YACs can carry the greatest amount of genomic DNA, with examples of constructs as large as one megabase. The use of yeast vectors has the additional benefit of combining the ease and precision of recombination in yeast cells with the ability to rebuild large sections of the human genome from individual YACs spanning overlapping portions of sequence. With careful manipulation, YACs can be engineered with appropriate markers for use as transgenes in mammalian cells (Davies *et al.*, 1992; Jakobovits, 1994).

In an initial study to test YACs as Ig transgenes, a clone was identified carrying 85 kb of human IgH sequence including two VH segments, all DH and JH segments, and the IgM and IgD constant region genes. To avoid further manipulation of the YAC, the construct was introduced into mouse embryonic stem (ES) cells by co-lipofection with a selectable marker plasmid. Of 1221 clones screened, two possessed the YAC-encoded IgM constant region DNA. Mice derived from one of these clones exhibited extremely low levels of human IgM in serum. Crossbreeding with mice possessing inactivated IgH loci increased human IgM levels in serum almost ten-fold, but the expression was still less than 200 ng/ml (Choi *et al.*, 1993).

In a similar approach with the light chain locus, a library of human genomic YACs was screened to isolate clones bearing the human IgK locus. A 300 kb clone carrying three VK, all JK, and the CK constant region in a contiguous sequence was introduced into murine ES cells via yeast protoplast fusion. Several clones from the resulting ES cells were shown to have integrated the unrearranged YAC-derived sequence, in some cases with additional yeast DNA. PCR assays demonstrated that V–J recombination of the transgenic locus utilizing all three of the VK segments occurred only in spleen. Human IgK chains were detectable in serum from the mice at a range of 1–3 μg/ml. These data showed that it was possible to use YACs to create functional transgenic Ig loci (Davies *et al.*, 1993).

To test whether the reduced representation of human IgK was due to a restricted variable repertoire, the same 300 kb human kappa YAC transgene was redesigned in a later study to incorporate 20 identical tandem repeats of human genomic cosmid bearing five human germline VK segments. The resulting construct had a length of 1.3 megabases, including 103 human VK segments, all JK segments and the CK constant region gene, and was inserted into mouse ES cells using protoplast fusion. The mice targeted with the transgene produced 1–60 μg/ml human IgK in serum, which was significantly greater than mice possessing the original unmodified 300 kb YAC transgene (Davies *et al.*, 1993), but less than 1% of the endogenous mouse IgK serum levels. When bred into mice bearing a deletion in the endogenous mouse IgK locus but intact IgL locus, the human IgK serum levels increased to 200–1200 μg/ml. Mouse IgL serum levels were elevated compared with wild-type mice and comparable to the human IgK serum levels. The differences in serum Ig levels were also observed in flow cytometric analysis of B cell populations expressing human IgK (Zou *et al.*, 1996).

A direct comparison of otherwise identical YAC transgenes bearing 3 versus 103 VK segments demonstrated that increasing the variable repertoire improved representation of human Ig in mice, even when the endogenous light chain loci are intact. Removal of competing mouse IgK chains further enhanced the usage of the human IgK transgene. In the transgenic mice lacking murine IgK, the mouse IgL levels also increased to levels almost equal to the transgenic human serum IgK. Because the mouse IgL variable repertoire is significantly less than that of the human IgK transgene, the equivalent expression levels suggest that the human transgene is less efficient in rearranging, expressing or competing with the murine IgL locus (Zou *et al.*, 1996).

Another approach using YACs combined human IgH and IgK transgenes to generate fully-human antibodies in mice. Both a human IgH YAC encoding five VH segments, all of the DH and JH segments, and extending through the IgM and IgD constant region genes, and a human IgK YAC incorporating three VK segments, all of the JK segments, and the CK constant region gene were isolated. Targeting mouse ES cells was accomplished by the fusion of yeast spheroplasts bearing the desired YACs. Resulting cells possessed intact YAC-derived human Ig loci sequences integrated randomly within the mouse genome. The targeted mice were bred with mice bearing mutations in the endogenous murine IgH and IgK loci to enhance the usage of human IgH and IgK transgenes. The mouse IgL locus was not eliminated. Studies with mice bearing only five VH and three VK segments expressed a diverse repertoire of VH–DH–JH and VK–JK rearrangements, with human antibodies observed in serum, albeit at lower levels compared with wild-type mice. Immunization of these mice led to antigen-specific humoral responses, and production and analysis of resulting hybridomas demonstrated that the antibodies possessed somatic hypermutations indicative of having undergone affinity maturation.

Flow cytometric analysis of B cell populations from transgenic mice in a wild-type background showed that the human YAC transgenes were poorly expressed relative to the endogenous mouse loci. In addition, significant levels of B cells co-expressing both endogenous mouse and human IgH and mouse and human IgK were observed, indicating that allelic exclusion is incomplete in these mice. Human transgenes are expressed much more efficiently in mice with inactivated endogenous IgH and IgK loci, restoring B cell development in bone marrow and the periphery, albeit at levels substantially lower than wild-type mice. The incomplete reconstitution of the B cell compartment was reflected in the decreased levels of serum Ig. In addition, the levels of mouse lambda light chain were greatly increased, further indicating that the human transgenes may not be operating efficiently *in vivo* (Green *et al.*, 1994).

A subsequent version of these mice was created using yeast recombination to build YAC constructs in a stepwise fashion, incorporating more of the IgH and IgK loci and increasing the variable repertoire to 66 VH segments and 32 VK segments. For the human IgH locus, four overlapping YACs were recombined to create a germline IgH sequence spanning a little over a megabase of sequence. Similarly, the human IgK locus was recreated by recombining four overlapping YACs covering 800 kb of human genomic DNA. The large YAC constructs were introduced into mouse ES cells using yeast spheroplast fusion. Careful analysis demonstrated that the YAC constructs were intact after integration into the mouse genome.

Analysis of mice possessing the larger Ig transloci using flow cytometry demonstrated that increasing the variable segment diversity allowed reconstitution of the B cell population to 40–60% of wild-type. The human IgH and IgK transgenes were similarly well expressed; however, there was a seven-fold increase in mouse IgL expressing cells, suggesting that the IgK transgene was not as efficiently utilized. Serum Ig levels were also substantially reconstituted, but at lower levels than wild-type mice.

Sequence analysis of the naïve antibody repertoire demonstrated usage of 11 of 34 functional VH segments, ten DH segments, and three JH segments in the human IgH transgene. In addition, eight of 25 functional VK segments and all five of the JK segments were used from the human IgK transgene. The proportion of usage of the individual segments resembled that of the normal human antibody repertoire. N-additions were observed in the heavy chain but not kappa light chain rearrangements, reflecting the greatly reduced terminal deoxynucleotidyl transferase (TdT) activity during kappa light chain recombination. The range and average lengths of the human IgH (8–19 amino acids) and IgK (9–10 amino acids) CDR3 regions were comparable to those of normal human antibodies.

Immunization of the transgenic mice resulted in production of sub-nanomolar affinity antibodies to IL-8, EGF and TNF-α. The sequences of the antibodies from the antigen-specific hybridomas showed that the responses were polyclonal, with both VH and VK sequences exhibiting somatic hypermutations. Biacore measurements showed that antibodies to each of the antigens had KAs in the range of 10^9 to 10^{10}/M (Mendez *et al.*, 1997).

Comparisons between Ig transgenes containing 5 VH × 3 VK (Green *et al.*, 1994) and 66 VH × 32 VK (Mendez *et al.*, 1997) variable repertoires showed that a greater repertoire size improved many aspects of antibody production. Transgenic mice possessing the larger potential primary repertoire were better able to reconstitute the B cell compartment in bone marrow, and in the periphery produced higher levels of serum Ig with greater variable region diversity and had improved humoral responses to antigens, resulting in antibodies of high affinity. Because B cell development requires productive rearrangement of first heavy then light chain Ig, the increase in variable segments should result in a greater diversity of productive rearrangement occurring at each locus. The increased repertoire diversity should in turn lead to the observed increase in serum Ig levels and improved response to immunization.

Because of the improved repertoire size possible through the use of YACs, a human IgK YAC containing over 400 kb of sequence was incorporated into the synthetic miniloci approach described previously (Lonberg *et al.*, 1994). As before, the transloci were introduced into mice bearing mutations in the endogenous heavy and kappa light chain loci. Based upon flow cytometric analysis of bone marrow, the addition of more human VK segments improved the reconstitution of B cell development. However, mature B cell populations were not significantly greater than in mice without the added YAC-derived sequences.

Immunization of mice with the larger human IgK repertoire with human CD4 antigen induced a specific humoral response. Approximately 1% of antigen-specific hybridomas from immunized mice expressed human heavy chains, paired with either human IgK or mouse IgL chains. Purified antibodies from the eight hybridomas expressing fully-human heavy and light chains were analyzed by BIAcore and all were shown to have KAs in the range of 10^9/M. The function of several of the antibodies was tested in mixed lymphocyte reactions, and two of three had inhibitory effects similar to the positive control antibody (Fishwild *et al.*, 1996).

The majority of efforts to introduce human Ig transloci into the mouse genome have focused upon human IgH and IgK loci, but not IgL. In normal humans, lambda chains are expressed in about 40–50% of antibodies, but are found in only 5% of mouse antibodies (McIntire and Rouse, 1970). This is in part due to the relatively limited repertoire diversity in mouse compared with human lambda loci. Mice possess few variable segments in a relatively small locus. In contrast, the human lambda locus is more than a megabase in size, with 70 VL segments and a series of 7 JL–CL pairs (Frippiat *et al.*, 1995). Data from mouse and human B cells suggests that light chain rearrangement is staged, with the kappa alleles recombining prior to lambda alleles (Hieter *et al.*, 1981). The specific mechanisms controlling the timing of recombination and the ratio of kappa:lambda expression are still not entirely clear, although there is evidence for differential locus accessibility and involvement of transcription control regions such as enhancers (Popov *et al.*, 1996).

In order to capture the increased repertoire diversity represented by the human lambda locus, a translocus was constructed using a YAC bearing half of the germline human VL segments combined with cosmids containing the VL–CL pairs and the 3′ enhancer region (Popov *et al.*, 1996). The resulting YAC carried a 380 kb section of the human germline IgL locus, and was inserted into mouse ES cells via yeast protoplast fusion. The efficiency of the human IgL was tested by breeding with mice possessing a mutation in the endogenous IgK locus. Serum from mice homozygous for the inactivated IgK locus exhibited significant levels of human lambda chain with little change in endogenous mouse lambda chain expression, indicating that the human IgL translocus competed effectively with the endogenous mouse alleles. In mice heterozygous for the inactivated IgK locus, human IgL levels were significantly lower than in the knockout mice but greater than endogenous mouse IgK expression. Flow cytometric analysis confirmed that the ratio of mouse IgK:human IgL expression resembled that of normal human serum.

Hybridomas were derived from the human IgL transloci mice heterozygous for the mouse IgK locus. The rearrangements of both IgK and IgL alleles were examined by Southern blotting techniques. No bias in rearrangements was observed in human IgL expressing hybridomas compared with mouse IgK expressors, indicating that mechanisms controlling the timing of light chain rearrangement were not retained in the translocus mice. Sequence analysis of rearranged human IgL transgenes showed that the majority of VL segments were used in rearrangements but with few N or P nucleotide additions, most likely reflective of the lack of TdT expression during light chain rearrangement in mice (Popov *et al.*, 1999).

Mice were produced carrying human IgH, IgK and IgL transloci in combination with mutations that inactivated the endogenous murine IgH and IgK loci in order to better reproduce the normal human antibody repertoire. The mature B cell population in spleen was reconstituted at 25% to 45% of wild-type mouse levels. Cells expressing human IgH, IgK and IgL were detectable with a ratio of human IgK to human IgL of about 1:3, in contrast to the 3:2 ratio found in humans. This may be in part due to the construction of the human IgK translocus, which possessed multiple identical repeats of a subset of five human VK segments instead of the genuine human germline IgK locus. Alternately, the altered ratios may have been due to a lack of appropriate genetic control elements or incompatibility with murine factors (Gorman *et al.*, 1996).

Human serum IgM levels in the human IgH, IgK, and IgL transloci mice ranged from 50–400 μg/ml, which is somewhat less than the IgM levels observed in wild-type mice maintained under pathogen-free conditions. Immunization further elevated human serum IgM, but was lower than the levels induced by immunization of wild-type mice. Hybridomas derived from the translocus mice produced 80% less IgM than typical mouse hybridomas. Sequences of the antigen-specific hybridomas exhibited predominant use of a single human VH segment with 13 of 27 DH segments and two of four JH segments in a variety of different rearrangements including N or P nucleotide additions, creating highly diverse CDR3 regions. The light chain sequences showed frequent usage of just two VK segments with two other VK, and a single VL segment also observed, with very little CDR3 diversity due to a lack of TdT expression during light chain rearrangements. Somatic hypermutation was found in the hybridoma sequences indicating that the sequences underwent affinity maturation. However, the low number of mutations may indicate that the transloci are not efficient templates for mutation (Nicholson *et al.*, 1999).

4.4.3.4 *Human minichromosome-based transloci*

In contrast to approaches that rebuild the human Ig loci with synthetic constructs or genomic fragments, the use of human chromosomes, or fragments thereof, has the potential to allow the introduction of the entire intact unrearranged human IgH, IgK, and IgL loci into mouse ES cells. Human fibroblasts were stably transfected with a construct bearing a G418 resistance gene (neo) which integrated randomly throughout the human genome. Microcells derived from the stably-transfected fibroblasts were fused with mouse ES cells and selected on G418. The resulting ES cells contained the human chromosome (or a portion thereof) bearing the neo gene. The modified ES cells were microinjected into mouse blastocysts and chimeras and screened for retention of human IgH, IgK, or IgL loci. The chimeric mice exhibited appropriate tissue-specific expression of the Ig loci and germline transmission of a minichromosome (Tomizuka *et al.*, 1997).

Mice possessing minichromosomes bearing the human IgH and IgK loci were bred with one another as well as with mice harboring deletions at the IgH and IgK loci to prevent endogenous murine antibody expression. The resulting mice exhibited significant reconstitution of the mature naïve B cell compartment, with overall serum levels of human IgM, IgG, and IgK equivalent to those in wild-type mice. Additionally, human Ig of isotypes IgG1, IgG2, IgG3, and IgG4 were all detected in mouse serum at comparable levels to those in human serum. Immunization with human serum albumin (HSA) or G-CSF resulted in a rapid antigen-specific human antibody response (Tomizuka *et al.*, 2000; Ishida *et al.*, 2002).

While the minichromosome approach allows incorporation of essentially the entire human Ig repertoire in mice, instability in the absence of drug selection can be problematic for both germline transmission and hybridoma production. Mice immunized with HSA exhibited substantial antigen-specific titers but produced ten-fold fewer hybridomas compared to wild-type mice. Although the human IgH minichromosome was maintained in the majority of hybridomas, the human IgK minichromosome was absent in almost half of the clones. This instability was similarly reflected in metaphase spreads of tail-derived fibroblasts as well as in germline transmission rates. The human minichromosome bearing the IgL locus was similarly unstable (Ishida *et al.*, 2002).

Because of the instability of the minichromosomes bearing human Ig light chain loci, mice carrying the human IgH minichromosome were bred with mice bearing the human IgK transloci (Fishwild *et al.*, 1996), combining the full human IgH locus with a more stable human IgK transgene. Immunization of these mice with either human CD4 or human G-CSF resulted in significantly higher titers compared with mice bearing human IgH or IgK minichromosomes alone. In addition, stable hybridomas were derived from the mice immunized with each antigen at a frequency similar to that observed with wild-type mice. Several of the monoclonal antibodies generated had affinities in the nanomolar range, which is equivalent to therapeutic antibodies to the same targets (Ishida *et al.*, 2002).

4.4.3.5 *Translocus summary*

Early work with rearranged transgenes demonstrated that production of human antibodies in mice was feasible. A progressive series of mice with increasingly large and complex transloci showed gradual improvements in function as a consequence of the greater potential diversity of the variable repertoire. These data revealed the importance of regulatory regions as well as the number of Ig variable segments present. Each of these approaches showed that it was possible for fully-human transloci to take advantage of *in vivo* processes of repertoire formation, diversification, selection and expression.

The use of human transloci in mice was not without drawbacks. Early versions of the transgenic mice derived from synthetic miniloci or YACs exhibited significant deficiencies in B cell development, expansion, and function. The addition of increasing numbers of variable segments partially overcame these deficiencies, suggesting that the diversity of the germline repertoire was proportional to the level of functional reconstitution. Because the progression of B cell development depends upon the production of functional heavy and light chains, an increased repertoire of variable segments should allow a greater chance for productive variable rearrangement and thus more efficient B cell differentiation. However, even with the most diverse human IgH and IgK germline repertoires, B cell differentiation and expansion is incomplete, suggesting there may be additional mechanistic limitations with the use of fully-human antibody sequences (Mendez *et al.*, 1997). Another complication was the observation of murine heavy chains in serum from mice which had inactivated endogenous IgH loci. The transplant of the human variables to mouse constant regions is presumably through inappropriate recombination or RNA splicing between the translocus and the endogenous alleles, which has been described (Durdik *et al.*, 1989; Knight *et al.*, 1995).

It was possible to overcome the various inefficiencies observed, with even the most complex human transloci, in a direct and ingenious way by functionally removing endogenous mouse Ig loci through gene inactivation. Without competition from two of the three murine Ig loci, B cells produced in translocus mice are substantially dependent upon rearrangement and expression of the human Ig genes. Although there may be contamination from murine IgL sequences, the majority of the B cells exclusively produce fully-human antibodies, and immunized mice produce genuine high affinity antibodies which may not require further modification or improvement for therapeutic applications.

Use of human Ig transloci mice has been incredibly successful, producing fully-human therapeutic antibodies directed against a diverse range of target antigens, including human ligands and receptors, human tumor markers, pathogenic proteins and polysaccharides, small molecules and many other relevant targets. It is especially important to highlight the effectiveness of mouse platforms in generating responses to human targets. The majority of human antibodies produced in mice have affinities between 0.1 nM and 10 nM, although there are also examples of antibodies with picomolar or even sub-picomolar affinities. This is similar to the range of affinities observed for mAbs isolated after immunization of wild-type mice, which perhaps reflects the limitations of affinity maturation *in vivo*. The majority of fully-human therapeutic antibodies in the clinic are derived from synthetic minilocus, minichromosome and YAC-based translocus approaches (Lonberg, 2008).

4.4.4 *Humanized mouse Ig loci*

A variety of translocus approaches have successfully introduced increasingly larger and more complex human Ig loci into the mouse genome, up to and including the entire genomic IgH and IgK loci as minichromosomes. While these approaches have progressively increased the repertoire diversity and functionality of the introduced human Ig loci, the efficiency of B cell development and expansion, allelic exclusion, and Ig expression remain lower than the natural murine immune response. These inefficiencies are somewhat ameliorated although never completely overcome with increased repertoire diversity, requiring inactivation of endogenous mouse Ig loci in order to enhance use of the human transloci.

One critical factor common to B cell development and function is signaling through the B cell receptor (BCR), which consists of the membrane form of antibody in a complex with specific signaling molecules and various positive and negative co-receptors (Brezski and Monroe, 2008). Most, if not all, of the signaling function of the antibody is through the heavy chain constant region. Therefore, if the interactions between the constant region and the signaling apparatus are inefficient, there could be effects upon B cell differentiation and antibody rearrangement, allelic exclusion and production. Although Ig constant regions are highly conserved, there are clear differences between mouse and human sequences. Therefore, although fully-human antibodies may be able to interact productively with the murine B cell machinery, these interactions may not be as efficient as those with wild-type mouse antibodies.

In addition to potential incompatibilities between specific mouse and human constant regions, most transloci, with the exception of the human IgH minichromosomes, do not have the full complement of Ig isotypes. Typically, synthetic and YAC IgH transloci are limited to human IgM and a single human IgG isotype, such as IgG1. The roles of the various isotypes in generation of an efficient humoral response are not fully clear, and thus exclusion of one or more may compromise immune function. More importantly, the antibody constant region must bind to a series of Fc receptors in order to coordinate responses of different immune cell populations and modulate the overall humoral response (Nimmerjahn and Ravetch, 2008). The IgH isotypes and the Fc receptors with which they interact differ between human and mouse. Not only are there fewer low affinity Fc gamma receptors in mice, the cell-specific expression patterns and specific isotypes with which they interact are different from those in humans (Nimmerjahn and Ravetch, 2006). By using exclusively the human constant regions, transloci mice may generate different responses through the Fc receptors. Antibody–antigen binding also mediates complement activation, which is similarly dependent upon interactions with antibody constant regions. These interactions may also be affected by species-specific differences in Ig constant region sequences (Sjöberg *et al.*, 2009). The complex interplay of antibodies, antigens, cell signaling, Fc receptor, and complement binding contribute to the overall productivity and specificity of the immune response *in vivo*.

In order to overcome the potential restrictions of using fully-human antibodies in mice, the benefits of a fully-human variable repertoire must be combined with the efficiency of the humoral response directed by murine constant regions. The most straightforward approach would be to link the germline human variable loci to the relevant murine constant regions. Class switching demonstrates that antibodies are modular, so that variable regions can be transferred between isotypes without significant loss of specificity (Stavnezer *et al.*, 2008). Typically, chimeric antibodies possess mouse variable regions linked to human constant regions which have been shown in many cases to function as well as the parental murine antibodies (Morrison *et al.*, 1984). While it would be possible to accomplish this using transloci, there is the risk that accessibility and expression control mechanisms may be excluded. The most direct method would be to target the murine Ig loci to eliminate the germline variables of the mouse and replace them with the equivalent human genomic sequences. The antibodies resulting from these "reverse chimeric" loci would have fully-human variables regions with mouse constant regions and would be predicted to maintain wild-type murine humoral responses. In addition, by targeting the genuine murine Ig loci, all of the accessibility and expression control regions outside of the variable region sequences will remain intact, allowing appropriate antibody expression throughout B cell development, activation and differentiation, whether the elements have been defined or not. Another advantage of directly targeting the genuine murine loci is that there is no need for separate inactivation of endogenous loci, reducing the complexity of breeding and avoiding the inappropriate trans-switching or trans-recombination events found in various transloci mice (Durdik *et al.*, 1989; Knight *et al.*, 1995).

Direct targeting of genomic loci takes advantage of the increased homologous recombination activity found in ES cells. By flanking a selectable marker with regions of sequence homology to specific regions in the mouse genome, homologous recombination replaces the sequences found between the regions of homology with the selectable marker at a low rate. For example, growing ES cells in appropriate selection media allows only targeted cells to survive. Incorporation of the selectable marker randomly in the genome as a transgene occurs with much higher frequency than specific targeting by homologous recombination. Therefore careful screening strategies, typically Southern blotting or genomic PCR, must be employed to ensure that properly targeted cells are recognized among the large number of transgenic ES cells.

Similarly to transloci approaches, the size and complexity of the Ig loci presents a significant technical challenge to removal of endogenous mouse sequences and replacement with human genomic loci. Although YACs provide the convenience of manipulation by recombination in yeast combined with extremely large insert sizes, Ig loci are extremely repetitive and at risk of inappropriate recombination events which may rearrange or remove critical sequences. Bacterial artificial chromosomes (BACs) are unable to carry as much DNA as YACs, but are much more easily handled and manipulated. Moreover, methods that take advantage of homologous recombination to allow specific targeting of BACs in bacteria have been developed. Much like targeting in ES cells, short regions of homology to BAC genomic DNA can be placed flanking a sequence of interest, including a selectable marker active in prokaryotes, to create a targeting vector that replaces the region between the areas of homology in the BAC. Growth in selectable media allows only bacteria bearing the targeted BAC to survive, and the resulting constructs can be screened for correct targeting (Zhang *et al.*, 1998). These modified BACs can in turn be used as targeting vectors in ES cells to make deletions of greater sizes than is possible with other vectors. The speed and precision of bacterial homologous recombination allows more complicated constructs to be created, including incorporation of heterologous sequences between the regions of homology that will be used to target ES cells. Thus, it is possible to create a large targeting vector that can specifically insert large sections of human genomic DNA into specific sites within the mouse genome *via* homologous recombination, which would be required to replace the mouse Ig loci with human sequences (Valenzuela *et al.*, 2003).

In typical targeting approaches using homologous recombination in mouse ES cells, the vectors are designed to allow screening *via* Southern blotting or long range genomic PCR. Due to the larger sizes of BACs, these strategies would be unable to easily span the distance across the homology regions used for targeting and therefore would be unable to distinguish targeted clones from transgenes. This potential limitation was overcome through the development of a quantitative PCR assay which distinguishes the reduction in copy number of probes in the targeted gene from the diploid copy number of a series of reference probes in non-targeted loci. This type of quantitative assay can also be employed to detect the appropriate insertion of selectable marker cassettes and heterologous DNA sequences, such as human genomic DNA (Valenzuela *et al.*, 2003).

The technical tools available to replace murine Ig variable loci with human germline variable sequences require multiple targeting events in ES cells to incorporate the entire Ig loci. The IgH and IgK loci are arranged similarly in mouse and humans, with variable segments upstream of a critical intronic enhancer sequence, followed by constant regions and additional enhancers. The IgH and IgK loci were targeted sequentially in separate ES cell lineages using similar approaches to first delete the murine genomic sequence and then to introduce the human germline Ig variable sequences. After initial targeting events at the 5′ and 3′ boundaries of the variable region loci, the entire mouse genomic sequence was removed, and initial BAC constructs with a minimal variable human variable locus with several V segments and all DH and JH segments (for the heavy chain locus) or all JK segments (for the IgK locus) were inserted. Mice were generated from the targeted ES cells and bred together as hetero- and homozygotes for both humanized alleles (Stevens, 2008).

The targeted mouse ES cells were derived from F1 heterozygotes of 129 × BL6 mice. By targeting the IgH allele of the 129 mouse strain with human sequences, use of the humanized locus was distinguishable by flow cytometry with antibodies specific for the IgMa allotype found in 129 compared with the IgMb allotype of the unmodified BL6 IgH allele. Analysis of wild-type, heterozygote, and homozygote mice demonstrated that the chimeric heavy chain allele was used efficiently in heterozygote mice which had normal B cell populations, although the reduced number of variable segments was reflected in the reduced number of B cells expressing the chimeric IgH chain. Interestingly, homozygote mice expressing only the relatively restricted human VH repertoire were phenotypically normal, exhibiting no differences from wild-type mice in bone marrow or splenic B cell populations or structural morphology. Further flow cytometric analysis of B cell differentiation in bone marrow demonstrated that all stages of B cell development were equivalent to wild-type mice. This was in direct contrast to transloci and minichromosome mice, which produced inefficiently reconstituted B cell populations even when the transloci possessed large variable repertoires (Stevens, 2009). Experiments with mice in which the murine antibody repertoire was deliberately restricted either by genetic manipulation (Xu and Davis, 2000; Cascalho *et al.*, 1996) or nuclear transfer from donor B cells (Hochedlinger and Jaenisch, 2002) suggest that limiting the variable repertoire does not prevent full B cell expansion in the periphery. Therefore, targeting the mouse Ig loci and expressing antibodies with genuine murine heavy chains allows full reconstitution of the B cell compartment despite having a relatively limited variable repertoire (Xu and Davis, 2000).

Additional flow cytometric studies in heterozygote mice demonstrated that humanized loci with relatively limited variable repertoires exhibited strong allelic exclusion between the humanized and wild-type IgH alleles. Similarly, mice homozygous for the humanized IgK allele effectively excluded the lambda locus. The ratio between expressed humanized IgK and wild-type mouse lambda loci was similar to wild-type mice, despite possessing fewer VK segments than wild-type mice. These results suggest that exclusion between IgK and IgL loci is not related to differences in repertoire diversity, and thus potential for productive rearrangements, but is intrinsic to the loci perhaps via enhancers or other regulatory elements.

All serum Ig isotypes from the doubly-targeted Ig mice were represented at similar levels to that of wild-type mice. Immunization of wild-type and homozygous humanized Ig littermates with protein antigens resulted in similar antigen-specific serum titers. Hybridomas from wild-type and humanized Ig mice yielded the same rate of productive fusions and analogous numbers antigen-specific mAbs. The range of affinities of the antigen-specific hybridomas from mice of both types was also comparable (Stevens, 2009).

Variable sequence analysis of non-immunized splenocyte and hybridoma antibodies exhibited an extremely diverse repertoire, with usage of nearly every functional VH, DH, and JH, or VK and JK segment introduced. N and P additions were observed at the IgH, but not IgK, CDR3, indicative of TdT function during only the heavy chain rearrangement process in mice. Comparison of IgM- and

IgG-derived sequences from non-immunized splenocytes and hybridoma sequences further demonstrated extensive somatic hypermutation at both the heavy and light chain variable regions, indicative of affinity maturation of the humanized alleles (Durdik *et al.*, 1989, Knight *et al.*, 1995).

Thus, direct replacement of the germline variable sequences at murine IgH and IgK loci with the equivalent human sequences successfully recapitulated wild-type humoral responses in mice. Use of murine constant regions expressed from genuine Ig loci appeared to overcome the inefficiencies of transloci approaches, making full use of endogenous immune response in mice. Importantly, humanized Ig loci were found functionally identical to mouse loci and produced antibodies with fully-human variable regions.

4.4.4.1 *Summary of humanized Ig loci*

The direct targeting of the murine Ig loci to replace genomic DNA encoding the unrearranged variable segments with the equivalent germline human Ig variable genomic sequences required a novel genomic engineering approach. Serial targeting of ES cells can create chimeric IgH and IgK loci bearing all of the germline human variable segments integrated into the appropriate mouse loci. The resulting antibodies possess fully-human variable regions joined to mouse constant regions, which appear to function similarly to the mouse antibodies. Phenotypic analyses also suggest that humanized Ig mice are identical to wild-type mice and reproduce the normal humoral response to antigens.

4.5 Alternate Translocus Approaches

The development of translocus approaches to produce human antibodies has been applied successfully to other species. Transgenic rats, rabbits, cows, chickens, and pigs bearing various human Ig locus constructs have been described (see, for example, Kuroiwa *et al.*, 2002). With the advent of induced pluripotent stem cell technologies and improved methods for genomic modification, there will be continued efforts to create novel human transloci for production of human antibody therapeutics in various species.

4.6 Conclusions

The use of translocus, minichromosome and humanized Ig loci in mice allows the rapid production and optimization of human antibody therapeutics by taking advantage of the genuine humoral responses *in vivo*. While there are substantial differences in the functionality and productivity of the immune response in transloci and humanized mice, all of these approaches have generated effective therapeutic antibodies to a large variety of targets. Combining human antibody production in mice with affinity maturation, Fc engineering, glycosylation modification, and other optimization technologies will lead to further improvements in treatment of human disease.

References

Almagro J.C., Fransson J. (2008). Humanization of antibodies. *Front Biosci* **13**: 1619–33.

Betz A.G., Milstein C., Gonzalez-Fernandez A., Pannell R., Larson T., Neuberger M.S. (1994). Elements regulating somatic hypermutation of an immunoglobulin kappa gene: critical role for the intron enhancer/matrix attachment region. *Cell* **77**: 239–48.

Brezski R.J., Monroe J.G. (2008). B-cell receptor. *Adv Exp Med Biol* **640**: 12–21.

Brinster R.L., Ritchie K.A., Hammer R.E., O'Brien R.L., Arp B., Storb U. (1983). Expression of a microinjected immunoglobulin gene in the spleen of transgenic mice. *Nature* **306**: 332–6.

Brüggemann M., Caskey H.M., Teale C., Waldmann H., Williams G.T., Surani M.A., Neuberger M.S. (1989). A repertoire of monoclonal antibodies with human heavy chains from transgenic mice. *Proc Natl Acad Sci USA* **86**: 6709–13.

Brüggemann M., Spicer C., Buluwela L., Rosewell I., Barton S., Surani M.A., Rabbitts T.H. (1991). Human antibody production in transgenic mice: expression from 100 kb of the human IgH locus. *Eur J Immunol* **21**: 1323–6.

Bucchini D., Reynaud C.A., Ripoche M.A., Grimal H., Jami J., Weill J.C. (1987). Rearrangement of a chicken immunoglobulin gene occurs in the lymphoid lineage of transgenic mice. *Nature* **326**: 409–11.

Cascalho M., Ma A., Lee S., Masat L., Wabl M. (1996). A quasi-monoclonal mouse. *Science* **272**: 1649–52.

Cedar H., Bergman Y. (2008). Choreography of Ig allelic exclusion. *Adv Immunol* **99**: 1–32.

Choi T.K., Hollenbach P.W., Pearson B.E., Ueda R.M., Weddell G.N., Kurahara C.G., Woodhouse C.S., Kay R.M., Loring J.F. (1993). Transgenic mice containing a human heavy chain immunoglobulin gene fragment cloned in a yeast artificial chromosome. *Nat Genet* **4**: 117–23.

Davies N.P., Rosewell I.R., Brüggemann M. (1992). Targeted alterations in yeast artificial chromosomes for inter-species gene transfer. *Nucleic Acids Res* **20**: 2693–8.

Davies N.P., Rosewell I.R., Richardson J.C., Cook G.P., Neuberger M.S., Brownstein B.H., Norris M.L., Brüggemann M. (1993). Creation of mice expressing human antibody light chains by introduction of a yeast artificial chromosome containing the core region of the human immunoglobulin kappa locus. *Biotechnology (NY)* **11**: 911–4.

Durdik J., Gerstein R.M., Rath S., Robbins P.F., Nisonoff A., Selsing E. (1989). Isotype switching by a microinjected mu immunoglobulin heavy chain gene in transgenic mice. *Proc Natl Acad Sci USA* **86**: 2346–50.

Eccles S., Sarner N., Vidal M., Cox A., Grosveld F. (1990). Enhancer sequences located 3′ of the mouse immunoglobulin lambda locus specify high-level expression of an immunoglobulin lambda gene in B cells of transgenic mice. *New Biol* **2**: 801–11.

Ferrier P., Krippl B., Blackwell T.K., Furley A.J.W., Suh H., Winoto A., Cook W.D., Hood L., Constantini F., Alt F.W. (1990). Separate elements control DJ and VDJ rearrangement in a transgenic recombination substrate. *EMBO J* **9**: 117–25.

Fishwild D.M., O'Donnell S.L., Bengoechea T., Hudson D.V., Harding F., Bernhard S.L., Jones D., Kay R.M., Higgins K.M., Schramm S.R., Lonberg N. (1996). High-avidity human IgG kappa monoclonal antibodies from a novel strain of minilocus transgenic mice. *Nat Biotechnol* **14**: 845–51.

Frippiat J.P., Williams S.C., Tomlinson I.M., Cook G.P., Cherif D., Le Paslier D., Collins J.E., Dunham I., Winter G., Lefranc M.P. (1995). Organization of the human immunoglobulin lambda light-chain locus on chromosome 22q11.2. *Hum Mol Genet* **4**: 983–91.

Giusti A.M., Coffee R., Manser T. (1992). Somatic recombination of heavy chain variable region transgenes with the endogenous immunoglobulin heavy chain locus in mice. *Proc Natl Acad Sci USA* **89**: 10321–5.

Gollahon K.A., Hagman J., Brinster R.L., Storb U. (1988). Ig lambda-producing B cells do not show feedback inhibition of gene rearrangement. *J Immunol* **141**: 2771–80.

Goodhardt M., Cavelier P., Akimenko M.A., Lutfalla G., Babinet C., Rougeon F. (1987). Rearrangement and expression of rabbit immunoglobulin kappa light chain gene in transgenic mice. *Proc Natl Acad Sci USA* **84**: 4229–33.

Gorman J.R., van der Stoep N., Monroe R., Cogne M., Davidson L., Alt F.W. (1996). The Ig(kappa) enhancer influences the ratio of Ig(kappa) versus Ig(lambda) B lymphocytes. *Immunity* **5**: 241–52.

Green L.L., Hardy M.C., Maynard-Currie C.E., Tsuda H., Louie D.M., Mendez M.J., Abderrahim H., Noguchi M., Smith D.H., Zeng Y., David N.E., Sasai D.H., Garza D., Brenner D.G., Hales J.F., McGuinness R.P., Capon D.J., Klapholz S., Jakobovits A. (1994). Antigen-specific human monoclonal antibodies from mice engineered with human Ig heavy and light chain YACs. *Nat Genet* **7**: 13–21.

Grosschedl R., Weaver D., Baltimore D., Costantini F. (1984). Introduction of a mu immunoglobulin gene into the mouse germ line: specific expression in lymphoid cells and synthesis of functional antibody. *Cell* **38**: 647–58.

Hieter P.A., Hollis G.F., Korsmeyer S.J., Waldmann T.A., Leder P. (1981). Clustered arrangement of immunoglobulin lambda constant region genes in man. *Nature* **294**: 536–40.

Hochedlinger K., Jaenisch R. (2002). Monoclonal mice generated by nuclear transfer from mature B and T donor cells. *Nature* **415**: 1035–8.

Hohn B., Koukolíková-Nicola Z., Lindenmaier W., Collins J. (1988). Cosmids. *Biotechnology* **10**: 113–27.

Hoogenboom H.R. (2005). Selecting and screening recombinant antibody libraries. *Nat Biotech* **23**: 1105–16.

Ishida I., Tomizuka K., Yoshida H., Tahara T., Takahashi N., Ohguma A., Tanaka S., Umehashi M., Maeda H., Nozaki C., Halk E., Lonberg N. (2002). Production of human monoclonal and polyclonal antibodies in TransChromo animals. *Cloning Stem Cells* **4**: 91–102.

Jakobovits A. (1994). Humanizing the mouse genome. *Curr Biol* **4**: 761–3.

Knight K.L., Kingzette M., Crane M.A., Zhai S.K. (1995). Transchromosomally derived Ig heavy chains. *J Immunol* **155**: 684–91.

Kohler G., Milstein C. (1975). Continuous cultures of fused cells secreting antibody of a predefined specificity. *Nature* **256**: 495–7.

Kuroiwa Y., Kasinathan P., Choi Y.J., Naeem R., Tomizuka K., Sullivan E.J., Knott J.G., Duteau A., Goldsby R.A., Osborne B.A., Ishida I., Robl J.M. (2002). Cloned transchromosomic calves producing human immunoglobulin. *Nat Biotechnol* **20**: 889–94.

Lanzavecchia A., Corti D., Sallusto F. (2007). Human monoclonal antibodies by immortalization of B cells. *Curr Opin Biotechnol* **18**: 523–8.

Lonberg N., Taylor L.D., Harding F.A., Trounstine M., Higgins K.M., Schramm S.R., Kuo C.C., Mashayekh R., Wymore K., McCabe J.G., Munoz-O'Regan D., O'Donnell S.L., Lapachet E.S.G., Bengoechea T., Fishwild D.M., Carmack C.E., Kay R.M., Huszar D. (1994). Antigen-specific human antibodies from mice comprising four distinct genetic modifications. *Nature* **368**: 856–9.

Lonberg N. (2008). Fully human antibodies from transgenic mouse and phage display platforms. *Curr Opin Immunol* **20**: 450–9.

McIntire K.R., Rouse A.M. (1970). Mouse immunoglobulin light chains: alterations of kappa:lambda ratio. *Fed Proc* **19**: 704–10.

Mendez M.J., Green L.L., Corvalan J.R., Jia X.C., Maynard-Currie C.E., Yang X.D., Gallo M.L., Louie D.M., Lee D.V., Erickson K.L., Luna J., Roy C.M., Abderrahim H., Kirschenbaum F., Noguchi M., Smith D.H., Fukushima A., Hales J.F., Klapholz S., Finer M.H., Davis C.G., Zsebo K.M., Jakobovits A. (1997). Functional transplant of megabase human immunoglobulin loci recapitulates human antibody response in mice. *Nat Genet* **15**: 146–56.

Morrison S.L., Johnson M.J., Herzenberg L.A., Oi V.T. (1984). Chimeric human antibody molecules: mouse antigen-binding domains with human constant region domains. *Proc Natl Acad Sci USA* **81**: 6851–5.

Nicholson I.C., Zou X., Popov A.V., Cook G.P., Corps E.M., Humphries S., Ayling C., Goyenechea B., Xian J., Taussig M.J., Neuberger M.S., Brüggemann M. (1999). Antibody repertoires of four- and five-feature translocus mice carrying human immunoglobulin heavy chain and kappa and lambda light chain yeast artificial chromosomes. *J Immunol* **163**: 6898–906.

Nimmerjahn F., Ravetch J.V. (2006). Fcgamma receptors: old friends and new family members. *Immunity* **24**: 19–28.

Nimmerjahn F., Ravetch J.V. (2008). Fc-receptors as regulators of immune responses. *Nat Rev Immunol* **8**: 34–47.

Nussenzweig M.C., Shaw A.C., Sinn E., Campos-Torres J., Leder P. (1987). Allelic exclusion in transgenic mice that express the membrane form of immunoglobulin mu. *Science* **236**: 816–9.

Nussenzweig M.C., Shaw A.C., Sinn E., Danner D.B., Holmes K.L., Morse H.C., Leder P. (1988). Allelic exclusion in transgenic mice carrying mutant human IgM genes. *J Exp Med* **157**: 613–27.

O'Brien R.L., Brinster R.L., Storb U. (1987). Somatic hypermutation of an immunoglobulin transgene in kappa transgenic mice. *Nature* **326**: 405–9.

Peled J.U., Kuang F.L., Iglesias-Ussel M.D., Roa S., Kalis S.L., Goodman M.F., Scharff M.D. (2008). The biochemistry of somatic hypermutation. *Annu Rev Immunol* **26**: 481–511.

Pinkert C.A., Manz J., Linton P.J., Klinman N.R., Storb U. (1989). Elevated PC responsive B cells and anti-PC antibody production in transgenic mice harboring anti-PC immunoglobulin genes. *Vet Immunol Immunopathol* **23**: 321–32.

Popov A.V., Bützler C., Frippiat J.P., Lefranc M.P., Brüggemann M. (1996). Assembly and extension of yeast artificial chromosomes to build up a large locus. *Gene* **177**: 195–201.

Popov A.V., Zou X., Xian J., Nicholson I.C., Brüggemann M. (1999). A human immunoglobulin lambda locus is similarly well expressed in mice and humans. *J Exp Med* **189**: 1611–20.

Ritchie K.A., Brinster R.L., Storb U. (1984). Allelic exclusion and control of endogenous immunoglobulin gene rearrangement in kappa transgenic mice. *Nature* **312**: 517–20.

Rusconi S., Kohler G. (1985). Transmission and expression of a specific pair of rearranged immunoglobulin mu and kappa genes in a transgenic mouse line. *Nature* **314**: 330–4.

Sjöberg A.P., Trouw L.A., Blom A.M. (2009). Complement activation and inhibition: a delicate balance. *Trends Immunol* **30**: 83–90.

Sohn J., Gerstein R.M., Hsieh C.L., Lemer M., Selsing E. (1993). Somatic hypermutation of an immunoglobulin mu heavy chain transgene. *J Exp Med* **177**: 493–504.

Stavnezer J., Guikema J.E., Schrader C.E. (2008). Mechanism and regulation of class switch recombination. *Annu Rev Immunol* **26**: 261–92.

Sternberg N. (1990). Bacteriophage P1 cloning system for the isolation, amplification, and recovery of DNA fragments as large as 100 kilobase pairs. *Proc Natl Acad Sci USA* **87**: 103–7.

Stevens S. (2008). Public Presentation:VelocImmune — A novel platform for human antibody discovery, 14 November 2008.

Stevens S. (2009). Public Presentation:VelocImmune — A novel platform for the discovery of human antibody therapeutics, 02 March 2009.

Storb U., Denis K.A., Brinster R.L., Witte O.N. (1985). Pre-B cells in kappa-transgenic mice. *Nature* **316**: 356–8.

Storb U., Pinkert C., Arp B., Engler P., Gollahon K., Manz J., Brady W., Brinster R.L. (1986). Transgenic mice with mu and kappa genes encoding antiphosphorylcholine antibodies. *J Exp Med* **164**: 627–41.

Tarlinton D.M. (2008). Evolution in miniature: selection, survival and distribution of antigen reactive cells in the germinal centre. *Immunol Cell Biol* **86**: 133–8.

Taylor L.D., Carmack C.E., Schramm S.R., Mashayekh R., Higgins K.M., Kuo C.C., Woodhouse C., Kay R.M., Lonberg N. (1992). A transgenic mouse that expresses a diversity of human sequence heavy and light chain immunoglobulins. *Nucleic Acids Res* **20**: 6287–95.

Tomizuka K., Yoshida H., Uejima H., Kugoh H., Sato K., Ohguma A., Hayasaka M., Hanaoka K., Oshimura M., Ishida I. (1997). Functional expression and germline transmission of a human chromosome fragment in chimaeric mice. *Nat Genet* **16**: 133–43.

Tomizuka K., Shinohara T., Yoshida H., Uejima H., Ohguma A., Tanaka S., Sato K., Oshimura M., Ishida I. (2000). Double trans-chromosomic mice: maintenance of two individual human chromosome fragments containing Ig heavy and kappa loci and expression of fully human antibodies. *Proc Natl Acad Sci USA* **97**: 722–7.

Tonegawa S. (1983). Somatic generation of antibody diversity. *Nature* **302**: 575–81.

Valenzuela D.M., Murphy A.J., Frendewey D., Gale N.W., Economides A.N., Auerbach W., Poueymirou W.T., Adams N.C., Rojas J., Yasenchak J., Chernomorsky R., Boucher M., Elsasser A.L., Esau L., Zheng J., Griffiths J.A., Wang X., Su H., Xue Y., Dominguez M.G., Noguera I., Torres R., Macdonald L.E., Stewart A.F., DeChiara T.M., Yancopoulos G.D. (2003). High-throughput engineering of the mouse genome coupled with high-resolution expression analysis. *Nat Biotechnol* **21**: 652–9.

Wagner S.D., Williams G.T., Larson T., Neuberger M.S., Kitamura D., Rajewsky K., Xian J., Brüggemann M. (1994a). Antibodies generated from human immunoglobulin miniloci in transgenic mice. *Nucleic Acids Res* **22**: 1389–93.

Wagner S.D., Popov A.V., Davies S.L., Xian J., Neuberger M.S., Brüggemann M. (1994b). The diversity of antigen-specific monoclonal antibodies from transgenic mice bearing human immunoglobulin gene miniloci. *Eur J Immunol* **24**: 2672–81.

Wagner S.D., Gross G., Cook G.P., Davies S.L., Neuberger M.S. (1996). Antibody expression from the core region of the human IgH locus reconstructed in transgenic mice using bacteriophage P1 clones. *Genomics* **35**: 405–14.

Xian J., Zou X., Popov A.J., Mundt C.A., Miller N., Williams G.T., Davis C.G., Neuberger M.S., Brüggemann M. (1998). Comparison of the performance of plasmid-based human IgK minilocus and YAC-based human IgK transloci for the production of human antibody repertoires in transgenic mice. *Transgenics* **2**: 333–43.

Xu J.L., Davis M.M. (2000). Diversity in the CDR3 region of VH is sufficient for most antibody specificities. *Immunity* **13**: 37–45.

Yamamura K., Kudo A., Ebihara T., Kamino K., Araki K., Kumahara Y., Watanabe T. (1984). Introduction of human gamma 1 immunoglobulin genes into fertilized mouse eggs. *J Biochem* **96**: 357–63.

Yamamura K., Kudo A., Ebihara T., Kamino K., Araki K., Kumahara Y., Watanabe T. (1986). Cell-type-specific and regulated expression of a human gamma 1 heavy-chain immunoglobulin gene in transgenic mice. *Proc Natl Acad Sci USA* **83**: 2152–6.

Yancopoulos G.D., Alt F.W. (1986). Regulation of the assembly and expression of variable-region genes. *Annu Rev Immunol* **4**: 339–68.

Zhang Y., Buchholz F., Muyrers J.P.P., Stewart A.F. (1998). A new logic for DNA engineering using recombination in Escherichia coli. *Nat Genet* **20**: 123–8.

Zou X., Xian J., Davies N.P., Popov A.V., Brüggemann M. (1996). Dominant expression of a 1.3 Mb human Ig kappa locus replacing mouse light chain production. *FASEB J* **10**: 1227–32.

5

Isotype Selection and Fc Engineering: Design and Construction of Fit-for-Purpose Therapeutic Antibodies

William R. Strohl

5.1 Introduction

Humans have five classes of immunoglobulins (Igs), IgA, IgD, IgE, IgG, and IgM (Fig. 5.1), all of which are secreted by activated B cells as complex glycoproteins possessing multiple binding activities (Table 5.1). All human Igs are comprised of a similar basic monomeric H2L2 structure consisting of a pair of heavy chains, each ranging from ca. 50–75 kDa in size, and each paired with a light chain of about 25 kDa. The heavy chains define the class of Ig, i.e. α, δ, ε, γ, and μ comprise the IgA, IgD, IgE, IgG, and IgM classes, respectively. The light chains are comprised of either κ or λ isoforms. Igs possess two major regions, Fc ("crystallizable fragment") and Fabs (antigen binding regions), both of which are built on domains consisting of β-pleated sheets. The Fc region is comprised only of the heavy chain, whereas Fabs are comprised of both heavy chain and light chain domains.

In the IgA, IgD, and IgG classes, the Fab is separated from the Fc by a flexible, non-structured hinge region, whereas in the IgE and IgM classes, an extra β-sheet constant domain replaces the hinge (Fig. 5.1), resulting in significantly less flexibility of the Fab domains in these Ig classes. Additionally, the IgA and IgM classes possess two additional unique structural features that confer additional

* Centocor R&D, Inc., USA.

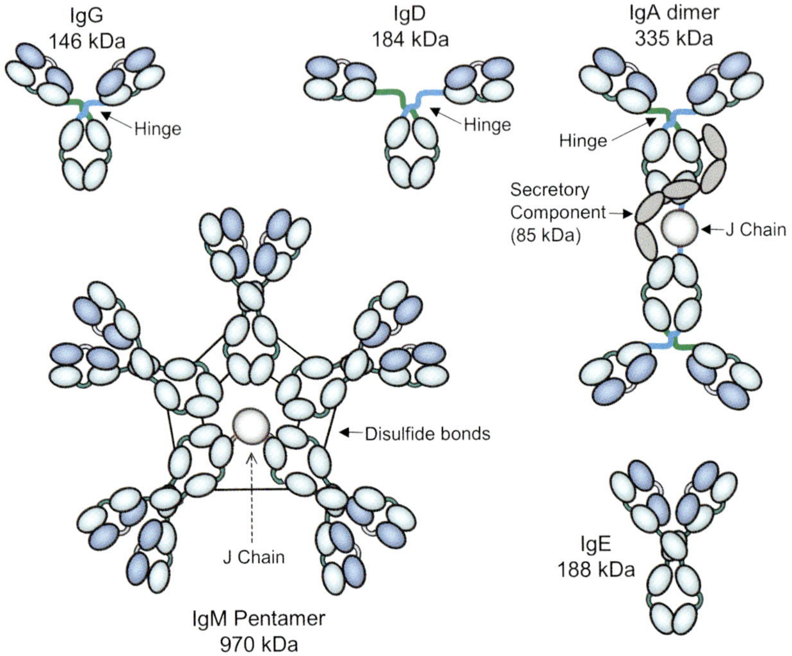

Figure 5.1 Depictions of the structures of human immunoglobulins, including IgG, IgD, IgA, IgM, and IgE. See Table 5.1 for more details.

activities upon them. The first is a C-terminal "tailpiece," which is an 18 amino acid non-structured sequence that confers the ability to multimerize through a critical cys residue (Sørensen *et al.*, 1999), and the second is an additional chain known as the joining chain (J chain), which confers the ability to bind to specific receptors such as the polymeric Ig receptor (pIgR) (Brandtzaeg and Prydz, 1984). IgDs also have a short tailpiece that allows them to be anchored to the surface of B cells (Sun *et al.*, 2005).

The light and heavy variable chains, which comprise the primary antigen binding sequences of Igs, are derived from families of V and J (and in heavy chains, D) genes that are common to all Ig classes. The antibody development process begins by expression of IgMs and IgDs on the surface of naïve B cells that, in response to antigen stimulation, then evolves through hypermutation and class-switching, as directed by the strength of antigen stimulation and cytokines, to the generation of high affinity IgGs.

The human IgG class is further subdivided into IgG1, IgG2, IgG3, and IgG4 isotypes (or subclasses), each possessing unique properties. The IgG subclasses appear to have arisen from gene duplication, which most likely occurred later in

Table 5.1 Characteristics of human antibody classes and isotypes

Characteristic or Function (Human Data Only)	Class or Isotype: IgG1	IgG2	IgG3	IgG4	Serum IgA	Dimeric sIgA1[a]	Dimeric sIgA2[a]	IgD	IgE	Pentameric IgM	Hexameric IgM
Approx. mol. wt. (kDa)	146	146	165	146	160	445	445	184	188	970	1140
Protein chains[b]	γ2L2	γ2L2	γ2L2	γ2L2	α2L2	(α2L2)2, J, SC	(α2L2)2, J, SC	δ2L2	ε2L2	(μ2L2)5, J	(μ2L2)6
Functional valency	2	2–4[c]	2	1[d]	2	4	4	2	2	5[e]	12
No. of Fc N-linked glycosylation sites (per monomer)[f]	1	1	1	1	2	3	3–5	3	7	5	5
Primary location(s)	serum	serum	serum	serum	serum[g]	external secretions[h]	external secretions[h]	B-cell surface; serum	mast cell, basophil surfaces	serum; B-cell surface	serum; B-cell surface
Hinge region	+	+	+	+	+	+	+	+	—	—	—
ADCC-NK cells	+	—	+	—		—	—	—	—	—	—
ADCP-MØ	+	+	+	+	?	+	+	—	—	—	—
ADCC/ADCP PMNs	—	—	—	—	+	—	—	—	—	—	—
FcRn binding	+	+	+[i]	+	—	—	—	—	—	—	—
Number of allotypes[j]	4	1	13	0	na[k]	na	na	na	na	na	na
Nominal serum half-life (days)	14–21	14–21	7	14–21	5	5	5	3	2	6	6
Approx. mean serum conc.	9 mg/mL	3 mg/mL	1 mg/mL	0.5 mg/mL	3.5 mg/mL[g]	—	—	35 μg/mL	300 ng/mL	1.4 mg/mL (95% of IgM)	0.1 mg/mL (5% of IgM)
Placental transfer	+ +	+	+ +	+ +	—	—	—	—	—	—	—
Produced by fetus during gestation in response to infection	—	—	—	—	—	—	—	—	—	+	+

(*Continued*)

Table 5.1 (*Continued*)

Characteristic or Function (Human Data Only)	Class or Isotype										
	IgG1	IgG2	IgG3	IgG4	Serum IgA	Dimeric sIgA1[a]	Dimeric sIgA2[a]	IgD	IgE	Pentameric IgM	Hexameric IgM
Hinge length (number of amino acids)	15	12	62	12	na	19	6	64	domain replaces hinge	domain replaces hinge	domain replaces hinge
J chain	—	—	—	—	—	+	+	—	—	+ (usually)[l]	—
Tailpiece	—	—	—	—	+	+	+	+	+/—	+	+
Secretory component (SC)[a]	—	—	—	—	—	+	+	—	—	—	—
pIgR-binding	—	—	—	—	—	+	+	—	—	+	—
Secreted into body fluids	—	—	—	—	—	+	+	—	—	—	—
Has function as B-cell-receptor (BCR)	+	+	+	+	—	—	—	+	—	+ (only as monomer)	+ (only as monomer)
C1q binding	++	+	++	—	—	—	—	—	—	+/—	+++
Complement-mediated cell lysis	++	+/—[m]	++	—	—	+[n]	+[n]	—	—	+/—	+++

Notes: **a**, Secretory component (SC) of secretory IgA (sIgA) is derived from pIg receptor (pIgR) during the process of secretion; **b**, Protein chains: L = light chain; J = joining chain; SC = secretory component; **c**, monomers behave like standard dimeric molecules, whereas covalent dimers should behave as tetrameric molecules (Salfeld, 2007); **d**, functionally unable to cross-link two antigens, and thus behaves as a monomer (van der Zee *et al.* 1986; Aalberse and Schuurman, 2002); **e**, while having ten V chains available for binding, the barrel-like structure appears to limit to five binding sites (Arnold *et al.*, 2005); **f**, not including J chain or secretory component; **g**, figure given is for all serum IgA, which is primarily monomer, but also contains minor dimeric and multimeric forms; **h**, e.g., saliva, urine, mucous, sweat, gastric fluids, tears; **i**, binding of IgG3 to FcRn is poorer than IgG1, apparently due to the H435R substitution in the IgG3 (Kim *et al.*, 1999); **j**, from Salfeld (2007); **k**, na = not applicable; **l**, J-chain-deficient pentameric IgM has also been described (Collins *et al.*, 2002); **m**, primarily via alternative pathway; **n**, through MBL-mediated mechanism.

the evolution of mammals (Vernersson *et al.*, 2004). This would explain why rabbits, for example, possess only a single major IgG subclass, whereas humans, rats, and mice have four IgG isotypes (Knight and Winstead, 1997). Recent evidence suggests that IgY, the isotype in bird species, is a predecessor to mammalian IgG and IgE isotypes (Taylor *et al.*, 2008).

The biological functionality of antibodies is divided into at least two major categories, i.e. the activity of the variable chains binding to the target antigen and the biology rendered by the Fc moiety of the antibody molecule. While the antigen binding activity is critical, a given variable chain will exert similar intrinsic affinities from one immunoglobulin (Ig) scaffold to the next while, on the other hand, each isotype exerts unique Ig-specific Fc-mediated immunological activities. Additionally, the Ig class or subclass can dramatically affect avidity. Finally, in recent years there has been a growing and more diverse effort to generate antibodies with engineered Fc regions, i.e. modified isotypes, to dial in or out desired Fc-mediated activities. The ultimate goal of these efforts is to generate therapeutic antibodies with Fc-mediated biological function that most closely matches that desired, i.e. development of "fit-for-purpose" antibodies (Chowdhury and Wu, 2005; Carter, 2006; Presta, 2006, 2008; Jefferis, 2007; Dimitrov and Marks, 2008; Strohl, 2009a, b).

This chapter describes the key properties of human antibody classes and isotypes and how these properties help define which isotypes or format should be used for a particular therapeutic intervention strategy. When antibodies are used as therapeutic proteins to carry out a desired functionality, it is important to focus the activity of that therapeutic antibody on the desired outcome. Superfluous activities not required to support the biology may be detrimental to the desired pharmacology or yield additional antibody-function related activities that may generate unwanted adverse events (AEs). These additional activities may be directed by domains present in the constant regions of the antibody, or may be due to the type, amount or pattern of antibody glycosylation. In discovering and optimizing an antibody therapeutic candidate, there are five factors that are typically considered by discovery biopharmaceutical scientists:

(1) nature of the target — e.g. whether the target is soluble, membrane-bound, located in the serum, in deep tissue, or a specialized organ; concentration or density of the target and distribution (narrow vs. broad). The nature of the target potentially impacts decisions on Fc functionality, size, affinity and dosing.

(2) affinity and/or avidity affects on the target in its natural environment — i.e. how tightly does the antibody need to bind its target?; does the antibody act monovalently, does it cross-link or not cross-link receptors, or are there other special binding attributes? The affinity/avidity of the therapeutic antibody to its target potentially impacts efficacy, dosing, tissue penetration, and internalization (if target is receptor).

(3) epitope — i.e. the portion of the target to which the antibody must bind to achieve the desired functionality. The epitope potentially impacts efficacy, mechanism of action, specificity (especially within protein families), intellectual property, and internalization (if target is receptor).

(4) desired activity directed by the Fc portion of the antibody:
   - FcγR binding leading to antibody-dependent cell cytotoxicity (ADCC) and/or antibody-dependent cell phagocytosis (ADCP);
   - binding to C1q, leading to complement-dependent cytotoxicity (CDC);
   - effect of glycosylation on Fc-mediated biological activities and/or clearance;
   - half-life (or extended half-life), as directed in part by the binding of IgG isotypes to FcRn;
   - Fc functionality potentially impacts efficacy, mechanism of action and toxicity;
   - FcγR-mediated avidity affects on target binding and clustering.

(5) the manufacturability or "developability" of the resultant optimized antibody molecules — i.e. does the candidate lead molecule possess the biophysical behavior (solubility, aggregation, oxidation, post translational modification) of the molecule well enough to develop? Developability impacts formulation, dosing, immunogenicity, development timelines, and ability to manufacture and package the drug economically in the desired format.

Antibody class and isotype selection are critical components of essentially all of the above-mentioned considerations, as will be described in the following sections.

5.2 Human Antibody Classes

Most isotype selection reviews to date have focused on IgGs, their ability to engage Fcγ receptors, and their glycosylation (Carter, 2006; Salfeld, 2007; Jefferis, 2007; Desjarlais *et al.*, 2007; Presta, 2008). Because there may be opportunities beyond IgGs, all classes and isotypes are at least considered here and reasons for consideration or rejection are given.

5.2.1 *IgD*

Human IgD antibodies are 184 kDa antibodies produced by mature, naïve B cells in parallel with the production of IgM. Most IgD remains anchored to B cells, although a small fraction is soluble. The structure of IgD is analogous to that of an IgG (Fig. 5.1), with a few distinctions. IgDs contain: a 7–8 amino acid tailpiece (sequence of YVTDHGPM) at the C-terminus of the constant chain that allows for anchorage onto the surface of B cells; an extra-long hinge region of 64 residues that is highly flexible, in part due to a tri-glycine stretch of residues; a T-like structure in solution; and three N-linked glycans, one of which is evolutionarily conserved with the N-297 of IgG, in the Fc region, as well as seven O-linked oligoglycans in the hinge region (Sun *et al.*, 2005). Membrane-bound IgD (mIgD) is a B cell receptor that is involved in class switching, B cell recruitment, peripheral B cell tolerance, and light chain rearrangement (Preud'homme *et al.*, 2000; Sun *et al.*, 2005). Additionally, a GPI-anchored form of IgD also has been described that contributes to the signaling pathways regulating the strength of B cell response to antigens (Chaturvedi *et al.*, 2002). Interestingly, mIgDs mostly have κ light chains, whereas secretory IgDs (secIgD) have λ light chains; the reason or purpose for this differentiation is not yet understood (Sun *et al.*, 2005).

The mean concentration of IgD in serum is approximately 35 μg/mL (Table 5.1), but the variability is quite high from person to person (Preud'homme *et al.*, 2000). The function of this circulatory IgD is not yet entirely clear. One possible hint to a role for soluble IgD is that there is a disproportionately high concentration of IgD in the middle ear (Samuelsson *et al.*, 2006). Proteins on invasive, encapsulated *Haemophilus influenzae*, but not on the less invasive non-typeable *H. influenzae*, bind the CH1 domain of human IgD antibodies (Samuelsson *et al.*, 2006). While these bacterially-expressed IgD binding proteins do not ascribe a role for IgD in protection from disease, they do raise questions about why microbes that occupy the niche in which IgD concentrations are disproportionately higher would evolve specific mechanisms to bind it.

Evolutionarily, IgD antibodies appear to be primordial, having arisen about 500 million years ago, similar to IgM antibodies (Ohta and Flajnik, 2006). The IgD class does not seem to offer any significant advantage over existing isotypes as a therapeutic scaffold. The naturally short half-life, involvement with B cell development and lack of in-depth understanding for the function of the soluble form likely make this class a "non-starter" for use in therapeutics. To date, there appears to be no consideration for use of the IgD isotype scaffold in a therapeutic setting.

5.2.2 *IgE*

IgE was shown in the late 1960s to be associated with the reagenic reaction that is responsible for natural defense against parasites such as helminths, including the roundworm ascaris (Ishizaka and Ishizaka, 1970). IgEs have a nominal soluble serum half-life of two days, and are normally found in serum in very low concentrations (300 ng/mL range) (Table 5.1). IgEs, which are approximately 188 kDa in size, consist of two heavy chain chains (ε chains) and two light chains (either κ or λ). IgEs have an extra C_H domain Cε2, similar to IgMs, which replace the position of the hinge found in IgGs (Fig. 5.1).

IgEs bind to two receptors, FcεRI, a high affinity, seven-transmembrane receptor located on the surface of mast cells, basophils and activated eosinophils, and a low affinity receptor, FcεRII (also known as CD23). The binding of IgE to the high affinity receptor, FcεRI, was shown to have a K_D of 3–5 × 10^{-9} M (Chan *et al.*, 2000), although other reports had put the interaction at about 10^{-10} M (Takai, 2002; Nakamura *et al.*, 2005; Harwood and McDonnell, 2007). IgE binds to FcεRI through the Cε3 domain (Harwood and McDonnell, 2007). IgE bound to the surface of mast cells or basophils through the FcεRI high affinity receptor has a bound half-life measured in several hours, and can reside in a bound state for many days. When FcεRI-bound IgEs encounter a multivalent antigen to which they bind, the IgEs and the FcεRI are clustered on the surface of the cells, causing degranulation of the mast cells or basophils and immediate release of histamines, leukotrienes and other molecules that trigger a hyper-reactive response and contribute significantly to bronchoconstriction (Gounni, 2006).

The low affinity (K_D ~ 1μM range) IgE receptor FcεRII is constitutively expressed on B cells and can be found on several non-B cell types, including T cells, monocytes, eosinophils, platelets, follicular dendritic cells, and Langerhan's cells. FcεRII is not a classical FcR, as it contains a C-terminal lectin-binding domain and belongs to the family of C-type lectins (Takai, 2002). Binding of IgE to its low affinity receptor, FcεRII, results in feedback negative regulation for the synthesis of IgE (Kilmon *et al.*, 2004).

To date, there are no reports of the IgE antibody being developed as a therapeutic platform, and due to its natural biological function, it is highly unlikely that an intact IgE scaffold would ever be used as a therapeutic. In fact, IgE itself is the target of the therapeutic IgG monoclonal antibody omalizumab (Xolair®), an IgG1 that blocks the ability of IgE to bind to its high affinity receptor, FcεRI, and as a result downregulates the receptor on the surface of mast cells (Chang and Shiung, 2006).

5.2.3 *IgM*

IgM antibodies, which are large polyvalent antibodies derived from the μ gene locus, are the first antibodies to be generated in response to an antigen *in vivo.* IgMs consist of heavy (Cμ1, Cμ2, Cμ3, Cμ4, VH) and light chains (CL plus Vκ or Vλ) that are built into monomeric units of about 190 kDa and polymerized into either a pentameric or hexameric format. A unique 18-amino acid sequence (PTLYNVSLVMSDTAGTCY), called the tailpiece, fused to the C-terminus of the heavy chain confers the ability to oligomerize (Sørensen *et al.*, 1999). The pentameric form also typically contains a 20 kDa J chain and comprises approximately 95% of all circulating IgM, whereas hexameric IgM (making up ~ 5% of circulating IgM) does not possess a J chain (Johansen *et al.*, 2000). J chain-deficient pentameric IgM has also been described (Collins *et al.*, 2002). Each multimeric state described above is also heterogeneous due to variations in disulfide bonds, making "IgM" a collection of very heterogeneous molecules. Finally, a small quantity of monomeric IgM also can be secreted, but this is usually associated with an inflammatory disease state (Xu *et al.*, 1994). Since the half-life values for pentameric and hexameric IgMs are both about 5 d, and the clearance rates appear to be similar, the more abundant pentameric form must be released into circulation at a higher rate (Hughey *et al.*, 1998).

The major characteristics of pentameric vs. hexameric IgMs are included in Table 5.1. The pentameric form of IgM also binds the pIg receptor (pIgR) through the binding activity of its J chain, which allows for transcytosis of pentameric IgMs from vascular space onto mucosal surfaces (Johansen *et al.*, 2000). The pentameric form of IgM has a barrel-shape which includes both an antigen-binding face and a non-antigen binding face (Arnold *et al.*, 2005), suggesting that while there are ten V chains, there may only be a valency of five (Table 5.1). The J chain, which occupies the back side of the barrel, appears to physically impede access to the V chains on that face (Arnold *et al.*, 2005). On the other hand, the hexameric form, which is physically a planar molecule (Arnold *et al.*, 2005), appears to have a full valency of 12, suggesting that all of its V chains are available for binding to antigen.

Hexameric IgM also strongly activates complement through the classical pathway and carries out complement-dependent cell lysis (Collins *et al.*, 2002). Pentameric IgM also activates the classical complement pathway *in vitro* using human complement with an efficiency of about 8–33% that of the hexameric form (Collins *et al.*, 2002; Arnold *et al.*, 2005). Pentameric IgM is also associated with the alternative complement pathway. The final major significant difference between pentameric and hexameric IgMs is that the hexameric form has been shown to be more likely involved in IgM-mediated autoimmune syndrome, probably due to its strong ability to activate complement (Hughey *et al.*, 1998).

Besides lectin receptors, IgM molecules interact with two receptors, polymeric Ig-receptor (pIgR; ca. 37 nM K_D) and Fcα/μR (ca. 1 nM K_D), and possibly a third (putative FcμR on NK cells) (Pricop *et al.*, 1993). The IgM-receptor interactions are relatively high affinity, with the binding to Fcα/μR at ca. 1 nM K_D (Ghumra *et al.*, 2009) and binding to pIgR at ca. 37 nM K_D (Norderhaug *et al.*, 1999). Since binding of IgM (and IgA) to pIgR (Braathen *et al.*, 2002) and Fcα/μR (Ghumra *et al.*, 2009) is J chain-dependent, it is likely that both receptors interact only with pentameric IgM containing intact J chain. Polymeric IgR is found on mucosal epithelial cells and is the mechanism by which IgAs and pentameric IgMs are translocated from circulation to mucosal surfaces (Kaetzel, 2005). Fcα/μR, on the other hand, is found in mice on B cells and macrophages, and in humans on a subpopulation of pre-germinal B cells and follicular dendritic cells, suggesting roles, albeit different, in immunology (Ghumra *et al.*, 2009). Thus, the biological activity of hexavalent IgMs, which make up only 5% of the circulating IgM population and strongly activate complement but do not bind IgM receptors, is substantially different from that of pentameric IgMs, which comprise 95% of the serum IgM species, and activate complement only moderately but bind both pIgR and Fcα/μR. These noteworthy differences could have a significant impact on production, purification and formulation of therapeutic pentameric IgMs, in which the hexameric forms might be considered as contaminating species with widely differing biology.

5.2.3.1 *IgMs as potential therapeutic monoclonal antibodies*

Several IgM therapeutic candidates have been placed into clinical development over the past two decades, but with no commercial products to date to show for the efforts. After Köhler and Milstein (1975) developed the hybridoma technology, potential therapeutic antibodies were produced by hybridomas, which were an excellent source of both IgGs and IgMs. Of the early therapeutic antibody candidates, four were IgMs that targeted blood-borne bacterial infections (Reichert and Dewitz, 2004). Two of these, edobacomab (XOMEN-55), a murine anti-lipid A IgM derived from mouse spleen cells, and nebacumab (Centoxin), a human anti-lipid A IgM derived from human spleen cells, made it into advanced clinical trails for sepsis in the early 1990s. Both were eventually discontinued for lack of efficacy (Reichert and Dewitz, 2004).

Additionally, early efforts to generate anti-tumor antibodies using hybridoma technology resulted in the isolation of several pentameric IgM monoclonal antibodies from cancer patients, but those possessed typical serum-IgM characteristics, including low affinity and specificity. Vollmers and Brändlein (2002) nonetheless have made the argument that IgMs might be the best biologics approach for fighting cancer, owing to their strong avidity affects as well as complement fixation. Some of the limitations for oncology, however, include poor tissue penetration due to the large size, as well as the inability of IgMs to cross the endothelial barrier, making it difficult for them to enter cancerous tissue (Vollmers and Brändlein, 2002). Even with these limitations, there have been several programs on IgM therapeutics for oncology, many of them targeting glyco-epitopes either unique to, or overexpressed on, cancerous tissues (i.e. compensate for intrinsically low affinity to glycans with avidity).

Researchers at Chugai Pharmaceutical Co. isolated an anti-GM3 IgM, L612, from EBV-transformed human B cells from patients with melanoma (Hoon *et al.*, 1993). The hybridoma cell line producing L612 made ca. 20% hexameric IgM, 74% pentameric IgM, and 6% of other minor IgM species (Azuma *et al.*, 2007). L612 binds to GM3, a prominent glyco-epitope on the surface of melanoma cells, and kills the GM3-expressing melanoma cells in a complement-dependent manner (Hoon *et al.*, 1993). Hybridoma-produced L612 was the first anti-tumor IgM reported to be placed into Phase 1 clinical trials, and it was found to produce no significant adverse side effects (Irie *et al.*, 2004). More recently, it has been shown that tumor-cell killing by L612 was mediated largely through the complement activation mediated by the hexameric fraction. Thus, an enriched hexameric recombinant form of L612, CA19, was generated for *in vitro* and *in vivo* analysis, as well as a potential therapeutic (Azuma *et al.*, 2007). CA19, which was produced by CHO cells in the absence of J chain, contained about 80% hexameric species and yielded much stronger complement-dependent killing of tumor cells than did the original L612 IgM (Azuma *et al.*, 2007). This demonstrates one possible manipulation of IgMs, i.e. to tune the effector function to meet the needs of the therapeutic.

Patrys Ltd. (Melbourne, Australia) has at least three lead anti-tumor IgMs in early development. SC-1 targets a specific glycoform on CD55 found only on the surface of stomach cancer cells and induces apoptosis of the targeted cancer cells *in vitro* and in *in vivo* models (Vollmers and Brändlein, 2006). It was demonstrated that neither complement activation nor ADCC were significant factors in its activity (Beutner *et al.*, 2008); thus, a pentameric IgM that does not induce either CDC or ADCC is a reasonable approach for this target. This IgM has been partnered with Astra-Zeneca and has already been in Phase I/IIa trials for cancer treatment. The second Patrys IgM, SM6, targets GRP78, a multi-functional glucose-regulated protein, and is in preclinical studies for oncology. The third Patrys IgM, LM-1, for which the target has not been disclosed publicly, is in late preclinical studies.

Other IgMs also are in early stages of development. Morphotek MorAb-028 IgM against G-28 glycolipid was reportedly in Phase I clinical trials in 2008 for metastatic melanoma, and Argenes ARG098, an anti-FAS (aka APO-1, CD95) agonist IgM for synovial injection, was placed into Phase 1 clinical trials in Belgium in 2008 for rheumatoid arthritis. Additionally, HMMC-1, an IgM being developed for ovarian cancer, binds an O-glycan found specifically on the surface of uterine cancer cells. HMMC-1 bound to about 55% of uterine endometrial adenocarcinoma, 77% of uterine cervical adenocarcinoma, and 75% of epithelial ovarian cancer tissues but not normal ovarian tissues (Nozawa *et al.*, 2004). Finally, the recombinant human IgM rHIgM22 that targets myelin on the surface of oligodendrites apparently promotes remyelination at those sites. Doses as low as 500 ng promoted spinal chord remyelination in murine demyelination models. The half-life of rHIgM22 in mice is 15 h, and in humans very little residual rHIgM22 is detected after 48 h, suggesting a short half-life clinically as well (Warrington *et al.*, 2007).

5.2.3.2 *When might an IgM be used as a therapeutic?*

Beyond being able to control the ratio of pentameric and hexameric IgM forms being produced in cell culture, there are several important requirements for IgM antibodies to be considered a potential standard platform, including *in vitro* optimization, production and purification. The variable chains of IgM antibody candidates have been manipulated successfully using phage display technologies *in vitro* to optimize

IgM therapeutic antibody candidates (Boel *et al.*, 2000; Pancook *et al.*, 2001). Additionally, several recombinant pentameric, J chain-fused IgMs have been produced in Crucell's Per.C6 human cell line at greater than 1 g/L in a protein-free medium (Tchoudakova *et al.*, 2009). Yields were from 0.5 g/L up to 2 g/L and productivities were ca. 20 pg/cell/day, suggesting sufficient production to be industrialized. Moreover, methods for purification have been developed using hydroxyapatite and cation exchange chromatography based on the fact that IgMs are typically more charged than IgGs (Gagnon *et al.*, 2008). Finally, conditions for stabilization and solubilization of IgMs have been worked out (Gagnon *et al.*, 2008). These advances form the basis on which an IgM therapeutic platform might be built.

Thus far, most of the IgMs being developed are J chain-bearing pentamers. If the J chain is not included as part of the IgM expression system, hexameric IgMs with significant complement-activation activity are the primary products obtained, as noted above for Chugai's CA-19. When the J chain is included, a small percentage of hexamer is still produced, and because even a small fraction of hexameric IgMs could generate such different activity than the majority pentameric IgMs, it is critical to have either no contamination of a pentameric IgM preparation with hexameric forms, or alternatively, a consistent batch-to-batch level of hexameric contaminant. Thus, control of the IgM product species is probably the greatest challenge, especially as production process changes occur.

Considering that the IgM platform is still not fully validated with a marketed product, as well as the inherent risk of developing a more complex compound, development and use of IgMs as therapeutics will probably remain relatively limited for at least the next decade. It is likely that for certain targets, particularly those such as glyco-epitopes on cancer cells, an IgM might ultimately be the best approach. Other targets that can be neutralized only through strong avidity-based responses, or where complement-mediated clearance is desired, may be targets for hexavalent IgMs as well. With the combination of medium-high affinity (e.g. ~ 1–5 nM), avidity, and potentially potent complement activation, one might end up with either a very potent drug, or alternatively, one that creates immune complexes and complement activation issues. These kinds of questions would need to be addressed in early phases of the IgM discovery process, since a decision would need to be made whether the risk was worth the value proposition of using an unproven platform.

5.2.4 *IgA*

IgAs, which are polyvalent antibodies derived from the alpha gene locus, are translocated to mucosal surfaces as the first line of defense against infectious microorganisms. IgAs consist of heavy (Cα1, Cα2, Cα3, hinge region, VH) and light chains (CL plus Vκ or Vλ) that are built into monomeric units of about 160,000 Da, generally similar to the structure of a standard IgG. IgA antibodies have from 2–5 N-linked glycosylation sites in the Cα chains, and an additional

3–6 O-linked glycosylation sites in the hinge region (Woof and Kerr, 2006). There are two isotypes of IgA, IgA1, and IgA2, generated from the α1 and α2 gene loci, respectively (Woof and Kerr, 2006). There are multiple sequence differences between IgA1 and IgA2 isotypes, but the most significant difference is found in the hinge region, in which the IgA1 hinges are typically 16–18 amino acids in length as compared with IgA2 hinges that are five amino acids long (Senior and Woof, 2005; Bonner *et al.*, 2009). The longer hinge of the IgA1 isoforms potentially allows for greater avidity, but also yields considerably greater sensitivity to proteolytic degradation (Senior and Woof, 2005). Furthermore, secretory IgA1 (sIgA1) isotypes are nearly planar in three-dimensional structure, whereas secretory IgA2 (sIgA2) isotypes are significantly more "bent"; these structural differences have been correlated with differential kinetics in response to receptor binding and IgA-dependent effector cell functionality (Bonner *et al.*, 2009). The IgA2 isotype has two allotypic forms, IgA2m(1) and IgA2m(2), which differ not only by amino acid residue changes, but also in the number of N-linked glycosylation sites (Woof and Kerr, 2006). Serum IgA, which is present in the ~ 3–5 mg/mL range and mostly (~ 90%) monomeric IgA1, is mostly produced in the bone marrow. The nominal half-life for serum IgA is about five days. Serum IgAs also can be the source of aberrant molecules that trigger IgA nephropathy (IgAN), which appears to be the result of accumulation of aberrant galactosylation of serum IgA1 species (Lin *et al.*, 2009).

Most of the polymeric IgA is produced locally in tissues. The overall production of IgA in mammals exceeds all other isotypes, with most of the IgA secreted across mucosal membranes to immerse the ~ 400 M^2 of mucosal surfaces, including respiratory, digestive, and genitourinary tracts, to protect against microorgamisms (Woof and Kerr, 2006) and maintain gut homeostasis (Cerutti and Rescigno, 2008). As with IgMs, IgAs possess a unique 18-amino acid tailpiece sequence (PTH-I/V-NVSVVMAEADGTCY) at the C-terminus of the heavy chain that is the key element for oligomerization (Sørensen *et al.*, 1999). IgA monomers are polymerized into a dimeric format containing a 15-kDa J chain that is required for binding pIgR, similar to IgMs (Braathen *et al.*, 2002). Higher order oligomers are also known to occur. The secreted form of IgA, sIgA, is dimeric and possesses an additional 80 kDa fragment called the "secretory component" (Fig. 5.1) that is derived from pIgR upon transcytosis from serum to the mucosa (Woof and Kerr, 2006).

The key importance of sIgA in microbial defense can be highlighted by the number of mechanisms used by mucosal microorganisms such as pneumococci, neisseriae, and *Haemophilus* spp., to defeat or neutralize sIgAs, including sIgA binding proteins and sIgA proteases (Senior and Woof, 2005; Woof and Kerr, 2006; Kazeeva and Shevelev, 2009). The most significant producers of the sIgA binding proteins are the group A and group B streptococci. The proteases produced by a variety of mucosal pathogens target a short sequence stretch in the hinge region shared by both sIgA1 and sIgA2 species, although the longer hinge of the sIgA1 isoforms makes them more susceptible to microbial proteases than are the sIgA2 isoforms (Senior and Woof, 2005).

Human IgA antibodies bind five different receptors: FcαRI (CD89), found primarily on neutrophils and monocytes; Fcα/μR, in humans found on pre-germinal B cells and follicular dendritic cells, pIgR, found primarily on epithelial cells, liver, lung and small intestine; asialoglycoprotein receptor, a lectin found primarily in the liver; and transferrin receptor, found on monocyte-derived dendritic cells (Takai, 2002; Monteiro and van de Winkel, 2003; Nakamura *et al.*, 2005; Wines and Hogarth, 2006; Woof and Kerr, 2006).

Monomeric human IgA binds human FcαRI with an affinity in the ~2–5 × 10^{-7} M range (Herr *et al.*, 2003). Efficient activation of neutrophil-mediated phagocytosis requires FcαRI cross-linking, similar to antigen-bound IgG cross-linking of low affinity receptors such as FcγRIIa and FcγRIIIa. Thus, antigen-complexed IgA forms appear to activate neutrophil and monocyte phagocytosis better than monomeric forms (Monteiro and van de Winkel, 2003; Woof and Kerr, 2006; Wines and Hogarth, 2006). One significant difference with IgA, however, is that a single IgA homodimeric molecule binds two FcαRIs, with the first IgA-FcαRI interaction at a K_D of 176 nM and the second interaction at 431 nM (Herr *et al.*, 2003). Binding of antigen-complexed IgA to FcαR triggers multiple cellular responses, including ADCC, ADCP, superoxide generation, calcium mobilization, degranulation, cytokine release and antigen presentation (Herr *et al.*, 2003; Woof and Kerr, 2006; Wines and Hogarth, 2006). Additionally, a form of FcαRI or unknown receptor appears to be present on NK cells that may be capable of triggering NK-cell-mediated ADCC (Mota *et al.*, 2003; Wines and Hogarth, 2006). Interestingly, mice do not have a functional FcαR (Takai, 2002; Woof and Kerr, 2006).

Human IgA binding to pIgR, which is high affinity (55 nM) interaction (Norderhaug *et al.* 1999; Table 5.2), requires the J chain, and is functional in translocating dimeric IgA across the endothelial barrier from the apical surface to the basolateral mucosal surface (Woof and Kerr, 2006; Wines and Hogarth, 2006). Upon engagement of IgA with pIgR, an 80 kDa polypeptide chain from pIgR, known as the "secretory component" (SC), is transferred to IgA and becomes a part of the secretory IgA (sIgA) molecule (Fig. 5.1; Table 5.1). The SC helps to protect sIgA from proteolytic degradation, but also interferes with the FcαRI binding site on the sIgA isoforms, leading some to call sIgA a "non-inflammatory immunoglobulin" (Monteiro and van de Winkel, 2003).

The function of IgA binding to Fcα/μR in humans has not yet been fully sorted out. In mice, where it is constitutively expressed on B cells and macrophages, Fcα/μR may function in endocytosis of antigen-complexed IgA (Woof and Kerr, 2006; Ghumra *et al.*, 2009). In humans, Fcα/μR is also present on follicular dendritic cells, although its function there with respect to IgA has not yet been elucidated. IgA binding to asialoglycoprotein receptor in the liver and transferrin receptor in the kidney appear to mediate clearance of IgA2 and IgA1, respectively (Monteiro and van de Winkel, 2003).

Table 5.2 Proteins that interact with human antibodies

Protein	Location or Cell Type	Human Antibody Isotype(s) That Bind in Physiologically Relevant Manner	Function
Human CD16 (FcγRIIIa)	NK cells, eosinophils, macrophages, monocytes, mast cells, dendritic cells, Langerhan's cells	IgG1, IgG3 (K_D ~ 100–800 nM)	Antibody-dependent activation; ITAM motif; antibody-dependent cellular cytotoxicity (ADCC) by NK cells
Human CD16B (FcγRIIIb)	Neutrophils, IFN-γ-induced eosinophils	IgG3 (K_D ~ 900–1000 nM); IgG1 (K_D ~ 5 μM)	Synergistic affects with other FcγRs; no signaling domain
Human CD32a (FcγRIIa)	Macrophages, neutrophils, eosinophils, platelets, dendritic cells, Langerhan's cells	IgG1 (K_D ~ 300 nM); IgG2, IgG3, IgG4 (K_D ~ 1–10 μM)	Low affinity receptor; phagocytosis by monocytes, macrophages, neutrophils; granule release by eosinophils
Human CD32b (FcγRIIb)	Macrophages, neutrophils, eosinophils, B cells, mast cells, basophils, dendritic cells	IgG1, IgG3, IgG4 (K_D ~ 5–10 μM)	Inhibition of stimulation response
Human CD32c (FcγRIIc)	NK cells	IgG3 >> IgG1 = IgG4 >> IgG2 (K_D ~5–10 μM)	Activation response in NK cells
CD64 (FcγRI)	Macrophages, eosinophils, dendritic cells, GCSF-activated neutrophils	IgG1, IgG3, IgG4 (K_D ~ 20–30 nM)	"High affinity receptor"; uptake; stimulation; activation of respiratory burst; killing
CD89 (FcαRI)	Neutrophils, monocytes, some macrophages, eosinophils, interstitial dendritic cells, Kuppfer cells	Monomeric IgA and dimeric sIgA (K_D ~ 176–431 nM[a]); dimeric IgA1 can bind two FcαRI molecules[a]	Neutrophil-mediated ADCC and ADCP by antigen-complexed IgA species

(Continued)

Table 5.2 (*Continued*)

Protein	Location or Cell Type	Human Antibody Isotype(s) That Bind in Physiologically Relevant Manner	Function
Fc α/μR	Subpopulation of pre-germinal B-cells and follicular dendritic cells (in humans)	Pentameric IgM (K_D ~ 1 nM); dimeric IgA1 (K_D ~ 10 nM)[b]	Potential role in binding IgM or IgA immune complexes and presenting antigens to B cells in germinal centers[b]
FcεRI	Mast cells, basophils, activated eosinophils	IgE (K_D ~ 3–5 nM; earlier reports were K_D ~ 100 pM)	Trigger allergic response
FcεRII (CD23)	B cells	IgE (K_D ~ 1 μM)	Regulation of IgE production
Asialoglycoprotein receptor	Hepatocytes	IgA (preferrably IgA2)	Clearance of IgA2
Transferrin R (CD71)	Epithelial cells; monocyte-derived dendritic cells; B lymphocytes	Dimeric or polymeric IgA1	Clearance of IgA1
MBL (mannose binding lectin or mannan binding lectin)	Serum protein; part of the innate immune system	IgG antibodies with G0 glycosylation; dimeric and polymeric IgA; 20% of IgM[c]	Recognition molecule of lectin pathway for complement activation[c]
CD206 (mannose receptor, MMR, CLE13D and MRC1)	Hepatic cells, astrocytes, epidermal Langerhan cells, perivascular microglia, glomerular microglia, macrophages, dendritic cells	IgG in G0 form (agalactosyl–IgG)[d]	Clearance of glycoproteins from the liver
C1q	Serum protein	IgG1, IgG3, hexameric IgM (30–80 μM range)[e]	First binder of the classical complement cascade

(*Continued*)

Table 5.2 (*Continued*)

Protein	Location or Cell Type	Human Antibody Isotype(s) That Bind in Physiologically Relevant Manner	Function
FcRn (neonatal receptor; Brambell receptor)	Epithelial cells, placenta, macrophages, monocytes, dendritic cells	IgG1, IgG2, IgG4 ($K_D \sim$ 1–2 μM or ~ 200 nM, depending on how measured) (IgG3, less so)	Maternal-fetal transfer; pH-dependent binding and recycling of IgGs and serum albumin, resulting in long half-life of these proteins
Polymeric Ig receptor (pIgR)	Mucosal epithelial cells	Dimeric IgA ($K_D \sim$ 55 nM); Pentameric IgM (J-chain required) ($K_D \sim$ 37 nM)[f]	Directional transcytosis of dimeric IgAs and pentameric IgMs across epithelial layers
Protein A	*Staphylococcus aureus*	IgG1, IgG2, IgG4 strongly; IgA, IgD, IgE, IgM weakly	Microbial defense mechanism
Protein G	*Streptococcus*	IgG1, IgG2, IgG3, IgG4	Microbial defense mechanism

References: **a**, Herr *et al.*, 2003; **b**, Ghumra *et al.*, 2009; **c**, Arnold *et al.*, 2006; **d**, Dong *et al.*, 1999; **e**, Schumaker *et al.*, 1976; **f**, Norderhaug *et al.*, 1999.
*Abbreviations: Nd = not known.

5.2.4.1 *IgAs as therapeutic antibodies*

Over the past decade, several investigators and companies have attempted to develop therapeutic sIgAs. The first sIgA molecule that advanced into clinical trials appears to be CaroRx™, a chimeric IgG–sIgA anti-*Streptococcus mutans* antibody to prevent recolonization of the plaque-forming microorganism (Wycoff, 2004). In that molecule, the Fabs including the V regions were derived from a murine IgG1 called Guy's 13; the J chain was mouse, and the secretory component was from rabbit. The chimeric sIgA/G was produced in tobacco plants with a reported yield of 200–500 mg/kg of plant material (Larrick *et al.*, 2001). CaroRx™, which reached Phase II clinical trials by 2004, was successful in preventing the recolonization of *S. mutans* over a four-month period (Ma *et al.*, 1998). No side effects or systemic anti-recombinant sIgA/G responses were reported, likely because the antibody was delivered to the oral cavity (Ma *et al.*, 1998).

Other companies too were trying to produce sIgA in plants during the same period of the late 1990s to early 2000s, but these efforts seem to have subsided in more recent years. Product quantity and quality, as well as potential regulatory hurdles for production, have dampened interest in using plants to produce sIgAs, especially if they are to be administered systemically (Stoger *et al.*, 2002). Moreover, reasonable cell culture methods have been devised by cloning the IgA sequences, the J chain and the secretory component into facile cell lines (Morrison and Chintalacharuvu, 2001; Beyer *et al.*, 2009). IgAs can be purified using either IgA-binding protein (IgAbp) or Jacalin, an IgA-binding lectin, as the affinity ligand for purification (Pack, 2000).

Over the past few years, the interest in developing sIgA therapeutics has picked up again. Dechant *et al.* (2007) reported on an anti-EGFR mouse–human chimeric IgA. They generated both IgA1 and IgA2 monomeric species and compared the activities of those antibodies to an IgG1 containing the same variable chains from mAb 225, the precursor of cetuximab. Both IgA1 and IgA2 constructs mediated ADCC on EGFR-bearing A431 cells using polymorphonuclear (PMN) cells but not mononuclear cells as effector cells, as would be expected by the presence of FcαRI on the PMNs (Dechant *et al.*, 2007). In both *in vitro* cell-based and whole blood assays, however, IgA2 was significantly more effective in killing EGFR-bearing A431 cells than IgA1; the IgA2 isotype also outperformed IgG1 in whole blood (Dechant *et al.*, 2007).

5.2.4.2 *When might an IgA be used as a therapeutic?*

There are two potential advantages for the use of an IgA class antibody as a therapeutic. First, if the desired site of action is mucosal rather than blood or tissue, then a dimeric sIgA may be an approach to try. The CaroRx™ therapeutic candidate was a great example of this, although the distinct advantage in that program was the local delivery to the oral cavity. If other candidates could be delivered locally to the desired mucosal site of action, then a dimeric sIgA approach may be a valid approach. A significant advantage of sIgAs in these mucosal environments is the protection from proteolytic cleavage conferred by the presence of the SC, which physically wraps around the Fcs of the sIgAs (Fig. 5.1). Systemic delivery for mucosal site of action may involve additional factors that could compromise success. For example, systemic delivery of a precursor IgA dimer, complete with J chain, but lacking the secretory component, would result in systemic binding to pIgR with transcytosis into a huge mucosal space (~ 400 M^2 of mucosal surfaces) (Woof and Kerr, 2006), potentially driving the need for high systemic doses to reach the desired mucosal surface.

The second potential advantage is the ability for IgAs to engage FcαRI, which is found on neutrophils, monocytes and other cytotoxic immune cells (Table 5.2), to drive neutrophilic and monocyte-driven phagocytosis (Woof and Kerr, 2006). Since neutrophils are amongst the most abundant white blood cells in circulation (~ 2.5–7.5 × 10^6/mL), the ability to recruit them, along with monocytes, to deplete certain targets is worth further investigation. There are multiple examples in the literature showing the value of targeting cell depletion through the activity of FcαRI. Several groups have demonstrated that bispecific antibodies that bind Her2, CD20, or fungal antigen with one V chain and FcαRI (CD89) with the other V chain very actively depleted the targeted cells in a variety of settings (Stockmeyer *et al.*, 2000). Pretreatment with G-CSF, which increases neutrophil count, is a possible adjunct to this form of therapy to increase the effector cell to target cell ratio (Stockmeyer *et al.*, 2000). This cytotoxic approach toward targeted cell depletion may be valid for at least a subset of targets in the oncology and infectious diseases field, although the size of a dimeric IgA might pose a limit to any significant tissue or tumor penetration. On the other hand, this approach may pose the potential for organ damage, particularly lung and kidney, caused by highly activated neutrophils, so a balance of efficacy and safety concerns would need to be taken into consideration for this type of an approach.

5.2.5 *IgG overview*

Human IgG antibodies are divalent antibodies derived from the gamma gene locus and are the most prevalent antibodies in the serum, with a mean concentration of about 13–14 mg/mL (Table 5.1). IgGs consist of heavy (CH1, CH2, CH3, hinge, VH) and light chains (CL plus Vκ or Vλ) that are built into units of about 146 kDa (except IgG3, which is ~ 165 kDa). There are four isotypes, IgG1, IgG2, IgG3, and IgG4, and all except IgG3 have been used to generate marketed therapeutic antibodies. Evolutionarily, the different IgG isotypes arose after diversification of mammalian species, so there are significant differences in isotype functionality between different species (Hsu *et al.*, 2006). Moreover, since subclasses arose after speciation, correlation of IgG subclasses and effector functions of non-primate mammals cannot be directly correlated with human IgG isotypes (Vernersson *et al.*, 2004).

The textbook half-life for IgG1, IgG2 and IgG4 isotypes is approximately 14–21 days, although an analysis of marketed and clinical candidate antibodies showed that both the mean and median half-life values for > 20 clinically studied antibodies in humans were in the 12–13 day range (Strohl, unpublished data). IgG3s have a shorter half-life, in the range of 7 days (Table 5.1). IgGs have a single glycosylation site in the constant region at N297 (based on EU numbering) (Figs. 5.2, 5.3). Additionally, it has been demonstrated that 20–30% of naturally occurring IgG antibodies also have N-linked glycosylation in the Fab region (Jefferis, 2007).

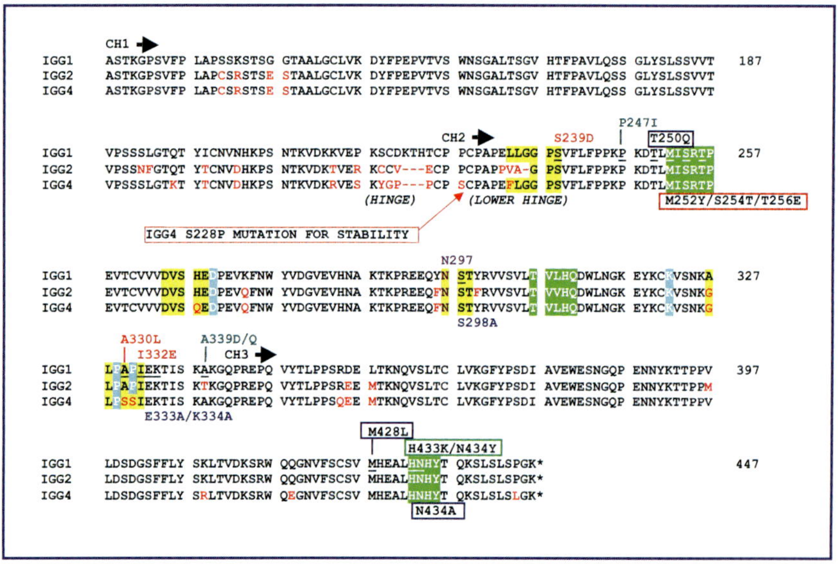

Figure 5.2 Sequences of human IgG1, IgG2, and IgG4 heavy chain contant regions, including residue locations on the linear sequences to which human Fcγ receptors (residues 234–239, 265–269, 297–299, 327–332; highlighted yellow) and C1q (residues 270, 322, 329, 331; highlighted blue), and FcRn (residues 252–257, 307–311, 433–436; highlighted green) bind. Differences in residues from IgG1 in IgG2 and IgG4 are noted in red. Several notable mutants resulting in increased ADCC (S239D, I332E, in red above sequence; S298A, E333A, K334A, in blue below sequence; P247I, A339D/Q, in green above sequence) or prolonged half-life (T250Q, M428L, in blue box above sequence; M252Y, S254T, T256E, in red box below sequence; H433K, N434Y, in green box above sequence) are noted.

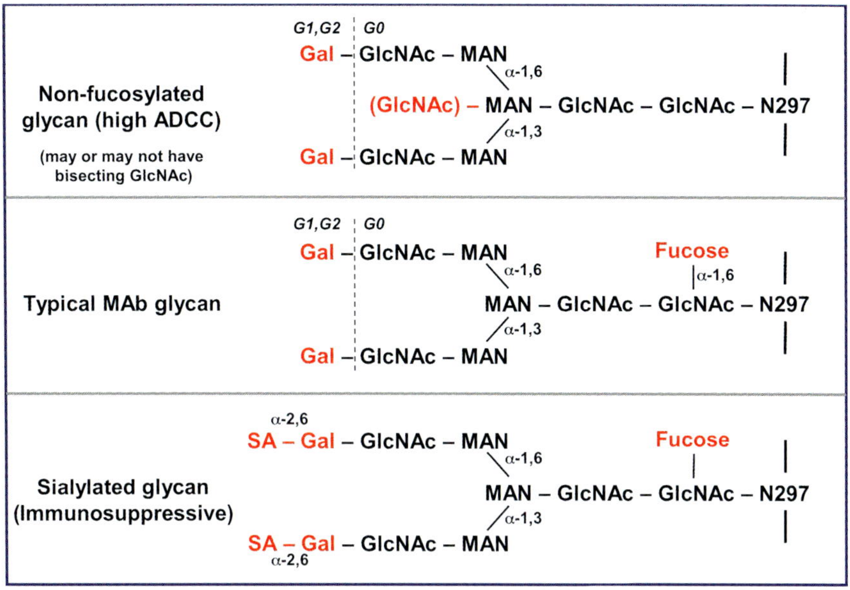

Figure 5.3 Structure of oligosaccharides attached to asparagine 297 (N297) of human IgG antibodies. The middle panel shows the major forms of IgG produced by CHO cell cultures, i.e. bi-antennary, fucosylated G0 (no GAL residues), G1 (one terminal GAL residue), and G2 (two terminal GAL residues) species (variable residues shown in red). The top panel shows a non-fucosylated oligosaccharide, which may be found either as a standard biantennary structure or with a bisecting GlcNAc residue (as shown in red), that confers at least ten-fold tighter binding to human FcγRIIIa and increased ADCC. The bottom panel shows a terminally α-2,6-sialylated IgG oligoglycan that confers immunosuppressive properties. Abbreviations: GlcNAc = *N*-acetyl-glucosamine; MAN = mannose; Gal = galactose.

Human IgG antibodies are multifunctional proteins that not only bind an antigen, but also elicit several types of responses through their interactions with complement and a variety of Fcγ receptors. Included in these responses are antibody-dependent cellular cytotoxicity (ADCC), antibody-dependent cellular phagocytosis (ADCP), as well as complement-mediated cytotoxicity (CMC). Domains within the antibody Fc regions are responsible for binding various receptors or complement components to effect these responses (Fig. 5.2). Human IgG isotypes, IgG1, IgG2, IgG3, and IgG4 have different binding to the initial complement component, C1q (Sontheimer *et al.*, 2005), and the Fcγ receptors due to the differences in their sequences, particularly in the hinge regions (Fig. 5.2; Jefferis, 2007). Another receptor, FcRn (neonatal receptor, also known as the Brambell receptor), binds internalized IgG in acidic endosomes, and then, through a pH-dependent step, recycles the antibodies back into the serum, giving IgG antibodies their prolonged half-life (Ghetie and Ward, 2000).

The following section delves into the details of isotype, glycoform, and engineered Fc selection.

5.3 Human IgG Isotypes and Their Receptors

5.3.1 *Human IgG isotypes in therapeutic antibodies and Fc fusion proteins*

Currently, there are more than 30 mAbs, FAbs, and Fc-fusion proteins on the market, and another 31 that are identified to be in advanced clinical trials (i.e. Phase III) (Strohl, unpublished). Of these, 67% (35 IgG1 antibodies and 6 Fc fusion proteins derived from natural IgG1) contain a natural, normally glycosylated IgG1 Fc domain. The remainder include 10% with modified Fcs, 6.5% IgG2s, another 6.5% as either Fabs or scFvs, 5% IgG4s, and another 5% were derived from murine IgGs. Salfeld (2007) lists additional clinical therapeutic candidates containing either IgG2 or IgG4 isotypes. The surprising fact that the "modified Fc" group, which includes IgGs possessing modifications either to increase or decrease effector functionality, was second to IgG1s in the advanced molecules shows the critical importance placed on developing antibodies with optimized Fc functionality for the desired pharmacology.

5.3.2 *IgG activities and binding to FcγRs*

Myeloid cells such as monocytes, macrophages, neutrophils, and dendritic cells express a variety of Fcγ receptors, including at least one activating FcγR (FcγRI, FcγRIIa, FcγRIIIa/b), as well as the inhibitor FcγRIIb. Neutrophils express FcγRIIa and FcγRIIIb (instead of FcγRIIIa), and also FcγRI when activated by G-CSF (Desjarlais *et al.*, 2007). It has long been known that binding of IgGs, through their Fc regions, to FcγRs on monocytes, macrophages and neutrophils could lead to phagocytosis of the target cells or particles (Clark, 1997; Ravetch and Bolland, 2001; Nakamura *et al.*, 2005). On the other hand, the only activating IgG receptor expressed on human natural killer cells is FcγRIIIa, and binding of cell-targeted IgGs to FcγRIIIa on NK cells leads to ADCC of those targeted cells (Ravetch and Bolland, 2001). Finally, binding of antibodies to the activating receptors, FcγRI, FcγRIIa, and FcγRIIIa on these immune cells results in the synthesis and release of reactive nitrogen and oxygen species that induce apoptosis, as well as inflammatory cytokines, which significantly affect the immune response toward the target of the antibodies (Li *et al.*, 2006b; Desjarlais *et al.*, 2007). Additionally, it has been demonstrated that polymorphisms of the key receptors, FcγRIIa, FcγRIIIa, and FcγRIIIb can have important implications in propensity to diseases (van Sorge *et al.*, 2003; Bournazos *et al.*, 2009) as well as efficacy of therapeutics binding those receptors (Cartron *et al.*, 2002; van Sorge *et al.*, 2003; Weng and Levy, 2003; Desjarlais *et al.*, 2007; Zhang *et al.*, 2007; Musolino *et al.*, 2008).

Bruhns *et al.* (2009) recently carried out a detailed comparative analysis of binding of human IgGs to FcγRs (Table 5.3). Below, a short description of each pertinent human FcγR is given. More information on human FcγRs can be found in various reviews (Clark, 1997; Ravetch and Bolland, 2001; Takai, 2002; van Sorge *et al.*, 2003; Woof and Burton, 2004; Nakamura *et al.*, 2005; Nimmerjahn and Ravetch, 2006, 2008).

5.3.2.1 *Human FcγRI*

Human FcγRI (CD64) is a 72 kDa protein known as the high affinity IgG receptor (Table 5.2; Fig. 5.4). FcγRI requires the FcR γ chain containing an intracellular immunoreceptor tyrosine-based activation motif (ITAM) for both stable cell surface expression and activity (van Sorge *et al.*, 2003; Nimmerjahn and Ravetch, 2008). FcγRI is widely distributed on myeloid-derived immune cells such as macrophages, neutrophils, eosinophils and dendritic cells. IgG1, IgG3, and IgG4 isotypes bind to it with an affinity in the range of 15–35 nM, whereas IgG2 does not bind it at all (Table 5.3; Bruhns *et al.*, 2009). Due to the relatively high affinity interaction, downstream signaling as a result of IgG binding to FcγRI appears to be triggered by a monomeric interaction of the antibody with the receptor. It appears that this signaling is a function of the antibody-binding α chain rather than the γ chain, although this needs further investigation (van Sorge *et al.*, 2003). Under normal physiological conditions, FcγRI would be occupied with a resident IgG1, IgG3, or IgG4 *in vivo*, ready to be displaced by a high-avidity immune complex (Bruhns *et al.*, 2009).

5.3.2.2 *Human FcγRIIa and FcγRIIc*

Human FcγRIIa (CD32a) is a 40 kDa low affinity, integral membrane receptor found on a wide variety of hematopoietic cells, including monocytes, macrophages, B cells, neutrophils, mast cells, and platelets (Takai, 2002; Nimmerjahn and Ravetch, 2008). Human FcγRIIa is a monomeric receptor in which the intracellular ITAM is part of the internal domain of the receptor (see Fig. 5.4; Nimmerjahn and Ravetch, 2008). FcγRIIa is not present on lymphocytes, including natural killer (NK) cells. Human IgGs bind FcγRIIa with an affinity in the μM range (Table 5.3); therefore, physiologically significant binding only occurs by antigen-complexed IgG capable of conferring avidity. The ITAM is capable of signaling, leading to cell activation including oxidative burst, phagocytosis and production of proinflammatory cytokines (Hart *et al.*, 2004). There is no direct equivalent of human FcγRIIa in mice or rats, although as discussed

Table 5.3 Comparison of binding activities for human IgGs and their receptors

	K_D Values for IgG Isotype Binding to FcγR (in nM)								
Fcγ Receptor	**XmAb5574* IgG1 Mutant[j]**	**Hu IgG1**	**Hu IgG2**	**Hu IgG3**	**Hu IgG4**	**Mu IgG1**	**Mu IgG2a**	**Mu IgG2b**	**Mu IgG3**
HuFcγRI	0.14[j]	15[a], 0.1[h], 8[i], 0.33[j]	Nb[a]	15[a]	30[a]	—	—	—	—
HuFcγRIIa – H131	170[j]	200[a], 690[g], 850[h], 1000[j]	2000[a]	1000[a]	6000[a]	800[g]	—	—	—
HuFcγRIIa – R131	80[j]	300[a], 910[h], 1000[j]	10000[a]	1000[a]	5000[a]	—	—	—	—
HuFcγRIIb/HuFcγRIIc[k]	100[j]	8000[a], 1670[g], 2300[h], 1400[j]	50000[a]	6000[a]	5000[a]	1370[g]	—	—	—
HuFcγRIIIa – F158	14[j]	800[a], 710[e], 77[f], 1570[g], 1000[h], 380[i], 2000[j]	30000[a]	130[a], 560[e], 51[f]	5000[a]	Nb[g]	—	—	—
HuFcγRIIIa – V158	5[j]	500[a], 280[h], 220[j]	15000[a]	100[a]	4000[a]	—	—	—	—
HuFcγRIIIb-NA1	—	5000[a]	Nb[a]	900[a]	Nb[a]	—	—	—	—
HuFcγRIIIb-NA2	—	5000[a]	Nb[a]	1000[a]	Nb[a]	—	—	—	—
HuFcγRIIIb-SH	—	5000[a]	Nb[a]	900[a]	Nb[a]	—	—	—	—
MuFcγRI	2[j]	600[i], 110[j]	—	—	—	Nb[d]	26[d]	Nb[d]	Nb[d]
MuFcγRII	100[j]	Nb[c], 1100[j]	—	—	—	300[c], 830[d]	2390[c], 1800[d]	450[c], 900[d]	Nb[c]
MuFcγRIII	—	Nb[c], Nb[i]	—	—	—	3200[c], 4800[d]	1460[c], 1800[d]	1550[c], 1600[d]	Nb[c]
MuFcγRIV	0.2[j]	536[b], 300[c], 30[j]	—	—	—	Nb[c], Nb[d]	34.5[c], 71[d]	59[c], 63[d]	Nb[c]

Symbols and abbreviations: Hu = human; Mu = mouse; Nb = no measureable binding; — = no available data.

* S239D = I332E mutant of IgG1 that increases binding to activating FcγRs and confers stronger ADCC and ADCP than wild-type IgG1.

Data are from the following references and recalculated from K_A or K_D values, and rounded where necessary: **a**, Bruhn *et al.*, 2009; **b**, Bruckheimer *et al.*, 2009; **c**, Nimmerjahn *et al.*, 2005; **d**, Baudino *et al.*, 2008b; **e**, monomeric IgG (Li *et al*, 2007); **f**, multimeric IgG binding with ca. ten-fold greater avidity than monomeric form (Li *et al*, 2007); **g**, Maenaka *et al.*, 2001; **h**, Richards *et al.*, 2008; **i**, Oflazoglu *et al.*, 2009 (cell-based data); **j**, Horton *et al.*, 2008; **k**, Human FcγRIIb and FcγRIIc have identical exodomains (Bruhns *et al.*, 2009), even though the inhibitor receptor, FcγRIIb, has an ITIM in its cytoplasmic domains and FcγRIIc has an activating ITAM in its cytoplasmic domain (Nimmerjahn and Ravetch, 2008).

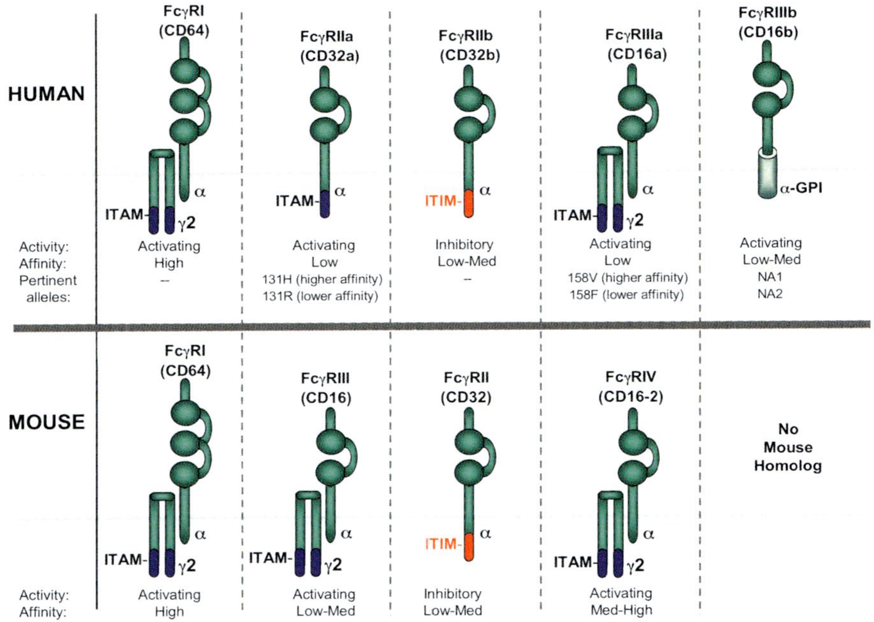

Figure 5.4 Model of the structures of human Fcγ receptors (top) and their mouse Fcγ receptor homologues, including comparative binding information, significant polymorphisms, and important domains. Abbreviations: γ2 = Fcγ receptor; α = alpha chain of receptor; ITAM = immunoreceptor tyrosine-based activation motif; ITIM = immunoreceptor tyrosine-based inhibition motif; α-GPI = glycosylphosphatidylinositol (linked receptor).

later in this chapter, mouse FcγRIII is most closely related in sequence, structure and function.

Human FcγRIIa has been strongly linked with macrophage-mediated phagocytosis, both by the disease association of its polymorphisms, as well as in *in vitro* functional assays (Richards *et al.*, 2008). Increased binding of IgG1 mutants to FcγRIIa strongly correlated with increased phagocytosis, whereas increased binding to either FcγRI or FcγRIIIa had a much weaker correlation with increased phagocytosis (Richards *et al.*, 2008).

There are two major polymorphisms of human FcγRIIa, the higher affinity H131 and a lower affinity R131. Individuals who are homozygous for the lower affinity R131 FcγRIIa polymorphism have been shown to be more susceptible to infections, particularly by encapsulated microorganisms (e.g. pneumococci, meningococci) (Platonov *et al.*, 1998; Yee *et al.*, 2000). This susceptibility is believed to be due to the inability of IgG2 to bind the lower affinity receptor form well enough to protect against encapsulated microorganisms (Yee *et al.*, 2000). Moreover, human FcγRIIa H131 and FcγRIIIb NA1 polymorphisms were correlated with optimal anti-pneumococcal phagocytic activity, implicating a role of FcγRIIa in neutrophil-mediated phagocytosis (Jansen *et al.*, 1999). Additionally, the higher affinity FcγRIIa polymorphism has been correlated with improved therapeutic outcome in treatment of neuroblastoma with an anti-GD2 murine antibody (Cheung *et al.*, 2006), and several similar correlations have been found with the FcγRIIa and FcγRIIIa polymorphisms both independent and/or together in cancer therapy (Weng and Levy, 2003; Zhang *et al.*, 2007; Desjarlais *et al.*, 2007; Musolino *et al.*, 2008; Bibeau *et al.*, 2009). In the Caucasian population, 24% of people have the 131HH homozygous allele, 42% have the 131HR heterozygous allele, and 34% have the 131RR lower affinity homozygous allele (van Schie and Wilson, 2000).

A variant of human FcγRII, FcγRIIc, is present on human NK cells of ca. 50% of the population (Veri *et al.*, 2007) and may function as a co-activation receptor along with FcγRIIIa (Ernst *et al.*, 2002; Nimmerjahn and Ravetch, 2008). Like FcγRIIa, FcγRIIc has an ITAM in its cytoplasmic domain. It appears that human FcγRIIc arose from genetic recombination with FcγRIIb, since the exodomain of FcγRIIc is identical to that of FcγRIIb (Bruhns *et al.*, 2009), while its intracellular activation domain is FcγRIIa-like (Nimmerjahn and Ravetch, 2008). There are multiple polymorphisms of FcγRIIc having differing levels of activation activity that may affect NK cell activation by IgG antibodies (Ernst *et al.*, 2002).

5.3.2.3 *Human FcγRIIb*

Human FcγRIIb (CD32b), which has an immunoreceptor tyrosine-based inhibitory motif (ITIM) embedded in its intracellular domain (Fig. 5.4), is the inhibitory receptor that helps to balance activation FcγR-mediated activities, as well as cytokine and antibody production (Baerenwaldt and Nimmerjahn, 2008). Blocked or impaired FcγRIIb signaling results in inflammatory and autoimmune conditions. FcγRIIb is also part of the B cell complex and plays a significant role in B cell maturation. Overexpression of FcγRIIb on B cells results in production of antibodies with higher affinity, demonstrating its important role as a negative regulator of B cell activation, as well as its role in balancing the antibody response toward antigens (Baerenwaldt and Nimmerjahn, 2008). Overexpression of FcγRIIb on macrophages, on the other hand, resulted in greater susceptibility to infection due to the dampening of the activating FcγR responses (Baerenwaldt and Nimmerjahn, 2008). Thus it appears that FcγRIIb expression on macrophages and B cells act to balance the humoral immune response. On the other hand, human FcγRIIb does not likely play a role in regulation of ADCC (Nakamura *et al.*, 2005), as it is not present on NK cells (Takai, 2002).

Three approaches have been used lately to target the inhibitory receptor, FcγRIIb, as part of a potential therapeutic strategy. In the first example, scientists at MacroGenics generated a highly specific anti-FcγRIIb antibody that does not bind to the activating receptor, FcγRIIa (Veri *et al.*, 2007). This anti-FcγRIIb antibody was shown to block immune complexes binding to the inhibitory receptor. Beyond use as an excellent research reagent, this antibody may find use therapeutically to block inhibitory responses resulting to improve the responsiveness of therapeutic antibodies that rely on ADCC and/or ADCP for their efficacy (Veri *et al.*, 2007). This could be particularly important in certain cancers such as melanoma, where it was recently shown that FcγIIb expression prevented NK-cell-mediated ADCC of human metastatic melanoma cells *in vitro* and protected B16F0 melanoma tumors from IgG-induced ADCC *in vivo* (Cassard *et al.*, 2008).

In a second approach to targeting FcγRIIb, Chu *et al.* (2008) engineered an anti-CD19 antibody with modified Fc functionality (IgG1-S267E/L328F) yielding 420-fold improved binding of the Fc to FcγRIIb with no substantial increase in binding to the activating FcγRs (and diminishment of FcγRIIIa binding). They demonstrated that co-engagement of both CD19 and FcγRIIb by this antibody could block B cell receptor (BCR) complex-mediated signaling (Chu *et al.*, 2008). A third path for modulating the activity of FcγRIIb is described in Section 4.6, in which highly sialylated IgGs in mice appear to function through a lectin to stimulate the expression of FcγRII on mouse macrophages to provide an immunosuppressive effect without affecting FcγRII on B cells (Anthony *et al.*, 2008a, b).

5.3.2.4 *Human FcγRIIIa*

Human FcγRIIIa (Fig. 5.4) is a low affinity receptor found on macrophages, NK cells, eosinophils and mast cells. Similar to FcγRI, FcγRIIIa requires the ITAM-containing FcR γ chain for stable cell surface expression and activity (van Sorge *et al.*, 2003; Nimmerjahn and Ravetch, 2008). FcγRIIIa is the activating receptor on human NK cells that is responsible for triggering ADCC when bound by complexed IgG1 or IgG3 antibodies. NK-cell-mediated ADCC via FcγRIIIa is one of the strongest cell depletion mechanisms in human serum. When human peripheral blood mononuclear cells (PBMCs) are used as the effector function source, NK cells bearing FcγRIIIa mediated most of the lysis by an anti-EpCAM IgG1 on $EpCAM^+$ LS180 cells; when NK cells were depleted or FcγRIIIa blocked by an antibody, most of the PBMC-mediated lytic activity decreased, indicating the key role played by FcγRIIIa-bearing NK cells (Richards *et al.*, 2008).

IgG1 and IgG3 bind FcγRIIIa with a K_D in the 100–800 nM range, but complexes of IgG1 or IgG3 bind with about a ten-fold greater affinity (Li *et al.*, 2007), leading to cell activation, cytokine release and ADCC. FcγRIIIa is glycosylated, with the receptor on NK cells having high-mannose- and complex-type oligosaccharides, resulting in a higher binding affinity for monomeric IgG on NK cells for FcγRIIIa on monocytes, which lack the high mannose glycosylation (van Sorge *et al.*, 2003).

There are two major polymorphisms of human FcγRIIIa, the higher affinity V158 polymorphism and a lower affinity F158 polymorphism (Nimmerjahn and Ravetch, 2006; Bournazos *et al.*, 2009). Individuals who are either heterozygous 158VF or homozygous 158FF for the lower affinity FcγRIIIa polymorphism have been shown to respond significantly more poorly to several anti-cancer therapeutic antibodies targeting receptors (e.g. Rituxan®, Herceptin®, Erbitux®) than individuals homozygous for the high affinity 158VV allele (Cartron *et al.*, 2002; van Sorge *et al.*, 2003; Weng and Levy, 2003; Desjarlais *et al.*, 2007; Zhang *et al.*, 2007; Musolino *et al.*, 2008). Taking a composite of the FcγRIIIa allelic distributions across several populations, tabulated by Chong *et al.* (2006), 12% of the composite population (n = 560) have the 158VV homozygous higher affinity form, 46% (n = 2200) have the 158VF heterozygous form, and 43% (n = 2051) have the 158FF lower affinity homozygous form.

5.3.2.5 *Human FcγRIIIb*

Human FcγRIIIb (CD16b) (Fig. 5.4) is a glycosylphosphatidyl inositol (GPI)-anchored receptor found on neutrophils and IFN-γ-induced eosinophils (Takai, 2002; Nimmerjahn and Ravetch, 2008). Since FcγRIIIb is GPI-anchored, it does not possess the same type of signaling domain as the other FcγRs, and it is generally believed that FcγRIIIb acts by synergizing with other FcγRs (Nimmerjahn and Ravetch, 2008). It also has been demonstrated that FcγRIIIb interacts with β2 integrins such as Mac-1 (CD11b/CD18), as well as other low affinity FcγRs on neutrophils to bind immune complexes, resulting in phagocytosis, degranulation and respiratory burst activity (Coxon *et al.*, 2001). More recently, FcγRIIIb has been shown to activate a unique signaling pathway via Syk, PI3K, and MEK-independent nuclear-restricted phosphorylation of ERK and Elk-1 (García-García *et al.*, 2009). Mice, rats, and non-human primates, including chimpanzees, appear to lack a structural equivalent of FcγRIIIb.

There are three different isoforms of FcγRIIIb, the NA1, NA2, and SH forms (Bruhns *et al.*, 2009). Of the two major isoforms of FcγRIIIb (NA1 and NA2), NA1 is correlated positively with neutrophil-mediated phagocytosis (Jansen *et al.*, 1999). While in biochemical assays, the binding of IgG1 (~ 5 μM range), and IgG3 (900–1000 nM range) did not vary significantly between the different isoforms (Bruhns *et al.*, 2009; Table 5.3), the NA1 form of FcγRIIIb had been shown previously to bind both IgG1 and IgG3 and confer phagocytic activity better than the NA2 isoform (Bredius *et al.*, 1994). In the Caucasian population, 14% of people have the NA1/NA1 allele, 48% are heterozygous NA1/NA2, and 38% are homozygous NA2/NA2 (van Schie and Wilson, 2000).

Neutrophils, considered as the first line of defense against invading microorganisms, utilize phagocytosis, oxidative burst, release of anti-microbial compounds via degranulation, and production of pro-inflammatory cytokines to kill. FcγRIIIb and FcγRIIa are constitutively expressed on neutrophils. It has been demonstrated by several groups that activation of human neutrophils by immune complexes requires both FcγRIIIb and FcγRIIa (Jakus *et al.*, 2008), resulting in phagocytosis and cytokine release.

5.3.2.6 *Complement activation*

The complement cascade is a highly complex, albeit phylogenetically ancient, component of our innate immune system comprising three different pathways and about 30 different proteins, which normally circulate in inactive forms that can be activated via proteolytic processing. The complement cascade, which is critical for both host defense as well as damaged tissue removal, has four major functions: (i) recognition and opsonization of targeted foreign object; (ii) phagocytosis of the foreign material; (iii) immune cell activation; and (iv) lysis (Gasque, 2004). The complement inflammatory cascade can be activated by direct interaction of C1q with antibody–antigen immune complexes (classical pathway), by the antibody-independent adsorption of complement factor C3 directly onto the surface of pathogens (alternative route), or by the lectin pathway, another antibody-independent pathway in which bacteria or other microbes bind the "recognition molecule," mannan binding lectin (MBL; also known as mannose binding lectin, mannan binding protein, mannose binding protein) (Sontheimer *et al.*, 2005; Arnold *et al.*, 2006; Ip *et al.*, 2009). Activation via any of these three pathways results in generation of the pro-inflammatory anaphylatoxins, C3a and C5a, as well as the pore-like membrane attack complex (MAC; C5b, C6, C7, C8, C9) that inserts into cell membranes causing lysis unless blocked by cell-surface inhibitors (Gasque, 2004).

The classical pathway for complement activation occurs when C1q, a 460 kDa glycoprotein found in human serum in the 70 μg/mL range, binds to IgM, IgG1, or IgG3s that are clustered upon antigen binding (Kishore and Reid, 2000). C1q interacts most avidly with the Fc of antigen-complexed IgG1, IgG3, hexameric IgM, and to a significantly lesser extent with IgG2. Using serum, human IgG1 and IgG3 were found to activate complement via the classical pathway whereas IgG2 was minimal and IgG4 did not activate at all (Seino *et al.*, 1993). While human IgG3 binds C1q better than IgG1, it has been shown that human IgG1 mediates complement-dependent lysis better than IgG3 (Brüggemann *et al.*, 1987). IgG4, IgA, pentameric IgM, IgE, or IgD do not bind C1q in a physiologically significant manner (Sontheimer *et al.*, 2005). Binding C1q alone is not enough to trigger complement activation. Additional factors including the epitope and target density are also important as to whether C1q binding results in complement activation (Bindon *et al.*, 1990; Lucisano-Valim and Lachman, 1991).

Human IgG1 binds C1q primarily through Fc residues D270, K322, P329, and P331 (Fig. 5.2; Idusogie *et al.*, 2000). Similarly, Thommesen *et al.* (2000) demonstrated that K322 was a critical residue for C1q binding and complement-mediated cell killing by human IgG3. The contact points for murine IgG2b binding to complement had previously been shown to be E318, K320 and K322 (Duncan and Winter, 1988). Of those, only K322 appears also to be applicable to human IgG binding to C1q (Idusogie *et al.*, 2000; Thommesen *et al.*, 2000).

The alternative pathway for complement activation makes up nearly 80% of all complement activation, so it is by far the most critical pathway (Ricklin and Lambris, 2007). As mentioned above, the alternate pathway is triggered by antibody-independent deposition of complement factor C3 on the antigen. Some antibodies appear to facilitate the alternate pathway, especially IgG2 (Dillman *et al.*, 1995). The lectin pathway, in which MBL is the key activator, is the third route to complement activation. MBL is a member of the collectin family that is produced by hepatocytes and circulates in serum at about 1.2 μg/mL (range < 50 ng/mL to 10 μg/mL) (Arnold *et al.*, 2006; Ip *et al.*, 2009). While the MBL pathway for complement activation is typically considered to be an antibody-independent pathway for complement activation, IgGs with the G0 form (agalactosyl form with terminal GlcNAc) (see Fig. 5.3) of glycosylation bind MBL in a Ca^{++}-dependent manner and complex with the surfaces of bacteria, viruses, fungi and other invading entities (Arnold *et al.*, 2006; Ip *et al.*, 2009). Additionally, dimeric IgA and about 20% of all IgMs bind MBL, triggering complement activation via the lectin pathway (Arnold *et al.*, 2006).

5.3.3 *Human IgG1*

Two thirds of all therapeutic antibodies and Fc fusion proteins employ an unmodified IgG1 isotype. Thus, more is probably known about the biology, functionality

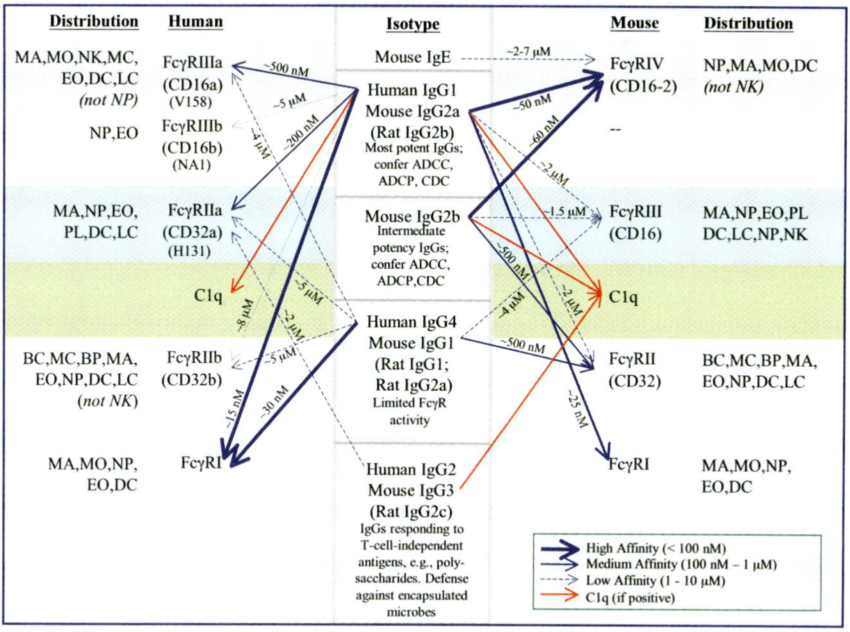

Figure 5.5 Comparison of human IgG isotypes used in therapeutic antibodies with the most similar IgG isotypes from mouse and rat. Included are the human and mouse Fcγ receptors and the approximate affinities of the IgGs for those receptors for the most prominent interactions. For the human receptors, data for the high affinity forms (see Table 5.3) were used. Binding of IgGs to C1q is noted by red lines. Key for distribution: MA = macrophages; MO = monocytes; NK = natural killer cells; MC = mast cells; EO = eosinophils; DC = dendritic cells; LC = Langhans cells; NP = neutrophils; PL = platelets; BC = B cells; BP = basophils.

and stability of IgG1 antibodies than any other isotype of antibody. Human IgG1 antibodies are typically generated as part of a Th1 type response to antigen stimulation. Human IgG1 antibodies bind to all forms of FcγRs (Table 5.3; Fig. 5.5) and are capable of triggering phagocytosis through multiple receptors (FcγRI, FcγRIIa, FcγRIIIa) on monocytes, macrophages and neutrophils. Binding of IgG1 to the high affinity receptor, FcγRI (K_D for IgG1, 15 nM), appears to require only monomeric interaction to induce activity, whereas for the low affinity receptors FcγRIIa (K_D for IgG1, 200–300 nM) and FcγRIIIa (K_D for IgG1, 500–800 nM), avidity affects are required to induce activity (Nimmerjahn and Ravetch, 2008; Bruhns *et al.*, 2009). Human IgG1s also can induce ADCC via binding to FcγRIIIa (K_D for IgG1, 5 μM) and possibly FcγRIIc (Ernst *et al.*, 2002) on NK cells. As noted above, human IgG1 antibodies also bind complement factor C1q and are capable of activating the complement cascade via the classic route as well as the alternative route (Brüggemann *et al.*, 1987; Seino *et al.*, 1993).

The IgG1 isotype has become the "default" isotype, especially for many of the early therapeutic antibody candidates. For therapeutic antibodies targeting cancer cell surface targets (e.g. rituximab, trastuzumab, cetuximab) or infectious disease targets on pathogens (e.g. palivizumab), the ability to bind FcγRs to trigger phagocytosis and/or ADCC, as well as activate the complement cascade may be viewed as positive or even critical attributes. IgG1s have also been used widely for soluble targets with which there is no apparent cellular interaction, e.g. blocking the binding of a cytokine to its receptor. Examples of this use include omalizumab (anti-IgE), bevacizumab (anti-VEGF), ustekinumab (anti-IL12-IL23p40), and canakinumab (anti-IL-1β). This approach appears to work well and has resulted in several blockbuster monoclonal antibody products.

5.3.4 *Human IgG2*

Human IgG2 antibodies are naturally produced in response to the complex carbohydrate antigens of microbial pathogens, and thus serve as one of the defense mechanisms against encapsulated microorganisms (Ferrante *et al.*, 1990). Human IgG2 binds with a five-fold greater affinity to the H131 isoform of FcγRIIa (K_D ~ 2 μM) than to the R131 isoform (K_D ~ 10 μM) (Table 5.3). This difference in binding is sufficient to trigger functionality only after cross-linking of antibodies and targets onto the surface of the effector cells. When binding to the higher affinity form of FcγRIIa, human IgG2 antibodies are capable of triggering complement-independent, PMN-mediated phagocytic activity, particularly against encapsulated microorganisms such as *Neisseria meningitidis*, *Haemophilus influenzae* and *Streptococcus pneumoniae* (Sanders *et al.*, 1995). Natural human IgG2 antibodies have been shown to demonstrate the least overall level of effector function of the four human IgG isotypes, which makes them attractive as the isotype of choice when little to no effector function is desired to drive therapeutic efficacy (Dillon *et al.*, 2008).

Human IgG2 antibodies weakly bind C1q, but typically activate complement via the alternative pathway rather than the classical pathway (Seino *et al.*, 1993). Due to the relatively weak binding of IgG2 to FcγR and components of the complement system, it has been shown that complex systems such as whole blood, which allow for immune complex formation, are required to evaluate the true effector functions of IgG2 antibodies (Jefferis, 2007). Lai *et al.* (2005) demonstrated that several human anti-pseudomonad IgG2 antibodies did not lyse the targeted *Pseudomonas* spp. in combination with monocytes or with serum alone, whereas addition of whole blood to the antibodies triggered lysis.

Human IgG2 antibodies have the capability of forming a covalent dimer through interaction of the cysteine residues in the hinge region (Yoo *et al.*, 2003).

It is hypothesized that this covalent dimer formation may confer an advantage to human IgG2s in binding polysaccharide antigens present on microorganisms (Yoo *et al.*, 2003, Salfeld, 2007). IgG dimer formation, however, is not unique to the IgG2 isotype, as IgG1s also have been shown to form both non-covalent and covalent dimers, albeit via different mechanisms (Yoo *et al.*, 2003).

There is currently one marketed antibody with the IgG2 isotype, the anti-EGFR mAb, panitumumab (Vectibix®). Other late stage clinical candidates with the IgG2 isotype include denosumab (AMG-162), an anti-RANK Ligand mAb, tremelimumab (CP-675,206), an anti-CTLA4 mAb, and figitumumab (CP-751,871), an anti-IGF-1R mAb. Salfeld (2007) lists additional IgG2 antibodies in human clinical trials.

With the development of panitumumab and the other late-stage clinical candidate IgG2 isotype mAbs, the behavior of the IgG2 has come under closer scrutiny. Besides the covalent dimer formation, the Fab moieties of the IgG2 isotype have been shown to undergo disulfide bond isomerization, resulting in a series of isoforms that appear to retain biological functionality (Dillon *et al.*, 2008; Guo *et al.*, 2008; Wypych *et al.*, 2008). While these structural isoforms need to be carefully characterized for each product candidate, they do not appear to undermine the value of an IgG2 isotype as a therapeutic candidate (Dillon *et al.*, 2008; Guo *et al.*, 2008; Wypych *et al.*, 2008). Salfeld (2007) pointed out that no preclinical or clinical data had yet been obtained that suggested that IgG2s posed any unique safety issues. Nevertheless, additional analytical assays and follow-up protocols have been proposed for IgG2 and other "novel" antibody isotypes or forms (Swann *et al.*, 2008).

An interesting "experiment" is underway currently that may help to pave the way for the future for differentiation of IgG1 vs. IgG2 antibodies, at least for cancer targets that are found on both healthy and cancer cells. For both EGFR and IGF-1R, one of the lead antibodies is an IgG1 and another lead is an IgG2; Panitumumab and cetuximab are IgG2 and IgG1 isotypes, respectively, of anti-EGFR antibodies, and figitumumab and Merck MK-0646 are IgG2 and IgG1 isotypes, respectively, of anti-IGF-1R antibodies. Moreover, for both of these targets, the receptor is found both on normal cells and in increased levels on cancer cells. It would be expected that in each case, the IgG1s would function not only through receptor binding and neutralization of ligand-mediated activity, but also via ADCC and potentially also by complement activation. These Fc-mediated functionalities could potentially lead to greater efficacy, but also to increased side effects (or possibly both). The IgG2s, on the other hand, might have some low level of macrophage or monocyte-mediated phagocytosis driven through FcγRIIa, but would not likely possess physiologically significant ADCC or CDC activities; thus, the primary functionality of the IgG2s would likely be via neutralization of ligand-mediated activity. It will be interesting to see how this plays out for both efficacy and safety, whether a trend appears across multiple targets or whether the data are target-specific.

5.3.5 *Human IgG3*

Similar to IgG1, IgG3 is capable of binding all of the human FcγRs (Bruhns *et al.*, 2009; Table 5.3), mediating strong ADCC (Michaelsen *et al.*, 1992) and of binding C1q and activating complement via both the classical and alternative routes (Seino *et al.*, 1993). As such, human IgG3 antibodies elicit one of the strongest cellular and complement-related responses of all the human isotypes. Human IgG3 is the isotype that stands out as significantly different from the others. While the other IgG isotypes possess short hinges encoded by single exons, IgG3 has a long hinge encoded by four exons, containing 62 amino acids and 11 disulfide bonds. Shortening of the hinge from 62 to 15 amino acids, or substituting the IgG4 hinge for the natural IgG3 hinge, did not substantially reduce the strong ADCC response conferred by IgG3 (Michaelsen *et al.*, 1992). Human IgG3 antibodies are highly susceptible to proteolysis (Jefferis, 2007), typically possess a short half-life (Kim *et al.*, 1999), and have multiple natural forms with extensive hinge length and constant domain polymorphisms within human populations (Jefferis, 2007). Kim *et al.* (1999) attributed the shorter half-life of human IgG3 to a H435R substitution which appears to influence its binding kinetics to FcRn. Other than the construction of an IgG3-based GM-CSF-fused or IL-2-fused Fc-fusion proteins targeting HER2 (Penichet and Morrison, 2001; De La Cruz *et al.*, 2006), there do not appear to be any significant efforts currently to make a therapeutic based on an intact human IgG3.

On the other hand, in an effort to enhance effector function activity, Natsume *et al.* (2008) constructed a set of different IgG1/IgG3 hybrid molecules. They found that variant 1133, which contained the CH1 and hinge of IgG1, and the CH2 and CH3 domains from IgG3, possessed the optimal set of characteristics, including the best complement-dependent cytotoxicity. This mutant form, however, did not bind Protein A, so the Protein A-binding sequences from IgG1 were grafted back into the hybrid 1133 molecule. The resultant molecule, 113F-minus-fucose, retained strong C1q-binding activity and complement-mediated lysis, as well as strong binding to FcγRs (Natsume *et al.*, 2008). *In vivo* experiments comparing the anti-CD20 versions of the 113F-minus-fucose mutant form *vs.* wild-type IgG1-minus-fucose demonstrated the superior CDC and B cell depleting activity of the 113F-minus-fucose mutant (Natsume *et al.*, 2008). It is possible that one of the IgG1/IgG3 hybrid isotypes such as 113F, or some variation thereof, might be worth consideration for therapeutic development.

5.3.6 *Human IgG4*

Human IgG4 is the least abundant IgG in human serum (Table 5.1). Along with IgE, human IgG4 is considered to be a part of the Th2 antibody response to antigens. Thus, typical allergens associated with a Th2 response, such as nematodes or pollen, stimulate not only IgE, but also IgG4 responses (Aalberse *et al.*, 2009). Additionally, when humans experience chronic exposure to protein antigens, including biopharmaceuticals such as erythropoietin, they typically undergo an isotype switch from IgG1 to IgG4 (Jefferis, 2007).

In the past, when a "benign blocker" was desired, it was often addressed by the use of scFv fragments, FAb fragments, or $F(Ab')_2$ fragments, often coupled to poly-ethylene glycol (PEG) to increase half-life. Cimzia® (certolizumab pegol) is a successful example of this type of strategy, resulting in a good TNF-α-blocking antibody that has a reasonable half-life of 14 days. These approaches have not been widely translated into therapeutics, however, at least in part due to the difficulty and costs in PEGylating specifically and consistently to form a uniform product. Finally, PEG itself is associated with vacuolization of renal cortical tubular epithelium cells (Bendele *et al.*, 1998). In the past, the human isotype IgG4 had been considered by some to be a good choice of an antibody isotype to be used as a "relatively benign blocker," because of its supposedly low affinity to FcγRIIa, FcγRIIIa and C1q. Key antibodies employing the human IgG4 isotype include Mylotarg® (gemtuzumab ozogamicin), the anti-VLA4 antibody Tysabri® (natalizumab), the Phase III candidate anti-CD22 antibody conjugate, inotuzumab ozogamycin, and the Phase II candidate anti-IL-13 antibody, CAT354.

IgG4, however, is known to be substantially different from its other human IgG counterparts (Aalberse and Schuurman, 2002; Aalberse *et al.*, 2009). It has long been known that IgG4s are unable to cross-link antigens *in vivo*, making them functionally monovalent (van der Zee *et al.*, 1986). Additionally, IgG4 molecules were long suspected of being able to bind multiple different antigens, functionally behaving like a bispecific molecule (Aalberse and Schuurman, 2002; Schuurman *et al.*, 1999). The mechanism for this behavior was linked to the ability of IgG4s to dissociate into two half molecules (Angal *et al.*, 1993; Aalberse and Schuurman, 2002; van der Neut Kolfschoten *et al.*, 2007), largely due to the flexibility of the hinge region promulgated by the "cys-pro-pro-cys" sequence that favored intrachain disulfide bond formation, leading to equilibrium with interchain disulfides and ultimately dissociation. It has been shown that reducing conditions activate this process and that the CH3 domain of IgG4 is involved as well (van der Neut Kolfschoten *et al.*, 2007). Once dissociated, the half-IgG4 molecules are free to pair indiscriminately with other half-IgG4 molecules of the same or different specificity, resulting in "natural" bispecific antibodies *in vivo* (van der Neut Kolfschoten *et al.*, 2007; Labrijn *et al.*, 2009). Not only does this Fab exchange occur under laboratory conditions, it has been recently shown that it can occur in serum with resident IgG4 molecules (Labrijn *et al.*, 2009); thus, a therapeutic IgG4 mAb could potentially end up sharing its functionality with a patient's endogenous IgG4 antibodies. Apparently, the equilibrium dictating self-reassociation vs. heteromolecular association is a stochastic process driven by concentration of each of the IgG4s in solution (van der Neut Kolfschoten *et al.*,

2007). The S228P mutant (EU numbering based on IgG1 sequence) of IgG4 (Fig. 5.2) was shown by Celltech to limit chain dissociation (Angal *et al.*, 1993), resulting in a better-behaved IgG4 antibody that possessed improved PK and tissue distribution properties.

Human IgG4 antibodies have several other properties that make them less than ideal as "benign blockers," i.e. antibodies with silenced Fc functionality. Human IgG4 antibodies, which bind FcγRI with a K_D of 30 nM, and all isoforms of FcγRIIa, FcγRIIb, FcγRIIc, and FcγRIIIa in the 4–6 μM range (Table 5.3), have been demonstrated to trigger cellular responses such as ADCP and ADCC through these activities (Ernst *et al.*, 2002; Nielsen *et al.*, 2007). While the level of ADCC exhibited by IgG4 antibodies is lower than by IgG1 isotypes, it is clearly not background. Additionally, non-fucosylated IgG4 demonstrated as much ADCC as did non-fucosylated IgG3 (Niwa *et al.*, 2005), so having a low percentage (e.g. 1–5%) of non-fucosylated IgG4 in a production preparation (as is usually the case), could lead to unwanted activities. Furthermore, an IgG4 version of the anti-CD52 mAb, CAMPATH, was shown in human trials to have substantial injection site reactions, to stimulate first injection-related cytokine release syndrome (CRS) and to deplete lymphocytes in serum (Isaacs *et al.*, 1996). Finally, IgG4 antibodies from serum of patients with Wegener's granulomatosis have also been shown to activate neutrophils by a mechanism requiring engagement of FcγRIIa and/or FcγRIIIb (Holland *et al.*, 2004). Thus, natural IgG4 antibodies are clearly not silent with respect to Fc functionality. Jefferis (2007) recently highlighted the investigation into TeGenero TGN1412, an anti-CD28 IgG4 agonist antibody that caused life-threatening cytokine storm in several volunteers in an aborted Phase I clinical study. He pointed out that investigators have concluded that the side effects of TGN1412 could not have been anticipated, apparently overlooking the fact that FcγR engagement by IgG4, monomerically (FcγRI) and/or via antigen-bound crosslinking (e.g. FcγRIIa, FcγRIIIa), which is well known to mediate ADCC and/or ADCP via multiple cell types, may have been a contributing factor.

High concentrations of serum IgG4 also have been associated with several potential pathological conditions, including Wegener's granulomatosis (Holland *et al.*, 2004), renal lesions (Saeki *et al.*, 2007) and autoimmune pancreatitis (Kamisawa and Okamoto, 2006). Given the variety of issues noted above with the natural IgG4 isotype, it is unlikely that too many additional antibodies requiring "silenced Fc functionality" (Labrijn *et al.*, 2008) will be developed using the wild-type IgG4 isotype. There are, however, several modified IgG4-related molecules, including the "IgG4-ala-ala" (see Section 4.4 on Fc-silenced antibodies), that are in development. Even given the caveats mentioned above, Salfeld (2007) suggested that no preclinical or clinical data had yet been obtained that suggested that IgG4 antibodies posed any unique safety issues. Nevertheless, additional analytical assays and follow-up protocols have been proposed for IgG4 and other "novel" antibody isotypes or forms (Salfeld, 2007; Swann *et al.*, 2008).

5.4. Modified IgG Antibodies as Therapeutics

5.4.1 *Current marketed mAbs and clinical candidates with modified Fc*

Four marketed therapeutic antibodies and Fc fusion proteins possess engineered Fc regions for the purpose of modulating Fc functionality, including abatacept (Orencia®), ecolizumab (Soliris®), romiplostim (Nplate®), and catumaxomab (Removab®). Additionally, there are at least three molecules in Phase III and another seven in Phase I and II clinical trials that have modified Fc functionality (Table 5.4). Romiplostim (NPlate®), a recently approved peptide-Fc fusion protein known as a peptibody, and the phase III anti-CD3 clinical candidate, otelixizumab, are examples of non-glycosylated Fc based therapeutics. The recently marketed anti-complement C5 mAb, ecolizumab, and CTLA4-Fc fusion protein, abatacept, as well as the Phase III anti-CD3 antibody, teplizumab, are examples of antibody-based therapeutics with modified Fc sequences designed to reduce or eliminate Fc effector functionality. There are at least five antibody candidates currently in Phase I or II clinical trials that have modified Fcs to increase Fc immune functionality including: two non- or low-fucosylated antibodies, the anti-CD20 antibody GA101 (Bello and Sotomayor, 2007), and the anti-IL-5 receptor antibody MEDI-563; the Fc sequence-modified anti-CD20 antibodies PRO-131921 and LY2469298 (AME-133v) (Bello and Sotomayor, 2007); and the anti-CD19 antibody Xmab-5574 (Zalevsky *et al.*, 2009). Finally, MedImmune has initiated clinical trials with MEDI-557, the Fc-modified "YTE" anti-respiratory syncytial virus (RSV) third generation antibody that has altered FcRn binding for extension of half-life (Dall'Acqua *et al.*, 2006). While modification of the Fc sequence can potentially yield neo-epitopes that could cause increased immunogenicity, this has yet to be a significant issue. Eculizumab, for example, which is a hybrid between IgG2 and IgG4 Fc domains, has very low immunogenicity (Alexion Pharmaceuticals, 2007). Nevertheless, additional *in vitro* assays and follow-up procedures have been proposed for any type of "novel" antibody isotype or form (Swann *et al.*, 2008).

5.4.2 Fc engineering for increased effector function

In some cases, as with the desired functionality of ADCC in treating lymphomas, the goal is to increase the interaction between the antibody and one or more receptors on specific or multiple cell types. Several groups have invested heavily in making and characterizing Fc mutations that modulate various Fc functionalities such as ADCC, ADCP, CDC, and half-life (Table 5.4). It is expected that many of these mutant Fc sequences (Table 5.5) will be applied to therapeutic candidates over the next several years, and that additional Fc-modified antibodies currently in the discovery stage or in preclinical development will soon enter clinical development.

Table 5.4 Examples of Fc-engineered mAb and Fc fusion proteins

Molecule	Company	Stage	Target	Indication	Fc Modification
Orencia® (Abatacept)	BMS	Approved in US 12/23/05	CD80/CD86	RA	CTLA4-Fc fusion protein – modified Fc to eliminate FcγR binding
Soliris® (Eculizumab)	Alexion Pharma	Approved in US 03/16/07	Complement C5	PNH (reduce hemolysis)	Humanized IgG2/4 hybrid – modified Fc to eliminate FcγR binding
Nplate® (Romiplostim)	Amgen	Approved in US 08/22/08	TPO-R	Thrombo-cytopenia	Fc-peptide fusion ("peptibody") with aglycosylated Fc, reducing FcγR binding
Removab® (Catumaxomab)	Fresenius/ Trion	Approved in EU 4/23/09	EpCAM and CD3	Malignant ascites; cancer	Rat IgG2b-mouse IgG2a hybrid
Teplizumab; HOKT3g1(Ala-Ala)	Macrogenics/ Eli Lilly	Phase III	CD3	Type 1 Diabetes	Humanized modified Fc – IgG1-ala-ala to reduce FcγR binding
Otelixizumab (ChAglyCD3; TRX4)	GSK/Tolerx	Phase III	CD3	Type 1 Diabetes	Aglycosylated, humanized IgG1 Fc to eliminate FcγR binding
Belatacept (LEA-29Y)	BMS	Phase II /III	CD80/CD86	Renal transplantation	CTLA4-Fc fusion protein-modified Fc to eliminate FcγR binding (affinity matured version of abatacept)
MEDI-563 (formerly BIW-8405)	Astra-Zeneca	Phase II	IL-5R	SLE	Afucosylated IgG based on BioWa's Potelligent technology to increase ADCC
PRO-131921	Genentech	Phase I/II	CD20	Oncology	IgG with modified Fc sequence for increased ADCC

(*Continued*)

Table 5.4 (*Continued*)

Molecule	Company	Stage	Target	Indication	Fc Modification
AME-133v (LY2469298)	Eli Lilly	Phase I/II	CD20	NHL	IgG with modified Fc sequence for increased ADCC
GA-101	Genentech/ Biogen/IDEC	Phase I	CD20	NHL, CLL	Afucosylated IgG1 generated using Glycart technology to increase ADCC
Ertumaxomab (Rexomun)	Fresenius/ Trion	Phase II/III	Her2/neu and CD3	Malignant ascites; cancer	Rat-murine hybrid Fc yielding strong ADCC; tri-functional bi-specific
MEDI-557	Astra-Zeneca	Phase I	F-protein on RSV	RSV	Fc sequence modified "YTE" mutant – longer half-life mAb
TRU-015	Trubion	Phase I	CD20	Oncology, RA	Fc sequence modified to reduce C1q binding and complement activation, with retention of normal ADCC functionality
R297	INSERM	Phase I	RhD	Hemolytic disease of fetus, newborn (HDFN)	Rat cell-derived IgG1 with low fucose results in high ADCC[a]
Xmab-5574	Xencor	Late preclinical	CD19	NHL	IgG – modified Fc sequence to increase ADCC and ADCP[b,c]

References: **a**, Beliard *et al.*, 2008; **b**, Horton *et al.*, 2008; **c**, Zalevsky *et al.*, 2009

Table 5.5 Examples of Fc sequence engineering for modified effector functionality or half-life*

Function	Company or Source	Mutations or Changes	Reference
Increased ADCC, ADCP	Genentech	IgG1-S298A, E333A, K334A (plus several other single and multiple mutations)	Shields *et al.*, 2001
Increased ADCC, ADCP	Xencor, exemplified by anti-CD19 XmAb5574	IgG1-S239D, I332E	Lazar *et al.*, 2006; Horton *et al.*, 2008; Zalevsky *et al.*, 2009
Increased ADCC, ADCP; with ablated CDC	Xencor	IgG1-S239D, A330L, I332E	Lazar *et al.*, 2006; Richards *et al.*, 2008; Bruckheimer *et al.*, 2009
Increased ADCC, ADCP	Applied Molecular Evolution (AME133)	IgG1-P247I, A339D or Q	Allan *et al.*, 2006
Increased ADCC, ADCP	Applied Molecular Evolution	IgG1-D280H, K290S with or without S298D or V	Watkins and Allan, 2004
Increased ADCC, ADCP	Macrogenics	Mutant 12: IgG1-F243L, R292P, Y300L; Mutant 16: IgG1-F243L, R292P, Y300L, P396L; Mutant 18: IgG1-F243L, R292P, Y300L, V305I, P396L	Stavenhagen *et al.*, 2007
Increased CDC	Kyowa Hakko	IgG1/IgG3 fusion versions 1133 and 113F	Natsumi *et al.*, 2008
Increased binding to FcγRs over wild-type	Dane Wittrup, MIT	IgG1-S298G, T299A, K290E (or N), and/ or K326E	Sazinsky *et al.*, 2008
Decreased ADCC, ADCP	GSK/Tolerx; Otelixizumab (ChAglyCD3; TRX4)	IgG1-N297A	Bolt *et al.*, 1993
Decreased ADCC, ADCP, cytokine storm	OKT3-ala,ala	IgG1-L234A, L235A	Alegre *et al.*, 1994
Decreased ADCC, ADCP, cytokine storm	PDL, Visilizumab (M291-ala,ala)	IgG2-V234A, G237A	Cole *et al.*, 1999
Decreased ADCC, ADCP	Xencor	IgG1-G236R, L328R	Horton *et al.*, 2008

(*Continued*)

Table 5.5 (*Continued*)

Function	Company or Source	Mutations or Changes	Reference
Decreased ADCC, ADCP	Campath variant	IgG4-L235A, G237A, E318A	Hutchins *et al.*, 1995
Decreased ADCC, ADCP	Clenoliximab (modified anti-CD4 mAb)	IgG4-S228P, L235E	Reddy *et al.*, 2000
Decreased ADCC, ADCP, CDC	Alexion (eculizumab IgG2/4)	IgG2 EU seq 118-260; IgG4 EU seq 261-447	Mueller *et al.*, 1997
Decreased ADCC, ADCP, CDC	Merck "IgG2m4"	IgG2; H268Q, V309L, A330S, A331S	Strohl, 2010
Decreased ADCC, ADCP	BMS (abatacept)	IgG1-C220S, C226S, C229S, P238S	Davis *et al.*, 2007
Decreased ADCC, ADCP, CDC	Seattle Genetics (anti-CD70 c1F6v)	IgG1-C226S, C229S, E233P, L234V, L235A	McEarchern *et al.*, 2007
Decreased ADCC, ADCP	Romiplostim	Produced in *Escherichia coli*, no glycosylation	Alexion, 2007 (prescribing information)
Decreased ADCC, ADCP, CDC	Medimmune	IgG1-L234F, L235E, P331S	Oganesyan *et al.*, 2008
Normal ADCC but with reduced CDC	Trubion (TRU-015)	Hinge mutant (possibly IgG1-C226S, P230S)	Hayden-Ledbetter *et al.*, 2009
Selective binding to inhibitory FcγRIIb receptor	Xencor; Anti-CD19	IgG1-S267E, L328F	Chu *et al.*, 2008
Increased half-life	MEDI	IgG1-M252Y, S254T, T256E	Dall'Acqua *et al.*, 2002, 2006
Increased half-life**	PDL	IgG1-T250Q, M428L	Hinton *et al.*, 2006; Datta-Mannan *et al.*, 2007a,b**
Increased half-life	Genentech	IgG1-N434A	Yeung *et al.*, 2009
Increased half-life	Sally Ward	IgG1-H433K, N434Y	Ward, 2006
Increased half-life	Derry Roopenian	IgG1-T307A, E380A, N434A	Petkova *et al.*, 2006
Increased half-life	Xencor Xtend™	IgG1-M428L, N434S	Zalevsky *et al.*, 2009

*All numberings are based on EU numbering system.
**Datta-Mannan *et al.*, 2007a, b were not able to reproduce the non-human primate PK data generated by Hinton *et al.*, 2006, using the variable chains of a different antibody (anti-TNF-α); thus, this set of mutations may not work for all antibodies.

Examples of Fc mutations resulting in increased ADCC, ADCP, and/or CDC are shown in Table 5.5 as well as in Shields *et al.* (2001), Lazar *et al.* (2006), Stavenhagen *et al.* (2007), and Richards *et al.* (2008). Shields *et al.* (2001) characterized several single and multiple mutations of IgG1 Fc that increased or decreased binding to Fcγ receptors. One mutant from that study, known as the "triple alanine mutant" (S298A, E333A, K334A; Table 5.5), has significantly increased binding to FcγRIIIa and ADCC activity. Lazar *et al.* (2006) engineered a double mutant, IgG1- S239D, I332E, that enhanced FcγR binding and increased ADCC and ADCP. Using a humanized anti-CD19 mAb, XmAb5574, Horton *et al.* (2008) measured the increase in binding to both human and mouse receptors conferred by the IgG1-S239D, I332E mutant. They found that the modified Fc antibody was from 8 to ~140-fold higher affinity than the wild-type IgG1, depending on the receptor (Table 5.3). Notably the greatest increase in affinity afforded by the S239D, I332E mutations were on the weaker-binding human FcγRIIIa-F158 (Table 5.3; Horton *et al.*, 2008).

With the addition of an A330L mutation to the IgG1-S239D, I332E eliminated C1q binding, thus giving the option to include or exclude that functionality (Lazar *et al.*, 2006). van der Kolk *et al.* (2001) suggested that CDC could be linked with injection site reactions, so there may be cases in which the A330L or similar mutation might be advantageous. Presta (2008) noted that the increases in binding to FcγRIII due to the IgG1-S239D, I332E, and IgG1-S239D, A330L, I332E mutants ranged from 30–170-fold, were at least in part target-dependent, and the increase in ADCC ranged from 10–100-fold. This is an important observation, as cellular activities conferred by identical Fcs may be very different depending on several other facts, including affinity of the antibody to its target, epitope, target density and internalization rate.

Most N297A-mutant non-glycosylated IgGs have been shown to have substantially decreased binding to FcγRs (e.g. Bolt *et al.*, 1993). Sazinsky *et al.*, (2008), however, demonstrated that a non-glycosylated IgG1-S298G, T299A, retained binding activity for both FcγRI and FcγRIIa, and activated FcγRIIa *in vitro*. Moreover, variants of IgG1-S298G, T299A containing K290E (or N), and/or K326E mutations (Table 5.5) also bound FcγRIIIa better than wild-type IgG1 (Sazinsky *et al.*, 2008). This approach offers a potentially separate path to obtaining antibodies with increased Fc functionality.

5.4.3 *Glyco-engineering for increased effector function*

A typical IgG antibody produced in CHO cells has a relatively predictable glycosylation pattern covalently attached to N297 in the CH2 domain of the Fc (Fig. 5.3; Sethuraman and Stadheim, 2006; Jefferis, 2007; Raju, 2008; Jefferis *et al.*, 2009a, b). The core IgG glycan structure consists of a bi-antennary oligosaccharide containing the structure: $(\text{Man-GlcNAc})_2$-Man-GlcNAc-GlcNAc-N297, where GlcNAc and Man represent *N*-acetyl-glucosamine and mannose, respectively. Fucose is typically attached by an α1, 6-linkage to the first GlcNAc residue (Fig. 5.3). This agalactosyl core structure is termed G0. The oligoglycans of antibody Fc domains typically display broad heterogeneity; galactose residues are often attached to one or both of the G0 terminal GlcNAc residues to generate G1 and G2 glycans, respectively (Fig. 5.3). Additionally, sialic acid residues may cap the galactose residues and a bisecting GlcNAc is also sometimes present (Fig. 5.3). Usually, glycans containing the bisecting GlcNAc are not fucosylated, since the substrate for the fucosyltransferase is a non-bisected glycan (Umana *et al.*, 1999; Shinkawa *et al.*, 2003). A typical CHO cell culture-produced IgG comprises 35% G0, 35% G1 (two isomers), and 16% G2 glycans, with minor structures making up the remainder of the isoforms (Arnold *et al.*, 2006). Usually less than 10% of these glycans are non-fucosylated, and another small portion of these glycans, typically less than 5%, is terminally sialylated (Raju, 2008). These minor forms, however, actually can play a large role in the biology exerted by the IgG, as discussed below.

Umana *et al.* (1999) demonstrated that engineering a cell line with β(1,4)-*N*-acetylglucosaminyl-transferase III, which resulted in bisected, non-fucosylated oligosaccharides possessing terminal galactose residues attached at N297, yielded significantly higher ADCC than the normal cell line. Shinkawa *et al.* (2003) later demonstrated that it was the non-fucosylation rather than the bisected oligosaccharide or the terminal galactose residues that was primarily responsible for the higher ADCC of the antibodies from that cell line. Shields *et al.* (2002) were the first to show definitively that human IgG1s possessing non-fucosylated glycans bind FcγRIIIa more strongly, resulting in higher ADCC. This increased binding to FcγRIIIa results in a substantially higher level of ADCC *in vitro* by all four IgG isotypes. Niwa *et al.* (2005) and Iida *et al.* (2009) demonstrated that non-fucosylated anti-CD20 IgG1 functioned by outcompeting both plasma IgG for FcγRIIIa binding and fucosylated antibodies for binding to the antigen. A non-fucosylated anti-CD20 IgG1 also was recently shown to possess greater neutrophil-mediated phagocytosis than standard IgG via enhanced activity binding to FcγRIIIb, extending the effects known to be mediated by non-fucosylated mAbs (Shibata-Koyama *et al.*, 2009).

Several different routes to obtaining non-fucosylated IgGs have been developed, including deletion of the fucosyltransferase in CHO or other host production cell lines (Yamane-Ohnuki *et al.*, 2004), generation of bisected glycans as noted above (Umana *et al.*, 1999; Shinkawa *et al.*, 2003), glyco-engineering of *Pichia pastoris* to generate human glycan-like non-fucosylated mAbs (Li *et al.*, 2006a) and addition of the inhibitor kifunensine to cell cultures, resulting in non-fucosylated oligomannose-containing IgG antibodies (Zhou *et al.*, 2008).

There are now at least two non- or low-fucosylated IgG1-based antibodies in clinical development, including MEDI-563 (formerly BIW-8405), an anti-IL-5 receptor antibody developed using BioWa's Potelligent™ technology (Shinkawa *et al.*, 2003), and GA-101 (Robak, 2009), a third generation anti-CD20 antibody from Roche/Genentech using the GlycArt technology. In both of these cases, the lack of fucosylation has been demonstrated to increase ADCC *in vitro*. *In vivo* improvement in non-fucosylated mAbs has been demonstrated using "humanized" *SCID* or *NOG* mice into which human PBMCs had been grafted (Niwa *et al.*, 2004; Ito *et al.*, 2009). Alternatively, an ALCL model was used with SCID mice to measure the improved *in vivo* efficacy of a non-fucosylated anti-CD30 mAb over its normal glycan counterpart, after a standard xenograft model had failed to show any difference (Cardarelli *et al.*, 2009). These complex models were thought to be required to measure the improvements in non-fucosylated human IgG1 antibody activities against cancer targets either because the non-fucosylated human IgG1 does not have substantially improved binding to murine Fcγ receptors (Niwa *et al.*, 2004) and/or the cell types and locations are not sufficient to work using subcutaneously placed tumors. Using a model system, however, Scallon *et al.* (2007a) were able to demonstrate a three- to five-fold difference in *in vivo* ADCC using immuno-competent Balb/c mice treated with hamster anti-mouse CD3 mAbs bearing low or high levels of fucosylation.

Within the next few years, we will learn whether the *in vitro* and non-clinical *in vivo* data translate into efficacy, dose-sparing expansion of patient pool (e.g. those patients with lower affinity polymorphisms of FcγRIIa and FcγRIIIa), or other medically-related outcomes in the clinic.

5.4.4 *Fc engineering for silenced effector function*

Approximately two thirds of all marketed and clinical candidate IgGs target cell-surface receptors or proteins (Strohl, 2009a, b). For many of these cellular targets, especially those on immune cells, ADCC, ADCP, CDC, and/or induced apoptosis may potentially pose a safety risk. In these cases, it may be desired to use IgGs that lack Fc immune functionality (Labrijn *et al.*, 2008). Soliris®, Orencia®, and Nplate® are marketed products that already incorporate this principle. An IgG2-IgG4 fusion design was used to largely reduce Fc effector function in Soliris® (eculizumab; Tables 5.4, 5.5), while the hinge region of Orencia® (abatacept), a CTLA4-Fc fusion product that binds CD80 and CD86 on T and B cells, was modified to greatly reduce effector function (Tables 5.4, 5.5; Davis *et al.*, 2007; Rother *et al.*, 2007).

Similarly, several other approaches have been used to generate IgGs with reduced complement activation and Fc-mediated cellular responses (Table 5.5). Muromomab-CD3 (Orthoclone OKT3), an IgG2a murine monoclonal antibody against the ε chain of T-cell receptors, has been used to treat graft vs. host disease (GVHD) since 1986. This antibody has long been shown to have two major liabilities, immunogenicity against the murine antibody and the stimulation of CRS

(Xu *et al.*, 2000). OKT3 was engineered to have a human IgG1 Fc and mutated in the hinge region (L234A, L235A) to reduce FcγR interactions (Alegre *et al.*, 1994; Xu *et al.*, 2000), significantly dampening CRS. Concurrently, another anti-CD3 murine mAb, visilizumab, was humanized and converted to a modified human IgG2 (V234A, G237A) as an approach to dampen CRS (Cole *et al.*, 1999). Perhaps the worst case of CRS occurred with CAMPATH, an anti-CD52 mAb (Wing *et al.*, 1996). As mentioned above, CRS, lymphocyte depletion and injection site reactions all were observed with a normal IgG4 variant of CAMPATH (Isaacs *et al.*, 1996). Thus an IgG4-L235A, G237A, E318A variant of CAMPATH was constructed to dampen Fc functionality and reduce CRS (Hutchins *et al.*, 1995). Other silenced Fc antibodies have also been described, including an IgG4-S228P, L235E (EU numbering, different from authors') anti-CD4 mAb, clenoliximab (Reddy *et al.*, 2000), and an IgG2m4 design (Strohl, 2009b; Strohl, 2010), which was applied to both anti-amyloid beta and anti-IL-13 receptor antibodies. In reviewing the need for therapeutic antibodies employing a silenced Fc, Jefferis (2007) concluded that perhaps a non-glycosylated IgG2 might be the best solution for effector function-ablated IgG antibodies.

5.4.5 *Non-glycosylated IgGs for silenced effector function*

Apart from the FcγR-binding non-glycosylated mutants described above (Sazinsky *et al.*, 2008), non-glycosylated mutant IgGs made by N297A/Q mutations have long been known to have reduced binding to Fcγ receptors and C1q (Tao and Morrison, 1989; Walker *et al.*, 1989; Bolt *et al.*, 1993). Non-glycosylated IgG1 still has some residual minor binding to FcγRI, although this does not appear to result in effector function activity (Davis *et al.*, 2007). Thus, this approach has been used by several groups to generate antibodies or Fc fusion proteins with minimized effector functions (Tao and Morrison, 1989; Walker *et al.*, 1989; Bolt *et al.*, 1993). Non-glycosylated IgGs have decreased thermostability (Ghirlando *et al.*, 1999) and are more susceptible to proteolytic degradation than wild-type IgG, but retain the overall PK characteristics, including FcRn interaction, of their normally glycosylated counterparts (Tao and Morrison, 1989). Two major differences were recently observed between the crystal structure of a mouse non-glycosylated Fc and wild-type glycosylated mouse Fc (Feige *et al.*, 2009). The non-glycosylated form adopted a "superclosed" conformation not conducive to binding to FcγRs, and the structure of the CH2 domain of the Fc was substantially altered. It is known that the C'E loop in CH2 of a non-glycosyl-Fc is destabilized, leading to an overall lower stability of CH2 (Feige *et al.*, 2009).

A few therapeutic antibodies and Fc fusion proteins either currently marketed or in late phase clinical trials have used the non-glycosyl-Fc approach to minimize effector functionality. Romiplostim (Nplate®), which is a recently approved thrombopoietin (TPO) receptor peptibody agonist constructed by fusing a TPO peptide mimetic to the C-terminus of an Fc moiety, has a non-glycosylated Fc moiety because it is produced in *Escherichia coli* (Kumagai *et al.*, 2007). A non-glycosyl-anti-CD3, otelixizumab (also known as ChAglyCD3; TRX4), is in Phase III clinical development by Tolerx and Glaxo Smith-Kline for treatment of Type I diabetes. Otelixizumab has an N297A mutation to make it a non-glycosylated IgG (Bolt *et al.*, 1993). The reasoning behind this design was identical to that for OKT3 and visilizumab described above.

5.4.6 *Sialylated IgG*

Intravenous immunoglobulin (IVIG), which has immunosuppressive activity, has been widely used to treat autoimmune diseases (Nimmerjahn and Ravetch, 2007). Recently, α-2,6-linked terminally sialylated IgGs (Fig. 5.3) were found to be the active fraction responsible for the anti-inflammatory action of high-dose IVIG (Kaneko *et al.*, 2006; Nimmerjahn and Ravetch, 2007; Anthony *et al.*, 2008a). Similarly, sialylated murine IgGs were found not only to have reduced binding to activating FcγRs, but also to bind mouse SIGN-R1, which was shown to upregulate inhibitory FcγRII on mouse macrophages leading to an anti-inflammatory response (Anthony *et al.*, 2008b). This confirmed the link between sialylated IVIG and FcγRII first proposed by Samuelsson *et al.* (2001), who had shown that sialylated IVIG lacked its immunosuppressive effect in either FcγRII knock-out mice or mice treated with an anti-mouse FcγRII-blocking antibody. Sialylated mouse IgG also bound human DC-SIGN, the human counterpart of mouse SIGN-R1 *in vitro* (Anthony *et al.*, 2008b), further suggesting a link with IVIG treatment in humans. In a separate study, three different highly sialylated human IgGs were shown to confer an anti-inflammatory response (Scallon *et al.*, 2007b) via decreased ADCC, by reducing the affinity of the antibodies to both human FcγRIIIa and their cell-surface target, suggesting that the anti-inflammatory properties of mouse sialylated antibodies (Kaneko *et al.*, 2006; Nimmerjahn and Ravetch, 2007; Anthony *et al.*, 2008a, b) may translate to humans.

5.4.7 *Fc engineering for half-life modification*

IgG antibodies bind to the neonatal receptor, FcRn, in a pH-dependent manner, which results in their recycling and long half-life (Dall'Acqua *et al*, 2002, 2006; Petkova *et al.*, 2006). IgG antibodies have been engineered to improve the pH-dependent binding to FcRn to prolong half-life, thereby reducing the frequency of administration (Table 5.5; Hinton *et al.*, 2006; Dall'Acqua *et al.*, 2006; Petkova *et al.*, 2006; Datta-Mannan *et al.*, 2007a, b). The best-known example is the "YTE" mutant (IgG1-M252Y, S254T, T256E) from MedImmune (Astra-Zeneca), which yielded a four-fold longer half-life in non-human primates (NHPs) as compared to the wild-type version of the same antibody (Dall'Acqua *et al.*, 2006). Phase I clinical trials on MEDI-557, a YTE-containing version of Numax (anti-RSV mAb), were initiated in December 2007, and clinical results on this mechanism should be made public soon. If prolonged half-life holds true in human trials of this mutant, with little or no mutant Fc-associated immunogenicity, this type of approach could result in generation of many enhanced half-life antibodies in the future. Correct pH-dependent binding characteristics alone may not be enough, however, to prolong half-life of an IgG. The T250Q, M428L mutant (Hinton *et al.*, 2006), for example, did not appear to prolong the half-life of an anti-TNF-α antibody (Datta-Mannan *et al.*, 2007a, b), even though it did prolong the half-life of an anti-HBV antibody (Hinton *et al.*, 2006), suggesting that some mutants may not have robust activity with all antibody sequences.

5.4.8 *Antibody engineering for size*

It has been demonstrated that the size, along with other factors such as turnover rate, affinity, hydrostatic pressure, and nature of the target, of an antibody-like molecule can alter tissue distribution, in particular with respect to tissue and tumor penetration (Yokota *et al.*, 1992; Thurber and Wittrup, 2008; Zhang *et al.*, 2009). Thus, there has been a widespread effort to devise monoclonal antibody-like molecules that retain the antigen binding and Fc functionalities of natural IgGs, yet possess the ability to penetrate tumors more deeply due to a small size and Stokes/hydrodynamic radius. Small antibody fragments (e.g. FAb, scFv, dAb, diabodies, BiTEs) do not possess an intact Fc region and are below the cut-off for filtration by kidney glomerulae, resulting in very short half-lives of typically less than an hour. Thus, there has been an effort to make small antibody-like molecules that are > 60 kDa and retain Fc functionality.

The most clinically advanced "small antibody" is the anti-TNF-α domain antibody, Arana ART621, which has a single variable domain antibody connected via a modified hinge to each arm of an Fc (Doyle *et al.*, 2007). Arana ART621, which is about half the size of a standard IgG, is currently in Phase II clinical trials for rheumatoid arthritis (Clinical Trial #NCT00854685). It is expected that ART621 will provide greater tissue penetration than a normal IgG. Additionally, Trubion has developed small modular immunopharmaceuticals (SMIPs), which consist of scFvs connected via a modified hinge to each arm of an Fc, with a total mass in the range of 100 kDa, about two thirds the size of a normal mAb. The hinge region of the anti-CD20 SMIP, TRU-015, has been modified to significantly reduce CDC while retaining wild-type ADCC activity (Hayden-Ledbetter *et al.*, 2009). Since CDC has been associated with a possible increase in injection site reactions (van der Kolk *et al.*, 2001), TRU-015 has been designed for size (potential advantage for tissue distribution), retention of desired Fc functionality (ADCC), and safety (lacking CDC).

5.5 NHP, Mouse and Rat IgG Isotypes and Their Use as Surrogate Species

5.5.1 *NHP FcγRs*

One of the key attributes for antibodies as therapeutic drugs is that they bind to their targets with very high specificity. Antibodies can easily be found that bind a target, but not to related proteins that differ by even just a few amino acids in the binding region. This highly desirable property of antibodies can, however, be problematic as well. Non-clinical pharmacology, safety, and reproductive studies all need to be carried out as part of an antibody development program (Foulkes, 2002; Tabrizi and Roskos, 2007; Longstaff *et al.*, 2009). These studies require antibodies that will bind the homologous target in pharmacologically relevant non-human species (Chapman *et al.*, 2007). In the rare, best-case scenario, the desired therapeutic would bind human, cynomologus monkey, mouse, and rat target equally, yielding approximately the same functionality in each case. On the other hand, if rat, mouse and/or NHP homologs of the human target of a therapeutic antibody candidate have substantially different sequences from the human target, the antibody may not cross-react with the same target from these species, as is the case with anti-TNF-α antibodies (Foulkes, 2002). For antibodies made from immunized mice, it is highly likely, in fact, that the antibody would not cross-react with the mouse homolog due to recognition as "self." In these cases, it is necessary to make surrogate antibodies that bind and function in a manner similar to the therapeutic antibody candidate (Wu *et al.*, 2006).

One factor that is often overlooked in making surrogate antibodies is the correct match with the human isotype. If a surrogate needs to be made for NHPs, it is most logical with the information we have today to use the same isotype as used in humans, since the FcγRs (Fig. 5.6) and IgGs (Shearer *et al.*, 1999; Scinicariello *et al.*, 2004) of NHPs are similar to those of humans. A significant caveat to this, however, is that differences in physiology in NHPs vs. humans may result is significantly different reactivity to a monoclonal antibody. For example, the CD28 agonist antibody, TGN1412, did not provoke the same life-threatening cytokine release syndrome in NHPs as it did in humans (Stebbings *et al.*, 2007; Longstaff *et al.*, 2009). Additionally, proteins may function differently or have different expression patterns in NHPs than in humans. As an example, NHPs have a homolog of human FcγRIIIa (CD16a) but not human FcγRIIIb (Rogers *et al.*, 2006). Moreover, while NHPs have the same antibody isotypes as humans, the reactivity of human IgG1 and IgG2 to NHP FcγRIIIa differs from the interaction with human FcγRIIIa (Rogers *et al.*, 2006). Thus, data from NHP studies with normal or modified human IgGs need to be viewed with these potential differences in mind.

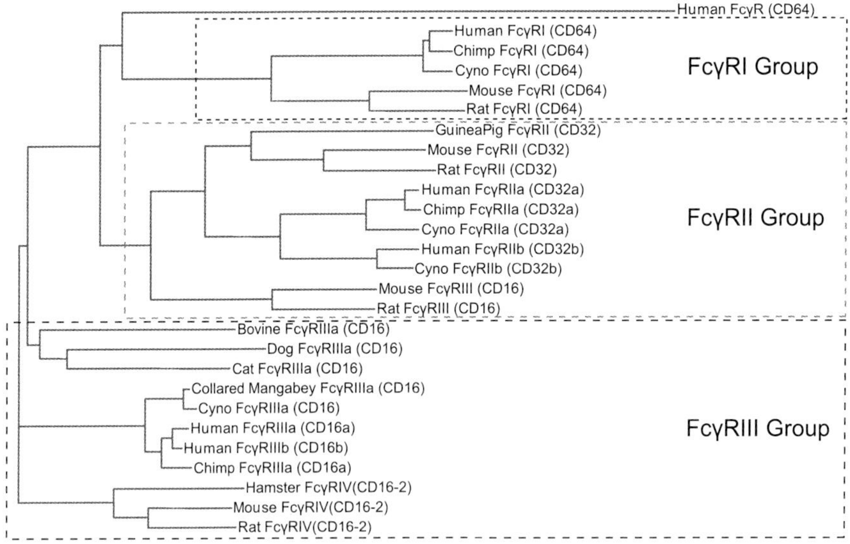

Figure 5.6 Dendrogram comparing the sequences of human vs. other mammalian Fcγ receptors. Human Fcα receptor was used as the outgroup. Based on the dendrogram data and information from the literature (Nimmerjahn *et al.*, 2005), the receptors are grouped into three major groups. This analysis also agrees with the groupings determined by Hirano *et al.* (2007).

5.5.2 *Rodent FcγRs*

The IgG isotypes and FcγR species of rodents are very different from human isotypes and FcγRs, although some broad similarities can be found, as depicted in Figs. 5.5 and 5.6. Mouse and rat immune cells express four different classes of Fcγ receptors, including FcγRI (CD64), FcγRII (CD32), FcγRIII (CD16), and FcγRIV (CD16–2) (Nimmerjahn *et al.*, 2005; Nimmerjahn and Ravetch, 2008). The activating receptors, FcγRI, FcγRIII, and the more recently described FcγRIV, interact with an ITAM-bearing common gamma chain (FcRγ) to make receptor complexes, analogous to those of human FcγRI and FcγRIIIa (Fig. 5.4; Nimmerjahn and Ravetch, 2008). The FcRγ chain with its ITAM sequence is essential for effector cell responses such as antibody-dependent cellular cytotoxicity (ADCC) and ADCP (Nimmerjahn and Ravetch, 2008). FcRγ knock-out mice, which do not express any of the activating receptors, FcγRI, FcγRIII, or FcγRIV, are resistant to autoimmune disorders such as glomerulonephritis and induced conditions such as collagen-induced arthritis (Nimmerjahn and Ravetch, 2008).

Both mice and rats have a high affinity, multi-subunit, activating FcγRI, similar in both sequence and function to that found in humans (Figs. 5.4–5.6). Mouse IgG2a, the only mouse isotype to bind mouse FcγRI, binds with an affinity of about 25 nM (Baudino *et al.*, 2008b), which is in the same range as human IgG1 binding to human FcγRI (Bruhns *et al.*, 2009). This allows the receptor to bind IgG in the monomeric state (Giorgini *et al.*, 2008). Mouse FcγRI is found on macrophages, monocytes, neutrophils, eosinophils and dendritic cells.

Both rats and mice possess an inhibitory FcγRII receptor that is structurally and functionally similar to human FcγRIIb (Figs. 5.4, 5.6; Nimmerjahn and Ravetch, 2008; Baerenwaldt and Nimmerjahn, 2008). Rodent FcγRII is a monomeric receptor in which the ITIM is harbored directly in the cytoplasmic domain of the receptor (Nimmerjahn and Ravetch, 2008). In mice, FcγRII is found on B cells, mast cells, basophils, macrophages, eosinophils, neutrophils, dendritic cells, and Langerhan's cells. FcγRII-knock-out mice are highly susceptible to a variety of autoimmune conditions, e.g. lupus erythematosus (Lin *et al.*, 2006), indicating the balancing role the inhibitory receptor plays in immune signalling homeostasis (Baerenwaldt and Nimmerjahn, 2008).

Neither mice nor rats possess a classic low affinity, activating FcγRIIa (CD32a) receptor, as found in humans and other primates (Figs. 5.4–5.6). The rodent FcγRIII receptor, which is a low affinity, multi-subunit activation receptor, most closely mimics that of human FcγRIIa (Figs. 5.4–5.6). Mouse and rat FcγRIII have been known widely in the literature as "CD16," although it is now clear that mouse and rat FcγRIII are not the functional equivalents of human CD16 (FcγRIIIa) (Nimmerjahn and Ravetch, 2008). Mouse IgG1, IgG2a, and IgG2b all bind this low affinity receptor in the 2–4 μM range, suggesting that activation requires formation of an antigen–antibody complex with the receptors, analogous to that observed with human FcγRIIa and FcγRIIIa (Jakus *et al.*, 2008). Mouse FcγRIII is found on macrophages, neutrophils, eosinophils, platelets, dendritic cells, Langerhan's cells, and NK cells (Nimmerjahn and Ravetch, 2008).

Mice and rats also have a novel receptor, called FcγRIV (CD16–2) (Figs. 5.4, 5.5), which is 63% identical in sequence to human FcγRIIIa (CD16a) (Fig. 5.6; Nimmerjahn *et al.*, 2005). FcγRIV has been termed an "intermediate affinity" receptor, although both mouse IgG2a and IgG2b isotypes bind FcγRIV in the 50–60 nM range, which is not so different from their binding to the mouse high affinity receptor FcγRI (25 nM range) (Fig. 5.5). Bruhns *et al.* (2009) defined a high affinity Fc receptor as having a K_A of at least 9×10^6/M (K_D ~ 110 nM) similar to the traditionally defined 100 nM cut-off. Additionally, Mancardi *et al.* (2008) recently demonstrated that FcγRIV could bind monomeric mouse IgG2a and IgG2b, thereby functioning as a high affinity receptor for those antibodies.

Mouse IgE also was recently found to bind mouse FcγRIV in the ~ 2–7 μM range (Mancardi *et al.*, 2008). Immune complexes of IgE bound to antigen were capable of displacing receptor-bound IgG on macrophages and inducing cytokine secretion (Hirano *et al.*, 2007; Mancardi *et al.*, 2008). Mancardi *et al.* (2008) further showed that human FcγRIIIa, the closest human receptor to mouse FcγRIV (Figs. 5.5, 5.6; Hirano *et al.*, 2007), does not bind human IgE, further demonstrating differences between human FcγRIIIa and mouse FcγRIV. There are several similarities between mouse FcγRIV and human FcεRI, including quaternary structural similarities and similar tissue distribution; therefore, it has been proposed that mouse FcγRIV is a low affinity receptor for IgE in mice (Mancardi *et al.*, 2008). This further demonstrates the uniqueness of rodent FcγRIV, which appears to be different from any other FcγR previously described (Hirano *et al.*, 2007; Mancardi *et al.*, 2008).

5.5.3 *Antibody-mediated cell killing in mice*

Steplewski *et al.* (1988) demonstrated early on that chimeric mouse–human antibodies containing either the human IgG1 or human IgG4 Fc were capable of interacting with mouse effector cells such as granulocytes and macrophages to effect cell lysis *in vitro* and inhibit tumor growth *in vivo* using human SW948 colon carcinoma cells in a xenograft nude mouse model. Interestingly, neither chimeric IgG2 nor chimeric IgG3 antibodies mediated cell lysis or inhibited tumor growth in those assays (Steplewski *et al.*, 1988). Oflazoglu *et al.* (2007) demonstrated that successful treatment of tumor xenografts in C.B-17 SCID mice using the human anti-CD30 IgG1, SGN-30, was not dependent on the presence of mouse NK cells. Upon treatment with SGN-30, survival was the same whether NK cells had been pre-depleted or not, whereas pre-depletion of macrophages prior to treatment led to significantly greater death (Oflazoglu *et al.*, 2007). In a follow-up study, Oflazoglu *et al.* (2009) showed that the cell-killing activity by a humanized IgG1 anti-CD40 mAb, SGN-40, in CB-17/lcr SCID mice was due to macrophages and not NK cells or neutrophils. Thus, human IgG1 antibodies kill tumors in mice via FcγR-mediated ADCP. Human IgG1 binds mouse FcγRI (K_D, 110 nM), FcγRII (K_D, 1.1 μM), and FcγRIV (K_D, 30 nM), but not mouse FcγRIII (Table 5.3; Nimmerjahn *et al.*, 2005). Since mouse FcγRIII is the only known Fcγ receptor on mouse NK cells (Nimmerjahn and Ravetch, 2006; Inoue *et al.*, 2007; Nimmerjahn and Ravetch, 2008), and depletion of NK cells does not significantly alter cell-mediated cytotoxicity in mice (Oflazoglu *et al.*, 2007, 2009), the biology of mouse effector cells is different than in humans (Nimmerjahn and Ravetch, 2008). Based on receptor binding and *in vivo* activites, mouse IgG2a and IgG2b isotypes interacting with FcγRIII and FcγRIV demonstrate the most potent combinations for immune function (Giorgini *et al.*, 2008). Macrophages and dendritic cells, armed with the full complement of mouse Fcγ receptors (Fig. 5.5), appear to be significant cell types in mice carrying out these immune functions (Inoue *et al.*, 2007; Minard-Colin *et al.*, 2008). Additionally, Jakus *et al.* (2008) recently demonstrated that antibody–antigen complexes involving both mouse FcγRIII and

FcγRIV could be formed to stimulate mouse neutrophils much in the same way that human neutrophils are activated by antibody–antigen complexes involving both human FcγRIIa and FcγRIIIb. Nimmerjahn and Ravetch (2008) recently concluded that, given current information, it appears that the IgG-dependent, cell-mediated activities in mice are mediated by granulocytes, monocytes or macrophages, not NK cells, primarily via the activity of FcγRIV.

5.5.4 *Rodent IgG isotypes*

Mice and rats both have IgA, IgD, IgE, IgG, and IgM classes of antibodies, similar to the repertoire in humans. They differ, however, in their IgG isotypes and the receptors the isotypes bind (Clark, 1997). Mice possess IgG isotype subclasses of IgG1, IgG2a, IgG2b, and IgG3; additionally, there are two versions of the IgG2a, known as IgG2a and IgG2a/c. Rats possess IgG1, IgG2a, IgG2b, and IgG2c isotypes. It is important to note that the numbering of IgG subtypes in rodents is completely unrelated to that with human IgGs. Thus, unlike human IgG1, mouse IgG1 is typically associated with Th2 responses, whereas a strong IgG2a response, sometimes including IgG2b and IgG3, is considered to be associated with Th1 (Slack, 1987).

The functions of mouse IgG1 antibodies are most closely related to those of human IgG4, which is generated as part of a Th2 response. In fact, Balb/c mice, which are predominantly Th2 driven, primarily produce a mouse IgG1 response to antigen, whereas Th1-driven C57Bl/6 mice generate predominantly an IgG2a response to antigens (Slack, 1987). Also, similar to human IgG4, mouse and rat IgG1 isotypes do not appear to bind C1q in a physiologically significant manner, and therefore do not activate the complement pathway via the classical route. Mouse and rat IgG1 antibodies, however, do not appear to carry out FAb arm exchange as shown for human IgG4 antibodies.

The most significant difference between human IgG4 and mouse IgG1 is that the mouse IgG1 does not bind the high affinity receptor FcγRI (Baudino *et al.*, 2008b). Mouse IgG1 mediates immune responses including phagocytosis *in vivo* through its interaction with FcγRIII (Hazenbos *et al.*, 1998; Hamaguchi *et al.*, 2006). As compared with the other mouse IgG isotypes, mouse IgG1 has a three amino acid deletion (N233, L234, E235) that has been shown to contribute to lack of FcγRIV and C1q binding (Baudino *et al.*, 2008b). Rat IgG1 and IgG2a share a deletion in this same relative position (Clark, 1997) and appear to possess activities most similar to mouse IgG1 (Brüggemann *et al.*, 1989). Furthermore, the sequence of rat IgG2a is most closely aligned with mouse IgG1 (Vernersson *et al.*, 2004). Finally, rat IgG2a has also been shown to be a poor inducer of ADCC (Clark, 1997).

It has long been recognized that mouse IgG2a is closest in activity to human IgG1 and, as mentioned above, Th1-driven C57Bl/6 mice generate an IgG2a response to most antigens (Slack, 1987). IgG2a binds mouse FcγRI and FcγRIV in the 25–50 nM range, FcγRIII and FcγRII in the 2 μM range, and C1q to activate the complement cascade via the classical pathway. The residues LLGGP at positions 234–238 unique to mouse IgG2a have been shown to be associated with the ability to bind the mouse FcγRI high affinity receptor (Baudino *et al.*, 2008b). Of significance is that the binding of mouse IgG2a to the higher affinity activating receptors is 50–100-fold stronger than the affinity to the inhibitor receptor FcγRII (Fig. 5.5). Induced anemia in mice using an anti-red blood cell mAb was driven by low dose anti-RBC IgG2a through its activity on the low affinity receptor, FcγRIII; more severe disease induced at high doses also required activities of the high affinity receptors FcγRI and FcγRIV (Baudino *et al.*, 2008a).

The rat IgG2b isotype most closely parallels the mouse IgG2a, both in activity (Brüggemann *et al.*, 1989; Clark, 1997; Fig. 5.4) and sequence similarity (Vernersson *et al.*, 2004). Both mouse IgG2a and rat IgG2b bind FcγRI (Clark, 1997), rat IgG2b has been shown to be a strong mediator of ADCC (Clark, 1997), and rat IgG2b is a strong complement activator using the classical pathway (Bindon *et al.*, 1988). Notably, one marketed product, Removab® (catumaxomab) takes advantage of a mouse IgG2a/rat IgG2b hybrid, which yields strong effector function activity contributed by both isotypes (Zeidler *et al.*, 1999). This "trifunctional" antibody, which is an anti-human CD3, anti-human EpCAM bispecific antibody with strong T-cell- and ADCC-driven effector functions, was isolated and is produced commercially by a quadroma cell line.

Similarly to mouse IgG2a, mouse IgG2b binds mouse FcγRIV in the 60 nM range, FcγRIII in the 2 μM range, and C1q to activate the complement cascade via the classical pathway (Table 5.3, Fig. 5.5). Some data support the conclusion that IgG2b has stronger complement activation activity than IgG2a, possibly through its interaction with C3 and C4 rather than C1q, which both isotypes bind well (Giorgini *et al.*, 2008). Different from mouse IgG2a, mouse IgG2b binds FcγRII in the 500 nM range and does not bind the high affinity receptor, FcγRI (Baudino *et al.*, 2008a). The ratio of binding to the only activation receptor it binds, FcγRIV, to the inhibitory receptor, FcγRII, is four-fold. Thus, based on binding characteristics, it would be expected that IgG2b would play an intermediate role in response to antigen stimulation, i.e. a milder type of Th1-type response (as compared with mouse IgG2a). Induced anemia in mice using an anti-red blood cell IgG2b mAb at both low and high doses required the interaction with the low affinity FcγRIII and the high affinity FcγRIV (Baudino *et al.*, 2008a). In several other models, however, *in vivo* mouse IgG2b activity appeared to be dependent on FcγRIV exclusively (Nimmerjahn and Ravetch, 2008).

Mouse IgG3 is the predominant isotype found to respond to T-independent antigens such as polysaccharides, similarly to human IgG2 (Díaz de Ståhl *et al.*, 2003). Consistent with this observation, IgG3-deficient mice are more susceptible to pneumococcal infections than wild-type mice (McLay *et al.*, 2002). The rat

IgG2c isotype is the most similar in sequence to mouse IgG3 and human IgG2, and also appears to function as the primary antibody response to T-cell-independent antigens such as carbohydrates (der Balian *et al.*, 1980; Yoo *et al.*, 2003). Mouse IgG3 antibodies are more potent activators than human IgG2 of both the classical and alternative complement pathways, which appears to be their primary mechanism of antigen neutralization and activation of the immune response system (Ding *et al.*, 2008). Rat IgG2c also binds C1q and activates complement (Bindon *et al.*, 1988). Thus, the primary difference between human IgG2 and mouse IgG3/rat IgG2c is that the human IgG2 is relatively poor at complement activation, and when it does so, it is more efficient via the alternative pathway (Seino *et al.*, 1993). As mentioned previously, human IgG2 can form covalent dimers as a potential mechanism to increase its avidity on repeat structure-type antigens (Yoo *et al.*, 2003). Similarly, both mouse IgG3 and rat IgG2c are known to form non-covalent self-aggregates, which are thought to help them in binding to polysaccharide antigens (Greenspan and Cooper, 1992; Clark, 1997; Yoo *et al.*, 2003). In the case of mouse IgG3, this effect is known to be an Fc-mediated, cooperative mechanism (Greenspan and Cooper, 1992).

5.6 Summary and Future Considerations

IgG isotype, Fc functionality and engineering possibilities are not the entire picture, as we are beginning to learn more fully. Antibodies of an identical isotype (e.g. IgG1) that are capable of binding FcγRs and C1q may also function in manners not explained by simple binding data, as demonstrated by the complexities of anti-CD20 and anti-TNF-α antibody biology. There are at least 11 known anti-CD20 mAbs in development for oncology and/or rheumatoid arthritis indications, which generally fall into either Type I and Type II anti-CD20 mAbs (Bello and Sotomayor, 2007; Glennie *et al.*, 2007; Beers *et al.*, 2008; Strohl, 2009a). Type I antibodies, rituximab, ofatumumab, ocrelizumab, and veltuzumab, are thought to function primarily by ADCC and CDC. These antibodies promote translocation into lipid rafts, but likely do not induce apoptosis. Type II anti-CD20 antibodies (e.g. GA-101) are thought to function via ADCC and apoptosis; however, these type II anti-CD20 antibodies apparently do not promote translocation into lipid rafts or use CDC to kill B cells (Bello and Sotomayor, 2007; Glennie *et al.*, 2007; Beers *et al.*, 2008).

Anti-TNF-α antibodies also have demonstrated a variety of cell-based effects through their activity on cell-bound TNF-α. The anti-TNF-α mAbs, infliximab, and adulimumab both bind C1q (Scallon *et al.*, 1995; Arora *et al.*, 2009; Mitoma *et al.*, 2008) and mediate CDC using recombinant cells expressing membrane bound TNF-α (Scallon *et al.*, 1995; Arora *et al.*, 2009; Mitoma *et al.*, 2008; Kaymakcalan *et al.*, 2009), whereas etanercept was marginal in both assays. Infliximab, adalimumab, and etanercept bind poorly to cells containing FcγRI,

FcγRIIa, or FcγRIIIa, and do not mediate ADCC in the absence soluble TNF-α. Addition of soluble TNF-α, however, apparently promotes cross-linking, leading to better binding of infliximab and adalimumab to the cells, and significantly increased ADCC activity (Arora *et al.*, 2009). Using recombinant cell lines expressing membrane-bound TNF-α, Mitoma *et al.* (2008) found that infliximab and adalimumab induced apoptosis, IL-10 production and G0/G1 cell cycle arrest, whereas etanercept did not. The anti-TNF-α antibodies were capable of generating an "outside-to-inside signaling cascade" through membrane-bound TNF-α, resulting in cell death. These combined mechanisms for both anti-CD20 and anti-TNF-α biologics reflect the complexity of antibody-target-effector interactions.

Thus, besides FcγR- and complement-mediated activities, it appears that affinity to the target, epitope, ability to cross-link, induction of receptor migration, outside-to-inside signaling, and perhaps even three-dimensional spatial considerations may play a role in mechanism of action. Additionally, isotype-independent factors, such as affinity, can have significant impacts on the activity (e.g. ADCC, internalization) mediated by the antibody (Tang *et al.*, 2007). Thus, even with a single isotype such as IgG1, significant differences in biology of different antibodies have now become apparent, demonstrating the need for greater depth in studies of antibody mechanism-of-action, no matter which isotype is chosen. Therefore, not only is isotype selection, the thesis of this chapter, critical to generating "fit-for-purpose" biologics, but so too are the nature of the target itself, the epitope on that target to which the antibody binds, the affinity of the antibody to that target and the level of cross-linking afforded by target density, local concentration of the target, target valency (e.g. dimers, trimers, etc.), and antibody behavior.

In summary, antibody class, isotype, and activity selection are critical to help tune a therapeutic antibody to the desired activity. While the first antibodies to be marketed were largely based on a standard IgG1 platform, the concept of engineering antibodies to go beyond what nature gives us (i.e. past what the natural IgG isotypes can give), has been around since the early 1980s (c.f., Neuberger *et al.*, 1984), so it is nothing new to speak of "fit-for-purpose" biologics. Many recent reviews have highlighted the need for making antibodies that are more tuned to the biology required of them; a few are listed here for reference (Chowdhury and Wu, 2005; Carter, 2006; Presta, 2006, 2008; Jefferis, 2007; Dimitrov and Marks, 2008). Moreover, several modified mAbs or Fc fusion proteins are already on the market or in the clinic, demonstrating the commitment the biopharmaceutical industry has made to the principle of generating "fit-for-purpose" biologics.

The next phase in understanding the activities of antibodies will likely include the push for a prospective deeper understanding of the interplay of antibody-mediated activities (e.g. CDC, ADCC, induction of apoptosis, cytokine secretion, internalization) and fate of the administered antibody (e.g. biodistribution, metabolism, elimination) as part of the selection of the ultimate candidate or antibody-like format for development. Additionally, development of better "humanized" mouse models bearing human effectors cells (Niwa *et al.*, 2004; Ito *et al.*, 2009) will help us to better understand the Fc functionality of normal and modified human IgG antibodies in disease models.

Acknowledgments

The author thanks Bernie Scallon for many insightful comments on this work, Jin Lu for assisting with the generation of Fig. 5.6, Jill Giles-Komar concerning the early history of IgM therapeutic candidates, Tom Nesspor concerning non-glycosylated IgGs, and Bernie Scallon and Omid Vafa for many discussions on therapeutic IgGs and their interactions with Fc receptors.

References

Aalberse R.C., Schuurman J. (2002). IgG4, breaking all the rules. *Immunology* **105**: 9–19.

Aalberse R.C., Stapel S.O., Schuurman J., Rispens T. (2009). Immunoglobulin IgG4: an odd antibody. *Clin Exp Allergy* **39**: 469–77.

Alegre M.L., Peterson L.J., Xu D., Sattar H.A., Jeyarajah D.R., Kowalkowski K., Thistlewaite J.R., Zivin R.A., Jolliffe L., Bluestone J.A. (1994). A non-activating "humanized" anti-CD3 monoclonal antibody retains innunosuppressive properties *in vivo*. *Transplantation* **57**: 1537–43.

Alexion Pharmaceuticals. (2007). Soliris® prescribing information. *http://www.accessdata.fda.gov/drugsatfda_docs/label/2007/125166lbl.pdf*

Allan B., Jiang W., Tang Y., Watkins J.D. (2006). Variant Fc regions. *WO 2006/020114* Publ. 2/23/2006.

Angal S., King D.J., Bodmer M.W., Turner A., Lawson A.D., Roberts G., Pedley B., Adair J.R. (1993). A single amino acid substitution abolishes the heterogeneity of chimeric mouse/human (IgG4) antibody. *Mol Immunol* **30**: 105–8.

Anthony R.M., Nimmerjahn F., Ashline D.J., Reinhold V.N., Paulson J.C., Ravetch J.V. (2008a). Recapitulation of IVIG anti-Inflammatory activity with a recombinant IgG Fc. *Science* **320**: 373–6.

Anthony R.M., Wermeling F., Karlsson M.C.I., Ravetch J.V. (2008b). Identification of a receptor required for the anti-inflammatory activity of IVIG. *Proc Natl Acad Sci USA* **105**: 19571–8.

Arnold J.N., Wormald M.R., Suter D.M., Radcliffe C.M., Harvey D.J., Dwek R.A., Rudd P.M., Sim R.B. (2005). Human serum IgM glycosylation. Identification of glycoforms that can bind to mannan-binding lectin. *J Biol Chem* **280**: 29080–7.

Arnold J.N., Dwek R.A., Rudd P.M., Sim R.B. (2006). Mannan binding lectin and its interaction with immunoglobulins in health and in disease. *Immunol Lett* **106**: 103–10.

Arora T., Padaki R., Liu L., Hamburger A.E., Ellison A.R., Stevens S.R., Louie J.S., Kohno T. (2009). Differences in binding and effector functions between classes of TNF antagonists. *Cytokine* **45**: 124–31.

Azuma Y., Ishikawa Y., Kawai S., Tsunenari T., Tsunoda H., Igawa T., Iida S.-i., Nanami M., Suzuki M., Irie R.F., Tsuchiya M., Yamada-Okabe H. (2007). Recombinant human hexamer-dominant IgM monoclonal antibody to ganglioside GM3 for treatment of melanoma. *Clin Cancer Res* **13**: 2745–50.

Baerenwaldt A., Nimmerjahn F. (2008). FcγRIIB — regulating the balance between protective and autoreactive immune responses. *Immunol Cell Biol* **86**: 482–4.

Baudino L., Nimmerjahn F., Azeredo da Silveira S., Martinez-Soria E., Saito T., Carroll M., Ravetch J.V., Verbeek J.S., Izui S. (2008a). Differential contribution of three activating IgG Fc receptors (FcγRI, FcγRIII, and FcγRIV) to IgG2a- and IgG2b-induced autoimmune hemolytic anemia in mice. *J Immunol* **180**: 1948–53.

Baudino L., Nimmerjahn F., Shinohara Y., Furukawa J.-I., Petry F., Verbeek J.S., Nishimura S.-I., Ravetch J.V., Izui S. (2008b). Impact of a three amino acid deletion in the CH2 domain of murine IgG1 on Fc-associated effector functions. *J Immunol* **181**: 4107–12.

Beers S.A., Chan C.H., James S., French R.R., Attfield K.E., Brennan C.M., Ahuja A., Shlomchik M.J., Cragg M.S., Glennie M.J. (2008). Type II (tositumomab) anti-CD20 monoclonal antibody out performs Type I (rituximab-like) reagents in B cell depletion reguardless of complement activation. *Blood* **112**: 4170–7.

Beliard R., Waegemans T., Notelet D., Massad L., Dhainaut F., de Romeuf C., Guemas E., Haazen W., Bourel D., Teillaud J.-L., Prost J.-F. (2008). A human anti-D monoclonal antibody selected for enhanced FcγRIII engagement clears RhD+ autologous red cells in human volunteers as efficiently as polyclonal anti-D antibodies. *Brit J Haematol* **141**: 109–19.

Bello C., Sotomayor E.M. (2007). Monoclonal antibodies for B cell lymphomas: rituximab and beyond. *Hematol Amer Soc Hematol Educ Program* **2007**: 233–42.

Bendele A., Seely J., Richey C., Sennello G., Shopp G. (1998). Short Communication: Renal tubular vacuolation in animals treated with polyethylene-glycol-conjugated proteins. *Toxicol Sci* **42**: 152–7.

Beutner U., Lorenz U., Illerti B., Rotti L., Timmermann W., Vollmers H.P., Muller-Hermelink H.K., Thiede A., Ulrichs K. (2008). Neoadjuvant therapy of gastric cancer with the human monoclonal IgM antibody SC-1: Impact on the immune system. *Oncol Rep* **19**: 761–9.

Beyer T., Lohse S., Berger S., Peipp M., Valerius T., Dechant M. (2009). Serum-free production and purification of chimeric IgA antibodies. *J Immunol Meth* **346**: 26–37.

Bibeau F., Lopez-Crapez E., Di Fiore F., Thezenas S., Ychou M., Blanchard F., Lamy A., Penault-Llorca F., Frebourg T., Michel P., Sabourin J.-C., Boissiere-Michot F. (2009). Impact of FcγRIIa-FcγRIIIa polymorphisms and KRAS mutations on the clinical outcome of patients with metastatic colorectal cancer treated with cetuximab plus irinotecan. *J Clin Oncol* **27**: 1122–9.

Bindon C.I., Hale G., Brüggemann M., Waldman H. (1988). Human monoclonal IgG isotypes differ in complement activating function at the level of C4 as well as C1q. *J Exp Med* **168**: 127–42.

Bindon C.I., Hale G., Waldmann H. (1990). Complement activation by immunoglobulin does not depend solely on C1q binding. *Eur J Immunol* **20**: 277–81.

Boel E., Verlaan S., Poppelier M.J., Nomdo J.G., Westerdaal A.C., Van Strijp J.A.G., Logtenberg T. (2000). Functional human monoclonal antibodies of all isotypes constructed from phage display library-derived single-chain Fv antibody fragments. *J Immunol Methods* **239**: 153–66.

Bolt S., Routledge E., Lloyd I., Chatenoud L., Pope H., Gorman S.D., Clark M., Waldmann H. (1993). The generation of a humanized, non-mitogenic CD3 monoclonal antibody which retains *in vitro* immunosuppressive properties. *Eur J Immunol* **23**: 403–11.

Bonner A., Almogren A., Furtado P.B., Kerr M.A., Perkins S.J. (2009). The nonplanar secretory IgA2 and near planar secretory IgA1 solution structures rationalize their different mucosal immune responses. *J Biol Chem* **284**: 5077–87.

Bournazos S., Woof J.M., Hart S.P., Dransfield I. (2009). Functional and clinical consequences of Fc receptor polymorphic and copy number variants. *Clin Exp Immunol* **157**: 244–54.

Braathen R., Sørensen V., Brandtzaeg P., Sandlie I., Johansen F.-E. (2002). The carboxyl-terminal domains of IgA and IgM direct isotype-specific polymerization and interaction with the polymeric immunoglobulin receptor. *J Biol Chem* **277**: 42755–62.

Brandtzaeg P., Prydz H. (1984). Direct evidence for an integrated function of J chain and secretory component in epithelial transport of immunoglobulins. *Nature* **311**: 71–3.

Bredius R.G.M., Fijen C.A.P., de Haas M., Kuijper E.J., Weening R.S., van de Winkel J.G.J., Out T.A. (1994). Role of neutrophil FcγRIIa (CD32) and FcγRIIIb (CD16) polymorphic forms in phagocytosis of human IgG1- and IgG3-opsonized bacteria and erythrocytes. *Immunology* **83**: 624–30.

Bruckheimer E.M., Fazenbaker C.A., Gallagher S., Mulgrew K., Furmann S., Coffman K.T., Walsh W., Ready S., Cook K., Damschroder M., Kinch M., Kiener P.A., Woods R., Gao C., Dall'Acqua W., Wu H., Coats S. (2009). Antibody-dependent cell-mediated cytotoxicity effector-enhanced EphA2 agonist monoclonal antibody demonstrates potent activity against human tumors. *Neoplasia* **11**: 509–17.

Brüggemann M., Williams G.T., Bindon C.I., Clark M.R., Walker M.R., Jefferis R., Waldmann H., Neuberger M.S. (1987). Comparison of the effector functions of human immunoglobulins using a matched set of chimeric antibodies. *J Exptl Med* **66**: 1351–61.

Brüggemann M., Teale C., Clark M., Bindon C., Waldmann H. (1989). A matched set of rat/mouse chimeric antibodies. Identification and biological properties of rat H chain constant regions μ, γ1, γ2a, γ2b, γ2c, ε, and α. *J Immunol* **142**: 3145–50.

Bruhns P., Iannascoli B., England P., Mancardi D.A., Fernandez N., Jorieux S., Daeron M. (2009). Specificity and affinity of human Fcγ receptors and their polymorphic variants for human IgG subclasses. *Blood* **113**: 3716–25.

Cardarelli P.M., Moldovan-Loomis M.-C., Preston B., Black A., Passmore D., Chen T.-H., Chen S., Liu J., Kuhne M.R., Srinivasan M., Assad A., Witte A., Graziano R.F., King D.J. (2009). *In vitro* and *in vivo* characterization of MDX-1401 for therapy of malignant lymphoma. *Clin Cancer Res* **15**: 3376–83.

Carter P.J. (2006). Potent antibody therapeutics by design. *Nature Rev Immunol* **6**: 343–57.

Cartron G., Dacheux L., Salles G., Solal-Celigny P., Bordos P., Colombat P., Watier H. (2002). Therapeutic activity of humanized anti-CD20 monoclonal antibody and polymorphism in igG Fc receptor FcgammaRIIIa gene. *Blood* **99**: 754–8.

Cassard L., Cohen-Solal J.F.G., Fournier E.M., Camilleri-Broët S., Spatz A., Chouaïb S., Badoual C., Varin A., Fisson S., Duvillard P., Boix C., Loncar S.M., Sastre-Garau X., Houghton A.N., Avril M.-F., Gresser I., Fridman W.H., Sautès-Fridman C. (2008). Selective expression of inhibitory Fcγ receptor by metastatic melanoma impairs tumor susceptibility to IgG-dependent cellular response. *Internat J Cancer* **123**: 2832–9.

Cerutti A., Rescigno M. (2008). The biology of intestinal immunoglobulin A responses. *Immunity* **28**: 740–50.

Chan L.A., Lyczak J.B., Zhang K., Morrison S.L., Saxon A. (2000). The novel human IgE epsilon heavy chain, epsilon tailpiece, is present in plasma as part of a covalent complex. *Mol Immunol* **37**: 241–52.

Chang T.W., Shiung Y.Y. (2006). Anti-IgE as a mast cell-stabilizing therapeutic agent. *J Allergy Clin Immunol* **117**: 1203–12.

Chapman K., Pullen N., Graham M., Ragan I. (2007). Preclinical safety testing of monoclonal antibodies: the significance of species relevance. *Nature Rev Drug Discovery* **6**: 120–6.

Chaturvedi A., Siddiqui Z., Bayiroglu F., Rao K.V.S. (2002). A GPI-linked isoform of the IgD receptor regulates resting B cell activation. *Nature Immunol* **3**: 951–7.

Cheung N.K., Sowers R., Vickers A.J., Cheung I.Y., Kushner B.H., Gorlick R. (2006). FCGR2A polymorphism is correlated with clinical outcome after immunotherapy of neuroblastoma with anti-GD2 antibody and granulocyte macrophage colony-stimulating factor. *J Clin Oncol* **24**: 2885–90.

Chong K.T., Ho W.F., Koo S.H., Thompson P., Lee E.J.D. (2006). Distribution of the FcγRIIIa 176 F/V polymorphism amongst healthy Chinese, Malays and Asian Indians in Singapore. *Brit J Clin Pharmacol* **63**: 328–32.

Chowdhury P.S., Wu H. (2005). Tailor-made antibody therapeutics. *Methods* **36**: 11–24.

Chu S.Y., Vostiar I., Kaki S., Moore G.L., Lazar G.A., Pong E., Joyce P.F., Szymkowski D.E., Desjarlais J.R. (2008). Inhibition of B cell receptor-mediated activation of primary human B cells by coengagement of CD19 and FcγRIIb with Fc-engineered antibodies. *Molec Immunol* **45**: 3926–33.

Clark M. (1997). Ig effector mechanisms. *Chem Immunol* **65**: 88–110.

Cole M.S., Stellrecht K.E., Shi J.D., Homola M., Hsu D.-H., Anasetti C., Vasquez M., Tso J.Y. (1999). HuM291, a humanized anti-CD3 antibody, is immunosuppressive to T cells while exhibiting reduced mitogenicity *in vitro*. *Transplantation* **68**: 563–71.

Collins C.W.L., Tsui F.W.L., Shulman M.J. (2002). Differential activation of human and guinea pig complement by pentameric and hexameric IgM. *Eur J Immunol* **32**: 1802–10.

Coxon A., Cullere X., Knight S., Sethi S., Wakelin M.W., Stavrakis G., Luscinskas F.W., Mayadas T.N. (2001). FcγRIII mediates neutrophil recruitment to immune complexes: A mechanism for neutrophil accumulation in immune-mediated inflammation. *Immunity* **14**: 693–704.

Dall'Acqua W.F., Woods R.M., Ward E.S., Palaszynski S.R., Patel N.K., Brewah Y.A., Wu H., Kiener P.A., Langermann S. (2002). Increasing the affinity of a human IgG1 for the neonatal Fc receptor: biological consequences. *J Immunol* **169**: 5171–80.

Dall'Acqua W.F., Kiener P.A., Wu H. (2006). Properties of human IgG1 engineered for enhanced binding to the neonatal Fc receptor (FcRn). *J Biol Chem* **281**: 23514–24.

Datta-Mannan A., Wichter D.R., Tang Y., Watkins J., Jiang W., Wroblewski V.J. (2007a). Humanized IgG1 variants with differential binding properties to the neonatal Fc receptor: Relationship to pharmacokinetics in mice and primates. *Drug Metab Disposition* **35**: 86–94.

Datta-Mannan A., Wichter D.R., Tang Y., Watkins J., Wroblewski V.J. (2007b). Monoclonal antibody clearance. Impact of modulating the interaction of IgG with the neonatal Fc receptor. *J Biol Chem* **282**: 1709–17.

Davis P.M., Abraham R., Xu L., Nadler S.G., Suchard S.J. (2007). Abatacept binds to the Fc receptor CD64 but does not mediate complement-dependent cytotoxicity or antibody-dependent cellular cytotoxicity. *J Rheumatol* **34**: 2204–10.

Dechant M., Beyer T., Schneider-Merck T., Weisner W., Peipp M., van de Winkel J.G.J., Valerius T. (2007). Effector mechanisms of recombinant IgA antibodies against epidermal growth factor receptor. *J Immunol* **179**: 2936–43.

De la Cruz Edmunds M.C., Tellers M., Chan C., Salmon P., Robinson D.K., Markussen J. (2006). Development of transfection and high-producer screening protocols for the CHOK1SV cell system. *Mol Biotechnol* **34**: 179–90.

der Balian G.P., Slack J., Clevinger B.L., Hervi-Bazin U., Davie J.M. (1980). Subclass restriction of murine antibodies. III. Antigens that stimulate IgG3 in mice stimulate IgG2 in rats. *J Exp Med* **152**: 209–18.

Desjarlais J.R., Lazar G.A., Zhukovsky E.A., Chu S.Y. (2007). Optimizing engagement of the immune system by anti-tumor antibodies: an engineer's perspective. *Drug Discovery Today* **12**: 898–910.

Díaz de Ståhl T., Dahlström J., Carroll M.C., Heyman B. (2003). A role for complement in feedback enhancement of antibody responses by IgG3. *J Exp Med* **197**: 1183–90.

Dillman S.L., Strelkauskas A.J., Sut H.R., Boakle R.J. (1995). Activation of human complement by totally human monoclonal antibodies. *Molec Immunol* **32**: 957–64.

Dillon T.M., Ricci M.S., Vezina C., Flynn G.C., Liu Y.D., Rehder D.S., Plant M., Henkle B., Li Y., Deechongkit S., Varnum B., Wypych J., Balland A., Bondarenko P.V. (2008). Structural and functional characterization of disulfide isoforms of the human IgG2 subclass. *J Biol Chem* **283**: 16206–15.

Dimitrov D.S., Marks, J.D. (2008). Therapeutic antibodies: current state and future trends — is a paradigm change coming soon? *Methods Mol Biol* **525**: 1–27.

Ding J.W., Zhou T., Zeng H., Ma L., Sjef Verbeek J., Yin D., Shen J., Chong A.S. (2008). Hyperacute rejection by anti-gal IgG_1, IgG_{2a}, and IgG_{2b} is dependent on complement and Fc-γ receptors. *J Immunol* **180**: 261–8.

Dong X., Storkus W.J., Salter R.D. (1999). Binding and uptake of agalactosyl IgG by mannose receptor on macrophages and dendritic cells. *J Immunol* **163**: 5427–34.

Doyle A.G., Woolven B.P., Tomlinson I.M., Lee J.A., Jennings P.A. (2007). Domain antibody construct. *WO 2007/087673 A1*.

Duncan A.R., Winter G. (1988). The binding site for C1q on IgG. *Nature* **332**: 738–40.

Ernst L.K., Metes D., Herberman R.B., Morel P.A. (2002). Allelic polymorphisms in the Fcgamma RIIc gene can influence its function on normal human natural killer cells. *J Mol Med* **80**: 248–57.

Ferrante A., Beard L.J., Feldman R.G. (1990). IgG subclass distribution of antibodies to bacterial and viral antigens. *Pediatr Infect Dis J* **9**: S16–24.

Feige M.J., Nath S., Catharino S.R., Weinfurtner D., Steinbacher S., Buchner J. (2009). Structure of the murine unglycosylated IgG1 Fc fragment. *J Mol Biol* **391**: 599–608.

Foulkes R. (2002). Preclinical safety evaluation of monoclonal antibodies. *Toxicology* **174**: 21–6.

Gagnon P., Hensel F., Richieri R. (2008). Purification of IgM monoclonal antibodies. *BioPharm Internat 26 March suppl*: 26–35.

García-García E., Nieto-Castañeda G., Ruiz-Saldaña M., Mora N., Rosales C. (2009). FcγRIIA and FcγRIIIB mediate nuclear factor activation through separate signaling pathways in human neutrophils. *J Immunol* **182**: 4547–56.

Gasque P. (2004). Complement: a unique innate immune sensor for danger signals. *Molec Immunol* **41**: 1089–98.

Ghetie V., Ward E.S. (2000). Multiple roles for the major histocompatibility complex Class I — related receptor FcRn. *Annu Rev Immunol* **18**: 739–66.

Ghirlando R., Lund J., Goodall M., Jefferis R. (1999). Glycosylation of human IgG-Fc: influences on structure revealed by differential scanning micro-calorimetry. *Immunol Lett* **68**: 47–52.

Ghumra A., Shi J., Mcintosh R.S., Rasmussen I.B., Braathen R., Johansen F.E., Sandlie I., Mongini P.K., Areschoug T., Lindahl G., Lewis M.J., Woof J.M., Pleass R.J. (2009). Structural requirements for the interaction of human IgM and IgA with the human Fc α/μ receptor. *Eur J Immunol* **39**: 1147–56.

Giorgini A., Brown H.J., Lock H.R., Nimmerjahn F., Ravetch J.V., Verbeek J.S., Sacks S.H., Robson M.G. (2008). FcγRIII and FcγRIV are indispensable for acute glomerular inflammation induced by switch variant monoclonal antibodies. *J Immunol* **181**: 8745–52.

Glennie M.J., French R.R., Cragg M.S., Taylor R.P. (2007). Mechanisms of killing by anti-CD20 monoclonal antibodies. *Mol Immunol* **44**: 3823–37.

Gounni A.S. (2006). The high-affinity IgE receptor (FcεRI): a critical regulator of airway smooth muscle cells? *Am J Physiol Lung Cell Mol Physiol* **291**: L312–21.

Greenspan N.S., Cooper L.J.N. (1992). Intermolecular cooperativity: a clue to why mice have IgG3? *Immunol Today* **13**: 164–8.

Guo A., Han M., Martinez T., Ketchem R.R., Novick S., Jochheim C., Balland A. (2008). Electrophoretic evidence for the presence of structural isoforms specific for the IgG2 isotype. *Electrophoresis* **29**: 2550–6.

Hamaguchi Y., Xiu Y., Komura K., Nimmerjahn F., Tedder T.F. (2006). Antibody isotype-specific engagement of Fcγ receptors regulates B lymphocyte depletion during CD20 immunotherapy. *J Exp Med* **203**: 743–53.

Hart S.P., Alexander K.M., Dransfield I. (2004). Immune complexes bind preferentially to FcγRIIA (CD32) on apoptotic neutrophils, leading to augmented phagocytosis by macrophages and release of proinflammatory cytokines. *J Immunol* **172**: 1882–7.

Harwood N.E., McDonnell, J.M. (2007). The intrinsic flexibility of IgE and its role in binding FcεRI. *Biomedicine Pharmacotherapy* **61**: 61–7.

Hayden-Ledbetter M.S., Cerveny C.G., Espling E., Brady W.A., Grosmaire L.S., Tan P., Bader R., Slater S., Nilsson C.A., Barone D.S., Simon A., Bradley C., Thompson P.A., Wahl A.F., Ledbetter J.A. (2009). CD20-directed small modular immunopharmaceutical, TRU-015, depletes normal and malignant B cells. *Clin Cancer Res* **15**: 2739–46.

Hazenbos W.L., Heijnen I.A., Meyer D., Hofhuis F.M., Renardel de Lavalette C.R., Schmidt R.E., Capel P.J., van de Winkel J.G., Gessner J.E., van den Berg T.K., Verbeek J.S. (1998). Murine IgG1 complexes trigger immune effector functions predominantly via FcγRIII (CD16). *J Immunol* **161**: 3026–32.

Herr A.B., White C.L., Milburn C., Wu C., Bjorkman P.J. (2003). Bivalent binding of IgA1 to FcαRI suggests a mechanism for cytokine activation of IgA phagocytosis. *J Mol Biol* **327**: 645–57.

Hinton P.R., Xiong J.M., Johlfs M.J., Tang M.T., Keller S., Tsurushita N. (2006). An engineered human IgG1 antibody with longer serum half-life. *J Immunol* **176**: 346–56.

Hirano M., Davis R.S., Fine W.D., Nakamura S., Shimizu K., Yagi H., Kato K., Stephan R.P., Cooper M.D. (2007). IgE^b immune complexes activate macrophages through FcγRIV binding. *Nature Immunol* **8**: 762–71.

Hoon D.S., Wang Y., Sze L., Kanda H., Watanabe Y., Morrison S.L., Morton D.L., Irie R.F. (1993). Molecular cloning of a human monoclonal antibody reactive to ganglioside GM3 antigen on human cancers. *Cancer Res* **53**: 5244–50.

Holland M., Hewins P., Goodal M., Adu D., Jefferis R., Savage C.O. (2004). Anti-neutrophil cytoplasm antibody IgG subclasses in Wegener's granulomatosis: a possible pathogenic role for the IgG4 subclass. *Clin Exp Immunol* **138**: 183–92.

Horton H.M., Bernett M.J., Pong E., Piepp M., Karki S., Chu S.Y., Richards J.O., Vostiar I., Joyce P.F., Repp R., Desjarlais J.R., Zhukovsky E.A. (2008). Potent *in vitro* and *in vivo* activity of an Fc-engineered anti-CD19 monoclonal antibody against lymphoma and leukemia. *Cancer Res* **68**: 8049–57.

Hsu E., Pulham N., Rumfelt L.L., Flajnik M.F. (2006). The plasticity of immunoglobulin gene systems in evolution. *Immunol Rev* **210**: 8–26.

Hughey C.T., Brewer J.W., Colosia A.D., Rosse W.F., Corley R.B. (1998). Production of IgM hexamers by normal and autoimmune B cells: implications for the physiologic role of hexameric IgM. *J Immunol* **161**: 4091–7.

Hutchins J.T., Kull F.C., Jr., Bynum J., Knick V.C., Thurmond L.M., Ray P. (1995). Improved biodistribution, tumor targeting, and reduced immunogenicity in mice with gamma 4 variant of Campath-1H. *Proc Natl Acad Sci USA* **92**: 11980–4.

Idusogie E.E., Presta L.G., Gazzano-Santoro H., Totpal K., Wong P.Y., Ultsch M., Meng Y.G., Mulkerrin, M.G. (2000). Mapping of the C1q binding site on Rituxan, a chimeric antibody with a human IgG1 Fc. *J Immunol* **164**: 4178–84.

Iida S., Kuni-Kamochi R., Mori K., Misaka H., Inoue M., Okazaki A., Shitara K., Satoh M. (2009). Two mechanisms of enhanced antibody-dependent cellular cytotoxicity (ADCC) efficacy of non-fucosylated therapeutic antibodies in human blood. *BMC Cancer* **9**: 58, 1–12.

Inoue Y., Kaifu T., Sugahara-Tobinai A., Nakamura A., Miyazaki J.-I., Takai T. (2007). Activating Fcγ receptors participate in the development of autoimmune diseases in NOD mice. *J Immunol* **179**: 764–74.

Ip W.K., Takahashi K., Ezekowitz R.A., Stuart L.M. (2009). Mannose binding lectin and innate immunity. *Immunol Rev* **230**: 9–21.

Irie R.F., Ollila D.W., O'Day S., Morton D.L. (2004). Phase I pilot clinical trial of human IgM monoclonal antibody to ganglioside GM3 in patients with metastatic melanoma. *Cancer Immunol Immunother* **53**: 110–7.

Isaacs J.D., Wing M.G., Greenwood J.D., Hazleman B.L., Hale G., Waldman H. (1996). A therapeutic human IgG4 monoclonal antibody that depletes target cells in humans. *Clin Exp Immunol* **106**: 427–33.

Ishizaka K., Ishizaka T. (1970). Biological function of γE antibodies and mechanisms of reaginic hypersensitivity. *Clin Exp Immunol* **6:** 25–42.

Ito A., Ishida T., Yano H., Inagaki A., Suzuki S., Sato F., Takino H., Mori F., Ri M., Kusumoto S., Komatsu H., Iida S., Inagaki H., Ueda R. (2009). Defucosylated anti-CCR4 monoclonal antibody exercises potent ADCC-mediated antitumor effect in the novel tumor-bearing humanized NOD/Shi-*scid*, IL-2R_null mouse model. *Cancer Immunol Immunother* **58**: 1195–206.

Jakus Z., Németh T., Verbeek J.S., Mócsai A. (2008). Critical but overlapping role of FcγRIII and FcγRIV in activation of murine neutrophils by immobilized immune complexes. *J Immunol* **180**: 618–29.

Jansen W.T.M., Breukels M.A., Snippe H., Sanders L.A.M., Verheul A.F.M., Rijkers G.T. (1999). Fcγ receptor polymorphisms determine the magnitude of *in vitro* phagocytosis of *Streptococcus pneumoniae* mediated by pneumococcal conjugate sera. *J Infectious Dis* **180**: 888–91.

Jefferis R. (2007). Antibody therapeutics: isotype and glycoform selection. *Expert Opin Ther* **7**: 1401–13.

Jefferis R. (2009a). Glycosylation as a strategy to improve antibody-based therapeutics. *Nature Rev Drug Disc* **8**: 226–34.

Jefferis R. (2009b). Recombinant antibody therapeutics: the impact of glycosylation on mechanism of action. *Trends Pharmacol Sci* **30**: 356–62.

Johansen F.E., Braathen R., Brandtzaeg P. (2000). Role of J chain in secretory immunoglobulin formation. *Scand J Immunol* **52**: 240–8.

Kaetzel C.S. (2005). The polymeric immunoglobulin receptor: bridging innate and adaptive immune responses at mucosal surfaces. *Immunol Rev* **206**: 83–99.

Kamisawa T., Okamoto A. (2006). Autoimmune pancreatitis: proposal of IgG4-related sclerosing disease. *J Gastroenterol* **41**: 613–25.

Kaneko Y., Nimmerjahn F., Ravetch J.V. (2006). Anti-inflammatory activity of immunoglobulin G resulting from Fc sialylation. *Science* **313**: 670–3.

Kaymakcalan Z., Sakorafas P., Bose S., Scesney S., Xiong L., Karaoglu Hanzatian D., Salfeld J., Sasso E.H. (2009). Comparisons of affinities, avidities, and complement activation of adalimumab, infliximab, and etanercept in binding to soluble and membrane tumor necrosis factor. *Clinical Immunology* **131**: 308–16.

Kazeeva T.N., Shevelev A.B. (2009). IgA-specific proteins of pathogenic bacteria. *Biochemistry (Moscow)* **74**: 12–21.

Kilmon M.A., Shelburne A.E., Chan-Li Y., Holmes K.L. Conrad D.H. (2004). CD23 trimers are preassociated on the cell surface even in the absence of its ligand, IgE. *J Immunol* **172**: 1065–73.

Kim J.K., Firan M., Radu C.G., Kim C.-H., Ghetie V., Ward E.S. (1999). Mapping the site on human IgG for binding of the MHC class I-related receptor, FcRn. *Eur J Immunol* **29**: 2819–25.

Kishore U., Reid K.B.M. (2000). C1q: structure, function, and receptors. *Immunopharmacology* **49**: 159–70.

Knight K.L., Winstead C.R. (1997). Generation of antibody diversity in rabbits. *Curr Opin Immunol* **9**: 228–32.

Köhler G., Milstein C. (1975). Continuous cultures of fused cells secreting antibody of predefined specificity. *Nature* **256**: 495–7.

Kumagai Y., Fujita T., Ozaki M., Sahashi K., Ohkura M., Ohtsu T., Arai Y., Sonehara Y., Nichol J.L. (2007). Pharmacodynamics and pharmacokinetics of AMG 531, a thrombopoiesis-stimulating peptibody, in healthy Japanese subjects: A randomized, placebo-controlled study. *J Clin Pharmacol* **47**: 1489–97.

Labrijn A.F., Aalberse R.C., Schuurman J. (2008). When binding is enough: nonactivating antibody formats. *Curr Opin Immunol* **20**: 479–85.

Labrijn A.F., Buijsse A.O., van den Bremer E.T.J., Verwilligen A.Y.W., Bleeker W.K., Thorpe S.J., Killestein J., Polman C.H., Aalberse R.C., Schuurman J., van de Winkel J.G.J., Parren P.W.H.I. (2009). Therapeutic IgG4 antibodies engage in Fab-arm exchange with endogenous human IgG4 *in vivo*. *Nature Biotechnol* **27**: 767–71.

Lai Z., Kimmela R., Petersen S., Thomas S., Pier G., Bezabeh B., Luod R., Schreiber J.R. (2005). Multi-valent human monoclonal antibody preparation against *Pseudomonas aeruginosa* derived from transgenic mice containing human immunoglobulin loci is protective against fatal pseudomonas sepsis caused by multiple serotypes. *Vaccine* **23**: 3264–71.

Larrick J.W., Yu L., Naftzger C., Jaiswal S., Wycoff K. (2001). Production of secretory IgA antibodies in plants. *Biomolec Eng* **18**: 87–94.

Lazar G.A., Dang W., Karki S., Vafa O., Peng J.S., Hyun L., Chan C., Chung H.S., Eivazi A., Yoder S.C., Veilmetter J., Carmichael D.F., Hayes R.J., Dahiyat B.I. (2006). Engineered antibody Fc variants with enhanced effector function. *Proc Natl Acad Sci USA* **103**: 4005–10.

Li H., Sethuraman N., Stadheim T.A., Zha D., Prinz B., Ballew N., Bobrowicz P., Choi B.K., Cook W.J., Cukan M., Houston-Cummings N.R., Davidson R., Gong B., Hamilton S.R., Hoopes J.P., Jiang Y., Kim N., Mansfield R., Nett J.H., Rios S., Strawbridge R., Wildt S., Gerngross T.U. (2006a). Optimization of humanized IgGs in glycoengineered *Pichia pastoris. Nature Biotechnol* **24**: 210–15.

Li J., Davis J., Bracht M., Carton J., Armstrong J., Gao W., Scallon B., Fung R., Emmel E., Zimmerman M., Griswold D.E., Li L. (2006b). Modulation of antigen-specific T cell response by a non-mitogenic anti-CD3 antibody. *Internat Immunopharmacol* **6**: 880–91.

Li P., Jiang N., Nagarajan S., Wohlhueter R., Selvaraj P., Zhu C. (2007). Affinity and kinetic analysis of Fcgamma receptor IIIa (CD16a) binding to IgG ligands. *J Biol Chem* **282**: 6210–21.

Lin Q., Xiu Y., Jiang Y., Tsurui H., Nakamura K., Kodera S., Ohtsuji M., Ohtsuji N., Shiroiwa W., Tsukamoto K., Amano H., Amano E., Kinoshita K., Sudo K., Nishimura H., Izui S., Shirai T., Hirose S. (2006). Genetic dissection of the effects of stimulatory and inhibitory IgG Fc receptors on murine lupus. *J Immunol* **177**: 1646–54.

Lin X., Ding J., Zhu L., Shi S., Jiang L., Zhao M., Zhang H. (2009). Aberrant galactosylation of IgA1 is involved in the genetic susceptibility of Chinese patients with IgA nephropathy. *Nephrol Dial Transplant* **24**: 3372–5.

Longstaff C., Whitton C.M., Stebbings R., Gray E. (2009). How do we assure the quality of biological medicines? *Drug Discovery Today* **14**: 50–5.

Lucisano-Valim Y.M., Lachmann P.J. (1991). The effect of antibody isotype and antigenic epitope density on the complement-fixing activity of immune complexes: A systematic study using chimaeric anti-NIP antibodies with human Fc regions. *Clin Exp Inmunol* **84**: l–8.

Ma J.K., Hikmat B.Y., Wycoff K., Vine N.D., Chargelegue D., Yu L., Hein M.B., Lehner T. (1998). Characterization of a recombinant plant monoclonal secretory antibody and preventive immunotherapy in humans. *Nature Med* **4**: 601–6.

Maenaka K., van der Merwe P.A., Stuart D.I., Jones E.Y., Sondermann P. (2001). The human pow affinity Fcγ receptors IIa, IIb, and III bind IgG with fast kinetics and distinct thermodynamic properties. *J Biol Chem* **276**: 44898–904.

Mancardi D.A., Iannascoli B., Hoos S., England P., Daeron M., Bruhns P. (2008). FcγRIV is a mouse IgE receptor that resembles macrophage FcεRI in humans and promotes IgE-induced lung inflammation. *J Clin Invest* **118**: 3738–50.

McEarchern J.A., Oflazoglu E., Francisco L., McDonagh C.F., Gordon K.A., Stone I., Klussman K., Turcott E., van Rooijen N., Carter P., Grewal I.S., Wahl A.F., Law C.-L.

(2007). Engineered anti-CD70 antibody with multiple effector functions exhibits *in vitro* and *in vivo* antitumor activities. *Blood* **109**: 1185–92.

McLay J., Leonard E., Petersen S., Shapiro D., Greenspan N.S., Schreiber J.R. (2002). γ3 gene-disrupted mice selectively deficient in the dominant IgG subclass made to bacterial polysaccharides. II. Increased susceptibility to fatal pneumococcal sepsis due to absence of anti-polysaccharide IgG3 is corrected by induction of anti-polysaccharide IgG1. *J Immunol* **168**: 3437–43.

Michaelsen T.E., Aase A., Norderhaug L., Sandlie I. (1992). Antibody dependent cell-mediated cytotoxicity induced by chimeric mouse-human IgG subclasses and IgG3 antibodies with altered hinge region. *Mol Immunol* **29**: 319–26.

Minard-Colin V., Xiu Y., Poe J.C., Horikawa M., Magro C.M., Hamaguchi Y., Haas K.M., Tedder T.F. (2008). Lymphoma depletion during CD20 immunotherapy in mice is mediated by macrophage FcγRI, FcγRIII, and FcγRIV. *Blood* **112**: 1205–13.

Mitoma H., Horiuchi T., Tsukamoto H., Tamimoto Y., Kimoto Y., Uchino A., Kentaro T., Harashima S.-i., Hatta N., Harada M. (2008). Mechanisms for cytotoxic effects of anti-tumor necrosis factor agents on transmembrane tumor necrosis factor alpha-expressing cells: comparison among infliximab, etanercept, and adalimumab. *Arthritis Rheum* **58**: 1248–57.

Monteiro R.C., van deWinkel J.G.J. (2003). IgA Fc receptors. *Annu Rev Immunol* **21**: 177–204.

Morrison S.L., Chintalacharuvu K.R. (2001). Secretory immunoglobulin produced by single cells and methods for making and using same. *US Patent 6,300,104 B1.*

Mota G., Manciulea M., Cosma E., Popescu I., Hirt M., Jensen-Jarolim E., Calugaru A., Galatiuc C., Regalia T., Tamandl D., Spittler A., Boltz-Nitulescu G. (2003). Human NK cells express Fc receptors for IgA which mediate signal transduction and target cell killing. *Eur J Immunol* **33**: 2197–205.

Mueller J.P., Giannoni M.A., Hartman S.L., Elliott E.A., Squinto S.P., Matis L.A., Evans M.J. (1997). Humanized porcine VCAM-specific monoclonal antibodies with chimeric IgG2/G4 constant regions block human leukocyte binding to porcine endothelial cells. *Molec Immunol* **34**: 441–52.

Musolino A., Naldi N., Bortesi B., Pezzuolo D., Capelletti M., Missale G., Laccabue D., Zerbini A., Camisa R., Bisagni G., Neri T.M., Ardizzoni A. (2008). Immunoglobulin G fragment C receptor polymorphisms and clinical efficacy of trastuzumab-based therapy in patients with HER-2/*neu*–positive metastatic breast cancer. *J Clin Oncol* **26**: 1789–96.

Nakamura A., Akiyama K., Takai T. (2005). Fc receptor targeting in the treatment of allergy, autoimmune diseases and cancer. *Expert Opin Therap Targets* **9**: 169–90.

Natsume A., In M., Takamura H., Nakagawa T., Shimizu Y., Kitajima K., Wakitani M., Ohta S., Satoh M., Shitara K., Niwa R. (2008). Engineered antibodies of IgG1/IgG3 mixed isotype with enhanced cytotoxic activities. *Cancer Res* **68**: 3863–72.

Neuberger M.S., Williams G.T., Fox R.O. (1984). Recombinant antibodies possessing novel effector functions. *Nature* **312**: 604–8.

Nielsen L.K., Hefsgaard Green T., Norderhaug L., Sandlie I., Hanefeld Dziegiel M. (2007). Functional *in vitro* studies of recombinant human immunoglobulin G and immunoglobulin A anti-D. *Transfusion* **47**: 306–15.

Nimmerjahn F., Bruhns P., Horiuchi K., Ravetch J.V. (2005). FcγRIV: a novel FcR with distinct IgG subclass specificity. *Immunity* **23**: 41–51.

Nimmerjahn F., Ravetch J.V. (2006). Fcγ Receptors: Old friends review and new family members. *Immunity* **24**: 19–28.

Nimmerjahn F., Ravetch J.V. (2007). The anti-inflammatory activity of IgG: the intravenous IgG paradox. *J Exp Med* **204**: 11–5.

Nimmerjahn F., Ravetch J.V. (2008). Fcγ receptors as regulators of immune responses. *Nature Rev Immunol* **8**: 34–47.

Niwa R., Hatanaka S., Shoji-Hosaka E., Sakurada M., Kobayashi Y., Uehara A., Yodoi H., Nakamura K., Shitara K. (2004). Enhancement of the antibody-dependent cellular cytotoxicity of low-fucose IgG1 is independent of FcgammaRllla functional polymorphism. *Clin Cancer Res* **10**: 6248–55.

Niwa R., Natsume A., Uehara A., Wakitani M., Iida S., Uchida K., Satoh M., Shitara K. (2005). IgG subclass-independent improvement of antibody dependent cellular cytotoxicity by fucose removal from Asn297-linked oligosaccharides. *J Immunol Meth* **306**: 151–60.

Norderhaug I.N., Johansen F.-E., Krajci P., Bradzaeg P. (1999). Domain deletions in the human polymeric Ig receptor disclose differences between its dimeric IgA and pentameric IgM interaction. *Eur J Immunol* **29**: 3401–9.

Nozawa S., Aoki D., Tsukazaki K., Susumu N., Sakayori M., Suzuki N., Suzuki A., Wakita R., Mukai M., Egami Y., Kojima-Aikawa K., Ishida I., Belot F., Hindsgaul O., Fukuda M., Fukuda M.N. (2004). A humanized monoclonal antibody with therapeutic potential against Müllerian duct-related carcinomas. *Clin Cancer Res* **10**: 7071–8.

Oflazoglu E., Stone I.J., Gordon K.A., Greal I.S., van Rooijen N., Law C.-L., Gerber H.-P. (2007). Macrophages contribute to the antitumor activity of the anti-CD30 antibody SGN-30. *Blood* **110**: 4370–2.

Oflazoglu E., Stone I.J., Brown L., Gordon K.A., van Rooijen N., Jonas M., Law C.-L., Grewal I.S., Gerber H.-P. (2009). Macrophages and Fc-receptor interactions contribute to the antitumor activities of the anti-CD40 antibody SGN-40. *Brit J Cancer* **100**: 113–7.

Oganesyan V., Gao C., Shirinian L., Wu H., Dall'Acqua W.F. (2008). Structural characterization of a human Fc fragment engineered for lack of effector functions. *Acta Cryst* **D64**: 700–4.

Ohta Y., Flajnik M. (2006). IgD, like IgM, is a primordial immunoglobulin class perpetuated in most jawed vertebrates. *Proc Natl Acad Sci USA* **103**: 10723–8.

Pack T.D. (2000). Purification of IgA. *Curr Protocols Immunol* **2**: 2.10B.1–2.10B.7

Pancook J.D., Beuerlein G., Pecht G., Tang Y., Nie Y., Wu H., Huse W.D., Watkins J.D. (2001). *In vitro* affinity maturation of human IgM antibodies reactive with tumor-associated antigens. *Hybridoma Hybridomics* **20**: 383–96.

Penichet M.L., Morrison S.L. (2001). Antibody–cytokine fusion proteins for the therapy of cancer. *J Immunol Methods* **248**: 91–101.

Petkova S.B., Akilesh S., Sproule T.J., Christianson G.J., Al Khabbaz H., Brown A.C., Presta L.G., Meng Y.G., Roopenian D.C. (2006). Enhanced half-life of genetically engineered human IgG1 antibodies in a humanized FcRn mouse model: potential application in humorally mediated autoimmune disease. *Internat Immunol* **18**: 1759–69.

Platonov A.E., Shipulin G.A., Vershinina I.V., Dankert J., van de Winkel J.G.J., Kuijper E.J. (1998). Association of human FcγRIIa (CD32) polymorphism with susceptibility to and severity of meningococcal disease. *Clin Infect Dis* **27**: 746–50.

Presta, L.G. (2006). Engineering of therapeutic antibodies to minimize immunogenicity and optimize function. *Adv Drug Deliv Rev* **58**: 640–56.

Presta, L.G. (2008). Molecular engineering and design of therapeutic antibodies. *Curr Opin Immunol* **20**: 460–70.

Pricop L., Rabinowich H., Morel P.A., Sulica A., Whiteside T.L., Herberman R.B. (1993). Characterization of the Fc μ receptor on human natural killer cells. Interaction with its physiological ligand, human normal IgM, specificity of binding, and functional effects. *J Immunol* **151**: 3018–29.

Preud'homme J.-L., Petit I., Barra A., Morel F., Lecron J.-C., Lelièvre E. (2000). Structural and functional properties of membrane and secreted IgD. *Molec Immunol* **37**: 871–87.

Raju T.S. (2008). Terminal sugars of Fc glycans influence antibody effector functions of IgGs. *Curr Opin Immunol* **20**: 471–6.

Ravetch J.V., Bolland S. (2001). IgG Fc receptors. *Annu Rev Immunol* **19**: 275–90.

Reddy M.P., Kinney C.A.S., Chaikin M.A., Payne A., Fishman-Lobell J., Tsui P., Dal Monte P.R., Doyle M.L., Brigham-Burke M.R., Anderson D., Reff M., Newman R., Hanna N., Sweet R.W., Truneh A. (2000). Elimination of Fc receptor-dependent effector functions of a modified IgG4 mAb to human CD4. *J Immunol* **164**: 1925–33.

Reichert J.M., Dewitz M.C. (2004). Anti-infective monoclonal antibodies: perils and promise of development. *Nature Rev Drug Disc* **5**: 191–5.

Richards J.O., Karki S., Lazar G.A., Chen H., Dang W., Desjarlais J.R. (2008). Optimization of antibody binding to FcγRIIa enhances macrophage phagocytosis of tumor cells. *Mol Cancer Therap* **7**: 2517–27.

Ricklin D., Lambris J.D. (2007). Complement-targeted therapeutics. *Nature Biotechnol* **25**: 1265–75.

Robak T. (2009). GA-101. A third-generation, humanized and glycol-engineered anti-CD20 mAb for the treatment of B cell lymphoid malignancies. *Curr Opin Investig Drugs* **10**: 588–96.

Rogers K.A., Scinicariello F., Attanasio R. (2006). IgG Fc receptor III homologues in non-human primate species: Genetic characterization and ligand interactions. *J Immunol* **177**: 3848–56.

Rother R.P., Rollins S.A., Mojcik C.J., Brodsky R.A., Bell L. (2007). Discovery and development of the complement inhibitor eculizumab for the treatment of paroxysmal nocturnal hemoglobinuria. *Nature Biotechnol* **25**: 1256–64.

Saeki T., Nishi S., Ito T., Yamazaki H., Miyamura S., Emura I., Imai N., Ueno M., Saito A., Gejyo F. (2007). Renal lesions in IgG4-related systemic disease. *Intern Med.* **46**: 1365–71.

Salfeld J.G. (2007). Isotype selection in antibody engineering. *Nature Biotechnol* **25**: 1369–72.

Samuelsson A., Towers T.L., Ravetch J.V. (2001). Anti-inflammatory activity of IVIG mediated through the inhibitory Fc receptor. *Science* **291**: 484–6.

Samuelsson M., Jendholm J., Amisten S., Morrison S.L., Forsgren A., Riesbeck K. (2006). The IgD CH1 region contains the binding site for the human respiratory pathogen *Moraxella catarrhalis* IgD binding protein MID. *Eur J Immunol* **36**: 2525–34.

Sanders L.A.M., Feldman R.G., Voorhorst-Ogink M.D., de Haas M., Rijkers G.T., Capel P.J., Zegers B.J., van de Winkel J.G. (1995). Human immunoglobulin G (IgG) Fc receptor IIA (CD32) polymorphism and IgG2-mediated bacterial phagocytosis by neutrophils. *Infect Immun* **63**: 73–81.

Sazinsky S.L., Ott R.G., Silver N.W., Tidor B., Ravetch J.V., Wittrup K.D. (2008). Aglycosylated immunoglobulin G1 variants productively engage activating Fc receptors. *Proc Natl Acad Sci USA* **105**: 20167–72.

Scallon B., Moore M.A., Trinh H., Knight D.M., Ghrayeb J. 1995. Chimeric anti-TNF-α monoclonal antibody cA2 binds recombinant transmembrane TNF-α and activates immune effector functions. *Cytokine* **7**: 251–9.

Scallon B., McCarthy S., Radewonuk J., Cai A., Naso M., Raju T.S., Capocasale R. (2007a). Quantitative *in vivo* comparisons of the Fcγ receptor-dependent agonist activities of different fucosylation variants of an immunoglobulin G antibody. *Internat Immunopharmacol* **7**: 761–72.

Scallon B.J., Tam S.H., McCarthy S.G., Cai A.N., Raju T.S. (2007b). Higher levels of sialylated Fc glycans in immunoglobulin G molecules can adversely impact functionality. *Molec Immunol* **44**: 1524–34.

Schumaker V.N., Calcott M.A., Spiegelberg H.L., Muller-Eberhard H.J. (1976). Ultracentrifuge studies of the binding of IgG of different subclasses to the Clq subunit of the first component of complement. *Biochemistry* **16**: 5175–81.

Schuurman J., Van Ree R., Perdok G.J., Van Doorn H.R., Tan K.Y., Aalberse R.C. (1999). Normal human immunoglobulin G4 is bispecific: it has two different antigen-combining sites. *Immunology* **97**: 693–8.

Scinicariello F., Engleman C.N., Jayashankar L., McClure H.M. Attanasio R. (2004). Rhesus macaque antibody molecules: sequence and heterogeneity of α and γ constant regions. *Immunology* **111**: 66–74.

Seino J., Eveleigh P., Warnaar S., Van Haarlem L.J.M., Van Es L.A., Daha M.R. (1993). Activation of human complement by mouse and mouse/human chimeric monoclonal antibodies. *Clin Exp Immunol* **94**: 291–6.

Senior B.W., Woof JM. (2005). Effect of mutations in the human immunoglobulin A1 (IgA1) hinge on its susceptibility to cleavage by diverse bacterial IgA1 proteases. *Infect Immun* **73**: 1515–22.

Sethuraman N., Stadheim T.A. (2006). Challenges in therapeutic glycoprotein production. *Curr Opin Biotechnol* **17**: 341–6.
Shearer M.H., Dark R.D., Chodosh J., Kennedy R.C. (1999). Comparison and characterization of Iimunoglobulin G subclasses among primate species. *Clin Diagnost Labor Immunol* **6**: 953–8.
Shibata-Koyama M., Iida S., Misaka H., Mori K., Yano K., Shitara K., Satoh M. (2009). Nonfucosylated rituximab potentiates human neutrophil phagocytosis through its high binding for FcγRIIIb and MHC class II expression on the phagocytotic neutrophils. *Exp Hematol* **37**: 309–21.
Shields R.L., Namenuk A.K., Hog K., Meng Y.G., Rae J., Briggs J., Xie D., Lai J., Stadlen A., Li B., Fox J.A., Presta L.G. (2001). High resolution mapping of the binding site on human IgG1 for Fc gamma RI, Fc gamma RII, Fc gamma RIII, and FcRn and design of IgG1 variants with improved binding to the Fc gamma R. *J Biol Chem* **276**: 6591–604.
Shields R.L., Lai J., Keck R., O'Connell L.Y., Hong K., Meng Y.G., Weikert S.H., Presta L.G. (2002). Lack of fructose on human IgG1 N-linked oligosaccharide improves binding to human Fcgamma RIII and antibody-dependent cellular toxicity. *J Biol Chem* **277**: 26733–40.
Shinkawa T., Nakamura K., Yamane N., Shoji-Hosaka E., Kanda Y., Sakurada M., Uchida K., Anazawa H., Satoh M., Yamasaki M., Hanai N., Shitara K. (2003). The absence of fucose but not the presence of galactose or bisecting N-acetylglucosamine of human IgG1 complex-type oligosaccharide shows the critical role of enhancing antibody-dependent cellular cytotoxicity. *J Biol Chem* **278**: 3466–73.
Slack J.H. (1987). Strain-dependent IgG subclass response patterns. *J Immunol* **139**: 3734–8.
Sontheimer R.D., Racila E., Racila D.M. (2005). C1q: Its functions within the innate and adaptive immune responses and its role in lupus autoimmunity. *J Invest Dermatol* **125**: 14–23.
Sørensen V., Sundvold V., Michaelsen T.E., Sandlie I. (1999). Polymerization of IgA and IgM: Roles of Cys309/Cys414 and the secretory tailpiece. *J Immunol* **162**: 3448–55.
Stavenhagen J.B., Gorlatov S., Tuaillon N., Rankin C.T., Li H., Burke S., Huang L., Johnson S., Bonvini E., Koenig S. (2007). Fc optimization of therapeutic antibodies enhances their ability to kill tumor cells *in vitro* and controls tumor expansion *in vivo* via low-affinity activating Fcγ receptors. *Cancer Res* **67**: 8882–90.
Stebbings R., Findlay L., Edwards C., Eastwood D., Bird C., North D., Mistry J., Dilger P., Liefooghe E., Cludts I., Fox B., Tarrant G., Robinson J., Meager T., Dolman C., Thorpe S.J., Bristow A., Wadhwa M., Thorpe R., Poole S. (2007) ."Cytokine storm" in the phase I trial of monoclonal antibody TGN1412: better understanding the causes to improve preclinical testing of immunotherapeutics. *J Immunol* **179**: 3325–31.
Steplewski Z., Sun L.K., Shearman C.W., Ghrayeb J., Daddona P., Koprowski H. (1988). Biological activity of human-mouse IgG1, IgG2, IgG3, and IgG4 chimeric monoclonal antibodies with antitumor specificity. *Proc Natl Acad Sci USA* **85**: 4852–6.

Stockmeyer B., Dechant M., van Egmond M., Tutt A.L., Sundarapandiyan K., Graziano R.F., Repp R., Kalden J.R., Gramatzki M., Glennie M.J., van de Winkel J.G.J., Valerius T. (2000). Triggering Fc α-receptor I (CD89) recruits neutrophils as effector cells for CD20-directed antibody therapy. *J Immunol* **165**: 5954–61.

Stoger E., Sack M., Fischer R., Christou P. (2002). Plantibodies: applications, advantages and bottlenecks. *Curr Opin Biotechnol* **13**: 161–6.

Strohl W.R. (2010). Non-immunostimulatory antibody and compositions containing the same. *US 7,700,009*, Issued 4/20/2010.

Strohl W.R. (2009a). Therapeutic Monoclonal Antibodies — Past, Present, and Future. In Z. An, (ed.), *Therapeutic Monoclonal Antibodies — From Bench to Clinic*, pp. 4–50. John Wiley & Sons, New York.

Strohl W.R. (2009b). Optimization of Fc-Mediated Effector Functions of Monoclonal Antibodies. *Curr Opin Biotechnol* **20**: 685–91.

Sun Z., Almogren A., Furtado P.B., Chowdhury B., Kerr M.A., Perkins S.J. (2005). Semi-extended solution structure of human myeloma immunoglobulin D determined by constrained X-ray scattering. *J Mol Biol* **353**: 155–73.

Swann P.G., Tolnay M., Mutukkumar S., Shapiro M.A., Rellahan B.L., Clouse K.A. (2008). Considerations for the development of therapeutic monoclonal antibodies. *Curr Opin Immunol* **20**: 493–9.

Tabrizi M.A., Roskos L.K. (2007). Preclinical and clinical safety of monoclonal antibodies. *Drug Discovery Today* **12**: 540–7.

Takai T. (2002). Roles of Fc receptors in autoimmunity. *Nature Rev Immunol* **2**: 580–92.

Tang Y., Lou J., Alpaugh R.K., Robinson M.K., Marks J.D., Weiner L.M. (2007). Regulation of antibody-dependent cellular cytotoxicity by IgG intrinsic and apparent affinity for target antigen. *J Immunol* **179**: 2816–23.

Tao M.-H., Morrison S.L. (1989). Studies of aglycosylated chimeric mouse-human IgG. Role of Carbohydrate in the Structure and Effector Functions Mediated by the Human IgG Constant Region. *J Immunol* **143**: 2595–601.

Taylor A.I., Gould H.J., Sutton B.J., Calvert R.A. (2008). Avian IgY binds to a monocyte receptor with IgG-like kinetics despite an IgE-like structure. *J Biol Chem* **283**: 16384–90.

Tchoudakova A., Hensel F., Murillo A., Eng B., Foley M., Smith L., Schoenen F., Hildebrand A., Kelter A.R., Ilag L.L., Vollmers H.P., Brandlein S., McIninch J., Chon J., Lee G., Cacciuttolo M. (2009). High level expression of functional human IgMs in human PER.C6® cells. *mAbs* **1**: 163–71.

Thommesen J.E., Michaelsen T.E., Løset G.A., Sandlie I., Brekke O.H. (2000). Lysine 322 in the human IgG3 CH2 domain is crucial for antibody dependent complement activation. *Molec Immunol* **37**: 995–1004.

Thurber G.M., Wittrup K.D. (2008). Quantitative spatiotemporal analysis of antibody fragment diffusion and endocytic consumption in tumor spheroids. *Cancer Res* **68**: 3334–41.

Umana P., Jean-Mairet J., Moudry R., Amstutz H., Bailey J.E. (1999). Engineered glycoforms of an anti-neuroblastoma IgG1 with optimized antibody-dependent cellular cytotoxic activity. *Nature Biotechnol* **17**: 176–80.

van der Kolk L.E., Grillo-López A.J., Baars J.W., Hack C.E., van Oers M.H. (2001). Complement activation plays a key role in the side-effects of rituximab treatment. *Brit J Haematol* **115**: 807–11.

van der Neut Kolfschoten M., Schuurman J., Losen M., Bleeker W.K., Martínez-Martínez P., Vermeulen E., den Bleker T.H., Wiegman L., Vink T., Aarden L.A., De Baets M.H., van de Winkel J.G.J., Aalberse R.C., Parren P.W.H.I. (2007). Anti-inflammatory activity of human IgG4 antibodies by dynamic Fab arm exchange. *Science* **317**: 1554–7.

van der Zee J.S., Van Swieten P., Aalberse R.C. (1986). Serologic aspects of IgG4 antibodies. II. IgG4 antibodies form small, nonprecipitating immune complexes due to functional monovalency. *J Immunol* **137**: 3566–71.

van Schie R.C.A.A., Wilson M.E. (2000). Evaluation of human FcγRIIA (CD32) and FcγRIIIB (CD16) polymorphisms in Caucasians and African-Americans using salivary DNA. *Clin Diagn Lab Immunol* **7**: 676–81.

van Sorge N.M., van der Pol W.L., van de Winkel J.G.J. (2003). FcγR polymorphisms: Implications for function, disease susceptibility and immunotherapy. *Tissue Antigens* **61**: 189–202.

Veri M.-C., Gorlatov S., Li H., Burke S., Johnson S., Stavenhagen J., Stein K.E., Bonvini E., Koenig S. (2007). Monoclonal antibodies capable of discriminating the human inhibitory Fcγ-receptor IIB (CD32B) from the activating Fcγ-receptor IIA (CD32A): biochemical, biological and functional characterization. *Immunology* **121**: 392–404.

Vernersson M., Aveskogh M., Hellman L. (2004). Cloning of IgE from the echidna (*Tachyglossus aculeatus*) and a comparative analysis of epsilon chains from all three extant mammalian lineages. *Dev Comp Immunol* **28**: 61–75.

Vollmers H.P., Brändlein S. (2002). Nature's best weapons to fight cancer. Revival of human monoclonal IgM antibodies. *Human Antibodies* **11**: 131–42.

Vollmers H.P., Brändlein S. (2006). Natural IgM antibodies: The orphaned molecules in immune surveillance. *Adv Drug Delivery Rev* **58**: 755– 65.

Walker M.R., Lund J., Thompson K.M., Jefferis R. (1989). Aglycosylation of human IgG1 and IgG3 monoclonal antibodies can eliminate recognition by human cells expressing FcγRI and/or FcγRII receptors. *Biochem J* **259**: 347–53.

Ward E.S. (2006). Immunoglobulin molecules with improved characteristics. *WO 2006/130834.*

Warrington A.E., Bieber A.J., Ciric B., Pease L.R., Van Keulen V., Rodriguez M. (2007). A recombinant human IgM promotes myelin repair after a single, very low dose. *J Neurosci Res* **85**: 967–76.

Watkins J.D., Allan B. (2004). Fc region variants. *WO 2004/074455 A2. Publ 9/2/2004.*

Weng W.K., Levy R. (2003). Two immunoglobulin G fragment C receptor polymorphisms independently predict response to rituximab in patients with follicular lymphoma. *J Clin Oncol* **21**: 3940–7.

Wines B.D., Hogarth P.M. (2006). IgA receptors in health and disease. *Tissue Antigens* **68**: 103–14.

Wing M.G., Moreau T., Greenwood J., Smith R.M., Hale G., Isaacs J., Waldmann H., Lachmann P.J., Compston A. (1996). Mechanism of first-dose cytokine-release syndrome by CAMPATH 1-H: involvement of CD16 (FcgammaRIII) and CD11a/CD18 (LFA-1) on NK cells. *J Clin Invest* **98**: 2819–26.

Woof J.M., Burton D.R. (2004). Human antibody-Fc receptor interactions illuminated by crystal structures. *Nature Rev Immunol* **4**: 89–99.

Woof J.M., Kerr M.A. (2006). The function of immunoglobulin A in immunity. *J Pathol* **208**: 270–82.

Wu B., Joshi A., Ren S., Ng C. (2006). The application of mechanism-based PK/PD modeling in pharmacodynamic-based dose selection of muM17, a surrogate monoclonal antibody for efalizumab. *J Pharm Sci* **95**: 1258–68.

Wycoff K.L. (2004). Secretory IgA Antibodies from Plants. *Curr Pharmaceut Design* **10**: 1381–6128.

Wypych J., Li M., Guo A., Zhang Z., Martinez T., Allen M.J., Fodor S., Kelner D.N., Flynn G.C., Liu Y.D., Bondarenko P.V., Speed Ricci M., Dillon T.M., Balland A. (2008). Human IgG2 antibodies display disulfide mediated structural isoforms. *J Biol Chem* **283**: 16194–205.

Xu H., Geddes R., Roberts-Thompson P.J. (1994). Low molecular weight IgM and CD5 B lymphocytes in rheumatoid arthritis. *Annals Rheum Dis* **53**: 383–90.

Xu D., Alegre M.L., Varga S.S., Rothermel A.L., Collins A.M., Pulito V.L., Hanna L.S., Dolan K.P., Parren P.W., Bluestone J.A., Jolliffe L.K., Zivin R.A. (2000). *In vitro* characterization of five humanized OKT3 effector function variant antibodies. *Cell Immunol* **200**: 16–26.

Yamane-Ohnuki N., Kinoshita S., Inoue-Urakubo M., Kusunoki M., Iida S., Nakano R., Wakitani M., Niwa R., Sakurada M., Uchida K., Shitara K., Satoh M. (2004). Establishment of *FUT8* knockout Chinese hamster ovary cells; an ideal host cell line for producing completely defucosylated antibodies with enhanced antibody-dependent cellular cytotoxicity. *Biotechnol Bioeng* **87**: 614–22.

Yee A.M.F., Phan H.M., Zuniga R., Salmon J.E., Musher D.M. (2000). Association between FcγRIIa-R131 allotype and bacteremic pneumococcal pneumonia. *Clin Infect Dis* **30**: 25–8.

Yeung Y.A., Leabman M.K., Marvin J.S., Qiu J., Adams C.W., Lien S., Starovasnik M.A., Lowman H.B. (2009). Engineering human IgG1 affinity to human neonatal Fc receptor: Impact of affinity improvement on pharmacokinetics in primates. *J Immunol* **182**: 7663–71.

Yokota T., Milenic D.E.,Whitlow M., Schlom J. (1992). Rapid tumor penetration of a single-chain Fv and comparison with other immunoglobulin forms. *Cancer Res* **52**: 3402–8.

Yoo E.M., Wims L.A., Chan L.A., Morrison S.L. (2003). Human IgG2 can form covalent dimers. *J Immunol* **170**: 3134–8.

Zalevsky J., Leung I.W.L., Karki S., Chu S.Y., Zhukovsky E.A., Desjarlais J.R., Carmichael D.F., Lawrence C.E. (2009). The impact of Fc engineering on an anti-CD19 antibody: increased Fcγ receptor affinity enhances B cell clearing in nonhuman primates. *Blood* **113**: 3735–43.

Zeidler R., Reisbach G., Wollenberg B., Lang S., Chaubel S., Schmitt B., Lindhofer H. (1999). Simultaneous activation of T cells and accessory cells by a new class of intact bispecific antibody results in efficient tumor cell killing. *J Immunol* **163**: 1246–52.

Zhang W., Gordon M., Schultheis A.M., Yang D.Y., Nagashima F., Azuma M., Chang H.-M., Borucka E., Lurje G., Sherrod A.E., Iqbal S., Groshen S., Lenz H.-J. (2007). FCGR2A and FCGR3A polymorphisms associated with clinical outcome of epidermal growth factor receptor– expressing metastatic colorectal cancer patients treated with single-agent cetuximab. *J Clin Oncol* **25**: 3712–8.

Zhang J., Liu X., Bell A., To R., Nath Baral T., Azizi A., Li J., Cass B., Durocher Y. (2009). Transient expression and purification of chimeric heavy chain antibodies. *Protein Express Purif* **65**: 77–82.

Zhou Q., Shankara S., Roy A., Qiu H., Estes S., McVie-Wylie A., Culm-Merdek K., Park A., Pan C., Edmunds T. (2008). Development of a simple and rapid method for producing non-fucosylated oligomannose containing antibodies with increased effector function. *Biotechnol Bioeng* **99**: 652–65.

6

Antibody Expression from Bacteria to Transgenic Animals

Paul Stephens and Berni M. Sweeney**

6.1 Introduction

Recombinant antibodies are the fastest growing class of protein based drugs. Since the first approval of Muromonab (OKT3, a murine IgG2a antibody to CD3) in 1986, there have been at least 29 monoclonal antibodies (mAbs) approved by the FDA for clinical use (Table 6.1), and more than 300 antibody candidates are in late phase clinical trials (Aires da *et al.*, 2008; Beck *et al.*, 2008; Jean-Mairet, 2007). It was predicted that growth of sales will continue and will reach 30 billion US dollars by 2010 (Evans and Das, 2005). As drugs, antibodies have many advantages, including high specificity and generally minimal toxicity. However, due to their predominant use in chronic indications and their relatively poor potency there has been a need to dose at the gram scale. This creates a burden on the biopharmaceutical industry to develop ways of expressing antibodies in a cost-effective manner. Cost of manufacture will vary depending on the expression system, titer of antibody, and scale of manufacture. A mammalian expression system expressing antibody at a concentration of 1 g/L will cost in the region of \$260/g when produced at the 250 kg scale, whereas a transgenic plant system at a 1,000 kg scale will cost \$14/g (Farid, 2007). The actual cost of protein generation constitutes 33% of the total cost of goods, with the cost of downstream processing activities such as purification and formulation contributing the remainder (Langer, 2009). This results in the cost of the drug to the patient being high, up to \$35,000 per annum (Farid, 2007), and therefore the drive to continually develop improved expression systems with higher expression titers, faster turnaround times and

* UCB Celltech, UK.

Table 6.1 Production cell lines of FDA approved therapeutic antibodies and fragments

Production Cell Line	Mab	Year	Company
CHO	Prolia/denosumab	2010	Amgen
CHO	Actemra/tocilizumab	2010	Roche
CHO	Stelara/ustekinumab	2009	Centocor/ OrthoBiotech
CHO	Arzerra/ofatumumab	2009	GSK
SP2/0	Ilaris/canatinumab	2009	Novartis
CHO	Simponi/golimumab	2009	Centocor/ OrthoBiotech
E. coli	Cimzia/certolizumab	2008	UCB
NS0	Soliris/eculizumab	2007	Alexion
CHO	Vectibix	2006	Amgen
E. coli	Lucentis	2006	Genentech
NS0	TheraCIM/nimotuzumab	2005	YM Biosciences
NS0	Tysabri/natalizumab	2005	Biogen-Idec/Elan Pharmaceuticals
CHO	Avastin/bevacizumab	2004	Genentech/Roche
SP2/0	Erbitux/cetuximab	2004	Imclone/Merck-Serono/BMS
CHO	Xolair/omalizumab	2003	Genentech/Novartis
CHO	Raptiva/efalizumab	2003	Genentech/Serono
Hybridoma	131I-Bexxar/tositumomab	2003	GSK
CHO	Humira/adalimumab	2002	Abbott
CHO	Zevalin/ibritumomab	2002	Biogen-Idec
CHO	MabCampath/alemtuzumab	2001	Millenium
NS0	Mylotarg/gemtuzumab	2000	Wyeth
CHO	Herceptin/trastuzumab	1998	Genentech
NS0	Synagis/palivizumab	1998	MedImmune/ AstraZeneca
SP2/1	Remicade/infliximab	1998	Centocor/Schering Plough
SP2/2	Simulect/basiliximab	1998	Novartis
CHO	Rituxan/rituximab	1997	Genentech/ Biogen-Idec
NS0	Zenapax/daclizumab	1997	PDL/Roche
SP2/3	Reopro/abciximab	1997	Centocor/Eli Lilly
Hybridoma	Orthoclone/muromonab	1986	Johnson&Johnson

Source: http://www.centerwatch.com/drug-information/FDA-approvals.

improved process economics is paramount. Of the 25 approved mAbs, three are Fab fragments and two of these, Lucentis and Cimzia, are expressed in *Escherichia coli*, with the remaining 23 all expressed in mammalian cells (Table 6.1). Despite the fact that the majority of approved antibodies are full-length, there is a great deal of interest in the industry to produce smaller antibody fragments such as Fab, $F(ab')_2$, single chain Fv, single domain Abs (sdAbs), as well as more complex bi-specific antibodies. Antibody fragments have a much shorter serum half life and hence are especially useful for diagnostic purposes and the treatment of acute diseases. In addition their small size offers the potential for enhanced tumor penetration. The ultimate use of the antibody will to a certain extent influence the expression system of choice. If a depleting antibody is required then a eukaryotic host, usually a mammalian cell line, producing an antibody with an optimal glycosylation pattern, will be appropriate. However, if an antibody fragment is acceptable, then expression in a bacterial system might be the best alternative.

The primary sequence of an antibody molecule is a critical factor in determining its expression level and downstream processing characteristics such as propensity to aggregation and stability. Considerable time may be spent "maturing" an antibody sequence to achieve the best characteristics for process development and manufacturing. The use of phage libraries and display technologies have greatly assisted in the selection of appropriate antibody sequences, some aspects of which will be reviewed in other chapters in this book. This chapter will review current antibody expression platforms, concentrating predominantly on mammalian cell expression systems and the development of improvements, which have given rise to the current industry standard of 1–2 g/L, with a trend towards increasing levels of productivity.

6.2 Antibody Expression in Prokaryotes

Expression of heterologous proteins in bacteria is well documented, with several proteins approved for clinical use (Walsh, 2006). The absence of complex post-translational modifications such as disulfide bridge formation and glycosylation has limited the use of bacteria for producing large recombinant proteins. However, their simplicity and ease of microbial fermentation has made them ideal for simple proteins or protein domains such as antibody fragments.

6.2.1 *Gram negative E. coli*

Even though there have only been limited reports of the expression of full-length antibodies in *E. coli* (Mazor *et al.*, 2009; Simmons *et al.*, 2002), this Gram negative prokaryote has been the system of choice for the production of antibody fragments, and there are now two antibody Fab fragments expressed in *E. coli* approved by the FDA: Cimzia® (certolizumab pegol) for the treatment of Crohn's disease (CD) and rheumatoid arthritis (RA), and Lucentis® (ranibizumab) for the treatment of age-related macular degeneration (AMD) (Keystone *et al.*, 2008; Rosenfeld *et al.*, 2006). The use of *E. coli* as an expression host provides many advantages, namely, ease of genetic manipulation, short process development times, simple fermentation, scalability and absence of safety issues associated with viral contamination.

The first reports of antibody expression in *E. coli* appeared in 1984 from research groups at Celltech (UK) and Genentech (USA). In both cases full-length mature antibody genes encoding heavy (HC) and light (LC) chain sequences were expressed simultaneously from the strong *E. coli trp* promoter in a single cell. Antibody HC and LC were produced as insoluble inclusion bodies in the cytoplasm, and antigen binding activity was reconstituted from refolded polypeptides (Boss *et al.*, 1984; Cabilly *et al.*, 1984). Cytoplasmic expression of antibody fragments often results in the production of mis-folded aggregates which accumulate as inclusion bodies. While the antibody chains can be re-folded, recovery levels of correctly folded proteins are generally poor. Strategies to improve the levels of expression of soluble fragments in the cytoplasm include removal of cysteine residues (Proba *et al.*, 1998; Worn and Pluckthun, 1998), or increasing the oxidizing conditions in the cytoplasm by co-expression of the periplasmic chaperone Skp in combination with mutations in the host cell's thioredoxin and glutathione reductase (*trxB* and *gor*) genes (Levy *et al.*, 2001).

In an attempt to overcome the issues with bacterial cytoplasmic expression, two groups described the production of functional soluble antibody fragments in the periplasm. In these cases the antibody protein was directed to the oxidizing environment of the periplasmic space by the use of the bacterial leader sequences from pectate lyase (PelB) or outer membrane protein A (OmpA) and alkaline phosphatase (PhoA) (Better *et al.*, 1988; Skerra and Pluckthun, 1988). It is now generally accepted that for optimal production, antibody fragment expression is directed to the periplasmic space, where yields in the range of 0.1 to 100 mg/L from shake flask cultures and up to 1–2 g/L in fermentors have been achieved (Better *et al.*, 1988; Carter *et al.*, 1992; Chen *et al.*, 2004; Skerra and Pluckthun, 1988). After expression the antibody can be recovered from the periplasmic space by osmotic shock, Tris/EDTA, or other extraction methods. The purity of the crude extract is greatly improved by performing the extraction at an elevated temperature of 60°C, which ensures the removal of mis-folded and proteolysed Fab fragments (Popplewell *et al.*, 2005).

Expression in *E. coli* is affected by numerous factors including gene and protein sequence, transcription and translation efficiencies, and protein folding, aggregation and degradation (Arbabi-Ghahroudi *et al.*, 2005; Donzeau and Knappik, 2007; Fernandez, 2004; Sorensen and Mortensen, 2005). Expression of antibody fragments can be influenced at the DNA level by codon usage and the presence of inhibitory 5′ mRNA hairpin loop structures (Humphreys *et al.*, 2000), the ratio of HC and LC (Humphreys *et al.*, 2002; Simmons *et al.*, 2002), and the stability and folding rate of the individual chains and their subsequent assembly (Oelschlaeger *et al.*, 2003; Worn and Pluckthun, 1998; Wulfing and Pluckthun, 1994). A variety of approaches have been taken to optimize the levels of expression of antibody fragments in the periplasmic space. Evolutionary approaches utilize the construction of recombinant libraries and subsequent selection via genetic screening or phage, yeast or ribosome display techniques. Humphreys *et al.* (2000) used enhanced expression of secreted alkaline phosphatase to identify improved signal peptides. A library of codon-optimized signal peptides from yeast glucanase BGL2 were screened, and a novel signal peptide sequence was isolated which, when linked to antibody heavy and light chains, yielded 260 mg/L of purified Fab from a fermentor. Martineau *et al.* (1998) took a different approach and expressed a library of scFv molecules that bound to and activated a mutant β-galactosidase. By plating on media containing limiting amounts of lactose, they were able to isolate a scFv with a 50-fold increase in expression. Phage display has also been used to isolate antibody binding domains with increased affinity. The isolation of antibodies with increased affinity is dependent on both the affinity of the phage antibody and its level of expression. It is therefore not surprising that when Deng *et al.* (1994) and Jackson *et al.* (1995) were attempting to isolate antibodies with improved affinity, they also found antibodies with increased levels of expression. The link between antibody fragment stability and its level of expression is also well established, and selection for antibody fragments with increased affinity under stress conditions such as low pH, high temperature, or high levels of chaotrope often results in increased levels of expression (Jespers *et al.*, 2004; Jung *et al.*, 1999).

While these library-based methods have proved useful, more rational approaches to optimization have also been attempted. Jung and Pluckthun describe CDR grafting from a non-soluble scFv into the framework of a well-expressed soluble scFv (Jung and Pluckthun, 1997). In other approaches individual amino acids have been changed to improve the yield of antibody fragments. Forsberg *et al.* (1997) changed multiple residues in the V_L (variable light) domain resulting in titers approaching 500 mg/L, while changes in V_H (variable heavy) prevented cell lysis and increased periplasmic accumulation up to 700 mg/L. Additionally, a single amino acid change at position 6 in V_H led to a 30-fold increase in expression of soluble scFv due to enhanced folding efficiencies (Kipriyanov *et al.*, 1997). Lien and Lowman (2008) also describe how modification of the primary sequence led to increased therapeutic benefit of their second generation anti-VEGF antibody Lucentis®. They performed both site-directed mutagenesis and phage display to derive their final variable region sequence, which had enhanced affinity and improved expression characteristics in *E. coli.*

The fate of expressed antibody fragments in the periplasm is a balance between mis-folded, non-disulfide bridged aggregates and soluble functional proteins. Strain engineering has been used to counteract the mis-folding, aggregation and proteolysis induced by high level expression of antibody fragments in both the periplasm and cytoplasm. Knappik *et al.* (1993) showed that overexpression of the periplasmic disulphide isomerase (DsbA) and proline cis-trans isomerase (PPIase) had a marginal benefit on the expression of scFv, Fab and Fv fragments and speculated that periplasmic folding occurred before disulfide bridge formation. However, Zhang *et al.* (2002) showed that overexpression of two different disulfide isomerases (DsbC and DsbG) did alter the ratio of soluble versus insoluble fragment, with the level of soluble functional scFv increasing to almost 50% of the total target protein. Furthermore, in two analogous studies, Bothmann and Pluckthun showed that both Skp and FkpA chaperones increased the level of soluble scFv by increasing the efficiency of folding and suppressing aggregation (Bothmann and Pluckthun, 1998; Bothmann and Pluckthun, 2000). This effect was independent of the enzymatic activity of PPIase. Interestingly, they showed that this effect was more marked in poorly folded proteins. Poorly folded antibodies have a lower rate of folding and disulfide bridge formation, leading to the accumulation of unfolded aggregates, whereas a better tolerated antibody folds more efficiently allowing rapid disulfide bridge formation and less aggregation. Therefore, factors that increase the rate of correct folding will help antibody production. Another factor that can affect antibody accumulation is proteolysis and *E. coli* strains which lack the major periplasmic proteases such as DegP and Prc have shown significant increases (two- to three-fold) in protein accumulation (Chen *et al.*, 2004).

The utility of *E. coli* for the production of therapeutic antibody fragments has been proven by the approval of Lucentis® and Cimzia®, and several other *E. coli* produced antibodies are in clinical or preclinical studies. The economics of *E. coli* fermentation and the relative simplicity of the downstream processing coupled to the power of phage display technologies and directed evolution to improve the antibody properties for their particular use make this prokaryotic organism an extremely attractive choice for antibody manufacture.

6.2.2 *Alternative bacterial expression systems*

There are limited reports on the expression of antibodies and antibody fragments in bacteria other than *E. coli.* Two independent groups describe the use of the L form of the Gram negative bacteria *Proteus mirabilis* to express scFvs. The L form of *P. mirabilis* lacks its outer membrane and is therefore devoid of any periplasmic space, consequently avoiding the accumulation of mis-folded and aggregated protein which could have potentially toxic side effects. Using *lac* promoter based expression vectors with an OmpA signal sequence, two groups purified active soluble scFv from the culture media with yields between 10 and 200 mg/L. Expression was enhanced by reducing the culture temperature during the production phase to

30°C (Kujau *et al.*, 1998; Rippmann *et al.*, 1998). A system based on the Gram negative *Pseudomonas fluorescens* was developed by DOW Chemical Company as an alternative expression system. Following elucidation of the entire genomic sequence of *P. fluorescens*, they developed a variety of alternate strains that could be utilized for the expression of different protein types. They have reported the successful expression of scFvs at levels up to 3 g/L (Squires *et al.*, 2006).

The use of various *Bacillus* and *Lactobacillus* species of Gram positive bacteria has also been reported for antibody expression. Like the L form of *P. mirabilis*, Gram positive bacteria lack an outer cell wall and have no periplasmic space, so secreted proteins accumulate in the culture media. There are several reports where *Lactobacilli* have been used for topical delivery of potentially therapeutic scFvs. The fact that *Lactobacilli* have GRAS (Generally Recognized As Safe) status and that they are commensal bacteria makes them ideal hosts for this purpose. Hammarstrom and colleagues developed *Lactobacillus* for the delivery of antibodies targeted to periodontal pathogens. They have directed expression of the scFvs to either the streptococcal antigen I/II of *Streptococcus mutans* or the RgpA protease of *Porphyromonas gingivalis* on to the surface of *Lactobacillus zeae* and *Lactobacillus paracasei*, respectively, and have seen some benefit in animal models of dental disease (Kruger *et al.*, 2002; Lee and Da Silva, 1996). Similarly, *Lactobacillus casei* has been used topically to deliver an anti-ICAM scFv to the vaginal mucosa of human PBL-SCID mice resulting in the reduction of infectivity of HIV-1 (Chancey *et al.*, 2006).

Expression of antibody fragments in three different strains of *Bacillus* has also been described. Levels of up to 100 mg/L of a Fab fragment directed against human urokinase-type plasminogen activator have been reported in *Bacillus brevis* using the promoter and signal sequence derived from the major cell-wall protein Mwp (Inoue *et al.*, 1997). While initial levels of expression of scFv in *Bacillus subtilis* were poor, improvements have resulted in a three- to four-fold increase in expression. Starting with a strain that was lacking six extracellular proteases, Wu and colleagues went on to delete the *hrcA* repressor gene, which controls the GroE and DnaK intracellular chaperone operons, and to overexpress the extracellular chaperone PrsA. The overexpression of both intra- and extracellular chaperones led to a 2.5-fold decrease in the level of insoluble aggregate and an increase in soluble secreted antibody fragment from 4 to 12 mg/L (Wu *et al.*, 1998; Wu *et al.*, 2002). A final variant in which the cell-wall protease WprA was deleted led to a final yield of 15 mg/L (Wu *et al.*, 1993). *Bacillus megaterium* does not express alkaline proteases and has exhibited high plasmid stability during its growth phase. Therefore, it is also an attractive host for heterologous protein expression. This strain has been used by Jordan *et al.* (2007, 2009) for the expression of both scFv and scFab. They reported levels of expression similar to that seen in *E. coli* but the specific activity was marginally higher, implying that more of the antibody fragments were correctly folded and active.

6.3 Antibody Expression in Eukaryotes

Unlike bacteria, eukaryotes possess complex protein folding and secretory pathways that not only make them effective in secretion of heterologously expressed proteins, but also allow them to carry out complex post-translational modifications. Eukaryotes are able to glycosylate secreted proteins, and this modification is absolutely essential for antibody effector functions such as antibody-dependent cell cytotoxicity (ADCC) and complement-dependent cytotoxicity (CDC). They are therefore the host of choice for the expression of full-length antibodies. However, alternative eukaryotic hosts generate different glycan structures, and an understanding of the nature of these structures and their subsequent effects is a key factor in the selection of an appropriate host.

The structure of the carbohydrate modifications can affect many of the biological properties of glycoproteins including pharmacokinetics, bioactivity, secretion, *in vivo* clearance, solubility, receptor recognition and antigenicity (Gerngross, 2005; Storring and Gaines Das, 1992; Takeuchi *et al.*, 1988; Wasley *et al.*, 1991; Wurm, 2004). The ability of an antibody to mediate its effector functions is dependent on its ability to bind FcγRIIIα (for ADCC) or C1q (for CDC). Both binding events are strongly influenced by the nature of the carbohydrate moieties on the antibody (Niwa *et al.*, 2005a; Niwa *et al.*, 2005b; Simmons *et al.*, 2002). The glycosylation machinery of the host cell along with culturing methods (Gawlitzek *et al.*, 1995; Jenkins and Curling, 1994; Schweikart *et al.*, 1999) can therefore have a major impact on the glycan content and potency of the recombinant antibody.

All IgGs possess a consensus N-linked glycosylation sequence (Asn-X-Ser/Thr: X = any amino acid except Pro) at position 297 in the CH2 domain of the heavy chain (Kanda *et al.*, 2007). The normal glycan structure found on polyclonal human IgGs is biantennary in nature and composed of two N-acetylglucosamine (GlcNAc) residues joined to the peptide backbone. Linked to the GlcNAc residues are three mannose (Man) residues that form a branched structure through their α1,3 and α1,6 linkages. Two additional GlcNAc residues are linked to the branched mannose residues (Fig. 6.1). In addition fucose (Fuc), galactose (Gal) and N-acetyl neuraminic acid (NANA) residues may be also be added. As varying glycan structures are attached to a single invariant peptide backbone, this leads to pools of different glycoforms of the protein. These variations may be found in the site-occupancy (macroheterogeneity) or in the structure of attached glycans (microheterogeneity) (Butler, 2006). It had been suggested by Jefferis that there is the potential for more than 400 different glycoforms to occur due to the random pairing of heavy chains with different glycan structures (Jefferis, 2005b; Walsh and Jefferis, 2006).

In addition to the N-glycosylation site found in the constant domain it is also estimated that 20% of naturally occurring IgGs have additional sites in their variable

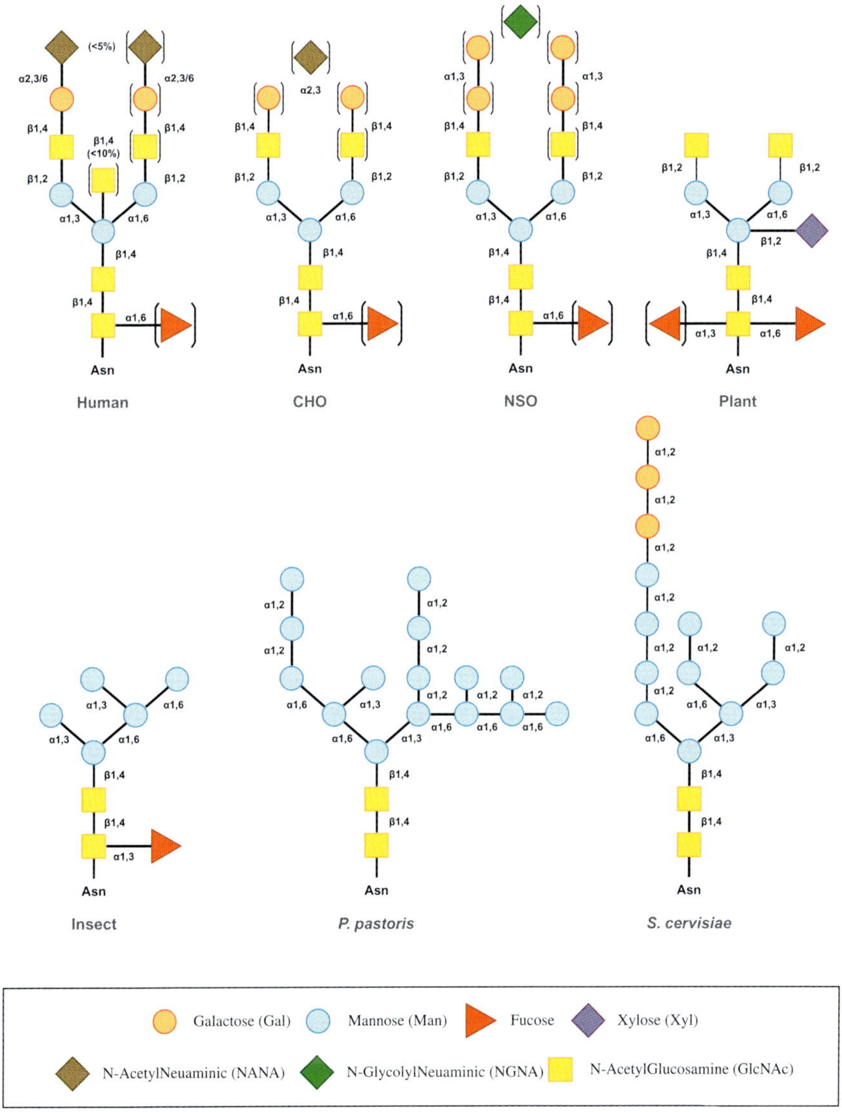

Figure 6.1 Typical complex N-glycan structures found on proteins produced by commercial production systems

domains (Lim *et al.*, 2008; Mimura *et al.*, 2007). However, this had been observed for only one approved therapeutic antibody, Erbitux® (cetuximab) (Holland *et al.*, 2006; Qian *et al.*, 2007; Li *et al.*, 2008). Unless it is required for activity, the removal of glycosylation sites from the variable domains by genetic engineering to ensure a more homogeneous antibody product is generally a preferred option.

6.3.1 *Yeast*

Yeast provides many advantages for the expression of recombinant antibodies and antibody fragments. Being eukaryotic, their sophisticated cellular machinery allows them to fold, glycosylate and secrete expressed proteins, yet they have the same simple fermentation requirements as single cellular prokaryotes. Their short generation time allows rapid scale-up and high biomass (100 g/L) using inexpensive chemically defined media. Yeast are also very well suited to downstream processing because they do not produce endotoxin, have very low levels of endogenously secreted proteins and are deemed to be safe, with some yeast strains having GRAS status. The most commonly used yeast for heterologous protein production are the traditional baker's yeast *Saccharomyces cerevisiae* (Yin *et al.*, 2007), *Pichia pastoris* (Cregg *et al.*, 2000; Zhou *et al.*, 2006) and *Schizosaccharomyces pombe* (Giga-Hama and Kumagai, 1999; Kumar and Singh, 2004).

Despite these advantages there have been relatively few reports of secretion of full-length functional antibodies from yeast. Wood *et al.* (1985) first reported the production of antibodies in *S. cerevisiae*. However, no full-length assembled antibody was detected in the culture supernatant, despite the fact the intact functional heterotetrameric antibody was found inside the cells. Using a *phosphoglycerate kinase* promoter and invertase leader sequence, Horwitz *et al.* (1998) produced both functional secreted full-length antibody and Fab fragment. The full-length antibody was able to exhibit ADCC, indicating that the glycosylation was sufficient for effective binding to FcγRIIIa, but the antibody did not show CDC. It was speculated that these differences in antibody effector function may be due to subtle differences in the carbohydrate structures on the expressed antibody. Ogunjimi *et al.* (1999) reported production of 36 mg/L of full-length antibody from *P. pastoris* using the *alcohol oxidase* 1 gene (*AOX1*) promoter and the *S. cerevisiae* mating factor α secretion signal. It has been speculated that the relatively low level expression of full-length antibodies may be due to the steric hindrance induced by the high mannose carbohydrate structures in the Fc domain (Gasser and Mattanovich, 2007).

A variety of reports describe the expression of antibody fragments in both *S. cerevisiae* and *P. pastoris* with levels between 70 mg/L and 1.2 g/L (Eldin *et al.*, 1997; Freyre *et al.*, 2000; Gurkan *et al.*, 2004). The higher levels were achieved using a combination of the *AOX1* promoter and the *S. cerevisiae* invertase leader sequence. Although high level expression of antibody fragments has been achieved in yeast, this can lead to an accumulation of mis-folded protein within the cell and the subsequent induction of the unfolded protein response (UPR). To overcome this issue Shusta *et al.* (1998) showed that co-expression of BiP (a member of the Hsp70 type chaperone family responsible for the protection of unfolded proteins from degradation) or protein disulfide isomerase (PDI) led to a two- to eight-fold increase in expression of five scFvs in *S. cerevisiae*. Similarly Gasser *et al.* (2006) showed that in *P. pastoris* intracellular retention of the Fab fragment could be alleviated to a certain extent by the expression of PDI and the UPR transcription factor HAC1, leading to a moderate increase in Fab expression of 1.3 and 1.9 fold, respectively.

Even though yeast can glycosylate secreted proteins, one of their major disadvantages is their inability to generate carbohydrate structures that correctly mimic those produced by mammalian cells (Fig. 6.1). Yeast are defective in the mannose trimming mechanism seen in higher eukaryotes, and this results in high mannose structures. In addition, they generate an excess of O-linked carbohydrates (Beck *et al.*, 2008; Li *et al.*, 2006). However, there have been extensive developments in the re-engineering of *P. pastoris* to produce more mammalian-like carbohydrate structures (Li *et al.*, 2006; Potgieter *et al.*, 2009). This glycoengineering was pioneered by GlycoFi Inc. (USA). They have eliminated enzymes involved in the endogenous yeast glycosylation pathways, while properly localizing five active eukaryotic proteins, including mannosidases I and II, N-acetylglucosaminyl transferases I and II, and uridine 5′-diphosphate (UDP)-N-acetylglucosamine transporter. Targeted localization of the enzymes enabled the generation of a synthetic *in vivo* glycosylation pathway, which produced the complex human heptasaccharide ($GlcNAc_2Man_3GlcNAc_2$) (Hamilton *et al.*, 2003; Hamilton and Gerngross, 2007). Li *et al.* (2006) also showed that, using these glycoengineered *P. pastoris* strains, the intact antibodies produced were capable of mediating high affinity FcγRIIIa binding and subsequent ADCC. They went on to produce these glycoengineered antibodies in bioreactors at levels in excess of 1 g/L (Potgieter *et al.*, 2009).

The most clinically advanced antibody produced in yeast is an anti-Interleukin 6 antibody (ALD518), which is expressed in *P. pastoris*. This is in a Phase II rheumatoid arthritis trial and a Phase I oncology trial conducted by Alder Biopharmaceutical Inc. (USA). They avoided the glycosylation issue by removing the N-linked glycosylation site in the antibody CH2 domain by site-directed mutagenesis. The results of the Phase I trial for treatment of cancer cachexia and fatigue indicated that it was safe and well tolerated, reversed fatigue, and increased hemoglobin and albumin levels (Clarke *et al.*, 2009).

As exemplified by the Alder clinical trial results and the recent acquisition of Glycofi by Merck and Co. (USA), yeast as a manufacturing platform for human therapeutic antibodies is gaining momentum.

6.3.2 *Filamentous fungi*

Filamentous fungi have been used widely in the food industry for the production of additives and have GRAS status. They achieve high levels of biomass in fermentation with densities up to 100 g/L of dry mass. In addition, production levels of 35 g/L of the host cell protein cellobiohydrolase have been reported (Durand *et al.*, 1988). There are now several reports of filamentous fungi being used to produce antibodies and antibody fragments (for reviews see Gasser and Mattanovich, 2007; Joosten *et al.*, 2003).

High levels of secretion from filamentous fungi are generally achieved by fusing the protein of interest to the carboxyl terminus of a highly secreted endogenous protein. Examples include glucoamylase from *Aspergillus niger* (Frenken *et al.*, 1998; *Ward et al.*, 2004) and cellobiohydrolase from *Trichoderma reesie* (Nyyssonen *et al.*, 1993). Exploiting the high level of endogenous secretion of glucoamylase, Ward *et al.* (2004) were able to achieve up to 0.9 g/L of a full-length antibody and 1.2 g/L of a Fab fragment from shake flask cultures over five days. Inclusion of a KexB cleavage site between the fusion protein and the antibody protein improved the efficiency of proteolytic removal of the fusion protein in the Golgi. However, removal is not always 100% efficient and can lead to heterogeneity at the N-terminus of the antibody chains. Processing was improved by including three glycine residues between the cleavage site and the beginning of the antibody sequence. They also observed that only half of the antibody heavy chains were glycosylated, and the sugars were of high mannose content and possessed a potentially immunogenic terminal galactose in the furanose form residue. Expression of an aglycosylated (Asn297Gln) antibody, however, resulted in reduced levels of expression. Interestingly, despite the differences in the glycosylation profile of the antibodies produced in the filamentous fungi, their circulating half life and ADCC function were similar to antibodies produced in mammalian cells.

Alternative antibody fragments such as scFv and llama V_{HH} have also been successfully expressed in *Aspergillus awamori* (Chancey *et al.*, 2006; Frenken *et al.*, 1998). A five-fold increase in expression (10 mg/L to 50 mg/L) was achieved in a shake flask by the use of the glucoamylase fusion partner, and levels increased further to 200 mg/L on fermentation. Potential issues with the proteolytic degradation of the V_{HH} domain were addressed with the addition of bovine serum albumin (BSA) to the media, which resulted in a five-fold increase in the level of V_{HH} accumulation. However, whether this was due solely to the proteolytic blocking activity of the BSA or the fact that BSA will enhance heterologous gene expression in cell culture is unclear.

Given their history of safe use and their potential to secrete high levels of protein, filamentous fungi may be useful for large scale antibody expression. However, more work is required to address the issues of product heterogeneity and inappropriate glycosylation. This may be advanced with the advent of engineered or newly selected strains with the ability to produce proteins with more human-like carbohydrate structures.

6.3.3 *Insect cells*

The use of insect cells for the expression of heterologous proteins has been greatly enabled by the use of viral vectors. Insect specific viruses, especially *Autographa californica* nuclear polyhedrosis virus (AcNPV), form the backbone of the majority of insect cell expression vectors. These viruses are deemed safe in that they have a limited cell tropism only infecting insect cells. The expression vectors utilize the powerful *polyhedrin* or *very late p10* promoters, which can lead to expression levels of up to 50% of total cell protein in insect cell lines such as Sf-9 and Sf-21 of *Spodoptera frugiperda*, or High Five™ of *Trichoplusia ni*. Proteins are produced transiently following a round of viral infection. The procedure can be lengthy and requires the generation of high viral titer stocks prior to the final production run. Heterologous proteins are then expressed 96 hours after infection. Cell lysis and subsequent release of proteases may affect protein quality, and protease inhibitors are often included in the production run. Alternative promoters such as the baculovirus *immediate early* (*IE1*) promoter have been used, directing expression of the heterologous gene to the early stage of the infection cycle when there is less cell lysis and more appropriate secretion and post-translational modification (Jarvis *et al.*, 1996). In addition to expression in insect cells using baculoviruses, there are reports of expression of antibodies and antibody fragments in stable insect cell lines. This avoids the need to generate infectious virus particles and moves the producer from a transient lytic system to a stable continuous system (Reavy *et al.*, 2000).

Despite the utility of insect cells and the baculovirus expression systems, their use for the expression of antibodies has proven somewhat limited. This may be due to the relatively low levels of observed expression. Liang *et al.* (2001) described the expression of a panel of full-length antibodies at levels of 6–18 mg/L, while Kretzschmar *et al.* (1996) achieved levels of 32 mg/L for the expression of a scFv from Sf-9 cells.

Expression of heterologous proteins in insect cells leads to the intracellular accumulation of mis-folded and aggregated protein with greater than 90% of expressed protein being retained within the cells (Shuler *et al.*, 1995). Various approaches to increase the level of antibody expression by co-expressing chaperones and foldases within insect cells have been attempted. Overexpression of BiP led to a 90% increase in antibody secretion (Hsu and Betenbaugh, 1997). Similarly, overexpression of HSP-70 led to an increase in soluble light chain precursor, which in turn led to an increase in soluble light chain and consequently an increase in accumulated soluble secreted antibody (Ailor and Betenbaugh, 1998).

Likewise co-expression of PDI led to an increase in the level of soluble IgG and subsequent increased accumulation of secreted antibody (Hsu *et al.*, 1996).

Examination of the sugar moieties on antibodies secreted into the culture media from High Five™ cells indicated an equal split of complex, hybrid and low mannose structures. The carbohydrate possessed terminal galactosyl residues but lacked terminal sialic acid. The majority of antibodies (84%) have α1,6 linked fucose with a lower proportion also possessing the potentially immunogenic α1,3 linked fucose moiety (Ailor and Betenbaugh, 1999; Hsu *et al.*, 1997). Despite the fact that the glycosylation pattern of antibodies produced in insect cells is somewhat different from that seen in mammalian cells, the insect-derived antibodies retained both CDC and ADCC (Edelman *et al.*, 1997; Jin *et al.*, 1995; zu Putlitz J. *et al.*, 1990). Attempts have been made to engineer insect cell host lines to produce more mammalian cell-like carbohydrates. The introduction of the mammalian genes encoding β1,4-galactosyltransferase, α2,6-sialyltransferase, sialic acid synthase and CMP-sialic acid synthase resulted in the production of sugar moieties with terminal sialic acid (Aumiller *et al.*, 2003; Harrison and Jarvis, 2006; Hollister *et al.*, 1998).

A potential option for industrializing the generation of antibodies from insect cells has been developed whereby a Fab fragment was produced from whole insect larvae. Recombinant baculoviruses encoding the Fab fragment were used to infect fifth instar larvae in an automated PERLXpress™ system; this resulted in the production of 1.1 g of Fab/kg of larvae (O'Connell *et al.*, 2007).

Despite the advantages of the baculovirus system for the expression of heterologous proteins, the production of antibodies in insect cells has not been widely adopted. This is presumably due to low levels of expression, issues with inappropriate glycosylation and the time taken to produce high titer infectious virus stocks.

6.3.4 *Transgenic plants and plant cells*

The primary attraction for the development of expression systems in plants is the potential for very large scale and cost-effective production. The scale of production is limited only by the availability of land acreage. Plants and algal systems are considered to have GRAS status and do not harbor pathogenic contaminants and pose little risk of viral, prion or bacterial endotoxin contamination. Methods for sowing, growing and harvesting are well established and do not involve high costs of labor, manufacturing facilities or complex media. A variety of different crops lend themselves to antibody production including expression in the leaves of plants such as tobacco (Valdes *et al.*, 2003) and alfalfa (Bardor *et al.*, 2003b), or in seeds such as rice, wheat (Stoger *et al.*, 2000), barley (Schunmann *et al.*, 2002), maize (Hood *et al.*, 2002), pea (Perrin *et al.*, 2000) and soya bean (Zeitlin *et al.*, 1998). The expression of heterologous proteins in seeds has the advantage of providing an inert environment for subsequent storage. One of the major issues for using recombinant plants is the environmental impact of genetically engineered crops. The public acceptance of plant-generated biotherapeutics may become an obstacle for the use of plants for antibody production. A potential solution could be provided by the expression of antibodies in the chloroplasts of transgenic algae such as *Chalamydomonas rienhardtii* (Fletcher *et al.*, 2007; Mayfield *et al.*, 2007). Plant cells have a similar endomembrane system to animal cells and consequently are able to carry out many of the same post-translational modifications including signal peptide processing and disulfide bridge formation. Plants are also capable of N-glycosylation but with a glycan composition distinct from that of animal cells.

The first report of full-length antibody expression in plant cells was by Hiatt *et al.* in 1989. There have since been numerous reports, and up to 20 companies are currently engaged in production of antibodies in plants (Farid, 2006). The most advanced is CaroRX^{TM}, produced in tobacco plants by Planet Biotech (Mountain View, California). This is a chimeric secretory IgA aimed at preventing the re-colonization of *Streptococcus mutans* in human plaque. However, it failed to show efficacy in a Phase II clinical trial (Weintraub *et al.*, 2005). Phase I clinical trials of anti-idiotypic antibodies produced in tobacco plants for the treatment of non-Hodgkin's lymphoma have also been reported (McCormick *et al.*, 2008).

Three basic methods are utilized to introduce DNA into plant cells depending on the final use of the expressed product: viral infection, gene transfer via *Agrobacterium tumefaciens*, and plasmid DNA coated particle bombardment. Viral infection is rapid and high level expression is achieved by systemic infection of the whole plant tissue. This has been exploited particularly for transient expression (Lico *et al.*, 2008). Viral infection requires the viral transcript to be inoculated on each plant tissue since systemic movement can be limited by the size of the transgene. The system is also prone to high mutation and deletion rates during viral RNA replication (Smith *et al.*, 1997). The use of agroinfection circumvents some of these issues, as viral replicons are delivered as DNA precursors (Marillonnet *et al.*, 2005; Vaquero *et al.*, 2002). Verch *et al.* (1998) expressed the mAb CO17-1A, directed to a colon cancer antigen, by independently introducing the genes encoding the heavy and light chains with a tobacco mosaic virus vector. More recently a co-viral infection system was developed by Giritich *et al.* (2006) which utilizes two sets of vectors derived from non-competing Tobacco Mosaic Virus (TMV) and Potato Virus-X (PVX), which each express a single antibody gene. This system allows the individual viruses to interact with different host cell proteins, allowing efficient viral replication and movement without interfering with each other. The HC and LC are therefore expressed in the same cells throughout the plant, resulting in yields of up to 0.5 g of assembled mAb/kg of fresh-leaf biomass.

For stable expression the methods of choice are either *A. tumefaciens*, which allows random DNA integration into the plant cell genome, or bombardment of plant cells with microprojectiles bearing DNA on their surface (Hansen and Wright, 1999). Antibody genes can either be co-transfected on two different vectors or linked on the same vector. For efficient antibody expression in plants it is important to use different promoter elements to avoid homology based gene silencing (De Neve M. *et al.*, 1999; De Wilde C. *et al.*, 2000). The strong constitutive *35S* promoter derived from Cauliflower Mosaic Virus (CaMV35S) is a popular choice in dicotyledons, whereas the maize *ubiquitin-1* promoter is preferred in monocotyledons (Hiatt *et al.*, 1989; Ma *et al.*, 1995). Ko *et al.* (2003) used a combination of the *CaMV35S* promoter to drive the heavy chain and the potato *proteinase inhibitor II* (*pin2*) promoter to drive the light chain for expression of an anti-rabies antibody in tobacco. Transgenic plants expressing individual antibody genes can be crossed to generate a plant producing an intact antibody. This method has been used to introduce other post-translational modifying genes such as secretory factor and J chain for IgAs or glycosylation modifying enzymes (Ma *et al.*, 1995).

Expression in the chloroplasts of algae has several advantages, not least of which is cost. It has been estimated that the cost of goods for antibodies made in *Chalamydomonas rienhardtii* is 6–8 USD/g compared to 150 USD/g in CHO (Chinese Hamster Ovary) cells (Franklin and Mayfield, 2005). In addition algae are grown in full containment, alleviating some of the environmental concerns about using plants in open fields; as algae do not produce pollen, there is also less risk of introducing transgenes into the food chain. Chloroplasts have a double membrane structure, and they can only be transformed efficiently by bombardment with tungsten or gold particles. However, once this has been achieved, the time from transformation to production is short. Stable cell lines can be generated in ten days, and scale-up can be achieved in only a few weeks. Each plant cell has 50–80 copies of the chloroplast genome. Consequently, there is the opportunity for a large gene dosage effect, and by utilizing alternative selection systems several different recombinant proteins can be expressed simultaneously.

Chloroplast expression is not prone to the gene silencing seen in other plant systems. Therefore, long-term stable expression can be maintained. In addition, algae are able to correctly fold and assemble complex proteins such as antibodies (Franklin and Mayfield, 2005). Through careful codon selection, the choice of appropriate 5′ untranslated regions and promoters, expression of up to 0.5% of total plant soluble protein has been expressed as a single chain antibody (Mayfield *et al.*, 2003; Mayfield *et al.*, 2007) The lack of any post-translational glycosylation in chloroplasts is a potential weakness and only antibodies or antibody fragments that do not require Fc effector function for their biological activity can be usefully made in chloroplasts. However, steps are being taken to introduce enzymes from the mammalian glycosylation pathway into chloroplasts (Ko *et al.*, 2009).

The glycosylation of antibodies produced in plants is different from that of mammalian cells. Both systems add the oligosaccharide precursor ($Glc_3Man_9GlcNAc_2$) onto the specified aspargine residues co-translationally in the endoplasmic reticulum, but differ in their late Golgi apparatus. Core α1,6 linked fucose residues and terminal sialic acids are added in mammalian cells, whereas bisecting β1,2 xylose and core α1,3 linked fucose residues are added in plants. In addition, β1,3 linked galactose and fucose are added to the terminal GlcNac residue to form a Lewis glycosylation epitope, which along with the β1,2 xylose and core α1,3 linked fucose form the glycol epitopes of some plant allergens (Saint-Jore-Dupas *et al.*, 2007; van Ree R. *et al.*, 2000). Expressed antibodies with these modifications were found to elicit an immune response in a majority of laboratory animals tested except BALB/c mice (Bardor *et al.*, 2003a). However, when these antibodies were administered to patients no adverse responses were observed (Ma *et al.*, 1998; Zeitlin *et al.*, 1998). Despite this, attempts have been made to either introduce heterologous or inactivate endogenous glycosyltransferases. Most notably the moss *Physcomitrella patens* has been engineered to prevent the addition of β1,2 xylose and α1,3 fucose (Koprivova *et al.*, 2004). The lack of α1,3 linked fucose on plant-generated antibodies leads to the same enhancement of ADCC as non-fucosylated antibodies produced in CHO cells without affecting the stability or affinity of the antibody (Cox *et al.*, 2006). The use of a carboxyl terminal H/KDEL tetra peptide tags to retain the antibody in the endoplasmic reticulum can also have a profound effect on expression levels and result in modified glycosylation patterns. Schillberg *et al.* (2003) noted a two- to ten-fold increase in expression in leaves and seeds of tobacco plants, as well as in the endosperm cells of rice, where scFvs accumulated in protein bodies and protein storage vacuoles (Torres *et al.*, 2001).

Clearly the production of antibodies in plants and plant cells has many attractions, the primary ones being the ease of scale-up and the consequent potential for relatively low cost of goods. Issues with the post-translational modification of plant-produced antibodies are being addressed. A recent paper described the production of 1.5 g of antibody/kg of fresh tobacco leaf (FW) (or 25% total soluble protein) utilizing the strong alfalfa *plastocyanin* promoter and 5′ and 3′ regulatory elements, the enhancing activity of the HcPro suppressor of silencing, a KDEL retention signal, and the co-expression of a chimeric β1,4 galactosyltransferase (Vezina *et al.*, 2009). This work shows what can be achieved by combining all of the most recent advances.

6.3.5 *Transgenic animals*

As with transgenic plants the major advantage of producing antibodies in transgenic animals is their scalability and the relatively low cost of production. Calculations of the cost per gram of antibody indicate that at the industrial scale

(100 kg/year) the cost of producing antibodies in the milk of transgenic goats is three to thirty times cheaper than mammalian cells (Farid, 2006). Herd size offers huge flexibility, where production can readily be tailored to demand without the need for large, expensive mammalian cell production plants. A mammalian cell culture facility costs millions of Euros and takes some years to build, whereas the cost of generating a founder animal is 30 to 50,000 Euros, and herd production time and costs are the same as that for any conventional animal husbandry (Kind and Schnieke, 2008). Transgenic animals also have an advantage over alternative non-mammalian systems in that the post-translational glycosylation carried out on the expressed antibody mimics more closely that produced in humans and mammalian cell culture (Kind and Schnieke, 2008). Drawbacks to transgenic animal production include the time required to produce a herd of sufficient size that is very species-specific (Houdebine, 2005; Houdebine, 2009), potential contamination with animal proteins and DNA, and contamination with prion and animal viruses leading to potential zoonotic diseases. These issues can be mitigated by the use of animals with well-defined pedigree, effective hygiene precautions and appropriate purification regimes.

The heterologous protein is generally targeted to the mammary gland and accumulated in the milk using milk protein specific promoters such as *β-casein, whey acidic protein*, and *beta lactoglobulin* (Castilla *et al.*, 1998; Limonta *et al.*, 1995; Tang *et al.*, 2008). The levels of expression seen in these systems is impressive, with up to 17 g/L in mice (Tang *et al.*, 2008) and 8 g/L in goats (Pollock *et al.*, 1999).

In addition to antibody production in transgenic mammals there have been several reports of antibody expression in chicken eggs. Egg white contains approximately 4 g of protein, half of which is ovalbumin, implying that a heterologous gene linked to the *ovalbumin* gene promoter could yield up to 1 gram of recombinant protein/egg. With hens laying up to 300 eggs per year, a single chicken could therefore produce approximately 300 g of recombinant protein/year (Ivarie, 2006). Zhu *et al.* (2005) reported a yield of 147.9 mg/L (or 3 mg/egg) when the antibody was expressed from the *ovalbumin* gene promoter in tubular gland cells of the oviduct, and Kamihira described levels of 20–150 mg/egg when expressed constitutively from the chicken *β-actin* promoter (Kamihira *et al.*, 2005; Kamihira *et al.*, 2009).

Expression of heterologous genes in transgenic animals often results in gene silencing due to the epigenetic characteristics of the host genome (Hofmann *et al.*, 2006). Attempts have been made to reduce this effect by the use of insulator elements such as the chicken *β-globin* gene insulator (Taboit-Dameron *et al.*, 1999; Tang *et al.*, 2008) or by the use of extended versions of the *α-lactalbumin* or *whey acidic protein* promoter gene control regions, which are believed to contain endogenous insulator elements (Fujiwara *et al.*, 1997; Rival-Gervier *et al.*, 2002; Stinnakre *et al.*, 1999).

Antibodies produced in transgenic animals exhibit a glycosylation pattern that is similar to that seen in man but with subtle exceptions. Antibodies expressed in animals are only partially sialylated like those expressed in CHO cells. There are two forms of sialic acid, N-acetyl neuraminic acid (NANA), the predominant form found on human proteins, and N-glycosyl neuraminic acid (NGNA), which is potentially immunogenic. Heterologous proteins expressed in transgenic chickens only possess NANA, whereas those expressed in ruminants only have NGNA, which becomes a potential drawback to their production (Raju *et al.*, 2000). Antibodies expressed in the tubular gland cells of the oviduct of chickens also lack fucose and consequently have the benefit of enhanced ADCC function (Zhu *et al.*, 2005).

6.3.6 *Mammalian cell expression systems*

Stably transfected mammalian cells expressing heterologous proteins were first generated by non-viral gene delivery in the early 1980s (Kaufman and Sharp, 1982; Ringold *et al.*, 1981; Scahill *et al.*, 1983), and this opened the door to the possibility of producing recombinant glycoproteins for the first time. This was followed by the adaptation of the bacterial cultivation process in stirred-tank bioreactors for large-scale cultivation of mammalian cells (Birch and Arathoon, 1990). Through this series of developments, the manufacture of recombinant therapeutic proteins in mammalian cells was born and led in 1987 to the FDA approval of tissue plasminogen activator (alteplase) as the first recombinant therapeutic glycoprotein. However, it was not until 1994 that the first recombinant antibody fragment, abciximab, came to the market, followed three years later by the first full-length antibodies, daclizumab and rituximab.

The initial processes developed in the 1980s using mammalian cells often resulted in low product yields. However, with improvements in all aspects of the process, product yields of 1–2 g/L are now commonplace, with claims of product titers as high as 27 g/L recently published (MacDonald, 2009). In addition to the factors that affect all recombinant protein expression, there are additional considerations for mammalian cells. These include the choice of host cell line, the site of integration, copy number, selection methods and cell culture conditions.

6.3.6.1 *Host cell lines*

The production of antibodies was once limited to hybridomas. However, hybridomas are not ideal for large-scale production processes due to their low levels of expression (<100 mg/L) and low integrated viable cell count (IVCC) in bioreactors. The advances in cloning and genetic engineering technologies now allows the isolation of the antibody genes from individual B-cells, which can then be cloned into optimized recombinant mammalian vectors for expression in high yielding host cell lines. For a production process to be economically viable it is not only essential that the cell line has high productivity but also the ability to grow in suspension, achieve high cell densities, and adapt to bioreactor conditions. Initial mammalian cell culture media had a requirement for complex growth factor cocktails and animal-derived components including serum albumin. However, with the advent of bovine spongiform encephalitis there has been a drive to replace these media formulations with simpler chemically defined, animal component-free (CDACF) media. This has in turn limited the choice of host cell lines to ones which can be adapted to these media. Mammalian cell lines typically employed for recombinant antibody production include CHO, NS0, baby hamster kidney (BHK), Human Embryonic Kidney 293 (HEK293), SP/2 cells and more recently the human-derived cell line PER.C6™.

From their humble beginning as a tool to understand somatic cell genetics, CHO cells are now the preferred mammalian host for the production of recombinant therapeutic proteins worldwide, with over 70% of approved mAbs expressed in these cells. This major share of the market is due to the recognized stability of chromosomally integrated transgenes (Schimke *et al.*, 1984), their relative ease of cultivation at large scale, and their adaptability to adherence-independent growth in serum- and protein-free media (Lee and Da Silva, 1996; Renner *et al.*, 1995; Zang *et al.*, 1995). CHO cells also have a very good safety profile, with the risk of adventitious viral agent reproduction shown to be extremely low. Of 44 human pathogenic viruses tested for replication in CHO cells, only seven were found to replicate (Wurm, 1997).

Another advantage of using CHO cells is that the glycoforms of CHO-produced IgGs have been shown to mimic closely those of human IgGs (Fig. 6.1). The main differences are the absence of the third N-acetylglucosamine bisecting arm, which is present in approximately 10% of human IgG glycoforms, and very low amounts of terminal N-acetyl neuraminic acid (NANA) (Jefferis, 2005a).

The original CHO cell line was derived from the ovary of the Chinese hamster (Kao and Puck, 1968). Subsequent mutagenic studies on these cells led to the establishment of the glycine-dependent strain, CHO-K1, and CHO-DXB11 (also referred to as CHO-DUKX or CHO-DUK-XB11), which lacks dihydrofolate reductase (DHFR) activity (Urlaub and Chasin, 1980). The ability of CHO cells to be cultivated *in vitro* allowed them to become a useful tool for early biochemical studies (Barendsen, 1979; Chang *et al.*, 1978; Galloway *et al.*, 1987a; Galloway *et al.*, 1987b; Glatt *et al.*, 1987) and subsequently led to their use as a host for recombinant protein expression using DNA transfection methods in the early 1980s (Kaufman and Sharp, 1982; Ringold *et al.*, 1981; Scahill *et al.*, 1983).

In order to meet economic and process requirements, intense optimization of these cell lines has since taken place. Most production CHO cell lines in use today have been selected to grow in suspension to high cell densities and in CDACF media. Genetically engineered CHO cell lines have also been generated to improve productivity by enhancing post-translational processing of complex proteins, altering glycosylation patterns, delaying the onset of apoptosis, and increasing or decreasing proliferation rates (see below).

Alternatives to CHO cells include the mouse B-cell lines, NS0 and SP2/0, which have also been widely used for antibody production. The NS0 cell line was derived from a plasmacytoma line (P3), and SP2/0 was derived from a subsequent P3/spleen cell hybrid (Kohler *et al.*, 1978). While both cell lines lack the ability to synthesize and secrete endogenous immunoglobulins, their differentiated B-cell origin make them particularly suitable for recombinant antibody production (Galfre and Milstein, 1981). The NS0 cell line has been used to produce mAbs, either as a fusion partner in hybridoma technology (Cazzola *et al.*, 1992; Novak *et al.*, 1986; Richards *et al.*, 1992) or as a host for transfection (Andersen and Krummen, 2002; Bebbington *et al.*, 1992). Both cell lines have been extensively characterized, adapt well to growth at large scale and can secrete high levels of recombinant proteins in solution.

The glycan structures of antibodies produced from NS0 are similar to human cells with the exception of the terminal sialic acid. The N-acetyl neuraminic acid that is found in humans and CHO cells is replaced with N-glycosyl neuraminic acid (NGNA) (Fig. 6.1; Hamilton and Gerngross, 2007; Jefferis, 2005a). While NGNA has been shown to be immunogenic in humans (Sheeley *et al.*, 1997), only 1–2% of the NS0-produced mAbs are terminally sialyated, and these glycoforms appear to be generally well tolerated.

NS0 cells have a requirement for exogenous cholesterol, which must be supplemented in the growth medium either on its own or via a natural or synthetic carrier (Gramer and Maas, 2003; Walowitz *et al.*, 2003). Successful attempts to isolate NS0 cells that grow in cholesterol-free and protein-free conditions, either by isolating a clonal variant (Birch *et al.*, 1994) or by adapting a population of cells (Lifely *et al.*, 1995), have been reported. However, to date, very little has been published on the use of cholesterol-independent NS0 cells in protein-free media for production of recombinant proteins. Over the last ten years there has been a significant decline in the use of NS0 cells, and the lack of commercially available CDACF media designed to support their growth to high cell densities may be a major contributing factor.

The human cell line HEK293, has been used as a host for small-scale recombinant protein production over the last 25 years. The ability of the cell line to be efficiently transfected in suspension at large scale using cost-effective methods such as polyethyleneimine (PEI) (Durocher *et al.*, 2002) or calcium-phosphate (Girard *et al.*, 2002) has meant that they are now widely used for the generation of gram scale recombinant proteins in transient systems (Pham *et al.*, 2006). The original HEK293 cell line was generated by the transformation of HEK cell cultures with sheared adenovirus 5 DNA (Graham *et al.*, 1977). For many years it was assumed that the HEK293 cells were of either a fibroblastic, endothelial or epithelial cell origin, all of which are abundant in the kidney. However, recent microarray analysis by Shaw *et al.* (2002) suggests that the cells may be neuronal in origin. Extensive optimization of this cell line has lead to the establishment of several variants with enhanced expression and growth profiles. Genetic engineering of the parental HEK293 cell line to express the simian virus 40 (SV40) Large T-antigen led to the establishment of the 293T cell line. Expression of the SV40 Large T-antigen allows episomal replication of transfected plasmids containing the SV40 origin of replication resulting in amplification of transfected plasmids and extended temporal expression of the desired gene products (Durocher *et al.*, 2002). The 293SFE cell line, another commonly used variant, was derived from a suspension adapted clonal cell line selected for its high specific growth rate (Schoofs *et al.*, 1998) and subsequently stably transfected with the *Epstein–Barr Nuclear Antigen 1* (*EBNA1*) gene. EBNA1 acts as a transactivator up-regulating many viral promoters resulting in enhanced protein expression and also allows episomal maintenance when the transfected plasmid DNA contains the origin of replication (*oriP*) from Epstein–Barr virus (EBV) (Baldi *et al.*, 2005).

PER.C6 cells are an alternative human cell line derived from embryonic retinoblastoma cells immortalized with Adenovirus5 E1A and E1B-encoding sequences (Fallaux *et al.*, 1998). While the cell line is well established for vaccine production (Maranga and Goochee, 2006; Xie *et al.*, 2002; Xie *et al.*, 2003), because of its human lineage and hence comparable glycosylation pattern, there is a growing interest in utilizing it to express recombinant IgGs (Jones *et al.*, 2003). While several therapeutic antibody companies are currently evaluating the Per.C6 cell line, no therapeutic antibody generated in this cell line has been approved to date.

The BHK21 cell line has also been used historically (Inoue *et al.*, 1996), but very little has been reported on its use as a production cell line in the last ten years.

6.3.6.2 *Vector systems*

For optimum stable and transient expression the design of the expression vector is critical. Key elements that must be included are promoter and enhancer elements, Kozak sequence and polyadenylation signals. Other factors to consider include selectable markers, origins of replication and chromatin remodeling elements.

In order to achieve efficient IgG expression, it is essential to control the balance of heavy chain and light chain production. On entry into the endoplasmic reticulum as unfolded polypeptides, both chains are modified and assembled. Only completely assembled molecules can bind to antigen and carry out effector functions. Light chains are synthesized 15–25% faster than heavy chains (Bergman *et al.*, 1981), and the isotype of the light chain has been shown to influence the kinetics of intracellular IgG assembly (Montano and Morrison, 2002).

Most light chains can be secreted as free monomers or homodimers (Kollmann *et al.*, 1966; Skvortsov and Gurvich, 1968). Whereas full-length heavy chains are only exported from the cell when combined with the light chain to form complete antibody molecules (Leitzgen *et al.*, 1997). The expression of both antibody chains is usually achieved either by co-transfecting cells with two independent monocistronic constructs or the use of a single vector with the light chain and heavy chain genes linked in series and transcription driven by two identical promoters. While the former is thought to be the least reproducible procedure (Fussenegger *et al.*, 1999), identical promoters on the same vector can introduce the risk of recombination events (Li *et al.*, 2007). However, the use of two different promoters in one vector may result in unbalanced expression of the two different chains. Houdebine and Attal (1999) attempted to control the relative levels of heavy chain and light chain by combining both genes on the same cistron with an internal ribosome entry site (IRES). However, a more recent comparative study of monocistronic and bicistronic expression constructs by Li *et al.* (2007) showed similar antibody expression levels and long-term stability in CHO cell lines independent of gene arrangement.

The vector systems commonly used to generate therapeutic cell lines usually contain strong promoters to drive expression of the antibody genes. The promoters are typically viral in origin, such as human cytomegalovirus (HCMV) *major immediate early* promoter, or they are derived from genes that are highly expressed in mammalian cells such as elongation factor 1α (EF1α) (Running Deer and Allison, 2004). For optimal RNA processing, it is essential to include one or more introns with appropriate splice sites and a polyadenylation signal. In addition, 5′ and 3′ untranslated regions are often optimized for improved transport and translation. Coding sequences themselves are now routinely optimized to enhance expression not only for codon optimization but also to remove cryptic splice sites, cryptic polyadenylation signals and regions with unfavorable mRNA secondary structures.

6.3.6.3 *Selection systems*

As well as the gene expression cassette, the vectors will usually contain a selection marker to allow stably transfected cells to be selected. The expression of the selection marker is normally controlled by a weak promoter to give selective pressure for integration into a transcriptionally active locus. As both the antibody genes and the selectable marker gene are linked, this should also result in the antibody genes being in a transcriptionally active region of the host genome.

Selection marker systems are based either on dominant or recessive selection. Recessive selection systems such as DHFR and glutamine synthetase (GS) restore a critical enzyme or protein to the cell that is required for survival. GS synthesizes glutamine, an essential amino acid, from glutamate and ammonia. Transfection of cells that lack endogenous GS with vectors containing the *GS* gene confers the ability to grow in glutamine free media. The genotype of NS0 cells makes them particularly suited for use with this system. Unlike other cell types, NS0 cells are glutamine auxotrophs and therefore require supplementation with glutamine for growth. Glutamine independence can be conferred upon transfection with a functional *GS* gene (Bebbington *et al.*, 1992). For cells such as CHO, which express endogenous GS, addition of methionine sulfoximine (MSX), a specific GS inhibitor, during the selection process can inhibit this endogenous activity, ensuring that only transfectants with high levels of transgene expression survive.

Similarly, transfection with the DHFR coding sequence can transform DHFR-deficient CHO cells to a $DHFR^+$ phenotype, thus allowing growth in glycine and nucleoside-deficient media. DHFR catalyses the conversion of folate to tetrahydrofolate. Tetrahydrofolate is involved in several biosynthetic pathways including the conversion of glycine to serine, the biosynthesis of thymidine monophosphate from deoxyuridine monophosphate, and for purine biosynthesis (Lucas *et al.*, 1996). When cultured in high levels of methotrexate (MTX), a folic acid analog that blocks DHFR activity, transfected cells can overcome the decrease in DHFR activity by spontaneous amplification of the *DHFR* locus and subsequent increase in expression. As amplification in cells usually occurs over large regions, the genetic linkage between *DHFR* and the transfected gene of interest ensures that the gene of interest is co-amplified. Repeated rounds of amplification with increasing concentrations of MTX can result in up to a 1000-fold increase in gene copy number, thus increasing the probability of securing a high producing transfectant (Kaufman *et al.*, 1978; Schimke *et al.*, 1978). However, gene amplification often necessitates lengthy cloning and screening steps and risks genetic instability leading to rearrangement and loss of expression of the gene of interest (Borth *et al.*, 1999; Kim *et al.*, 2001).

Bianchi and McGrew (2003) described a novel approach to improve this labor-intensive process. They determined that the DHFR protein can be divided into two separate domains and subsequently reassembled as a functioning protein using leucine zippers. They linked each *DHFR* gene fragment with an intervening IRES to either the light chain or heavy chain genes on separate vectors. Survival in selective media requires assembly in a 1:1 ratio of the expressed DHFR fragments. Linkage between the DHFR fragments and the individual antibody chains should also result in the 1:1 expression of each chain leading to optimal antibody assembly and accumulation. They also claim that this process of DHFR protein complementation leads to concurrent gene amplification resulting in a two- to five-fold increase in expression of the linked antibody genes compared to the standard methods. To our knowledge, this approach has not been used for antibody manufacture.

Dominant systems have generally relied on expression vectors containing genes that code for enzymes that confer antibiotic resistance to the cells and therefore ensure cell survival in the presence of antibiotics. Neomycin, hygromycin, blasto-cidin, puromycin, and zeocin resistance genes are commonly used dominant selection markers. Further stringency can be introduced by impairment of the expression of the selection gene by reducing the efficiency of the Kozak sequence or by altering the specific enzyme activity (Sautter and Enenkel, 2005). This approach is the basis for the BiHEX (Boehringer Ingelheim Gmbh) selection system. This system is based on the impairment of the specific activity of neomycin-phosphotransferase (NPT) by introduction of a point mutation in a conserved catalytic domain (Sautter and Enenkel, 2005).

6.3.6.4 *Integration events*

The site of integration of the transfected genes in the mammalian genome has a significant influence on levels of transcription. The DNA of the host genome is organized into euchromatin and hetrochromatin. Euchromatin is the open transcriptionally active portion of the genome and is characterized by high levels of histone acetylation. Constitutively expressed genes such as *glyceraldehyde-3-phosphate dehydrogenase* (*GAPDH*) and other housekeeping genes are predominantly located in these regions. By contrast, heterochromatin is characterized by extensive histone deacetylation and high levels of histone methylation, resulting in tightly packed, inactive regions with limited access to the transcription machinery. Integration into regions of heterochromatin therefore results in either low levels of expression or gene silencing. Integration of a recombinant gene into the host genome at a transcriptionally active site is extremely rare with only 1 in 10,000 events resulting in high levels of expression (Gorman and Bullock, 2000). As current transfection techniques rely on random integration, this requires laborious transfection campaigns that screen large numbers of clones to be certain of identifying high expressing transfectants.

There is now a great desire to increase the probability of obtaining stable, highly expressing transfectants, and several approaches that modify the chromatin structure or target a specific site of integration are now being investigated. Considerable effort has been focused on identifying elements that locally modify the chromatin structure or protect the transgene from epigenetic effects. For example, *HS4* insulators (from the chicken *β globin* locus) (Pikaart *et al.*, 1998), locus control regions (LCRs) (Li *et al.*, 1999), scaffold or matrix attachment regions (S/MARs) (Bode *et al.*, 2000), Ubiquitous Chromatin Opening Elements (UCOEs) (Antoniou *et al.*, 2003), and conserved anti-repressor elements (STARs) (Kwaks *et al.*, 2003) have all shown some utility (Table 6.2). While there is a desire to differentiate between these elements, they all appear to have an "insulator effect," providing protection from epigenetic factors, while some also have the ability to create regions of euchromatin, thus allowing high level expression regardless of the site of transgene integration. Aldrich *et al.* (1998) have also

Table 6.2 Commercially available elements that locally modify the chromatin structure

Element	Company	Feature	Function	Advantages
STAR (Star Select)	Crucell		Elements that counteract genetic gene repression Flank transgene 5′ and 3′	Increases number of high expressing clones (25–50 pcd*) As majority of clones are high expressing, only need to screen 30–50 clones
S/MARS (Scaffold/ matrix attachment regions)	Selexis	Contains AT rich nucleotide motifs	Influences gene expression by anchoring active chromatin domains to the nuclear matrix	20-fold increase in expression (80 pcd)
UCOEs (Ubiquitous chromatin opening elements)	Millipore	From genetic regions surrounding housekeeping genes	Maintain an open chromatin structure Flank transgene 5′ and 3′	10–25-fold increase in expression 15–40 pcd (1.5–2 g/L)

* pcd = per cell per day.

reported on an alternative expression augmenting sequence element (EASE) identified in the CHO genome. While EASE is not classed as a promoter/enhancer/SARS/MARS or LCR element and has no open reading frame (therefore, it does not encode a transactivator), it appears to facilitate integration into the genome resulting in high copy numbers. In addition, EASE-containing vectors appear to allow greater amplification of the transgene with increasing concentrations of MTX, as well as enhance viability and stability. Alternative approaches that target preselected transcriptionally active sites by site-specific integration include the Cre/loxP (Sauer, 1998) and Flp/FRT (Zhu *et al.*, 1995) systems. This approach has not been adopted to date for manufacturing cell lines.

Another approach is the use of artificial chromosomes. Artificial chromosomes allow the introduction of large amounts of DNA into the cell in an autonomously replicating, non-integrating format. The ACE system (Artificial Chromosome system) developed by Chromos Molecular Systems Inc. (USA), is an artificial mouse chromosome that contains multiple recombination acceptor sites. ACE host cells are co-transfected with an ACE targeting plasmid containing the gene of interest, a selection marker gene and a helper plasmid containing the *ACE integrase* gene. While this system removes the concerns over the influence of the integration site, it has been reported that expression levels tend to decrease over time (Williams and Cranenburgh, 2008). From a regulatory point of view, this issue of stability will have to be addressed before the system can be used for manufacture.

6.3.6.5 *Host cell line engineering*

There is a continuing desire to improve the characteristics of the host cell lines in order to enhance the activity of the antibody as well as increase its productivity. A variety of approaches have been adopted, including modifications of the endogenous glycosylation enzymes, modifications to genes controlling cell growth, and apoptosis and enhancements to the secretory machinery. In most cases these modifications have been carried out in CHO cells.

As discussed above the nature of the attached carbohydrate can affect the antibody effector function. Observations have shown enhanced ADCC by non-fucosylated human IgG1 antibodies (Niwa *et al.*, 2005b) or antibodies bearing a bisecting N-acetylglucosamine (Shinkawa *et al.*, 2003). This has led to the development of the Potelligent® (BioWA, Japan) and the GlycoMAb (Glycart, Switzerland) technologies based around cell lines engineered with altered glycosylation machinery. In the Potelligent® cell line, both *FUT8* gene alleles, which encode α1,6-fucosyltransferase, were disrupted by sequential homologous recombination (Yamane-Ohnuki *et al.*, 2004). The absence of α1,6-fucosyltransferase, which catalyzes the transfer of fucose from GDP-fucose to GlcNAc, results in the production of non-fucosylated antibodies. GlycoMAb cell lines are stably transfected with the gene encoding 1,4-N-acetylglucosaminyltransferase III (GnTIII), resulting in the expressed antibodies bearing bisecting N-acetylglucosamine oligosaccharides (Ferrara *et al.*, 1987). Several antibodies generated using the Potelligent® Technology and GlycoMAb technology are now in clinical trials.

An alternative approach to generating antibodies with increased ADCC involves the addition of inhibitors for enzymes involved in the glycosylation pathway during the culture process. Zhou *et al.* (2008) showed that kifunensine, a glycosidase inhibitor, is effective in blocking complex-type glycan formation in recombinant humanized IgG1s secreted by either hybridoma or CHO cells. The addition of kifunensine to batch cultures had no significant effect on cell growth and viability, and antibodies expressed contained mainly oligomannose-type glycans and demonstrated increased ADCC activity and affinity for FcγRIIIa but reduced C1q binding. Similar effects were also reported by Kanda *et al.* (2007) on addition of kifunenesine to the Potelligent® cell line expressing an anti-human CD20 chimeric antibody. While the potential of using inhibitors, which is a much cheaper option than the costly engineering solution described above, has been demonstrated in research, no data is yet available on its application in large-scale culturing for clinical grade material.

In an effort to maintain cell viability and prolong productivity, some success has been reported with host cells engineered to overexpress genes found to be up-regulated in cancer cells. Overexpression of Bcl-2 and Bcl-xl (anti-apoptotic proteins known to suppress the mitochondrial apoptotic pathway) have been shown to increase host cell survival in culture (Tey and Al-Rubeai, 2004). Other Bcl-2 family members and homologs (E1B-19K- the adenovirus Bcl-2 homolog and Aven-Bcl-2), which are involved in disruption or inhibition of the apoptosome complex, were also shown to have a similar effect (Figueroa, Jr. *et al.*, 2007; Majors *et al.*, 2008). More recently, Dorai *et al.* (2009) co-expressed E1B-19K, Aven, and a mutated form of XIAP (X-linked Inhibitor of Apoptosis Protein), and demonstrated a significant reduction in lactate and ammonia production, allowing prolonged cultivation with increasing IVCC and a resultant increase in product yield.

Others have addressed the issue of achieving higher productivity by modifying the translational and secretory pathways. The endoplasmic reticulum (ER) plays an essential role in secretion, and several studies have shown that there is a significant relationship between protein secretion and the regulation of ER components in the host cells. To date the most common approach to increase secretion rates has been to overexpress ER chaperones such as grp78 and protein disulphide isomerase. While significant increases in protein production have been observed in yeast and insect cell systems (Ailor and Betenbaugh, 1998; Hsu *et al.*, 1997) reports of similar increases in mammalian cells have been inconsistent (Borth *et al.*, 2005; Davis *et al.*, 2000; Kitchin and Flickinger, 1995). Dinnis and James proposed that since chaperones and other proteins involved in protein folding and expression exist as large multi-protein complexes, global expansion of all components of the secretory pathway is required for generic improvement of antibody secretion rather than overexpression of selected proteins. Overexpression of proteins known or suspected of having a key role in controlling ER expansion such as BLIMP-1, or the initiation of ER expansion (such as XBP-1 and ATF6), has now become the major area of focus (Dinnis and James, 2005; Ku *et al.*, 2008).

Overall, while benefits of overexpression and/or deletion of genes involved in apoptosis and protein secretion have been shown at a research level, their application for production of therapeutics is still under review.

6.3.6.6 *Stable versus transient expression*

Development of a mAb therapeutic, from early-stage research through preclinical and clinical development, requires ever-increasing amounts of antibody. The choice of expression system used in the different stages is dependent on a number of factors including the number of antibody variants, quantity and quality of material required, and the speed of delivery. For large-scale production of therapeutic antibodies, stable gene expression is the preferred option as it allows for greater process consistency and control of final product quality. Expression level and cell density are two critical factors in clonal cell line selection, with expression levels ranging from 20 to 60 pg/cell/day (15–18 pg/cell/day roughly equates to ~1 g/L) and cell densities of 10–150 million cells/ml required for an optimal commercial process (Coco-Martin J.M., 2008). To achieve this, the costly and time-consuming process of extensive host cell line development is required. Currently, the stable cell line selection process as described above necessitates the screening of hundreds if not thousands of individual cell lines to obtain a sufficient number with the desired high protein productivity and cell growth rate. Automation has begun to revolutionize the clonal cell line recovery process and allow even larger numbers of clones to be screened, with some of the major biopharma companies developing automated platforms that both screen for colonies and expand cultures to the T75 static flask stage (Lindgren *et al.*, 2009).

Rather than investing in large-scale automation for the traditional limiting dilution method of single cell cloning, some companies are investigating other ways of isolating high producing clones using FACs based methods (Carroll S., 2007). Alternatively, clonal cell lines can be isolated using systems such as ClonePIX (Genetix Ltd. UK), CellCellector (Aviso Gmbh Germany) and CellXpress (Sigma Aldrich Corp.) These systems enable automated detection, picking and transfer of high expressing clones when transfected cells are plated in a semisolid medium and stained with a fluorescently labeled antibody.

Transfectant selection is primarily based on expression levels but other characteristics contributing to productivity such as cell growth and viability over the culture time as well as stability and product quality also need to be assessed. Screening is most often carried out in shake flask cultures. A final panel of transfectants is then screened in a predictive model of the manufacturing process, which includes the feeding regimes used in the final process.

During the initial selection process when only research quantities (10–1000 mg) are sufficient, faster methods of recombinant protein production are required, and large-scale transient expression methods are now routinely being used. Both academic and industrial groups have now begun to invest in optimization of transient processes in order to supply material for late-stage research and early-stage clinical use (Morrow, 2008). In contrast to stable gene expression, transient gene expression results in short-term recombinant protein production, typically ten days from DNA transfection to harvest. HEK293 and CHO cells are the main cell lines used for this process, with HEK293 cells routinely achieving higher transfection rates and subsequent yields. The HEK293 cells also appear to be more amenable to transfection with PEI, which is inexpensive and easily scalable, resulting in lower overall production costs (Pham *et al.*, 2006). Recent reports from Florian Wurm's laboratory have described an improved PEI transient transfection protocol that yields titers exceeding 1 g/L using HEK293E cells (Backliwal *et al.*, 2008).

While HEK293 cells are capable of carrying out most of the post-translational folding and processing required to generate functional mature proteins, the nature of the attached glycan structure is somewhat different from that seen in CHO cells (Van den Nieuwenhof *et al.*, 2000). As CHO cells are the preferred manufacturing host, it would therefore be preferable to develop a high yielding CHO transient expression system. We may therefore see a decline in the use of HEK293 cells over the coming years as transient production levels in CHO cells increase. To date, however, this has proved to be challenging, as yields are much lower and optimization of transfection methods have proved to be more difficult (Derouazi *et al.*, 2004; Galbraith *et al.*, 2006; Haldankar *et al.*, 2006).

Several different approaches have been tried to improve the transient expression levels in CHO cells, including use of anti-mitotic microtubule disrupting agents to arrest CHO cells in the G2/M phase (Tait *et al.*, 2004), development of an Epi-CHO system by utilizing *Py* elements for plasmid replication, and *EBNA-1* and *oriP* for plasmid maintenance and segregation (Kunaparaju *et al.*, 2005). The Tait group also improved expression levels by addition of recombinant insulin-like growth factor and reducing the culture temperature to 32°C (Galbraith *et al.*, 2006). Wurm's group, alongside their HEK293 work, has recently reported development of a proprietary CHO/PEI system with mAb expression titers of 90 μg/ml (Wulhfard, 2009). We have also developed a CHO transient system using electroporation, which routinely yields expression levels of 200–300 μg/ml at the one- to five-liter scale (unpublished).

With the advent of large-scale transient production of recombinant mAbs for preclinical use and the current level of investment in optimization of the host cell and technology for improved clonal identification and selection, mammalian cell expression is and probably will continue to be the dominant bioproduction vehicle for the foreseeable future.

6.4 Conclusions and Future Directions

There have been major advances in the production of antibodies over the last 15 years since the approval of the first recombinant antibodies for therapeutic use. Levels of 1–2 g/L are industry standard with much higher titers being reported in specialized fermentation processes (MacDonald, 2009). There has also been a realization that antibody expression is not simply a function of the expression system of choice but is highly dependent on the antibody itself, with different antibodies demonstrating large differences in their inherent levels of expression. Therefore, in addition to optimizing antibody sequences at the DNA level, optimization of antibody sequences at the protein level can also have an impact on expression levels (Cain, 2009).

Despite the fact that large amounts of resources have been spent on alternative expression systems, the use of mammalian cells predominates. This conservatism within the industry is slowly being challenged by yeast and plant expression platforms, and antibodies derived from these systems have been produced for clinical trials (Clarke *et al.*, 2009; McCormick *et al.*, 2008; Weintraub *et al.*, 2005). The general reluctance within the industry to move to alternative platforms presumably reflects the concerns that antibodies produced in alternative systems will come under additional scrutiny and delays in regulatory approval (Rader, 2008).

The future for antibody-based drug treatments is bright. As our ability to increase the robustness of production systems improves, the result will be a significant reduction in the cost of goods, and these exquisite drugs will become more accessible to the patients who are in critical need of them.

Acknowledgments

David Humphreys, Katharine Cain and Fay Saunders are gratefully acknowledged for their critical reading and valuable comments on the manuscript.

References

Ailor E., Betenbaugh M.J. (1998). Overexpression of a cytosolic chaperone to improve solubility and secretion of a recombinant IgG protein in insect cells. *Biotechnol Bioeng* **58**: 196–203.

Ailor E., Betenbaugh M.J. (1999). Modifying secretion and post-translational processing in insect cells. *Curr Opin Biotechnol* **10**: 142–5.

Aires da S.F., Corte-Real S., Goncalves J. (2008). Recombinant antibodies as therapeutic agents: pathways for modeling new biodrugs. *BioDrugs* **22**: 301–14.

Aldrich T.L., Thomas J.N., Morris A.E. (1998). Improved bicistronic mammalian expression vectors using expression augmenting sequence element (EASE). *Cytotechnology* **28**: 9–17.

Andersen D.C., Krummen L. (2002). Recombinant protein expression for therapeutic applications. *Curr Opin Biotechnol* **13**: 117–23.

Antoniou M., Harland L., Mustoe T., Williams S., Holdstock J., Yague E., Mulcahy T., Griffiths M., Edwards S., Ioannou P.A., Mountain A., Crombie R. (2003). Transgenes encompassing dual-promoter CpG islands from the human TBP and HNRPA2B1 loci are resistant to heterochromatin-mediated silencing. *Genomics* **82**: 269–79.

Arbabi-Ghahroudi M., Tanha J., MacKenzie R. (2005). Prokaryotic expression of antibodies. *Cancer Metastasis Rev* **24**: 501–19.

Aumiller J.J., Hollister J.R., Jarvis D.L. (2003). A transgenic insect cell line engineered to produce CMP-sialic acid and sialylated glycoproteins. *Glycobiology* **13**: 497–507.

Backliwal G., Hildinger M., Hasija V., Wurm F.M. (2008). High-density transfection with HEK-293 cells allows doubling of transient titers and removes need for a priori DNA complex formation with PEI. *Biotechnol Bioeng* **99**: 721–7.

Baldi L., Muller N., Picasso S., Jacquet R., Girard P., Thanh H.P., Derow E., Wurm, F.M. (2005). Transient gene expression in suspension HEK-293 cells: application to large-scale protein production. *Biotechnol Prog* **21**: 148–53.

Bardor M., Faveeuw C., Fitchette A.C., Gilbert D., Galas L., Trottein F., Faye L., Lerouge P. (2003a). Immunoreactivity in mammals of two typical plant glyco-epitopes, core alpha(1,3)-fucose and core xylose. *Glycobiology* **13**: 427–34.

Bardor M., Loutelier-Bourhis C., Paccalet T., Cosette P., Fitchette A.C., Vezina L.P., Trepanier S., Dargis M., Lemieux R., Lange C., Faye L., Lerouge P. (2003b). Monoclonal C5-1 antibody produced in transgenic alfalfa plants exhibits a N-glycosylation that is homogenous and suitable for glyco-engineering into human-compatible structures. *Plant Biotechnol J* **1**: 451–62.

Barendsen G.W. (1979). Influence of radiation quality on the effectiveness of small doses for induction of reproductive death and chromosome aberrations in mammalian cells. *Int J Radiat Biol Relat Stud Phys Chem Med* **36**: 49–63.

Bebbington C.R., Renner G., Thomson S., King D., Abrams D., Yarranton, G.T. (1992). High-level expression of a recombinant antibody from myeloma cells using a glutamine synthetase gene as an amplifiable selectable marker. *Biotechnology (NY)* **10**: 169–75.

Beck A., Wagner-Rousset E., Bussat M.C., Lokteff M., Klinguer-Hamour C., Haeuw J.F., Goetsch L., Wurch T., Van D.A., Corvaia N. (2008). Trends in glycosylation, glycoanalysis and glycoengineering of therapeutic antibodies and Fc-fusion proteins. *Curr Pharm Biotechnol* **9**: 482–501.

Bergman B., Holm S.E., Holmlund D., Kaijser B. (1981). Serum antibodies against Escherichia coli and Proteus mirabilis after different types of operations. *Acta Chir Scand* **147**: 93–7.

Better M., Chang C.P., Robinson R.R., Horwitz, A.H. (1988). Escherichia coli secretion of an active chimeric antibody fragment. *Science* **240**: 1041–3.

Bianchi A.A., McGrew J.T. (2003). High-level expression of full-length antibodies using trans-complementing expression vectors. *Biotechnol Bioeng* **84**: 439–44.

Birch J.R., Arathoon R. (1990). Suspension culture of mammalian cells. *Bioprocess Technol* **10**: 251–70.

Birch J.R., Boraston R.C., Metcalfe H., Brown M.E., Bebbington C.R., Field R.P. (1994). Selecting and designing cell lines for improved physiological characteristics. *Cytotechnology* **15**: 11–6.

Bode J., Benham C., Knopp A., Mielke C. (2000). Transcriptional augmentation: modulation of gene expression by scaffold/matrix-attached regions (S/MAR elements). *Crit Rev Eukaryot Gene Expr* **10**: 73–90.

Borth N., Mattanovich D., Kunert R., Katinger H. (2005). Effect of increased expression of protein disulfide isomerase and heavy chain binding protein on antibody secretion in a recombinant CHO cell line. *Biotechnol Prog* **21**: 106–11.

Borth N., Strutzenberger K., Kunert R., Steinfellner W., Katinger H. (1999). Analysis of changes during subclone development and ageing of human antibody-producing heterohybridoma cells by northern blot and flow cytometry. *J Biotechnol* **67**: 57–66.

Boss M.A., Kenten J.H., Wood C.R., Emtage J.S. (1984). Assembly of functional antibodies from immunoglobulin heavy and light chains synthesised in E. coli. *Nucleic Acids Res* **12**: 3791–806.

Bothmann H., Pluckthun A. (1998). Selection for a periplasmic factor improving phage display and functional periplasmic expression. *Nat Biotechnol* **16**: 376–80.

Bothmann H., Pluckthun A. (2000). The periplasmic Escherichia coli peptidylprolyl cis,trans-isomerase FkpA. I. Increased functional expression of antibody fragments with and without cis-prolines. *J Biol Chem* **275**: 17100–5.

Butler M. (2006). Optimisation of the cellular metabolism of glycosylation for recombinant proteins produced by Mammalian cell systems. *Cytotechnology* **50**: 57–76.

Cabilly S., Riggs A.D., Pande H., Shively J.E., Holmes W.E., Rey M., Perry L.J., Wetzel R., Heyneker H.L. (1984). Generation of antibody activity from immunoglobulin polypeptide chains produced in Escherichia coli. *Proc Natl Acad Sci USA* **81**: 3273–7.

Cain K. (2009). "The effect of primary antibody sequence" proceedings of the 5th IBC Cell Line Development and Engineering, June, 2009.

Carroll S., N.M.A.-R.M. (2007). Monitoring of growth, physiology and productivity of animal cells by flow cytometry in Portner R. (Ed.), Animal Cell Biotechnology, Methods and Protocols. *Methods in Biotechnology* **24**, pp. 223–37, Springer.

Carter P., Kelley R.F., Rodrigues M.L., Snedecor B., Covarrubias M., Velligan M.D., Wong W.L., Rowland A.M., Kotts C.E., Carver M.E. (1992). High level Escherichia coli expression and production of a bivalent humanized antibody fragment. *Biotechnology (NY)* **10**: 163–7.

Castilla J., Sola I., Pintado B., Sanchez-Morgado J.M., Enjuanes L. (1998). Lactogenic immunity in transgenic mice producing recombinant antibodies neutralizing coronavirus. *Adv Exp Med Biol* **440**: 675–86.

Cazzola F., Saggin L., Callegaro L. (1992). Production of novel monoclonal antibodies against rabbit platelet factor four. *Hybridoma* **11**: 61–9.

Chancey C.J., Khanna K.V., Seegers J.F., Zhang G.W., Hildreth J., Langan A., Markham R.B. (2006). Lactobacilli-expressed single-chain variable fragment (scFv) specific for intercellular adhesion molecule 1 (ICAM-1) blocks cell-associated HIV-1 transmission across a cervical epithelial monolayer. *J Immunol* **176**: 5627–36.

Chang C.C., Castellazzi M., Glover T.W., Trosko J.E. (1978). Effects of harman and norharman on spontaneous and ultraviolet light-induced mutagenesis in cultured Chinese hamster cells. *Cancer Res* **38**: 4527–33.

Chen C., Snedecor B., Nishihara J.C., Joly J.C., McFarland N., Andersen D.C., Battersby J.E., Champion K.M. (2004). High-level accumulation of a recombinant antibody fragment in the periplasm of Escherichia coli requires a triple-mutant (degP prc spr) host strain. *Biotechnol Bioeng* **85**: 463–74.

Clarke S.J., Smith J.T., Gebbie C., Sweeney C., Olszewski N. (2009). A phase I, pharmacokinetic (PK), and preliminary efficacy assessment of ALD518, a humanized anti-IL-6 antibody, in patients with advanced cancer. *J Clinical Oncology* **27**: 3025.

Coco-Martin J.M., Harmsen M.M. (2008). The impact of cell culture mediam on cell line and process — A review of therapeutic protein expression by Mammalian cells. *BioProcess International* **6**: 28–33.

Cox K.M., Sterling J.D., Regan J.T., Gasdaska J.R., Frantz K.K., Peele C.G., Black A., Passmore D., Moldovan-Loomis C., Srinivasan M., Cuison S., Cardarelli P.M., Dickey L.F. (2006). Glycan optimization of a human monoclonal antibody in the aquatic plant Lemna minor. *Nat Biotechnol* **24**: 1591–7.

Cregg J.M., Cereghino J.L., Shi J., Higgins D.R. (2000). Recombinant protein expression in Pichia pastoris. *Mol Biotechnol* **16**: 23–52.

Davis R., Schooley K., Rasmussen B., Thomas J., Reddy P. (2000). Effect of PDI overexpression on recombinant protein secretion in CHO cells. *Biotechnol Prog* **16**: 736–43.

De Neve M., De B.S., De Wilde C., Van H.H., Strobbe I., Jacobs A., Van M., Depicker A. (1999). Gene silencing results in instability of antibody production in transgenic plants. *Mol Gen Genet* **260**: 582–92.

De Wilde C., Van H.H., De B.S., Angenon G., De J.G., Depicker A. (2000). Plants as bioreactors for protein production: avoiding the problem of transgene silencing. *Plant Mol Biol* **43**: 347–59.

Deng S.J., MacKenzie C.R., Sadowska J., Michniewicz J., Young N.M., Bundle D.R., Narang S.A. (1994). Selection of antibody single-chain variable fragments with improved carbohydrate binding by phage display. *J Biol Chem* **269**: 9533–8.

Derouazi M., Girard P., Van T.F., Iglesias K., Muller N., Bertschinger M., Wurm F.M. (2004). Serum-free large-scale transient transfection of CHO cells. *Biotechnol Bioeng* **87**: 537–45.

Dinnis D.M., James D.C. (2005). Engineering mammalian cell factories for improved recombinant monoclonal antibody production: lessons from nature? *Biotechnol Bioeng* **91**: 180–9.

Donzeau M., Knappik A. (2007). Recombinant monoclonal antibodies. *Methods Mol Biol* **378**: 14–31.

Dorai H., Nemeth J.F., Cammaart E., Wang Y., Tang Q.M., Magill A., Lewis M.J., Raju T.S., Picha K., O'Neil K., Ganguly S., Moore G. (2009). Development of mammalian production cell lines expressing CNTO736, a glucagon like peptide-1-MIMETIBODY: factors that influence productivity and product quality. *Biotechnol Bioeng* **103**: 162–76.

Durand H., Clanet M., Tiraby G. (1988). Genetic improvement of Trichoderma reesi for large scale cellulose production. *Enzyme and Microbial Technology* **10**: 341–6.

Durocher Y., Perret S., Kamen A. (2002). High-level and high-throughput recombinant protein production by transient transfection of suspension-growing human 293-EBNA1 cells. *Nucleic Acids Res* **30**: E9.

Edelman L., Margaritte C., Chaabihi H., Monchatre E., Blanchard D., Cardona A., Morin F., Dumas G., Petres S., Kaczorek M. (1997). Obtaining a functional recombinant anti-rhesus (D) antibody using the baculovirus-insect cell expression system. *Immunology* **91**: 13–9.

Eldin P., Pauza M.E., Hieda Y., Lin G., Murtaugh M.P., Pentel P.R., Pennell C.A. (1997). High-level secretion of two antibody single chain Fv fragments by Pichia pastoris. *J Immunol Methods* **201**: 67–75.

Evans D., Das R. (2005). Monoclonal Antibody therapies evolving into a $30 billion market. DataMonitor.

Fallaux F.J., Bout A., van d. V., I, van den Wollenberg D.J., Hehir K.M., Keegan J., Auger C., Cramer S.J., van O.H., van der Eb A.J., Valerio D., Hoeben R.C. (1998). New helper cells and matched early region 1-deleted adenovirus vectors prevent generation of replication-competent adenoviruses. *Hum Gene Ther* **9**: 1909–17.

Farid S.S. (2006). Established bioprocesses for producing antibodies as a basis for future planning. *Adv Biochem Eng Biotechnol* **101**: 1–42.

Farid S.S. (2007). Process economics of industrial monoclonal antibody manufacture. *J Chromatogr B Analyt Technol Biomed Life Sci* **848**: 8–18.

Fernandez L.A. (2004). Prokaryotic expression of antibodies and affibodies. *Curr Opin Biotechnol* **15**: 364–73.

Ferrara P., Pecceu F., Marchese E., Vita N., Roskam W., Lupker J. (1987). Characterization of recombinant glycosylated human interleukin 2 produced by a recombinant plasmid transformed CHO cell line. *FEBS Lett* **226**: 47–52.

Figueroa B., Jr., Ailor E., Osborne D., Hardwick J.M., Reff M., Betenbaugh M.J. (2007). Enhanced cell culture performance using inducible anti-apoptotic genes E1B-19K and Aven in the production of a monoclonal antibody with Chinese hamster ovary cells. *Biotechnol Bioeng* **97**: 877–92.

Fletcher S.P., Muto M., Mayfield S.P. (2007). Optimization of recombinant protein expression in the chloroplasts of green algae. *Adv Exp Med Biol* **616**: 90–8.

Forsberg G., Forsgren M., Jaki M., Norin M., Sterky C., Enhorning A., Larsson K., Ericsson M., Bjork P. (1997). Identification of framework residues in a secreted recombinant antibody fragment that control production level and localization in Escherichia coli. *J Biol Chem* **272**: 12430–6.

Franklin S.E., Mayfield S.P. (2005). Recent developments in the production of human therapeutic proteins in eukaryotic algae. *Expert Opin Biol Ther* **5**: 225–35.

Frenken L.G., Hessing J.G., van den Hondel C.A., Verrips C.T. (1998). Recent advances in the large-scale production of antibody fragments using lower eukaryotic microorganisms. *Res Immunol* **149**: 589–99.

Freyre F.M., Vazquez J.E., Ayala M., Canaan-Haden L., Bell H., Rodriguez I., Gonzalez A., Cintado A., Gavilondo J.V. (2000). Very high expression of an anti-carcinoembryonic antigen single chain Fv antibody fragment in the yeast Pichia pastoris. *J Biotechnol* **76**: 157–63.

Fujiwara Y., Miwa M., Takahashi R., Hirabayashi M., Suzuki T., Ueda M. (1997). Position-independent and high-level expression of human alpha-lactalbumin in the milk of transgenic rats carrying a 210-kb YAC DNA. *Mol Reprod Dev* **47**: 157–63.

Fussenegger M., Bailey J.E., Hauser H., Mueller P.P. (1999). Genetic optimization of recombinant glycoprotein production by mammalian cells. *Trends Biotechnol* **17**: 35–42.

Galbraith D.J., Tait A.S., Racher A.J., Birch J.R., James D.C. (2006). Control of culture environment for improved polyethylenimine-mediated transient production of recombinant monoclonal antibodies by CHO cells. *Biotechnol Prog* **22**: 753–62.

Galfre G., Milstein C. (1981). Preparation of monoclonal antibodies: strategies and procedures. *Methods Enzymol* 73: 3–46.

Galloway S.M., Armstrong M.J., Reuben C., Colman S., Brown B., Cannon C., Bloom A.D., Nakamura F., Ahmed M., Duk S. (1987a). Chromosome aberrations and sister chromatid exchanges in Chinese hamster ovary cells: evaluations of 108 chemicals. *Environ Mol Mutagen 10 Suppl* **10**: 1–175.

Galloway S.M., Deasy D.A., Bean C.L., Kraynak A.R., Armstrong M.J., Bradley M.O. (1987b). Effects of high osmotic strength on chromosome aberrations, sister-chromatid exchanges and DNA strand breaks, and the relation to toxicity. *Mutat Res* **189**: 15–25.

Gasser B., Mattanovich D. (2007). Antibody production with yeasts and filamentous fungi: on the road to large scale? *Biotechnol Lett* **29**: 201–12.

Gasser B., Maurer M., Gach J., Kunert R., Mattanovich D. (2006). Engineering of Pichia pastoris for improved production of antibody fragments. *Biotechnol Bioeng* **94**: 353–61.

Gawlitzek M., Conradt H.S., Wagner R. (1995). Effect of different cell culture conditions on the polypeptide integrity and N-glycosylation of a recombinant model glycoprotein. *Biotechnol Bioeng* **46**: 536–44.

Gerngross T. (2005). Production of complex human glycoproteins in yeast. *Adv Exp Med Biol* **564**: 139.

Giga-Hama Y., Kumagai H. (1999). Expression system for foreign genes using the fission yeast schizosaccharomyces pombe. *Biotechnol Appl Biochem* **30** (**Pt 3**): 235–44.

Girard P., Derouazi M., Baumgartner G., Bourgeois M., Jordan M., Jacko B., Wurm F.M. (2002). 100-liter transient transfection. *Cytotechnology* **38**: 15–21.

Giritch A., Marillonnet S., Engler C., van E.G., Botterman J., Klimyuk V., Gleba Y. (2006). Rapid high-yield expression of full-size IgG antibodies in plants coinfected with noncompeting viral vectors. *Proc Natl Acad Sci USA* **103**: 14701–6.

Glatt H., Gemperlein I., Turchi G., Heinritz H., Doehmer J., Oesch F. (1987). Search for cell culture systems with diverse xenobiotic-metabolizing activities and their use in toxicological studies. *Mol Toxicol* **1**: 313–34.

Gorman C., Bullock C. (2000). Site-specific gene targeting for gene expression in eukaryotes. *Curr Opin Biotechnol* **11**: 455–60.

Graham F.L., Smiley J., Russell W.C., Nairn R. (1977). Characteristics of a human cell line transformed by DNA from human adenovirus type 5. *J Gen Virol* **36**: 59–74.

Gramer M.J., Maas J. (2003). Optimal NS0 cell growth in a hollow fiber bioreactor requires increased serum concentration or a cholesterol supplement on the cell side of the fiber. *Biotechnol Prog* **19**: 1762–6.

Gurkan C., Symeonides S.N., Ellar D.J. (2004). High-level production in Pichia pastoris of an anti-p185HER-2 single-chain antibody fragment using an alternative secretion expression vector. *Biotechnol Appl Biochem* **39**: 115–22.

Haldankar R., Li D., Saremi Z., Baikalov C., Deshpande R. (2006). Serum-free suspension large-scale transient transfection of CHO cells in WAVE bioreactors. *Mol Biotechnol* **34**: 191–9.

Hamilton S.R., Bobrowicz P., Bobrowicz B., Davidson R.C., Li H., Mitchell T., Nett J.H., Rausch S., Stadheim T.A., Wischnewski H., Wildt S., Gerngross T.U. (2003). Production of complex human glycoproteins in yeast. *Science* **301**: 1244–6.

Hamilton S.R., Gerngross T.U. (2007). Glycosylation engineering in yeast: the advent of fully humanized yeast. *Curr Opin Biotechnol* **18**: 387–92.

Hansen G., Wright M.S. (1999). Recent advances in the transformation of plants. *Trends Plant Sci* **4**: 226–31.

Harrison R.L., Jarvis D.L. (2006). Protein N-glycosylation in the baculovirus-insect cell expression system and engineering of insect cells to produce "mammalianized" recombinant glycoproteins. *Adv Virus Res* **68**: 159–91.

Hiatt A., Cafferkey R., Bowdish K. (1989). Production of antibodies in transgenic plants. *Nature* **342**: 76–8.

Hofmann A., Kessler B., Ewerling S., Kabermann A., Brem G., Wolf E., Pfeifer A. (2006). Epigenetic regulation of lentiviral transgene vectors in a large animal model. *Mol Ther* **13**: 59–66.

Holland M., Yagi H., Takahashi N., Kato K., Savage C.O., Goodall D.M., Jefferis R. (2006). Differential glycosylation of polyclonal IgG, IgG-Fc and IgG-Fab isolated from the sera of patients with ANCA-associated systemic vasculitis. *Biochim Biophys Acta* **1760**: 669–77.

Hollister J.R., Shaper J.H., Jarvis D.L. (1998). Stable expression of mammalian beta 1,4-galactosyltransferase extends the N-glycosylation pathway in insect cells. *Glycobiology* **8**: 473–80.

Hood E.E., Woodard S.L., Horn M.E. (2002). Monoclonal antibody manufacturing in transgenic plants — myths and realities. *Curr Opin Biotechnol* **13**: 630–5.

Horwitz A.H., Chang C.P., Better M., Hellstrom K.E., Robinson R.R. (1988). Secretion of functional antibody and Fab fragment from yeast cells. *Proc Natl Acad Sci USA* **85**: 8678–82.

Houdebine L.M. (2005). Use of transgenic animals to improve human health and animal production. *Reprod Domest Anim* **40**: 269–81.

Houdebine L.M. (2009). Production of pharmaceutical proteins by transgenic animals. *Comp Immunol Microbiol Infect Dis* **32**: 107–21.

Houdebine L.M., Attal J. (1999). Internal ribosome entry sites (IRESs): reality and use. *Transgenic Res* **8**: 157–77.

Hsu T.A., Betenbaugh M.J. (1997). Coexpression of molecular chaperone BiP improves immunoglobulin solubility and IgG secretion from Trichoplusia ni insect cells. *Biotechnol Prog* **13**: 96–104.

Hsu T.A., Takahashi N., Tsukamoto Y., Kato K., Shimada I., Masuda K., Whiteley E.M., Fan J.Q., Lee Y.C., Betenbaugh M.J. (1997). Differential N-glycan patterns of secreted and intracellular IgG produced in Trichoplusia ni cells. *J Biol Chem* **272**: 9062–70.

Hsu T.A., Watson S., Eiden J.J., Betenbaugh M.J. (1996). Rescue of immunoglobulins from insolubility is facilitated by PDI in the baculovirus expression system. *Protein Expr Purif* **7**: 281–8.

Humphreys D.P., Carrington B., Bowering L.C., Ganesh R., Sehdev M., Smith B.J., King L.M., Reeks D.G., Lawson A., Popplewell A.G. (2002). A plasmid system for optimization of Fab′ production in Escherichia coli: importance of balance of heavy chain and light chain synthesis. *Protein Expr Purif* **26**: 309–20.

Humphreys D.P., Sehdev M., Chapman A.P., Ganesh R., Smith B.J., King L.M., Glover D.J., Reeks D.G., Stephens, P.E. (2000). High-level periplasmic expression in Escherichia coli using a eukaryotic signal peptide: importance of codon usage at the 5′ end of the coding sequence. *Protein Expr Purif* **20**: 252–64.

Inoue Y., Lopez L.B., Kawamoto S., Arita N., Teruya K., Seki K., Shoji M., Kamei M., Hashizume S., Shiozawa Y., Tachibana H., Ohashi H., Yasumoto K., Nagashima A., Nakahashi H., Suzuki T., Imai T., Nomoto K., Takenoyama M., Katakura Y., Shirahata S. (1996). Production of recombinant human monoclonal antibody using ras-amplified BHK-21 cells in a protein-free medium. *Biosci Biotechnol Biochem* **60**: 811–7.

Inoue Y., Ohta T., Tada H., Iwasa S., Udaka S., Yamagata H. (1997). Efficient production of a functional mouse/human chimeric Fab′ against human urokinase-type plasminogen activator by Bacillus brevis. *Appl Microbiol Biotechnol* **48**: 487–92.

Ivarie R. (2006). Competitive bioreactor hens on the horizon. *Trends Biotechnol* 24: 99–101.

Jackson J.R., Sathe G., Rosenberg M., Sweet R. (1995). *In vitro* antibody maturation. Improvement of a high affinity, neutralizing antibody against IL-1 beta. *J Immunol* **154**: 3310–9.

Jarvis D.L., Weinkauf C., Guarino L.A. (1996). Immediate-early baculovirus vectors for foreign gene expression in transformed or infected insect cells. *Protein Expr Purif* **8**: 191–203.

Jean-Mairet J. (2007). 'How to set up a successful start-up.' Proceedings of the Canarias BioInvest Forum, Canary Islands, Jan, 2007.

Jefferis R. (2005a). Glycosylation of natural and recombinant antibody molecules. *Adv Exp Med Biol* **564**: 143–8.

Jefferis R. (2005b). Glycosylation of recombinant antibody therapeutics. *Biotechnol Prog* **21**: 11–6.

Jenkins N., Curling E.M. (1994). Glycosylation of recombinant proteins: problems and prospects. *Enzyme Microb Technol* **16**: 354–64.

Jespers L., Schon O., Famm K., Winter G. (2004). Aggregation-resistant domain antibodies selected on phage by heat denaturation. *Nat Biotechnol* **22**: 1161–5.

Jin B.R., Ryu C.J., Kang S.K., Han M.H., Hong H.J. (1995). Characterization of a murine-human chimeric antibody with specificity for the pre-S2 surface antigen of hepatitis B virus expressed in baculovirus-infected insect cells. *Virus Res* **38**: 269–77.

Jones D., Kroos N., Anema R., van M.B., Vooys A., van der Kraat S., van der Helm E., Smits S., Schouten J., Brouwer K., Lagerwerf F., van B.P., Opstelten D.J., Logtenberg T., Bout A. (2003). High-level expression of recombinant IgG in the human cell line per.c6. *Biotechnol Prog* **19**: 163–8.

Joosten V., Lokman C., van den Hondel C.A., Punt P.J. (2003). The production of antibody fragments and antibody fusion proteins by yeasts and filamentous fungi. *Microb Cell Fact* **2**: 1–15.

Jordan E., Al-Halabi L., Schirrmann T., Hust M. (2009). Antibody production by the gram-positive bacterium Bacillus megaterium. *Methods Mol Biol* **525**: 509–16, xiv.

Jordan E., Hust M., Roth A., Biedendieck R., Schirrmann T., Jahn D., Dubel S. (2007). Production of recombinant antibody fragments in Bacillus megaterium. *Microb Cell Fact* **6**: 2.

Jung S., Honegger A., Pluckthun A. (1999). Selection for improved protein stability by phage display. *J Mol Biol* **294**: 163–80.

Jung S., Pluckthun A. (1997). Improving *in vivo* folding and stability of a single-chain Fv antibody fragment by loop grafting. *Protein Eng* **10**: 959–66.

John Morrow K., Jr. (2008). Optimizing transient gene expression: applications expected to move beyond discovery and the preclinic to clinical realm. *Genetic Engineering and Biotechnology News* Vol. 28 No. 5.

Kamihira M., Kawabe Y., Shindo T., Ono K., Esaka K., Yamashita T., Nishijima K., Iijima S. (2009). Production of chimeric monoclonal antibodies by genetically manipulated chickens. *J Biotechnol* **141**: 18–25.

Kamihira M., Ono K., Esaka K., Nishijima K., Kigaku R., Komatsu H., Yamashita T., Kyogoku K., Iijima S. (2005). High-level expression of single-chain Fv-Fc fusion protein in serum and egg white of genetically manipulated chickens by using a retroviral vector. *J Virol* **79**: 10864–74.

Kanda Y., Yamada T., Mori K., Okazaki A., Inoue M., Kitajima-Miyama K., Kuni-Kamochi R., Nakano R., Yano K., Kakita S., Shitara K., Satoh M. (2007). Comparison of biological activity among nonfucosylated therapeutic IgG1 antibodies with three different N-linked Fc oligosaccharides: the high-mannose, hybrid, and complex types. *Glycobiology* **17**: 104–18.

Kao F.T., Puck T.T. (1968). Genetics of somatic mammalian cells, VII. Induction and isolation of nutritional mutants in Chinese hamster cells. *Proc Natl Acad Sci USA* **60**: 1275–81.

Kaufman R.J., Bertino J.R., Schimke R.T. (1978). Quantitation of dihydrofolate reductase in individual parental and methotrexate-resistant murine cells. Use of a fluorescence activated cell sorter. *J Biol Chem* **253**: 5852–60.

Kaufman R.J., Sharp P.A. (1982). Amplification and expression of sequences cotransfected with a modular dihydrofolate reductase complementary dna gene. *J Mol Biol* **159**: 601–21.

Keystone E., Heijde D., Mason D., Jr., Landewe R., Vollenhoven R.V., Combe B., Emery P., Strand V., Mease P., Desai C., Pavelka K. (2008). Certolizumab pegol plus methotrexate is significantly more effective than placebo plus methotrexate in active rheumatoid arthritis: findings of a fifty-two-week, phase III, multicenter, randomized, double-blind, placebo-controlled, parallel-group study. *Arthritis Rheum* **58**: 3319–29.

Kim N.S., Byun T.H., Lee G.M. (2001). Key determinants in the occurrence of clonal variation in humanized antibody expression of CHO cells during dihydrofolate reductase mediated gene amplification. *Biotechnol Prog* **17**: 69–75.

Kind A., Schnieke A. (2008). Animal pharming, two decades on. *Transgenic Res* **17**: 1025–33.

Kipriyanov S.M., Moldenhauer G., Martin A.C., Kupriyanova O.A., Little M. (1997). Two amino acid mutations in an anti-human CD3 single chain Fv antibody fragment that affect the yield on bacterial secretion but not the affinity. *Protein Eng* **10**: 445–53.

Kitchin K., Flickinger M.C. (1995). Alteration of hybridoma viability and antibody secretion in transfectomas with inducible overexpression of protein disulfide isomerase. *Biotechnol Prog* **11**: 565–74.

Knappik A., Krebber C., Pluckthun A. (1993). The effect of folding catalysts on the *in vivo* folding process of different antibody fragments expressed in Escherichia coli. *Biotechnology (NY)* **11**: 77–83.

Ko K., Brodzik R., Steplewski Z. (2009). Production of antibodies in plants: approaches and perspectives. *Curr Top Microbiol Immunol* **332**: 55–78.

Ko K., Tekoah Y., Rudd P.M., Harvey D.J., Dwek R.A., Spitsin S., Hanlon C.A., Rupprecht C., Dietzschold B., Golovkin M., Koprowski H. (2003). Function and glycosylation of plant-derived antiviral monoclonal antibody. *Proc Natl Acad Sci USA* **100**: 8013–8.

Kohler G., Hengartner H., Shulman M.J. (1978). Immunoglobulin production by lymphocyte hybridomas. *Eur J Immunol* **8**: 82–8.

Kohler G., Milstein C. (2005). Continuous cultures of fused cells secreting antibody of predefined specificity. 1975. *J Immunol* **174**: 2453–5.

Kollmann G., Friedman H., Shapiro B. (1966). Chemical protection against radiation damage to an immune response in mouse spleen cells *in vitro*. SAM-TR-66-69. *Tech Rep SAM-TR*, 1–8.

Koprivova A., Stemmer C., Altmann F., Hoffmann A., Kopriva S., Gorr G., Reski R., Decker E.L. (2004). Targeted knockouts of Physcomitrella lacking plant-specific immunogenic N-glycans. *Plant Biotechnol J* **2**: 517–23.

Kretzschmar T., Aoustin L., Zingel O., Marangi M., Vonach B., Towbin H., Geiser M. (1996). High-level expression in insect cells and purification of secreted monomeric single-chain Fv antibodies. *J Immunol Methods* **195**: 93–101.

Kruger C., Hu Y., Pan Q., Marcotte H., Hultberg A., Delwar D., van Dalen P.J., Pouwels P.H., Leer R.J., Kelly C.G., van D.C., Ma J.K., Hammarstrom L. (2002). In situ delivery of passive immunity by lactobacilli producing single-chain antibodies. *Nat Biotechnol* **20**: 702–6.

Ku S.C., Ng D.T., Yap M.G., Chao S.H. (2008). Effects of overexpression of X-box binding protein 1 on recombinant protein production in Chinese hamster ovary and NS0 myeloma cells. *Biotechnol Bioeng* **99**: 155–64.

Kujau M.J., Hoischen C., Riesenberg D., Gumpert J. (1998). Expression and secretion of functional miniantibodies McPC603scFvDhlx in cell-wall-less L-form strains of Proteus mirabilis and Escherichia coli: a comparison of the synthesis capacities of L-form strains with an *E. coli* producer strain. *Appl Microbiol Biotechnol* **49**: 51–8.

Kumar R., Singh J. (2004). Expression and secretion of a prokaryotic protein streptokinase without glycosylation and degradation in schizosaccharomyces pombe. *Yeast* **21**: 1343–58.

Kunaparaju R., Liao M., Sunstrom N.A. (2005). Epi-CHO, an episomal expression system for recombinant protein production in CHO cells. *Biotechnol Bioeng* **91**: 670–7.

Kwaks T.H., Barnett P., Hemrika W., Siersma T., Sewalt R.G., Satijn D.P., Brons J.F., van B.R., Kwakman P., Kruckeberg A.L., Kelder A., Otte A.P. (2003). Identification of anti-repressor elements that confer high and stable protein production in mammalian cells. *Nat Biotechnol* **21**: 553–8.

Langer E.S. (2009). Expression system royalties battle. *BioPharm International* 22. Accessed online via http://biopharminternational.findpharma.com/biopharm/Expression+Systems/Expression-System-Royalties-Battle/ArticleStandard/Article/detail/590347.

Lee F.W., Da Silva N.A. (1996). Ty1-mediated integration of expression cassettes: host strain effects, stability, and product synthesis. *Biotechnol Prog* **12**: 548–54.

Leitzgen K., Knittler M.R., Haas I.G. (1997). Assembly of immunoglobulin light chains as a prerequisite for secretion. A model for oligomerization-dependent subunit folding. *J Biol Chem* **272**: 3117–23.

Levy R., Weiss R., Chen G., Iverson B.L., Georgiou G. (2001). Production of correctly folded Fab antibody fragment in the cytoplasm of Escherichia coli trxB gor mutants via the coexpression of molecular chaperones. *Protein Expr Purif* **23**: 338–47.

Li H., Sethuraman N., Stadheim T.A., Zha D., Prinz B., Ballew N., Bobrowicz P., Choi B.K., Cook W.J., Cukan M., Houston-Cummings N.R., Davidson R., Gong B., Hamilton S.R., Hoopes J.P., Jiang Y., Kim N., Mansfield R., Nett J.H., Rios S., Strawbridge R., Wildt S., Gerngross T.U. (2006). Optimization of humanized IgGs in glycoengineered Pichia pastoris. *Nat Biotechnol* **24**: 210–5.

Li J., Menzel C., Meier D., Zhang C., Dubel S., Jostock T. (2007). A comparative study of different vector designs for the mammalian expression of recombinant IgG antibodies. *J Immunol Methods* **318**: 113–24.

Li Q., Zhang M., Duan Z., Stamatoyannopoulos G. (1999). Structural analysis and mapping of DNase I hypersensitivity of HS5 of the beta-globin locus control region. *Genomics* **61**: 183–93.

Li S., Kussie P., Ferguson K.M. (2008). Structural basis for EGF receptor inhibition by the therapeutic antibody IMC-11F8. *Structure* **16**: 216–27.

Liang M., Dubel S., Li D., Queitsch I., Li W., Bautz E.K. (2001). Baculovirus expression cassette vectors for rapid production of complete human IgG from phage display selected antibody fragments. *J Immunol Methods* **247**: 119–30.

Lico C., Chen Q., Santi L. (2008). Viral vectors for production of recombinant proteins in plants. *J Cell Physiol* **216**: 366–77.

Lien S., Lowman H.B. (2008). Therapeutic anti-VEGF antibodies in Chernajovsky Y., Nissim, A. (Eds). *Therapeutic Antibodies. Handb Exp Pharmacol* **181**: 131–50, Springer-Verlag, Berlin.

Lifely M.R., Hale C., Boyce S., Keen M.J., Phillips J. (1995). Glycosylation and biological activity of CAMPATH-1H expressed in different cell lines and grown under different culture conditions. *Glycobiology* **5**: 813–22.

Lim A., Reed-Bogan A., Harmon B.J. (2008). Glycosylation profiling of a therapeutic recombinant monoclonal antibody with two N-linked glycosylation sites using liquid chromatography coupled to a hybrid quadrupole time-of-flight mass spectrometer. *Anal Biochem* **375**: 163–72.

Limonta J., Pedraza A., Rodriguez A., Freyre F.M., Barral A.M., Castro F.O., Lleonart R., Gracia C.A., Gavilondo J.V., de la Fuente J. (1995). Production of active anti-CD6 mouse/human chimeric antibodies in the milk of transgenic mice. *Immunotechnology* **1**: 107–13.

Lindgren K., Salmen A., Lundgren M., Bylund L., Ebler A., Faldt E., Sorvik L., Fenge C., Skoging-Nyberg U. (2009). Automation of cell line development. *Cytotechnology* **59**: 1–10.

Lucas B.K., Giere L.M., DeMarco R.A., Shen A., Chisholm V., Crowley C.W. (1996). High-level production of recombinant proteins in CHO cells using a dicistronic DHFR intron expression vector. *Nucleic Acids Res* **24**: 1774–9.

Ma J.K., Hiatt A., Hein M., Vine N.D., Wang F., Stabila P., van D. C., Mostov K., Lehner T. (1995). Generation and assembly of secretory antibodies in plants. *Science* **268**: 716–9.

Ma J.K., Hikmat B.Y., Wycoff K., Vine N.D., Chargelegue D., Yu L., Hein M.B., Lehner T. (1998). Characterization of a recombinant plant monoclonal secretory antibody and preventive immunotherapy in humans. *Nat Med* **4**: 601–6.

MacDonald G. (2008). PerC.6 platform achieves highest ever yield. *In-Pharma Technologist.*

Majors B.S., Arden N., Oyler G.A., Chiang G.G., Pederson N.E., Betenbaugh M.J. (2008). E2F-1 overexpression increases viable cell density in batch cultures of Chinese hamster ovary cells. *J Biotechnol* **138**: 103–6.

Maranga L., Goochee C.F. (2006). Metabolism of PER.C6 cells cultivated under fed-batch conditions at low glucose and glutamine levels. *Biotechnol Bioeng* **94**: 139–50.

Marillonnet S., Thoeringer C., Kandzia R., Klimyuk V., Gleba Y. (2005). Systemic Agrobacterium tumefaciens-mediated transfection of viral replicons for efficient transient expression in plants. *Nat Biotechnol* **23**: 718–23.

Martineau P., Jones P., Winter G. (1998). Expression of an antibody fragment at high levels in the bacterial cytoplasm. *J Mol Biol* **280**: 117–27.

Mayfield S.P., Franklin S.E., Lerner R.A. (2003). Expression and assembly of a fully active antibody in algae. *Proc Natl Acad Sci USA* **100**: 438–42.

Mayfield S.P., Manuell A.L., Chen S., Wu J., Tran M., Siefker D., Muto M., Marin-Navarro J. (2007). Chlamydomonas reinhardtii chloroplasts as protein factories. *Curr Opin Biotechnol* **18**: 126–33.

Mazor Y., Van B. T., Iverson B.L., Georgiou G. (2009). Isolation of full-length IgG antibodies from combinatorial libraries expressed in Escherichia coli. *Methods Mol Biol* **525**: 217–39, xiv.

McCormick A.A., Reddy S., Reinl S.J., Cameron T.I., Czerwinkski D.K., Vojdani F., Hanley K.M., Garger S.J., White E.L., Novak J., Barrett J., Holtz R.B., Tuse D., Levy R. (2008). Plant-produced idiotype vaccines for the treatment of non-Hodgkin's lymphoma: safety and immunogenicity in a phase I clinical study. *Proc Natl Acad Sci USA* **105**: 10131–6.

Mimura Y., Ashton P.R., Takahashi N., Harvey D.J., Jefferis R. (2007). Contrasting glycosylation profiles between Fab and Fc of a human IgG protein studied by electrospray ionization mass spectrometry. *J Immunol Methods* **326**: 116–26.

Montano R.F., Morrison S.L. (2002). Influence of the isotype of the light chain on the properties of IgG. *J Immunol* **168**: 224–31.

Niwa R., Natsume A., Uehara A., Wakitani M., Iida S., Uchida K., Satoh M., Shitara K. (2005a). IgG subclass-independent improvement of antibody-dependent cellular cytotoxicity by fucose removal from Asn297-linked oligosaccharides. *J Immunol Methods* **306**: 151–60.

Niwa R., Sakurada M., Kobayashi Y., Uehara A., Matsushima K., Ueda R., Nakamura K., Shitara K. (2005b). Enhanced natural killer cell binding and activation by low-fucose IgG1 antibody results in potent antibody-dependent cellular cytotoxicity induction at lower antigen density. *Clin Cancer Res* **11**: 2327–36.

Novak M., Borecky L., Kontsekova E., Kontsek P., Novakova M. (1986). Preparation and characterization of hybridomas producing monoclonal antibodies against human alpha interferon. *Acta Virol* **30**: 228–33.

Nyyssonen E., Penttila M., Harkki A., Saloheimo A., Knowles J.K., Keranen S. (1993). Efficient production of antibody fragments by the filamentous fungus Trichoderma reesei. *Biotechnology (NY)* **11**: 591–5.

O'Connell K.P., Kovaleva E., Campbell J.H., Anderson P.E., Brown S.G., Davis D.C., Valdes J.J., Welch R.W., Bentley W.E., van Beek N.A. (2007). Production of a recombinant antibody fragment in whole insect larvae. *Mol Biotechnol* **36**: 44–51.

Oelschlaeger P., Lange S., Schmitt J., Siemann M., Reuss M., Schmid R.D. (2003). Identification of factors impeding the production of a single-chain antibody fragment in Escherichia coliby comparing *in vivo* and *in vitro* expression. *Appl Microbiol Biotechnol* **61**: 123–32.

Ogunjimi A.A., Chandler J.M., Gooding C.M., Recinos A.I., Choudary P.V. (1999). High-level secretory expression of immunologically active intact antibody from the yeast Pichia pastoris. [Article]. *Biotechnology Letters* **21**: 561–7.

Perrin Y., Vaquero C., Gerrard I., Sack M., Drossard J., Stoger E., Christou P., Fischer R. (2000). Transgenic pea seeds as bioreactors for the production of single chain Fv antibody fragment (scFv) used in cancer diagnosis and therapy. *Molecular Breeding* **6**: 345–52.

Pham P.L., Kamen A., Durocher Y. (2006). Large-scale transfection of mammalian cells for the fast production of recombinant protein. *Mol Biotechnol* **34**: 225–37.

Pikaart M.J., Recillas-Targa F., Felsenfeld G. (1998). Loss of transcriptional activity of a transgene is accompanied by DNA methylation and histone deacetylation and is prevented by insulators. *Genes Dev* **12**: 2852–62.

Pollock D.P., Kutzko J.P., Birck-Wilson E., Williams J.L., Echelard Y., Meade H.M. (1999). Transgenic milk as a method for the production of recombinant antibodies. *J Immunol Methods* **231**: 147–57.

Popplewell A.G., Sehdev M., Spitali M., Weir A.N. (2005). Expression of antibody fragments by periplasmic secretion in Escherichia coli. *Methods Mol Biol* **308**: 17–30.

Potgieter T.I., Cukan M., Drummond J.E., Houston-Cummings N.R., Jiang Y., Li F., Lynaugh H., Mallem M., McKelvey T.W., Mitchell T., Nylen A., Rittenhour A.,

Stadheim T.A., Zha D., d'Anjou M. (2009). Production of monoclonal antibodies by glycoengineered Pichia pastoris. *J Biotechnol* **139**: 318–25.

Proba K., Worn A., Honegger A., Pluckthun A. (1998). Antibody scFv fragments without disulfide bonds made by molecular evolution. *J Mol Biol* **275**: 245–53.

Qian J., Liu T., Yang L., Daus A., Crowley R., Zhou Q. (2007). Structural characterization of N-linked oligosaccharides on monoclonal antibody cetuximab by the combination of orthogonal matrix-assisted laser desorption/ionization hybrid quadrupole-quadrupole time-of-flight tandem mass spectrometry and sequential enzymatic digestion. *Anal Biochem* **364**: 8–18.

Raju T.S., Briggs J.B., Borge S.M., Jones A.J. (2000). Species-specific variation in glycosylation of IgG: evidence for the species-specific sialylation and branch-specific galactosylation and importance for engineering recombinant glycoprotein therapeutics. *Glycobiology* **10**: 477–86.

Reavy B., Ziegler A., Diplexcito J., Macintosh S.M., Torrance,L., Mayo M. (2000). Expression of functional recombinant antibody molecules in insect cell expression systems. *Protein Expr Purif* **18**: 221–8.

Renner W.A., Lee K.H., Hatzimanikatis V., Bailey J.E., Eppenberger H.M. (1995). Recombinant cyclin E expression activates proliferation and obviates surface attachment of chinese hamster ovary (CHO) cells in protein-free medium. *Biotechnol Bioeng* **47**: 476–82.

Richards C.M., Aucken H.A., Tucker E.M., Hannant D., Mumford J.A., Powell J.R. (1992). The production of equine monoclonal immunoglobulins by horse-mouse heterohybridomas. *Vet Immunol Immunopathol* **33**: 129–43.

Ringold G., Dieckmann B., Lee F. (1981). Co-expression and amplification of dihydrofolate reductase cDNA and the Escherichia coli XGPRT gene in Chinese hamster ovary cells. *J Mol Appl Genet* **1**: 165–75.

Rippmann J.F., Klein M., Hoischen C., Brocks B., Rettig W.J., Gumpert J., Pfizenmaier K., Mattes R., Moosmayer D. (1998). Procaryotic expression of single-chain variable-fragment (scFv) antibodies: secretion in L-form cells of Proteus mirabilis leads to active product and overcomes the limitations of periplasmic expression in Escherichia coli. *Appl Environ Microbiol* **64**: 4862–9.

Rival-Gervier S., Viglietta C., Maeder C., Attal J., Houdebine L.M. (2002). Position-independent and tissue-specific expression of porcine whey acidic protein gene from a bacterial artificial chromosome in transgenic mice. *Mol Reprod Dev* **63**: 161–7.

Rosenfeld P.J., Brown D.M., Heier J.S., Boyer D.S., Kaiser P.K., Chung C.Y., Kim R.Y. (2006). Ranibizumab for neovascular age-related macular degeneration. *N Engl J Med* **355**: 1419–31.

Running Deer J., Allison D.S. (2004). High-level expression of proteins in mammalian cells using transcription regulatory sequences from the Chinese hamster EF-1alpha gene. *Biotechnol Prog* **20**: 880–9.

Saint-Jore-Dupas C., Faye L., Gomord V. (2007). From planta to pharma with glycosylation in the toolbox. *Trends Biotechnol* **25**: 317–23.

Sauer B. (1998). Inducible gene targeting in mice using the Cre/lox system. *Methods* **14**: 381–92.

Sautter K., Enenkel B. (2005). Selection of high-producing CHO cells using NPT selection marker with reduced enzyme activity. *Biotechnol Bioeng* **89**: 530–8.

Scahill S.J., Devos R., Van der Heyden J., Fiers W. (1983). Expression and characterization of the product of a human immune interferon cDNA gene in Chinese hamster ovary cells. *Proc Natl Acad Sci USA* **80**: 4654–8.

Schillberg S., Fischer R., Emans N. (2003). Molecular farming of recombinant antibodies in plants. *Cell Mol Life Sci* **60**: 433–45.

Schimke R.T., Alt F.W., Kellems R.E., Kaufman R.J., Bertino J.R. (1978). Amplification of dihydrofolate reductase genes in methotrexate-resistant cultured mouse cells. *Cold Spring Harb Symp Quant Biol* **42** (**Pt 2**): 649–57.

Schimke R.T., Beverley S., Brown P., Cassin R., Federspiel N., Gasser C., Hill A., Johnston R., Mariani B., Mosse E. (1984). Gene amplification and drug resistance in cultured animal cells. *Cancer Treat Rev* **11 Suppl A**: 9–17.

Schoofs G., Monica T.J., Ayala J., Horwitz J., Montgomery T., Roth G., Castillo F. J. (1998). A high-yielding serum-free, suspension cell culture process to manufacture recombinant adenoviral vectors for gene therapy. *Cytotechnology* **28**: 81–9.

Schunmann P.H.D., Coia G., Waterhouse P.M. (2002). Biopharming the SimpliRED™ HIV diagnostic reagnet in barley, potato and tobacco. *Molecular Breeding* **9**: 113–21.

Schweikart F., Jones R., Jaton J.C., Hughes G.J. (1999). Rapid structural characterisation of a murine monoclonal IgA alpha chain: heterogeneity in the oligosaccharide structures at a specific site in samples produced in different bioreactor systems. *J Biotechnol* **69**: 191–201.

Shaw G., Morse S., Ararat M., Graham F.L. (2002). Preferential transformation of human neuronal cells by human adenoviruses and the origin of HEK 293 cells. *FASEB J* **16**: 869–71.

Sheeley D.M., Merrill B.M., Taylor L.C. (1997). Characterization of monoclonal antibody glycosylation: comparison of expression systems and identification of terminal alpha-linked galactose. *Anal Biochem* **247**: 102–10.

Shinkawa T., Nakamura K., Yamane N., Shoji-Hosaka E., Kanda Y., Sakurada M., Uchida K., Anazawa H., Satoh M., Yamasaki M., Hanai N., Shitara K. (2003). The absence of fucose but not the presence of galactose or bisecting N-acetylglucosamine of human IgG1 complex-type oligosaccharides shows the critical role of enhancing antibody-dependent cellular cytotoxicity. *J Biol Chem* **278**: 3466–73.

Shuler M.L., Hammer D.A., Granados R.R., Wood, H.A. (1995). Overview of baculovirus-insect culture system. In M. L. Shuler, H. A. Wood, R. R. Granados and D. A. Hammer (Eds.) *Baculovirus expression systems and biopesticides*, pp. 1–12. Wiley-Liss, New York.

Shusta E.V., Raines R.T., Pluckthun A., Wittrup K.D. (1998). Increasing the secretory capacity of Saccharomyces cerevisiae for production of single-chain antibody fragments. *Nat Biotechnol* **16**: 773–7.

Simmons L.C., Reilly D., Klimowski L., Raju T.S., Meng G., Sims P., Hong K., Shields R.L., Damico L.A., Rancatore P., Yansura D.G. (2002). Expression of full-length immunoglobulins in Escherichia coli: rapid and efficient production of aglycosylated antibodies. *J Immunol Methods* **263**: 133–47.

Skerra A., Pluckthun A. (1988). Assembly of a functional immunoglobulin Fv fragment in Escherichia coli. *Science* **240**: 1038–41.

Skvortsov V.T., Gurvich A.E. (1968). Relative rates of synthesis of immunoglobulins and light chains in rabbit spleen cells during secondary response. *Nature* **218**: 377–8.

Smith G., Walmsley A., Polkinghorne I. (1997). Plant-derived immunocontraceptive vaccines. *Reprod Fertil Dev* **9**: 85–9.

Sorensen H.P., Mortensen K.K. (2005). Advanced genetic strategies for recombinant protein expression in Escherichia coli. *J Biotechnol* **115**: 113–28.

Squires C.H., Retallack D.M., Chew L.C., Ramseier T.M., Schneider J.C., Talbot H.W. (2006). Heterologous protein production in P.fluorescens. *Bioprocess Inernational*, pp. 54–8.

Stinnakre M.G., Soulier S., Schibler L., Lepourry L., Mercier J.C., Vilotte J.L. (1999). Position-independent and copy-number-related expression of a goat bacterial artificial chromosome alpha-lactalbumin gene in transgenic mice. *Biochem J* **339** (**Pt 1**): 33–6.

Stoger E., Vaquero C., Torres E., Sack M., Nicholson L., Drossard J., Williams S., Keen D., Perrin Y., Christou P., Fischer R. (2000). Cereal crops as viable production and storage systems for pharmaceutical scFv antibodies. *Plant Mol Biol* **42**: 583–90.

Storring P.L., Gaines Das R.E. (1992). The International Standard for Recombinant DNA-derived Erythropoietin: collaborative study of four recombinant DNA-derived erythropoietins and two highly purified human urinary erythropoietins. *J Endocrinol* **134**: 459–84.

Taboit-Dameron F., Malassagne B., Viglietta C., Puissant C., Leroux-Coyau M., Chereau C., Attal J., Weill B., Houdebine L.M. (1999). Association of the 5′HS4 sequence of the chicken beta-globin locus control region with human EF1 alpha gene promoter induces ubiquitous and high expression of human CD55 and CD59 cDNAs in transgenic rabbits. *Transgenic Res* **8**: 223–35.

Tait A.S., Brown C.J., Galbraith D.J., Hines M.J., Hoare M., Birch J.R., James D.C. (2004). Transient production of recombinant proteins by Chinese hamster ovary cells using polyethyleneimine/DNA complexes in combination with microtubule disrupting anti-mitotic agents. *Biotechnol Bioeng* **88**: 707–21.

Takeuchi M., Takasaki S., Miyazaki H., Kato T., Hoshi S., Kochibe N., Kobata A. (1988). Comparative study of the asparagine-linked sugar chains of human erythropoietins purified from urine and the culture medium of recombinant Chinese hamster ovary cells. *J Biol Chem* **263**: 3657–63.

Tang B., Yu S., Zheng M., Ding F., Zhao R., Zhao J., Dai,Y., Li N. (2008). High level expression of a functional human/mouse chimeric anti-CD20 monoclonal antibody in milk of transgenic mice. *Transgenic Res* **17**: 727–32.

Tey B.T., Al-Rubeai M. (2004). Suppression of apoptosis in perfusion culture of Myeloma NS0 cells enhances cell growth but reduces antibody productivity. *Apoptosis* **9**: 843–52.

Torres E., Gonzalez-Melendi P., Stoger E., Shaw P., Twyman R.M., Nicholson L., Vaquero C., Fischer R., Christou P., Perrin Y. (2001). Native and artificial reticuloplasmins co-accumulate in distinct domains of the endoplasmic reticulum and in post-endoplasmic reticulum compartments. *Plant Physiol* **127**: 1212–23.

Urlaub G., Chasin L.A. (1980). Isolation of Chinese hamster cell mutants deficient in dihydrofolate reductase activity. *Proc Natl Acad Sci USA* 77: 4216–20.

Valdes R., Gomez L., Padilla S., Brito J., Reyes B., Alvarez T., Mendoza O., Herrera O., Ferro W., Pujol M., Leal V., Linares M., Hevia Y., Garcia C., Mila L., Garcia O., Sanchez R., Acosta A., Geada D., Paez R., Luis V.J., Borroto C. (2003). Large-scale purification of an antibody directed against hepatitis B surface antigen from transgenic tobacco plants. *Biochem Biophys Res Commun* **308**: 94–100.

Van den Nieuwenhof I. M., Koistinen H., Easton R.L., Koistinen R., Kamarainen M., Morris H.R., Van D., I, Seppala M., Dell A., Van den Eijnden D.H. (2000). Recombinant glycodelin carrying the same type of glycan structures as contraceptive glycodelin-A can be produced in human kidney 293 cells but not in chinese hamster ovary cells. *Eur J Biochem* **267**: 4753–62.

van Ree R., Cabanes-Macheteau M., Akkerdaas J., Milazzo J.P., Loutelier-Bourhis C., Rayon C., Villalba M., Koppelman S., Aalberse R., Rodriguez R., Faye L., Lerouge P. (2000). Beta(1,2)-xylose and alpha(1,3)-fucose residues have a strong contribution in IgE binding to plant glycoallergens. *J Biol Chem* **275**: 11451–8.

Vaquero C., Sack M., Schuster F., Finnern R., Drossard J., Schumann D., Reimann A., Fischer R. (2002). A carcinoembryonic antigen-specific diabody produced in tobacco. *FASEB J* **16**: 408–10.

Verch T., Yusibov V., Koprowski H. (1998). Expression and assembly of a full-length monoclonal antibody in plants using a plant virus vector. *J Immunol Methods* **220**: 69–75.

Vezina L.P., Faye L., Lerouge P., D'Aoust M.A., Marquet-Blouin E., Burel C., Lavoie P.O., Bardor M., Gomord V. (2009). Transient co-expression for fast and high-yield production of antibodies with human-like N-glycans in plants. *Plant Biotechnol J* 7: 442–55.

Walowitz J.L., Fike R.M., Jayme D.W. (2003). Efficient lipid delivery to hybridoma culture by use of cyclodextrin in a novel granulated dry-form medium technology. *Biotechnol Prog* **19**: 64–8.

Walsh G. (2006). Biopharmaceutical benchmarks 2006. *Nat Biotechnol* **24**: 769–76.

Walsh G., Jefferis R. (2006). Post-translational modifications in the context of therapeutic proteins. *Nat Biotechnol* **24**: 1241–52.

Ward, M., Lin, C., Victoria, D.C., Fox, B.P., Fox, J.A., Wong, D.L., Meerman, H.J., Pucci, J.P., Fong R.B., Heng, M.H., Tsurushita, N., Gieswein, C., Park, M., Wang, H. (2004) *Appl Environ Microbiol* **70**: 2567–76.

Wasley L.C., Timony G., Murtha P., Stoudemire J., Dorner A.J., Caro J., Krieger M., Kaufman, R.J. (1991). The importance of N- and O-linked oligosaccharides for the biosynthesis and *in vitro* and *in vivo* biologic activities of erythropoietin. *Blood* 77: 2624–32.

Weintraub J.A., Hilton J.F., White J.M., Hoover C.I., Wycoff K.L., Yu L., Larrick J.W., Featherstone J.D. (2005). Clinical trial of a plant-derived antibody on recolonization of mutans streptococci. *Caries Res* **39**: 241–50.

Weir A.N., Nesbitt A., Chapman A.P., Popplewell A.G., Antoniw P., Lawson A.D. (2002). Formatting antibody fragments to mediate specific therapeutic functions. *Biochem Soc Trans* **30**: 512–6.

Williams S., Cranenburgh R. (2008). Improving Clone Production for Increased Protein Yield from Mammalian Cell Lines. *Innovations in Pharmaceutical Technology* **26**: 53–6.

Wood C.R., Boss M.A., Kenten J.H., Calvert J.E., Roberts N.A, Emtage J.S. (1985). The synthesis and *in vivo* assembly of functional antibodies in yeast. *Nature* **314**: 446–9.

Worn A., Pluckthun A. (1998). Mutual stabilization of VL and VH in single-chain antibody fragments, investigated with mutants engineered for stability. *Biochemistry* **37**: 13120–7.

Wu S.C., Ye R., Wu X.C., Ng S.C., Wong S.L. (1998). Enhanced secretory production of a single-chain antibody fragment from Bacillus subtilis by coproduction of molecular chaperones. *J Bacteriol* **180**: 2830–5.

Wu S.C., Yeung J.C., Duan Y., Ye R., Szarka S.J., Habibi H.R., Wong S.L. (2002). Functional production and characterization of a fibrin-specific single-chain antibody fragment from Bacillus subtilis: effects of molecular chaperones and a wall-bound protease on antibody fragment production. *Appl Environ Microbiol* **68**: 3261–9.

Wu X.., Ng S.C., Near R.I., Wong S.L. (1993). Efficient production of a functional single-chain antidigoxin antibody via an engineered Bacillus subtilis expression-secretion system. *Biotechnology (NY)* **11**: 71–6.

Wulfing C., Pluckthun A. (1994). Protein folding in the periplasm of Escherichia coli. *Mol Microbiol* **12**: 685–92.

Wulhfard S. (2009). Transient recombinant protein expression in mammalian cells: The role of mRNA level and stability. PhD Thesis, EPFL, Lausanne.

Wurm F.M. (1997). Aspects of Gene Transfer and Gene Amplification in Recombinant Mammalian Cells. In H. R. W. R. Hauser and R. Wagner (Eds.) *Mammalian Cell Biotech in protein production*, pp. 87–114. W. de Gruyter, Berlin, New York.

Wurm F.M. (2004). Production of recombinant protein therapeutics in cultivated mammalian cells. *Nat Biotechnol* **22**: 1393–8.

Xie L., Metallo C., Warren J., Pilbrough W., Peltier J., Zhong T., Pikus L., Yancy A., Leung J., Aunins J.G., Zhou W. (2003). Large-scale propagation of a replication-defective adenovirus vector in stirred-tank bioreactor PER.C6 cell culture under sparging conditions. *Biotechnol Bioeng* **83**: 45–52.

Xie L., Pilbrough W., Metallo C., Zhong T., Pikus L., Leung J., Aunins J.G., Zhou W. (2002). Serum-free suspension cultivation of PER.C6(R) cells and recombinant adenovirus production under different pH conditions. *Biotechnol Bioeng* **80**: 569–79.

Yamane-Ohnuki N., Kinoshita S., Inoue-Urakubo M., Kusunoki M., Iida S., Nakano R., Wakitani M., Niwa R., Sakurada M., Uchida K., Shitara K., Satoh M. (2004). Establishment of FUT8 knockout Chinese hamster ovary cells: an ideal host cell line for producing completely defucosylated antibodies with enhanced antibody-dependent cellular cytotoxicity. *Biotechnol Bioeng* **87**: 614–22.

Yin J., Li G., Ren X., Herrler G. (2007). Select what you need: a comparative evaluation of the advantages and limitations of frequently used expression systems for foreign genes. *J Biotechnol* **127**: 335–47.

Zang M., Trautmann H., Gandor C., Messi F., Asselbergs F., Leist C., Fiechter A., Reiser J. (1995). Production of recombinant proteins in Chinese hamster ovary cells using a protein-free cell culture medium. *Biotechnology (NY)* **13**: 389–92.

Zeitlin L., Olmsted S.S., Moench T.R., Co M.S., Martinell B.J., Paradkar V.M., Russell D.R., Queen C., Cone R.A., Whaley K.J. (1998). A humanized monoclonal antibody produced in transgenic plants for immunoprotection of the vagina against genital herpes. *Nat Biotechnol* **16**: 1361–4.

Zhang Z., Li Z.H., Wang F., Fang M., Yin C.C., Zhou Z.Y., Lin Q., Huang,H.L. (2002). Overexpression of DsbC and DsbG markedly improves soluble and functional expression of single-chain Fv antibodies in Escherichia coli. *Protein Expr Purif* **26**: 218–28.

Zhou J.M., Tang Y.X., Fang D.Y., Zhou J.J., Liang Y., Guo H.Y., Jiang L.F. (2006). Secreted expression and purification of dengue 2 virus full-length nonstructural glycoprotein NS1 in Pichia pastoris. *Virus Genes* **33**: 27–32.

Zhou Q., Shankara S., Roy A., Qiu H., Estes S., McVie-Wylie A., Culm-Merdek K., Park A., Pan C., Edmunds T. (2008). Development of a simple and rapid method for producing non-fucosylated oligomannose containing antibodies with increased effector function. *Biotechnol Bioeng* **99**: 652–65.

Zhu L., van de Lavoir M.C., Albanese J., Beenhouwer D.O., Cardarelli P.M., Cuison S., Deng D.F., Deshpande S., Diamond J.H., Green L., Halk E.L., Heyer B.S., Kay R.M., Kerchner A., Leighton P.A., Mather C.M., Morrison S.L., Nikolov Z.L., Passmore D.B., Pradas-Monne A., Preston B.T., Rangan V.S., Shi M., Srinivasan M., White S.G., Winters-Digiacinto P., Wong S., Zhou W., Etches R.J. (2005). Production of human monoclonal antibody in eggs of chimeric chickens. *Nat Biotechnol* **23**: 1159–69.

Zhu X.D., Pan G., Luetke K., Sadowski P.D. (1995). Homology requirements for ligation and strand exchange by the FLP recombinase. *J Biol Chem* **270**: 11646–53.

zu Putlitz J., Kubasek W.L., Duchene M., Marget M., von Specht B.U., Domdey H. (1990). Antibody production in baculovirus-infected insect cells. *Biotechnolopgy (NY)* **8**: 651–4.

7

Current Trends in Antibody Purification and Lead Selection

Jie Chen and Andrew Nixon**

7.1 Introduction

Biologic drugs are fast becoming one of the dominant approaches for developing new therapeutics for a variety of diseases. For the past decade, there has been continuous growth in the development of biotherapeutics, among which more than half are monoclonal antibodies (Reichert, 2009). Since their creation in the early 1970s, monoclonal antibodies have opened up a wide range of new therapies in diverse indication areas. The discovery of monoclonal antibody technology is recognized as one of the most significant scientific accomplishments in biotechnology. The number of approved antibody drugs further underscores the clinical effectiveness of this class of therapeutics in oncology and inflammatory diseases (Reichert and Pavlou, 2004). Previous chapters have discussed various technologies that facilitate antibody drug discovery. How to translate these technologies into high quality drugs is a major challenge in the development of antibody-based therapeutics, with both effective product design and process development strategies being critical for success. This chapter will provide an overview of evolving strategies in lead selection for monoclonal antibody drug discovery and discuss commonly used and recently emerged separation technologies in the process of monoclonal antibody purification.

7.2 Current Trends in Antibody Purification

The manufacturing process for biologics is not a passive supply chain. Instead it plays one of the most active roles in the drug development life cycle. This is because the

---

* Dyax Corp., USA.

bioactivities of biological drugs are embedded in their tertiary structures and complex relationship with the surrounding environment, something that is often heavily influenced by the manufacturing processes. "The process is the product" has been the perception in the development of biological therapeutics (Kozlowski, 2007).

All immunoglobulins, independent of their antigen-binding specificity, have a common structure with four polypeptide chains: two identical heavy (H) chains and two identical light (L) chains connected by inter-chain disulfide bonds and non-covalent interactions. This common structure can be dissected into two regions of biological function: Fab for antigen binding and Fc for mediating biological function. Because of their unique structural similarities, the production of recombinant antibodies is one of the most successful industrial manufacturing processes that implement a platform strategy. In the common generic downstream purification platform, Protein A (ProA) affinity chromatography and ion exchange chromatography, plus adequate viral clearance procedures, are usually sufficient to meet the purification demands for most of the therapeutic antibody candidates delivered by discovery pipelines (Shukla *et al.*, 2007b). In addition, separation techniques such as Hydrophobic Interaction Chromatography (HIC) and Mixed-Mode Chromatography have demonstrated high aggregate reduction capability (Gagnon, 2008; Chen *et al.*, 2008; Chen *et al.*, 2009), and are able to provide versatile purification tools in response to the product quality demands by the Food and Drug Administration (FDA).

Monoclonal antibodies are among the most complex and expensive pharmaceutical products to manufacture, in which the downstream purification process (DSP) represents over 50% of the total manufacturing cost (Farid, 2007; Farid, 2009) The past decade has brought substantial improvements in the upstream process yields of monoclonal antibodies produced in mammalian cell expression systems with productivities of 3–4 g/L routinely achieved. The shift of the manufacturing bottleneck from the upstream to the downstream process has resulted in an industry-wide drive for high throughput, low cost and high purity downstream purification process (Moreira, 2007; Jones *et al.*, 2007; Gottschalk, 2006; Gottschalk, 2009a). The purification process development strategies, from platform design to individual separation technology, are consequently undergoing revolutionary changes. Resins based on new separation mechanisms or new generations of well-known chromatographic technologies are being introduced to increase DSP efficiencies (Sofer, 2006; Langer, 2010; Shukla and Thommes, 2010). Traditional bead-based chromatography processes are being challenged by membrane-based separation technologies that present significant benefits in terms of process speed and operation associated cost reductions. As feed titers continue to increase towards and beyond 5 g/L, other purification processes such as precipitation methods could also offer significant economic and throughput advantages (Przybycien *et al.*, 2004; Curling and Gottschalk, 2007; Gottschalk, 2009b). For monoclonal antibody manufacturing, the critical goal has shifted from searching for state-of-the art technologies capable of producing sufficient quantities of antibodies with desired purity to meet clinical needs, to identifying and developing strategies to improve process throughput and efficiency, consequently reducing cost, while continuing to fulfill increasingly demanding quality criteria set by the FDA (Trexler-Schmidt *et al.*, 2009; Morrow, 2007a; Morrow, 2008). Our evolving understanding of upstream and downstream processes in this new era of biopharmaceutical manufacturing will pave the road for the rational design of purification processes for each new monoclonal antibody product.

7.2.1 *Chromatographic technology*

7.2.1.1 *Protein A affinity chromatography*

The history of ProA affinity chromatography has been linked to the development of therapeutic antibody drugs from the beginning. Protein A, a cell wall component of *Staphylococcus aureus*, is widely used as an affinity ligand for the purification of antibodies because of its five (E, D, A, B, and C) highly homologous antibody-binding domains. Each of the five domains has a similar ability to bind to the Fc region of IgG molecule primarily through hydrophobic interactions (Moks *et al.*, 1986). The main binding site is at the groove between the C_H2 and C_H3 domains of IgG (Diesenhofer, 1981). In addition to the general understanding of the ProA binding region in the Fc region, alternative binding sites within the variable heavy chain region of the V_H3 sub-family antibodies have also been identified (Sasso *et al.*, 1989; Sasso *et al.*, 1991; Graille *et al.*, 2000; Fig. 7.1).

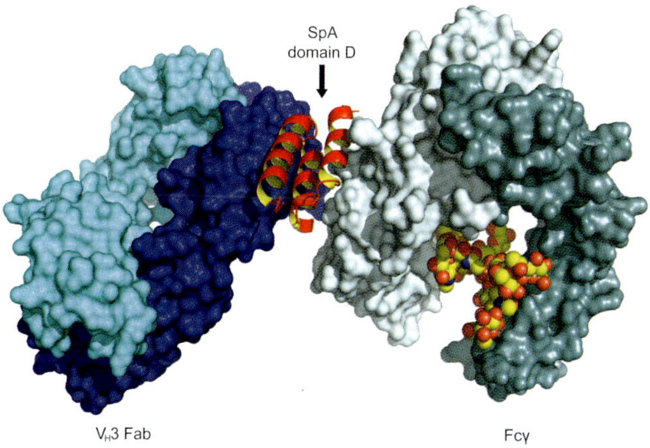

Figure 7.1 Protein A-IgG co-crystal structure

Protein A of *Staphylococcus aureus* has been shown to interact with antibody molecules at two binding sites identified: one is at the groove between the C_H2 and C_H3 domains of Fc region and the other is within the Fab region of the V_H3 subfamily antibodies (Reproduced from Graille *et al.*, 2000).

Therefore, ProA-based separation can also be used for the purification of Fabs with V_H3 variable domain. It has been reported that IgGs in the V_H3 sub-family may interact with ProA primarily through the variable region rather than the classic Fc-binding region, causing variations in elution behavior during ProA chromatographic purification (Ghose *et al.*, 2005). IgG adsorption to ProA chromatography media occurs under neutral physiological conditions, which allows IgG products to be directly captured from various native and recombinant sources. When the IgG binds to ProA, a conformational change, resulting from an induced fit binding mechanism, occurs contributing to the strikingly low pH conditions required for elution. Structural studies have revealed that there are highly conserved histidyl residues on both ProA and IgG at the binding interface, which are positively charged during low pH elution conditions to produce electrostatic repulsion for IgG desorption (Diesenhofer, 1981; Haake *et al.*, 1982). One benefit of the typically low pH (< 3.8) ProA elution step has been the demonstration of significant viral inactivation (Valdes *et al.*, 2002; Brorson *et al.*, 2003). Therefore, a dedicated holding time post-elution from ProA is generally built into the DSP to enhance the viral clearance capabilities of the process. The major problem associated with such acidic treatment of antibody molecules is aggregation and subsequent risk of loss of activity (Shukla *et al.*, 2005; Shukla *et al.*, 2007a). Conformationally changed antibody variants (acid conformer) have been identified when IgG molecules are purified using ProA chromatography with low pH exposure. Despite no changes detected in antigen-binding ability, the acid conformer could contribute to aggregate formation, increased susceptibility to proteolysis, and present a new epitope for immunogenicity (Sakamoto *et al.*, 2009). In addition, ProA is sensitive to proteolysis, the major cause of leaching during the purification process (Carter-Franklin *et al.*, 2007). Because Protein A is derived from bacteria, and as such possesses high antigenicity and immunomodulatory effects (Silverman *et al.*, 2005; Gomez *et al.*, 2006), leached ligand from the ProA chromatographic purification step must be removed in subsequent purification processes as a clinical safety consideration.

The high specificity of binding between ProA and IgG allows this single step of affinity chromatography to remove over 97% of the impurities derived from upstream mammalian cell culture process. After the Protein A affinity purification step, only trace impurities such as residual host cell proteins, DNA, leached ProA, and aggregates remain, all of which can be removed with one or two subsequent chromatographic purification steps (Shukla *et al.*, 2007b). Single-step purification with ProA affinity chromatography is often sufficient to support the initial *in vitro* study for antibody lead selection and has been the dominant capture step methodology used from laboratory-scale to commercial-scale manufacturing due to its powerful and generic process with straightforward development path.

Despite the high purification resolution of ProA affinity separation, the continuous increase of productivity in upstream mammalian cell culture has resulted in a challenge to the ProA-based traditional downstream purification platform. The cost and scale-up problems, together with issues associated with ligand leaching, instability under alkaline sanitization condition, low pH-induced aggregation, and dysfunction have promoted extensive studies in searching for alternative strategies to replace ProA in the monoclonal antibody purification process. Alternative affinity ligands or processes that can overcome the drawbacks of ProA chromatography are of great interest for the antibody production industry. Such alternatives would be especially beneficial for those antibody products that generate high levels of aggregates resulting from the ProA purification procedure.

Various strategies have been explored in search of alternative affinity ligands and some promising progresses have been made using synthetic peptide-based technologies that can mimic the ProA–Fc binding mechanism. However, there has not yet been success in finding a less expensive and more effective affinity separation replacement with a similar specificity to ProA (Ghose *et al.*, 2006; Hober *et al.*, 2007; Low *et al.*, 2007, Liu *et al.*, 2009). Meanwhile, ProA resin has also improved through both matrix engineering and advances in ligand chemistry in response to the demands of monoclonal antibody drug manufacturing. New generation ProA resins possess improved binding capacity and process efficiency for better process throughput. Among them, MabselectXtra from GE healthcare, Prosep vA Ultra from Millipore, and Mab Capture A from Applied Biosystems, all of which exhibit significantly higher dynamic binding capacity and/or increased process speed and reduced elution volume (Hahn *et al.*, 2003; Hahn *et al.*, 2005; Hahn *et al.*, 2006; Swinnen *et al.*, 2007). In particular, MabselectSuRe from GE healthcare was designed with an engineered ProA ligand that is built with multi-Z domain, a functional analog of B domain of Protein A. MabselectSuRe presents not only stronger resistance to alkaline sanitization conditions, longer cycle time and decreased ligand leaching (GE Healthcare, 11001164AA; GE Healthcare, 11001165AB; Jagschies *et al.*, 2006; Hahn *et al.*, 2006; Hober *et al.*, 2007), but also reduces the variable region binding, resulting in less elution pH variation, thereby enhancing the robustness of the ProA step process (Jansson *et al.*, 1998; Ghose *et al.*, 2005). With small ProA columns used in a multi-cycle strategy being widely adopted for industrial-scale manufacturing of monoclonal antibodies, products such as MabselectSuRe Protein A resin would certainly provide an enhancement to this old, but powerful, affinity separation tool as the primary capture process for antibody purification (Shukla and Thommes, 2010).

7.2.1.2 *Ion exchange chromatography*

Ion exchange (IEX) chromatography is one of the most popular and powerful bioseparation chromatography techniques. The IEX resin is defined based on the charged counter-ions available for exchange. Anion exchange resins have positively charged ligands while cation exchange resins have negatively charged ligands. The interaction between the charged ligands on the resin and oppositely charged solute molecules provides the basic ion exchange separation mechanism. Proteins are amphoteric molecules and the charged residues on the surface of a correctly folded functional protein change as a function of pH in the surrounding buffer environment. When the buffer pH is below the isoelectric point (pI), the pH at which a protein's net charge is zero, the protein carries a net positive charge and can bind to a cation exchanger. Similarly, when the buffer pH is above the pI, the protein carries a net negative charge and can bind to an anion exchanger (GE Healthcare, 11000421AA). The charge properties, including differences in charged residues and their distribution on the surface, are unique for each specific protein molecule. Proteins with charged amino acid differences can therefore be separated by ion exchange chromatography. As a result, ion exchange chromatography has been proven to have high-resolution power in separating the desired product from various impurities.

Besides being classified as a cation or anion exchanger based on the charge property of their counter-ions, each ion exchange chromatography resin within a category can further be divided into a strong or weak subclass based on its ligand ionizing capability over pH ranges, rather than on strength of binding (Desai, 2000). For example, GigaCap S, a strong cation exchange resin, and GigaCap CM, a weak cation exchange resin from Tosoh Bioscience, have reported resin-binding capacity of 100 mg/mL resin or above for tested polyclonal human IgG (Tosoh Bioscience, product literature). These different types of ion exchange resins form a wide spectrum of ion exchange chromatography tools to meet specific separation process needs. Their selectivity and resolution are further enhanced through a diversity of resin support matrices and linker spacer arms that offer resins with different chemical and physical properties, that in turn influence ligand accessibility and potential secondary interactions at the resin surface. Therefore, among the same ion exchange category, resins offered from different vendors or different subclass can all present different selectivity of impurity separation, which makes resin screening an important variable in ion exchange process development (Staby *et al.*, 2000; Staby *et al.*, 2001; Staby *et al.*, 2004; Staby *et al.*, 2005; Staby *et al.*, 2006; Staby *et al.*, 2007; Wu and Walter, 1992; Hardin *et al.*, 2009; DiLeo *et al.*, 2007).

Studies have shown that in monoclonal antibody purification processes, anion and cation exchange resins possess different selectivity and resolution for removal of impurities. Anion exchange is known to effectively reduce acidic impurities such as DNA, endotoxin, viruses, and the acidic population of host cell proteins. Because of the relative high pI for most monoclonal antibodies, anion exchange chromatography can be easily operated in a flow-through mode. When used in such a manner, the antibody product is not bound by the resin, while the impurities are retained on the column until cleaning and regeneration.

Conversely, cation exchange is effective for host cell protein removal and purifying monomeric IgG from product related species such as aggregates and fragments (Gagnon, 1996; Norling *et al.*, 2005; Stein *et al.*, 2007; Follman and Fahrner, 2004). Because the separation is based on individual protein specific charge properties, unlike affinity chromatography, the ion exchange purification process is not as generic as the ProA affinity process and requires optimization for each monoclonal antibody product. Fortunately, the process development strategy is not complicated. Conductivity and pH are the two major variables controlling optimal binding, wash and elution conditions (Basey and Blank, 2002; Lebreton *et al.*, 2009). Low conductivity is generally required for loading, which often results in undesired sample manipulation for feed material such as dilution or in-process ultrafiltration/diafiltration (UFDF). In recent years, several new ion exchange resins have shown improved resistance against feedstock with mildly elevated conductivity while having a minimal decrease in loading capacity. Binding pH is usually determined by protein pI, but the local charge distribution of the protein at the site of interaction with the ligand is more important and more relevant than the net charge of the protein as a whole. Therefore, optimal binding pH cannot be simply predicted by protein pI and needs to be determined empirically. A wash step using a buffer of intermediate pH and with a conductivity that is usually higher than that used for binding, but lower than elution condition, can offer substantial impurity removal. Finally, either pH dominant, salt dominant, or a combination strategy by increasing buffer pH, ionic strength, or both, can be used for elution of therapeutic antibodies from ion exchange resins (Mhatre *et al.*, 1995; Tishchenko *et al.*, 1998; Zhou *et al.*, 2007a). The various strategies may provide different resolution but should be applied based on product stability.

In response to the increased demand for process throughput, a variety of new generation cation and anion exchange resins were launched including Capto resins from GE, TMAE Hicap resins from EMD/Merck Biosciences, and Toyopearl GigaCap resins from Tosoh Bioscience. The most distinctive achievement in ion exchange resin design and engineering is the highly improved dynamic binding capacity of near or beyond 100 mg IgG/mL resin. This dramatically increased capacity has made ion exchange chromatography stand out from all other chromatography separation techniques for process efficiency and cost benefits. Additionally, it has had a significant impact on reshaping monoclonal antibody purification process development strategies. Cation exchange chromatography has now become an attractive alternative capture step, to replace affinity ProA chromatography for high titer IgG batch mammalian cell culture, to achieve scale-up practicality and cost efficiency. For the first time, there is now a good argument for antibody purification processes without ProA affinity chromatography. The DSP with cation exchange as an initial capture step instead of the conventional Protein A affinity chromatography has begun to be implemented in practice. This is a noteworthy turn of events for the conservative biomanufacturing industry, and the CEX capture, non-ProA platform approach is gaining momentum along with an industry-wide drive for high throughput and low cost antibody drug manufacturing (Arunakumari *et al.*, 2007; Ferreira *et al.*, 2007; Arunakumari, 2009; Lain *et al.*, 2009; Arunakumari *et al.*, 2009a).

7.2.1.3 *Hydrophobic interaction chromatography*

The discovery of hydrophobic interaction not only established the foundation for adsorption chromatography separation but also extended the theoretical understanding of solute–solute, solute–solvent, and solvent–solvent interactions in general (GE Healthcare, 11001269AA; Pratt and Chandler, 1980). Hydrophobic interactions are both a physical–chemical process and a biological process, critical for protein folding and maintaining protein functional structural integrity (Dyson *et al.*, 2006). Despite extensive studies on the mechanism of hydrophobic interaction and its broad applications in biological molecule purification, the universal rule of hydrophobic interaction that guides protein folding, protein–solvent and protein–protein interactions has not been fully elucidated. The mechanism of protein adsorption/desorption on hydrophobic interaction chromatography (HIC) remains multilateral. One of the most widely accepted theories involves thermodynamically favorable dehydration and associated effects. Water molecules form a hydration shell that surrounds the hydrophobic ligand and non-polar groups on protein surfaces. The number of highly ordered water molecules decreases when the protein interacts with hydrophobic ligands as it binds to the HIC column. This displacement of the surrounding ordered water molecules leads to an increase in entropy; therefore, adsorption on an HIC column is thermodynamically favorable (Queiroz *et al.*, 2001).

Hydrophobic interaction chromatography is a well-studied and powerful separation tool that separates biomolecules according to their hydrophobicity. As is also the case with charge properties, the hydrophobicity characteristics of a protein are unique and are the driving force for HIC separation. Both the distribution of non-polar amino acid residues and the effective hydrophobic surface area of a globular protein influence the affinity of binding on an HIC column. The non-polar groups of a protein are normally buried within the tertiary structure during the folding process as a result of hydrophobic interactions. Slight changes in the protein's native structure, through incorrect folding or denaturation processes, would therefore alter the distribution of hydrophobic residues, something to which the HIC method is sensitive and would be sufficient to allow for separation. Thus, HIC can be complementary to other types of chromatographic techniques, such as ion exchange and size exclusion chromatography, making it one of the most widely used separation methods in the purification of biotherapeutics. An HIC step is included in the manufacturing processes for many marketed monoclonal antibody products, principally for its capability of removing less active, mis-folded and aggregated forms of protein product, and of cleaning DNA, viruses, and host cell protein contaminants. Patents have been issued for the use of HIC in both bind-elute and flow-through mode to purify the monomeric form of monoclonal antibody products (Shadle *et al.*, 1995; Rinderknecht and Zapata, 2006).

The success of developing an HIC purification process for monoclonal antibodies requires the optimization of a complex set of variables to allow binding and elution with high recovery and purity. The degree of hydrophobicity of the HIC resin ligand, choice of salt and ionic strength, buffer pH environment, operation temperature, and various modifiers can all contribute to the desired selectivity (Gagnon, 1996; GE Healthcare, 11001269AA). The type of ligand and its density determine the ultimate hydrophobicity of the resin, which sets up the primary adsorption attributes for the proteins to be purified. GE Healthcare and Tosoh Bioscience offer a large range of HIC resins with different degrees of hydrophobicity strength to facilitate the HIC method development. The effect of salt on hydrophobic interactions has been known since the early 1900s and is a dominant factor that influences the binding and selectivity of HIC separation. Salt in the solution interacts with water molecules, leaving fewer molecules to shield hydrophobic areas on both protein and ligand, thereby promoting HIC adsorption. The principal parameters that determine the effect of salt on protein retention by the HIC column are the chemical properties of the salt and its molality. An increase in salt molality results in an increased retention of proteins in hydrophobic chromatography; similarly, a decrease in salt concentration will increase the elution strength for protein desorption (Xia *et al.*, 2004; Mahn *et al.*, 2007). In general, the ability of various salts to promote adsorption follows the ranking of the Hofmeister series of lyotropic ions (Kunz *et al.*, 2004; Bostrom *et al.*, 2004), while the magnitude of the effect of the salt depends on the characteristics of the amino acid residues at the binding site. Therefore, the same salt may have a different strength of effect on different proteins.

Generally, hydrophobic interaction chromatography utilizes mobile phases with high salt concentrations to promote adsorption of monoclonal antibodies; therefore, the protein solubility in high salt concentration needs to be carefully investigated prior to studying the binding capacity. The effect of pH on binding and selectivity depends on the pI of the protein, or more importantly on the titration of charged groups at the local binding site. Because of the variation of charged group distribution of different proteins, the effect of pH is usually not predictable and is an important process variable that needs to be optimized for maximal selectivity (Hjerten *et al.*, 1986; Xia *et al.*, 2005). Since hydrophobic interactions are entropy-driven processes, temperature increases will enhance hydrophobic interactions. However, high temperature can also cause conformational changes in the protein and expose embedded non-polar groups. Altering the temperature factor for process optimization must therefore be done cautiously based on protein stability.

Finally, many additives have been reported to have the ability to enhance or weaken hydrophobic interactions. Ethylene glycol/propylene glycol and mild organic hydrophobic competitors such as ethanol are often used in HIC processes to facilitate elution. Similarly, caution should be taken when using the additives since they may denature or otherwise impact the biological activity of the protein (Gagnon 1996; GE Healthcare, 11001269AA; Shadle *et al.*, 1995).

Even though HIC is a very powerful tool in monoclonal antibody purification processes, limitations such as relatively complicated process method development, low binding capacity and the requirement for high salt concentrations in the binding mobile phase, often result in process throughput restriction and complicate process operation during large-scale manufacturing. Efforts have been made to improve the properties of HIC resins. Tosoh Bioscience introduced a new generation of HIC resins with improved binding capacities achieved by facilitating mass transport of monoclonal antibody molecules into the ligand binding sites through pore size optimization (Chen *et al.*, 2008; O'Donnell *et al.*, 2006). In addition, mobile phase strategies have also been explored for optimization of hydrophobic interaction chromatography separation. A dual salt system has been reported to significantly enhance the dynamic binding capacity of antibody molecules on the HIC column and to increase protein solubility and stability during separation (Senczuk *et al.*, 2009; Nunes *et al.*, 2009). Finally, a new branch of chromatography separation tools called mixed-mode chromatography has been developed, which include a hydrophobic interaction element to achieve high resolution power. As we enter a new era of biotherapeutic manufacturing with high expectation for purity, hydrophobic interaction and related separation techniques will continue to be a favorite choice for monoclonal antibody purification processes.

7.2.1.4 *Mixed-mode chromatography*

Mixed-mode chromatography has evolved to become a very important chromatography method for protein purification (Majors *et al.*, 2009). Protein interactions with mixed-mode chromatographic media are complicated because of the complex chemistry of both the resins and the proteins being purified. Proteins are macromolecules with positively and negatively charged, hydrophobic and hydrophilic residues arranged in a specific way to form the functional tertiary structure. On the other hand, mixed-mode chromatographic media utilize multi-modal operating ligands functioning through various adsorption or desorption mechanisms (Rios, 2007; Zhou, 2009; Chung *et al.*, 2010).

The different mechanisms such as hydrophobic and electrostatic interactions can have either a synergistic or repulsive effect. The complexity of the protein–ligand interaction in mixed-mode chromatography provides distinctive opportunities for selectivity and resolution to separate desired product from various impurities. Great reduction for aggregates is one of the most recognized beneficial features of mixed-mode chromatography. Mixed-mode resins are also known for their tolerance of challenging feedstock conditions and their contribution to process operation flow improvement (Gagnon, 2008; Chen *et al.*, 2010; Kaleas *et al.*, 2010; Oehme and Peters, 2010). However, the mixed-modal interaction also poses a challenge in terms of identifying the optimal process parameters and developing robust industrial-scale processes. The effects of pH, salt type and ionic strength on the physical–chemical properties of the ligand and individual proteins, especially at the local interaction site, are critical for the success of mixed-mode separation. Given this complexity, the use of Design of Experiments (DOE) studies to optimize binding and elution conditions are likely to be more important in mixed-mode chromatography process development than for traditional chromatography methods. Commercially available mixed-mode resins that are commonly used in monoclonal antibody purification include Mercapto-Ethyl-Pyridine (MEP) hydrophobic charge induction resin, Capto adhere multi-modal anion exchange resin, and Ceramic Hydroxyapatite CHT resins.

Ceramic Hydroxyapatite (CHT) is a special resin in which both the ligand and matrix are constructed from the same chemical material, calcium phosphate. CHT is categorized as a mixed-mode resin based on its two functional groups: positively charged calcium ions (C-site) and negatively charged phosphate groups (P-sites). The electrostatic interactions of both C- and P- sites with charged moieties at the protein surface, and the metal affinity attraction between carboxyl groups on the protein and C-sites on the CHT support are the two major characterized separation mechanisms of the CHT resin (Biorad; Ng *et al.*, 2006). The binding conditions for CHT have been well defined as low levels of phosphate (5–10 mM) at pH over 6.5. Two different strategies are used for CHT elution: a sodium chloride salt gradient (cation exchange dominant method) and increasing phosphate molality (mixed-mode combinational method). Sodium chloride gradients at constant low phosphate levels have been reported to have better impurity reduction (Gagnon *et al.*, 2006; Ng *et al.*, 2006), although the sole increase of sodium phosphate is still a commonly used elution method. The application of CHT in monoclonal antibody purification was initially recognized for its superior ability to reduce aggregates (Franklin, 2003; Sun *et al.*, 2004). Now CHT has been incorporated into industrial-scale processes for biologics manufacturing. In monoclonal antibody purification processes, CHT mostly serves as a polishing step to selectively reduce impurities such as aggregates, leached ProA, host cell proteins, residual DNA, endotoxins, and viruses (Gagnon *et al.*, 2006; Gagnon, 2009).

Mercapto-Ethyl-Pyridine (MEP) was designed as a pseudo-affinity ligand of a hydrophobic charge induction resin, MEP HyperCel, in an attempt to replace ProA (Schwartz *et al.*, 2001) and capture antibodies from clarified mammalian cell

SEC has been used as the final polishing step in antibody purification processes where it is mainly used to purify monomers from product related impurities such as aggregates and fragments (Azevedo *et al.*, 2008; Suzanne and Bonnerjea, 2007). However, in order to achieve the desired resolution in SEC separation, sample loading volumes are restricted (no more than 2–5% of column volume is recommended), and flow rates need to be slow. With these process restrictions, SEC as a purification tool presents an insurmountable hurdle in scale-up and therefore has not been widely adopted for purification of antibody products at industrial-scale manufacturing (Ibere *et al.*, 2001; Suzanne and Bonnerjea, 2007). By contrast, as an analytical tool, SEC is the industry standard for the assessment of soluble aggregate levels in antibody products at every stage of product development and characterization (Kurz, 2007; Flatman *et al.*, 2007).

7.2.2 *Membrane separation technology*

Traditionally, membrane technologies that are commonly used in antibody purification processes are membrane filtration-based techniques. Membrane filtration uses a physical barrier to separate solutes based on their size and shape through the use of pressure, and it can be operated either with a dead-end flow or a cross-flow format. Two factors determine the effectiveness of a membrane filtration process: selectivity and productivity. Selectivity is expressed as retention factor, while productivity is expressed as flux. Membrane filtration is widely used in antibody purification processes including bioreactor harvesting, protein concentration, buffer exchange, virus filtration, and sterile filtration (van Reis and Zydney, 2001; Dixit, 2008; Saxena *et al.*, 2009). In addition to the unique function of each membrane filtration, much more is often accomplished in the process including batch volume reduction, feed conditioning for subsequent chromatographic steps, and achievement of partial purification. All of these processes make use of the same size-based separation principle, but the actual operation methods could vary significantly.

As high density and large-scale mammalian cell culture becomes increasingly important for the production of complex recombinant monoclonal antibodies, the first downstream operation unit of clarification plays a critical role in overall process throughput and product quality. Membrane filtration operations, tangential flow microfiltration (TFF) and depth filtration in particular, are favorably employed in clarification processes to remove cells and cellular debris from mammalian cell culture suspensions (van Reis *et al.*, 1991; Singhvi *et al.*, 1996; Shukla and Kandula, 2008; Wang *et al.*, 2008; Roush and Lu, 2008). Virus clearance has become a standard component of modern biologic purification processes and is essential for the safety of drug products (ICH and FDA, 1998). To eliminate the risk of viral contamination, orthogonal viral inactivation and removal methods are usually incorporated to provide robust and reliable viral clearance through the process design in which the membrane process, nanofiltration in particular, contribute the critical part in viral clearance (Ray and Tarrach, 2008; Dutton, 2009). Nanofiltration is one of the most effective virus reduction methods that can provide a multi-$\log_{10}$ reduction of virus contamination including small non-enveloped viruses (van Holten *et al.*, 2002; Yokoyama *et al.*, 2004; Zhou and Tressel, 2008). Various nanofiltration devices, including Planova virus filters from Asahi Kasei Medical Co. LTD., Viresolve Pro filters from Millipore, Novsip/Ultipo filters from Pall Life Sciences, and Virisart CPV filters from Sartorius, are all designed for effective removal of low levels of viruses as small in size as tens of nanometers (Hongo-Hirasaki *et al.*, 2006; Brough *et al.*, 2002; Brandwein and Aranha-Creado, 2000; Sartorius). Finally, UFDF is used in practically every biologics manufacturing process, at in-process steps to reduce the batch volume and to prepare load material for the subsequent chromatography separation step, and at the final step to transfer the purified product into the final formulation buffer at the defined concentration for drug delivery (van Reis and Zydney, 2001; Schwartz, 2003; Luo *et al.*, 2006).

Beyond general membrane filtration applications, membrane separation technologies have become increasingly important in the purification of biological products. In the past decade, as monoclonal antibody productivity in mammalian cell culture increased towards 5 g/L or higher, the column chromatography dominant purification strategy has faced enormous challenges for throughput, scalability and economics. Membrane adsorption or membrane chromatography technologies are gaining momentum in their application to the purification of biological products from laboratory to industrial manufacturing scale, mainly due to the unique and efficient process properties, disposable single-use engineering design, and economical benefits (Gottschalk, 2008; Glynn and Boyle, 2010).

Compared to conventional column chromatography with packed bead-based resin as the stationary support, membrane chromatography technology uses synthetic microporous or macroporous thin membranes to fulfill the same function. The mechanical properties of the two separation technologies contribute to the critical difference in solute transportation and process characteristics. Because the majority of potential binding sites on bead-based chromatography media are located inside of pores of the bead matrix, the transport of macromolecules in traditional column chromatography depends mainly on diffusion to reach the binding sites. By contrast, for the adsorptive membrane, binding sites are located in openly accessible pores on the membrane and the transportation of solutes to their binding sites takes place predominantly by convection. The significant reduction in mass transfer resistance greatly enhances process efficiency by decreasing process time and recovery liquid volume (Ghosh, 2002; Charcosset, 2006; Saxena *et al.*, 2009; Zhou, 2009).

A wide range of separation chemistries have been employed in membrane adsorbers including ion exchange, affinity, and hydrophobic interaction (Klein,

2000; Knudsen *et al.*, 2001; Boi, 2007; Fraud *et al.*, 2008). However, despite the long history of membrane separation technology, until recently, relative low binding capacity, low resolution, and lack of scale-up devices have prevented membrane adsorbtion approaches from replacing column chromatography separation in large-scale manufacturing applications. Innovations in membrane engineering technology, especially in the single-use/disposable module design, have allowed membrane separation to be practically adopted by the biopharmaceutical manufacturing industry (Saxena *et al.*, 2009; Zydney, 2009; Mora *et al.*, 2006; Zhou *et al.*, 2007b). Although the short-term vs. long-term cost and the environmental impact remains controversial, the advantages of membrane processes are well recognized for overcoming scale-up issues related to packed bed columns, significantly reducing process time, decreasing buffer consumption, and eliminating costs associated with column packing and cleaning (Lim *et al.*, 2007a; Lim *et al.*, 2007b; Gottschalk, 2008).

Positively charged depth filters are commonly utilized in monoclonal antibody harvest processes to enhance the impurity clearance of host cell proteins, DNA, and endotoxins, while clarifying cells and cellular debris (Roush and Lu, 2008). It has been demonstrated that reduced levels of host cell contaminants prior to ProA affinity chromatography can decrease the turbidity problem in the ProA elution product pool (Yigzaw *et al.*, 2006). Anion exchange separation has been widely used in a flow-through mode to remove trace impurities following the ProA affinity capture step in monoclonal antibody manufacturing processes. Single use, disposable Q membrane adsorbers present improved process efficiency and cost-saving benefits while retaining similar or better impurity and contaminant removal capabilities, especially for large biomolecules such as viruses and nucleic acids due to the larger pore size in the membrane (Phillips *et al.*, 2005; Zhou and Tressel, 2008; Clutterbuck *et al.*, 2007; Lim *et al.*, 2007b; Giovannoni *et al.*, 2009). The recently introduced HIC membranes further extend membrane separation when applied to antibody purification by enhancing impurity clearance, especially aggregates (Kuczewski and Zarbis-Papastoitsis, 2009; Fraud *et al.*, 2009; Zhao, 2009). Overall, to achieve the binding capacity and outstanding selectivity of column chromatography separation, several limitations with membrane separation technologies still remain to be overcome. As the membrane separation technology expands its applications, it is generally agreed in the industry that both bead- and membrane-based separation technologies will continue to advance to meet biotherapeutic drug manufacturing challenges for more effective and efficient processes with lower costs (Kelley, 2007).

7.2.3 *Precipitation*

Precipitation was the first effective separation method applied to the purification of antibodies. In 1851, Dr. Peter Ludvig Panum discovered serum-globulin through precipitation by dilution and pH adjustment using acetic acid (Haslam, 1913). This separation method matured and became very popular with Edwin Cohn's blood plasma product work during World War II (Cohn, 1947). Since then, various precipitation procedures have been developed and employed as a fast and inexpensive method to isolate antibodies from complex feed streams including serum, ascitic fluid and culture media. The purification process using precipitation and fractionation methods remains widely used in the blood product industry. The mechanism of immunoglobulin precipitation is fairly simple. Immunoglobulins in solution are surrounded by a tightly bound hydration shell. Certain chemicals including salts, acids, cationic detergents, organic solvents, and organic polymers can disturb the hydration shell and cause the immunoglobulin molecules to interact with each other, thereby precipitating from solution and forming visible insoluble aggregates. The precipitation process, when controlled properly, is reversible. Precipitated antibody product can then be collected and resolubilized in an appropriate buffer. Despite operating by the same mechanism as hydrophobic interaction chromatography, fractional salt precipitation has a much lower degree of precision. In general, precipitation separation is considered to be a mature but low resolution technology.

By the time monoclonal antibody technology was developed in the 1970s, chromatography technology dominated the separation field. Precipitation methods have yet to be used as mainstream purification methods in industrial-scale manufacturing processes for recombinant antibodies. This, however, may change. Precipitation methods, as cheap, fast, and easy-to-operate separation techniques, seem ideally suited to address downstream processing concerns regarding the cost and facility limitations associated with the use of ProA to initially capture antibodies from the high titer and large volume mammalian cell culture feed streams (Thommes and Etzl, 2007; Glynn, 2008; Gottschalk, 2008; Arunakumari *et al.*, 2009).

In recent years, the application of precipitation methods, as a cheap and effective primary recovery tool, prompted an exploration of the potential advantages of the implementation of these methods in very large-batch-scale monoclonal antibody production. Precipitation methods could be used to capture monoclonal antibody product from clarified cell culture supernatant with high recovery yields. This would then allow reconstitution of the captured antibody into a manageable batch size for subsequent chromatography separation. In addition, various precipitation techniques have also been tested to enhance impurity reduction which is especially beneficial for non-ProA DSP based purification processes. Lyotropic salts such as ammonium and sodium sulfate are the most commonly used additives for precipitating immunoglobulins because of the well-understood protein–salt interaction mechanism and no concerns regarding clearance of the additive (Habeeb and Francis, 1976; Thommes and Etzl, 2007; Glynn, 2008; Wang *et al.*, 2006; Grodzki and Berenstein, 2010).

Polyethylene glycol precipitation is also applied in antibody purification processes and has been demonstrated to be a powerful product precipitation tool for antibody recovery and host cell protein removal (Kuczewski *et al.*, 2010). Precipitation with caprylic acid under pH-controlled conditions has been developed as a scalable process and presents significant benefits for both host-cell contaminant and virus removal in non-ProA based purification platforms (Wang *et al.*, 2008; Wang *et al.*, 2009a). Flocculation is another application that utilizes a precipitation mechanism and is often developed as a pre-treatment method for mammalian cell culture prior to a filtration step. This contaminant removal precipitation procedure improves the efficiency of the subsequent filtration membrane process by significantly decreasing filter membrane area requirements (Kim *et al.*, 2001; Westoby *et al.*, 2010). Finally, precipitation is reported to facilitate high concentration formulation processes when compared to traditional ultrafiltration approaches for antibody drug products (Matheus *et al.*, 2009). As the applications of precipitation technology in monoclonal antibody purification continue to evolve, revisiting and exploring this old separation technique offers a great deal of potential in the development of low cost and high efficiency industrial-scale manufacturing processes.

The clinical success of antibody therapeutics depends on not only the desired efficacy of the drug product but also on effective manufacturing. As innovations continue in expression systems and bioseparation technologies, the concerns of capacity and the ability of downstream processes to handle growing production demands of monoclonal antibody drugs will subside. Systematic and effective studies can be designed using tools such as DOE, to allow accurate mapping of the optimal operation conditions associated with each separation step, which will help to develop robust manufacturing processes. Intelligently designing flexible platform approaches and utilizing various expression and separation technologies will maximize product quality and process efficiencies and in turn will provide cost savings for future monoclonal antibody drug manufacturing.

7.3 Candidate Selection

Monoclonal antibodies have proven to be extremely valuable biotherapeutic products. Worldwide initiatives have been made to ensure innovative and efficient discovery and development of promising antibody drug products. One such initiative has been the Quality by Design (QbD) approach supported by the FDA. According to the FDA guidelines described in "Quality Systems Approach to Pharmaceutical cGMP Regulations", "quality should be built into the product, testing alone cannot be relied on to ensure product quality". The principles of the QbD concept require that strategies are developed and implemented to select lead candidates with the desired drug-like properties. By taking such an approach, a thorough understanding of the product characteristics and manufacturing

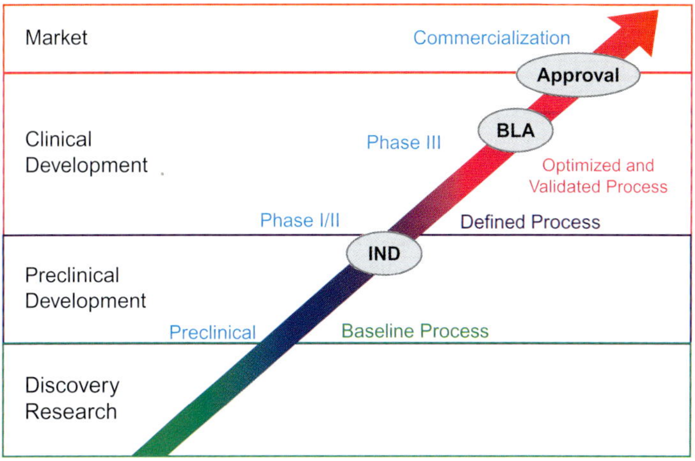

Figure 7.2 Drug development life cycle

Parallel product and process development map in monoclonal antibody drug development life cycle.

processes should be established that will then ensure the consistent delivery of high quality drug products and subsequently result in reduced regulatory oversight (FDA, 2006). Even though the discovery efforts occupy only a small part in a drug's development life cycle (Fig. 7.2), it is the most critical phase because the future therapeutic drug candidate is selected at that time, and the inherent quality of the selected molecule will impact the rest of the drug developmental path. By fully leveraging previous experiences with the discovery and development of antibody therapeutics, it is possible to design a comprehensive strategy that will help to build the desired high-quality drug-like properties into the selected therapeutic candidates. A comprehensive assessment of the suitability of a therapeutic antibody to advance into clinical development can be achieved while still in the discovery phase to facilitate the program moving from research to commercialization in a timely and cost-effective manner. This section will discuss the lead selection strategies and early CMC (chemistry, manufacture, control) development plans with the goal of identifying and advancing the best antibody drug candidates through the drug development life cycle.

7.3.1 *Lead selection for affinity and specificity*

To drive antibody-based therapeutics towards the clinic, hierarchical selection strategies are usually designed and employed. The therapeutic antibody discovery process begins with a disease target and an associated rationale of mechanism of action, which to be appropriately investigated require that the antibody leads resulting from the discovery efforts meet established criteria regarding affinity and specificity. Affinity to the target molecule defines the future monoclonal antibody drug's efficacy and influences the drug dose required to obtain the desired effectiveness. Both of these factors are especially important for monoclonal antibody drugs because the typical doses required for treatment are usually significantly higher than most other biologics such as hormone or cytokine therapy (Farid, 2007; Farid, 2009). Currently, monoclonal antibodies with affinities ranging from 0.1–10 nM, can be routinely obtained through advanced selection technologies (Stockwin *et al.*, 2003). In addition to affinity, absolute target specificity is a desirable quality of the drug because lack of specificity can lead to significant and unacceptable side effects (Pavlaki and Zucker, 2003; Walsh, 2004). Unlike small molecule drugs, which often are unable to distinguish targets with similar structures, the antigen-binding domain, Fab, in monoclonal antibodies is composed of a complex amino acid sequence and unique tertiary structure that permits antibodies to target antigens with high selectivity. Monoclonal antibodies can exhibit high affinity and specificity towards a single epitope on the target molecule. This kind of specific binding property is very sensitive to change. Even small changes of the amino acid sequence in the variable region of the Fab can increase affinity and/or specificity, making it possible to differentiate between the slightest differences in target antigen epitopes. The high specificity of monoclonal antibodies plays an important role in therapeutic mechanisms, especially cancer indications since many cancer targets are modified or subforms of normal regulatory agents in the body (Pavlaki and Zucker, 2003; Devy *et al.*, 2009). On the other hand, this high specificity feature, while a benefit clinically, is a potential issue pre-clinically, since it is highly desirable to have a lead that is able to cross-react with the animal species that will be used for pre-clinical efficacy and safety evaluation. A surrogate of the drug candidate may have to be used for animal studies when a lead has narrow species selectivity.

The technology platforms for developing fully human monoclonal antibody drugs, including transgenic animals and phage display, have generally been adapted to modern high throughput target selection and screening systems using methodologies such as ELISA, FACS, or surface plasmon resonance. While the ideal scenario would incorporate a functional assay as the primary screen of candidates, it is more practical to use simple binding studies to reduce the pool of candidates that may then be evaluated in the appropriate functional assays. At the screening phase, it is often possible to devise competition-based binding assays that allow further screening of those antibodies that are able bind to the region of the target thought to mediate function. This combination of binding and competition assay data is then able to provide a superior evaluation of an antibody drug candidate's affinity and specificity, and to facilitate advancement of suitable candidates into the relevant *in vitro* and *in vivo* experiments (Stockwin *et al.*,

2003). Previous chapters in this volume have discussed various state of the art selection, screening, and protein engineering methodologies that are used to identify potent leads and to enhance binding kinetics and improve specificity of monoclonal antibodies as needed. The design of appropriate selection strategies to meet previously established affinity and selectivity criteria is the first step in the lead selection process and in ensuring that the lead drug candidates advanced possess the properties essential for both desired safety and efficacy.

7.3.2 *Lead selection for* in vitro/in vivo *efficacy and safety*

Since most monoclonal antibodies are initially identified on the basis of the ability to bind the target antigen, it is crucial that monoclonal antibody candidates are evaluated as early in the process as possible using a functional assay that is dependent on the activity of the target of interest (Stockwin *et al.*, 2003; Tabrizi *et al.*, 2009). Antibodies identified in the initial selection may function via interfering, blocking, or stimulating effects. Cell-based bioassays, enzymatic assays and animal-based potency or safety model studies are intended to act as a measure of the functional capability of a biological product. In turn these pre-clinical studies are used to help build the necessary translational knowledge to design future clinical studies to eventually determine the true clinical "potency" and effective dose ranges. The better the established correlation between the expected clinical responses and the activity in pre-clinical biological assays and testing, the lower the risk of drug candidates failing during clinical studies due to inadequate efficacy, poor pharmacokinetics, and unacceptable toxicity. Although monoclonal antibodies generally display favorable safety profiles and are well tolerated at high doses, the dreadful clinical consequence from the Phase I trial of TGN 1412, a humanized IgG4 monoclonal antibody drug, has radically changed this view. TGN1412 is an agonistic anti-CD28 antibody that induced catastrophic systemic organ failure in healthy volunteers (TeGenero, 2005; Liedert *et al.*, 2007). Therefore, the development of appropriate bioassays and the selection of appropriate animal efficacy and toxicity models are important elements in the antibody drug lead selection and discovery plan. Characterization of safety in relevant species, one having a similar target antigen expression profile and functional integrity, is pivotal to effectively identifying potential benefits and risks to humans. When evaluating the biological function of a lead drug candidate, it is essential to ensure that the selected lead drug candidate, or its surrogate, demonstrate a comparable binding affinity profile in both human and in pre-clinical test species. Although there is no guarantee for the predictability of pre-clinical studies for the evaluation of efficacy and safety, the data obtained from biophysical experiments, animal pharmacokinetic (PK)/pharmacodynamic (PD) studies and the drug efficacy/safety profile in animal models can provide guidance for drug dose and a recommendation for safety monitoring in clinical studies. Therefore, the data regarding the efficacy and safety profile of lead drug candidates resulting from pre-clinical proof of concept and toxicity studies are among the most important requirements to be addressed in the research and development (R&D) drug discovery phase in order to advance therapeutic candidates to the clinic (Devy *et al.*, 2009; Tabrizi *et al.*, 2009).

7.3.3 *Lead selection for minimal immunogenicity potential*

Unlike vaccine-based drugs that are intended to induce an immune response to achieve drug bioactivities, the immune responses generated in response to biologic-based therapeutics can affect efficacy and safety in treated patients. Immune responses are expected with administration of large and complex proteins, such as biotherapeutics. Drug-related immune responses can result in clinical consequences that range from severe hypersensitivity and deficiency syndromes to neutralization of the drug's therapeutic effects and/or alteration of the drug's PK/PD profiles. Anti-drug antibodies are one of the most common results of immunogenicity in protein-based drug therapy. Since many endogenous proteins are biologically redundant, generation of antibodies that result in neutralization of therapeutic protein products does not always produce an obvious clinical effect. However, the effects can sometimes be devastating when there is cross-reactivity with essential endogenous proteins, especially when the endogenous protein serves a biologically unique function (Shankar *et al.*, 2006; Thorpe and Wadhwa, 2003). The examples of adverse anti-drug antibody responses include autoimmune thrombocytopenia caused by recombinant thrombopoietin treatment and pure red cell aplasia associated with administration of a particular formulation of erythropoietin (Regenmortel *et al.*, 2005; Li *et al.*, 2001). Because the effects of immunogenicity can be quite severe, and pre-clinical animal studies may not accurately predict the reaction in humans, regulatory agencies are developing risk-based guidelines for the evaluation of drug immunogenicity. The FDA recommendation for immunogenicity minimization at early stages of product development deals with mostly theoretical factors, primarily focusing on product origin, and three process control areas: protein folding, formulation, and impurities (Rosenberg and Worobec, 2004a; Rosenberg and Worobec, 2004b; Rosenberg and Worobec *et al.*, 2005).

The first generation of monoclonal antibodies that entered clinical trials in the early 1980s were of murine origin. These rodent antibodies, such as OKT3, the only approved murine monoclonal antibody for human therapeutic use, are highly immunogenic in patients and trigger a human anti-mouse antibody response that results in rapid clearance of the drug product from the body and limits its therapeutic benefits (Jones *et al.*, 2007; Adair, 2000). The key factor to reducing the inherent immunogenicity of a product candidate is to make the therapeutic monoclonal antibody as similar as possible to naturally occurring human immunoglobulin sequences without losing the changes that result in the desired functional properties. Therapeutic antibody product design has gone through multiple stages of development as the technology has evolved to facilitate the discovery efforts, from entirely mouse sequences, to chimeric mouse–human fusions, to humanized versions where only small regions of the mouse sequence remain in the CDRs, and eventually to today's fully human sequences (Winter and Milstein, 1991; Jones *et al.*, 2007). Theoretically the closer the sequence of a monoclonal antibody product is to the human sequence the less immunogenic it should be. However, results from clinical trials have not yet proven this concept. All currently marketed antibodies exhibit some level of immunogenicity including "fully human" antibodies (Adair, 2000; Adair *et al.*, 2002; Vaughan *et al.*, 1998). The immunogenicity level is not consistently correlated with the level of murine sequence components. Why are "fully human" monoclonal antibodies still immunogenic? First of all, "fully human" sequences do not represent completely natural human germline sequences, and some antibodies containing these sequences may diverge from the germline through accumulation of somatic mutations (Clark *et al.*, 2006). Mutation of the framework regions of selected lead antibody candidates to the corresponding germline sequence through protein engineering technology could help to minimize potential immunogenicity. Additionally, allotypes of immunoglobulins represent genetic differences between members of the same species. Anti-drug antibodies have been reported to induce different clinical responses among different allotype patients, indicating allotype mismatches could be another source of immunogenicity (Jefferis and Lefranc, 2009). Future personalized medicine approaches could result in a possible solution for this problem by producing specific allotype monoclonal antibody drug products for specific groups of patients.

Immunogenicity may also be a result of factors other than product sequence-related issues. Factors related to the product and process including aggregation, molecular mimicry, modification of the protein, adjuvant in the formulation, delivery route and dosing, and patient immune system status all affect the potential immunogenicity of antibody products. Monoclonal antibodies are produced from living organisms, purified through complex downstream purification processes and are preserved in the final formulation buffer until delivery to the site of action. Ensuring the protein's structural integrity, while minimizing physical and chemical modifications during the manufacturing process, is also crucial for immunogenicity control.

The current industry standard expression systems for monoclonal antibodies are mammalian cells, such as Chinese hamster cell line (CHO) and the murine myeloma cell line, NS0. It is well known that post-translational modifications, glycosylation in particular, are important for biological function of biotherapeutics (Doyle and Mamula, 2001; Walsh and Jefferis, 2006; Morrow, 2007b; Jefferis, 2006). Proteins expressed in mammalian cells may not yield a fully human post-translational modification profile and thus can be recognized as foreign by the body and trigger immune responses in patients. In addition, impurities derived from animal host cell systems can also be a source of immunogenic agents. Therefore, developing and using human cell expression systems such as PER.C6, a human retina-derived cell line, may provide better microenvironments for correct protein folding and more natural human modifications. A cause for concern with the human cell-derived host cell systems is the potential for contamination of the product with adventitious agents. In addition, the impurities derived from human host cell expression systems may exhibit molecular mimicry problems and more easily trigger autoimmune responses in humans. Several antibody products produced from the PER.C6 cell line are currently in clinical studies. The upcoming clinical results should reveal whether human cell-produced monoclonal antibodies present less risk of immunogenicity than those produced using rodent cell lines.

Physical and chemical changes of the product structure during the manufacturing process including oxidation, deamidation, disulfide bond shuffling, degradation, and aggregation could also promote immune responses to the drug products. This is especially true when there are factors that can further provoke an immune response, such as aggregates and materials with adjuvant-like activity in the formulation. Aggregates are not only a source of novel epitopes but also have the ability to cross-link B cell receptors, and enhance antigen processing and presentation, thereby facilitating the generation of immune responses (Hermeling *et al.*, 2004). Traditionally, costly downstream purification processes and formulation studies have been relied upon to remove and reduce the levels of misfolded and aggregated components to minimize the potential clinical impact. It would be ideal to minimize the potential immunogenicity of lead candidates with better stability characteristics during the early drug discovery phase. This is indeed the approach that is now being taken through an emphasis, during the discovery and development phase, on reducing antibody aggregation and structural instability. Strategies such as identifying and correcting aggregation hot spot sequence motifs using protein engineering approaches, and implementing screening and evaluation platforms during lead selection and process development to predict the intrinsic aggregation propensity of antibody drug products are being introduced to help build and select structural stability into antibody drug candidates (Voynov *et al.*, 2009; Wang *et al.*, 2007; Wang *et al.*, 2009b; Ionescu *et al.*, 2007; Dudgeon *et al.*, 2009; Arbabi-Ghahroudi *et al.*, 2008).

The growing concerns regarding the safety of biotherapeutics demand new technologies to identify direct predictive tools relevant to clinical responses, and discover critical immunogenic epitopes and remove them during the therapeutic molecule design. Advanced protein sciences research and novel protein engineering/selection methodologies are working together towards the design of improved molecules with reduced potential immunogenicity. The ultimate goal is to implement these new strategies to predict and resolve potential immunogenicity issues during the discovery phase so that the selected therapeutic candidates possess the desired safety attributes before advancement into clinical studies.

7.3.4 *Lead selection for molecule stability and manufacturability — CMC plan at early stage of drug development*

The FDA is moving forward with new regulatory guidelines regarding quality control for biological drugs (FDA, 2006), while biopharmaceutical companies are devoting time to reduce the discovery and development timeline. The key to achieving the goals from both sides is to clearly define the quality criteria at a project's initiation rather than to continuously attempt to rescue a failing project. The complex nature of an antibody structure leaves an opportunity for alteration of the molecule during the expression, purification, handling, storage, and transportation processes. To date, all monoclonal antibodies, regardless of their source, are heterogeneous products, which include various post-translational modifications and chemical–physical property differences (Fig. 7.3; Chirino and

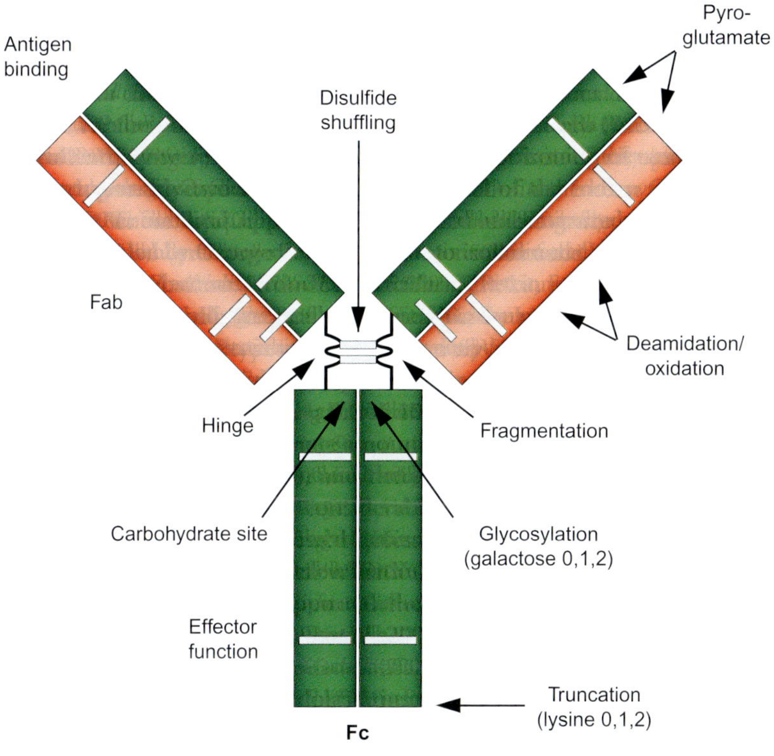

Figure 7.3 Sources of IgG molecular heterogeneity

Monoclonal antibodies produced in mammalian cell culture host systems are heterogeneous products that are prone to chemical and physical structural alteration during the expression and production process, handling, storage and transportation (Adapted from Chirino and Mire-Sluis, 2004).

Mire-Sluis, 2004). The inherent physical and chemical instability of monoclonal antibodies is the root of this heterogeneity. Many lead molecules fail during development not only as a result of inadequate efficacy, unacceptable toxicity, or immunogenicity, but also due to molecular instability, poor manufacturability, and difficulty in formulation (Zabrecky, 2008). One way to reduce the potential risk of failure in later-stage development is to be conscientious and thoughtful during the early stage of lead selection. The physicochemical properties of a given antibody molecule will determine how it can be expressed, purified, and what type of formulation will best suit it. Therefore, it is very important to take the potential opportunities of building high quality attributes such as stability and manufacturability into the selected lead molecule at the early drug discovery phase using the knowledge resulting from decades of research on monoclonal antibody molecules and their behavior during production processes.

In response to the FDA's emphasis on CMC for biologics, the strategic plan should be incorporated into the drug development program as early as possible to facilitate the overall success of therapeutic drug advancement. Molecule stability, manufacture process control, formulation development, and product characterization are the four major programs that make up CMC development (FDA, 2003a; FDA, 2003b; Shah, 2005). In the research stage, the aim of the CMC plan is to ensure the delivery of a sufficient supply of lead candidates with a quality suitable for pre-clinical efficacy and safety studies, and to evaluate the stability and feasibility of lead candidates for their potential success during the future development activities.

The first CMC program is to examine the intrinsic stability of the top antibody therapeutic candidates and eliminate the ones with potential structural instability problems. After a panel of antibodies is generated through antigen target binding selection, candidates are ranked based on binding affinity and selectivity, and narrowed down to a few top candidates. At this stage it is important to evaluate the stability attributes of the top candidates including both chemical and physical stability. Chemical instability involves the protein amino acid sequence alteration while physical instability describes the changes in higher order structures of the protein. At this early drug development stage, there are limitations on material availability and tight project timelines. Well-planned and realistic strategies for such studies are extremely important. There are many analytical testing assays developed for antibody drug characterization including various HPLC methods, capillary electrophoresis, Analytical Ultracentrifugation (AUG), Mass Spectrometry (MS), Fourier Transform Infrared spectroscopy (FTIR), Circular Dichroism spectroscopy (CD), and Differential Scanning Calorimetry (DSC) (Kurz, 2007; Wang *et al.*, 2007). Simple, rapid, and highly sensitive methods that can detect the molecular and structural changes of the selected antibodies are highly desirable at this early stage. Various studies have reported using DSC to assess molecular thermal stability as a predictor of antibody drug candidate stability (Garber and Demarest, 2007; Ahrer *et al.*, 2006; Wen *et al.*, 2008). A quick stress degradation study can be highly valuable because the comparative stability profile generated under the same stress conditions among different lead candidates can help identify the preferred lead drug candidate. However, it is important to note that degradation derived from forced stress studies may not necessarily occur under normal conditions. The designed forced degradation stress studies can only be used to assess molecular stability of selected candidates comparatively, and may be valuable in facilitating pre-clinical process and pre-formulation development (Kats, 2005; Chi *et al.*, 2003). In addition to establishing effective evaluation strategies for making the best choice of lead candidate, protein engineering methodologies may also be applied, when necessary, to construct more potent and more stable variants of selected therapeutic candidates (Walsh, 2007; Jefferis, 2001; Lundblad and Bradshaw, 2006; Angal *et al.*, 1993; Banks *et al.*, 2008; Pepinsky *et al.*, 2010). When structure related stability issues dictate either protein engineering or, in the extreme case, additional lead selection, it is much easier and more cost effective to identify and correct these issues in the early discovery phase rather than in the later drug development stages.

The second CMC program is pre-clinical scale bioprocess development. This program is typically responsible for not only providing material supply for various *in vitro* and *in vivo* proof of concept pre-clinical studies, but also provides the evaluation of the manufacturability of selected lead therapeutic candidates. Because of the relatively large doses that are required for biopotency using antibodies as therapeutics, it is very important to predict the success of a production process for new drug development programs intended to meet the requirements for large-scale manufacturing. Even though a lead candidate may demonstrate high affinity for the target antigen and the specified selectivity profile anticipated for potential clinical efficacy, if it fails to be cost effective in manufacturing at the scale required to meet the demands of commercialization, the drug will remain an interesting research tool without realizing the benefit to patients. Good pre-clinical process development programs address potential product quality/quantity concerns, as well as operability, scalability, economic and regulatory risks.

In pre-clinical process development programs, it is vital for any new lead drug candidate moving from discovery into development to be supported by a clonal cell line that can produce the desired quantity of the therapeutic candidate with the desired quality. The success of an antibody drug production program relies on the clonal cell line used for antibody production to be thoroughly characterized, and the expressed antibody to meet expected product quality criteria. Changing the cell line used for the manufacturing process could lead to the requirement for a clinical comparability study, which can be costly and time-consuming depending upon when the change is made. Therefore, the earlier that a clonal cell line can be identified for use in the manufacturing process, the lower the risk that pre-clinical and clinical bridging studies would be required at a later stage. It is an achievable goal to embed the timeline of stable clonal cell line generation into that of other pre-clinical study activities to ensure that the lead therapeutic candidate can be expressed at a high level and with the quality desired for late-stage manufacturing.

The pre-clinical process development program should establish an efficient and flexible platform for downstream purification (DSP) processes, which should be able to address any anticipated major issues in future large-scale manufacturing. A well-understood and well-controlled DSP process is critically responsible for final product quality. DSP development programs aim to identify the most efficient route from the starting materials to pure, active final products with the highest yield. The evaluation of a lead candidate's performance in response to the downstream purification platform adopted by the manufacturing facility is very important. Despite the structural similarities between monoclonal antibodies, there are surprisingly diverse behaviors reported for even the most commonly used traditional ProA purification scheme (Shukla *et al.*, 2007b). With a good understanding of the behavior of the lead molecule it is possible to craft a platform purification process that is well suited to the specific therapeutic candidate. This will result in shorter development times, reduced costs, and lower scale-up risks in later development stages. Although R&D laboratories only use small-scale processes to make the quantities needed for proof of concept pre-clinical testing, this process establishes the baseline for future optimization efforts and larger-scale cGMP manufacturing. To ensure a process can deliver a consistently high quality product, problems are best identified and addressed early on during the pre-clinical laboratory-scale production, which is much simpler and more effective than trying to make changes in more expensive pilot- and commercial-scale operations.

The third CMC program is formulation development. The basic concerns of biotherapeutic formulation are stability and structural integrity of the active molecule. The components of the formulation buffer are determined based on their ability to preserve the native conformation of the protein (Wang *et al.*, 2007). In addition, product concentration is also a determining factor for successfully delivering large quantities of a monoclonal antibody therapeutic to patients in the least painful and most effective manner. A generic pre-formulation program with a focus on a few key quality attributes of the product can be effective and efficient. In general, taking advantage of platform approaches in monoclonal antibody production processes to pick the lead candidate from discovery programs that best fit the platform process can significantly reduce process development timelines and ultimately reduce the time to patient (Copmann *et al.*, 2001; Apostol *et al.*, 2008; Rathore *et al.*, 2008–2009).

The fourth CMC program is product characterization. It is a central tenet of the overall product control strategy, since it is this characterization that decides what features of the lead molecule are important to ensure the criteria regarding the therapeutic candidate's efficacy and safety are met. Molecular derivatives of the antibody or product-related impurities are classified into four major categories: mass variants, charge variants, post-translational modification variants, and disulfide bond variants. Therefore, characterization of the heterogeneity of a monoclonal antibody product is not trivial for the product manufacturing process and quality control of the drug. However, product characterization is an evolving process that involves contributions at different stages of the product life cycle (Faulkner, 2000; Schenerman *et al.*, 2004). Characterization of lead candidate molecules at the discovery stage is the starting point for understanding a drug product and its process. The information that can be obtained from molecule stability studies and from the behavior of the product molecule during pre-clinical process development studies provides important first-hand knowledge that helps define the critical attributes of the therapeutic candidate. This knowledge will then be drawn upon throughout the rest of the drug development lifecycle. In addition, analyzing the primary sequence and tertiary structure as they relate to stability behavior among the selected lead candidates can also reveal important sequence motifs and characteristics required for target binding and favorable drug-like properties. Establishing an understanding of the correlation between behavior phenotype and the product molecular structure can further enhance future antibody discovery programs. In summary, the manufacturability assessment studies and early CMC development program can facilitate the lead selection process, and greatly assist the translation of drug candidates from discovery into development, and ultimately to commercialization.

The process of therapeutic candidate selection is a science based practice that incorporates a balance of risk and benefit into the decision making procedure. As part of this decision making process it is critical to consider parameters that would suggest likely efficacy in the patient population such as target binding affinity/selectivity, *in vitro* and *in vivo* animal pharmacokinetic and safety data, together with those properties that would indicate good drug-like properties such as physicochemical stability and manufacturability. Figure 7.4 depicts a decision tree of items that might be considered in a systematic approach to the therapeutic candidate selection process. The more predictive information that is available for lead candidates and the more integrated that the pre-clinical process program is in the discovery process, the more effective will be the translation of high tech to high quality drugs, and the sooner patients will realize the benefits.

Implementation of intelligent lead selection strategies can reduce the discovery time and cost, and increase the chance of later stage development and clinical success. Monoclonal antibodies constitute a major class of approved therapeutic proteins, and they are likely to continue to be one of the most significant and flexible biopharmaceutical products in the future. The increasing sophistication of antibody discovery and characterization technologies will drive constant changes

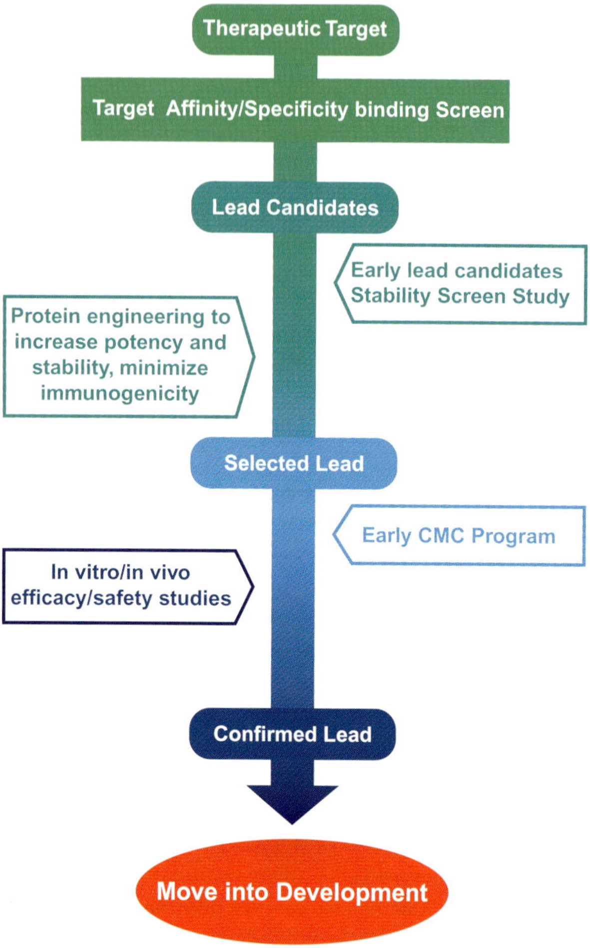

Figure 7.4 A hierarchical selection scheme for lead selection during monoclonal antibody drug discovery

to our assessments and decisions for designing, analyzing and controlling lead selection towards the goal of ensuring the candidates with the highest probability of development are advanced.

Acknowledgments

Crystal structure image (Fig. 7.1) was kindly provided by Dr. Enrico A. Stura (Laboratoire de Structure des Protéines, Département d'Ingénierie et d'Etude des

Protéines, Commissariat à l'Energie Atomique, 91191 Gif-sur-Yvette, France). We wish to thank Dr. Enrico A. Stura, Dr. Gregg J. Silverman (Department of Medicine, University of California at San Diego, USA), and the *Proceedings of the National Academy of Sciences* for their permission for reproduction of the image. In addition, we wish to thank the authors of the article by Chirino and Mire-Sluis, 2004 as well as Macmillan Publishers Ltd. (*Nature Biotechnology*) for permission to adapt the antibody image (Fig. 7.3).

References

Adair F. (2000). Immunogenicity: the last hurdle for clinically successful therapeutic antibodies. *BioPharm Int* **13**: 42–6.

Adair F., Ozanne D. (2002). The immunogenicity of therapeutic proteins. *BioPharm Int* **15**: 30–6.

Ahrer K., Buchacher A., Ibere G., Jungbauer A. (2006). Thermodynamic stability and formation of aggregates of human immunoglobulin G characterized by differential scanning calorimetry and dynamic light scattering. *J Biochem Biophys Methods* **66**: 73–86.

Angal S., King D.J., Bodme, M.W., Turner A., Lawson A.D.G., Roberts G., Pedley B., Adair J.R. (1993). A single amino acid substitution abolishes the heterogeneity of chimeric mouse/human (IgG4) antibody. *Mol Immunol* **30**: 105–8.

Apostol I., Schofield T., Koeller G., Powers S., Stawicki M., Wolfe R.A. (2008). A rational approach for setting and maintaining specifications for biological and biotechnology-derived products: Parts 1–3. *BioPharm Int* **21**: 16–22, 17–21, 28–36.

Arbabi-Ghahroudi M., To R., Gaudette N., Hirama T., Ding W., MacKenzie R., Tanha J. (2008). Aggregation-resistant VHs selected by *in vitro* evolution tend to have disulfide-bonded loops and acidic isoelectric points. *Prot Eng Des Sel* **22**: 59–66.

Arunakumari A., Wang J., Ferreira G. (2007). Alternatives to Protein A: improved downstream process design for human monoclonal antibody production. *BioPharm Int Suppl* **20**: 36–40.

Arunakumari A. (2009). Implementing cost reduction strategies for HumAb manufacturing processes. *BioPharm Int Suppl* **7**: 48–54.

Arunakumari A., Wang J., Ferreira G. (2009). Advances in non-Protein A purification processes for human monoclonal antibodies. *BioPharm Int* **22**: 22–6.

Azevedo A.M., Rosa R.A.J., Ferreira I.F., Aires-Barros M.R. (2008). Integrated process for the purification of antibodies combining aqueous two-phase extraction, hydrophobic interaction chromatography and size-exclusion chromatography. *J Chromatogr A* **1213**: 154–61.

Banks D.D., Gadgil H.S., Pipes G.D., Bondarenko P.V., Hobbs V., Scavezze J.L., Kim J., Jiang X.R.J., Mukku V., Dillon T.M. (2008). Removal of cysteinylation from an unpaired sulfhydryl in the variable region of a recombinant monoclonal IgG1 antibody improves homogeneity, stability, and biological activity. *J Pharm Sci* **97**: 764–79.

Basey C., Blank G. (2002). US Patent 6417335/6489447.

Biorad, CHT Ceramic Hydroxyapatite Instruction Manual.

Boi C. (2007). Membrane adsorbers as purification tools for monoclonal antibody purification tools for monoclonal antibody purification. *J Chromatogr B Analyt Technol Biomed Life Sci* **848**: 97–127.

Boschetti E. (2002). Antibody separation by hydrophobic charge induction chromatography. *Trends Biotech* **20**: 333–7.

Bostrom M., Williams D.R.M., Ninham B.W. (2004). Why the properties of proteins in salt solutions follow a Hofmeister series. *Curr Opin Colloid and Interface Sci* **9**: 48–52.

Brandwein H., Aranha-Creado H. (2000). Membrane filtration for virus removal. *Dev Biol* **102**: 157–63.

Brorson K., Krejci S., Lee K., Hamilton E., Stein K., Xu Y. (2003). Bracketed generic inactivation of rodent retroviruses by low pH treatment for monoclonal antibodies and recombinant proteins. *Biotechnol Bioeng* **82**: 321–9.

Brough H., Antoniou C., Carter J., Jakubik J., Xu Y., Lutz H. (2002). Performance of a novel Viresolve NFR virus filter. *Biotechnol Prog* **18**: 782–95.

Burton S.C., Harding D.R.K. (1998). Hydrophobic charge induction chromatography: salt independent protein adsorption and facile elution with aqueous buffers. *J Chromatogr A* **814**: 71–81.

Carter-Franklin J.N., Victa C., McDonald P., Fahrner R. (2007). Fragments of protein A eluted during protein A affinity chromatography. *J Chromatogr A* **1163**: 105–11.

Charcosset C. (2006). Membrane processes in biotechnology: an overview. *Biotechnol Bioeng* **24**: 482–92.

Chen J., Tetrault J., Ley A. (2008). Comparison of standard and new generation hydrophobic interaction chromatography resins in the monoclonal antibody purification process. *J Chromatogr A* **1177**: 272–81.

Chen J., Jetrault J., Zhang Y., Wasserman A., Conley G., DiLeo M., Haimes E., Nixon A., Ley A. (2010). Mixed-mode chromatography in the monoclonal antibody purification process. *J Chromatogr A* **1217**: 216–24.

Chi E.Y., Krishnan S., Randolph T.W., Carpenter J.F. (2003). Physical stability of protein in aqueous solution: mechanism and driving forces in nonnative protein aggregation. *Pharmaceut Res* **20**: 1325–36.

Chirino A.J., Mire-Sluis A. (2004). Characterizing biological products and assessing comparability following manufacturing changes. *Nat Biotechnol* **22**: 1383–91.

Chung W.K., Hou Y., Holstein M., Freed A., Makhatadze G.I., Cramer S.M. (2010). Investigation of protein binding affinity in multimodal chromatographic systems using a homologous protein libray. *J Chromatogr A* **1217**: 191–8.

Clark L.A., Ganesan S., Papp S., van Vlijmen H.W.T. (2006). Trends in antibody sequence changes during the somatic hypermutation process. *J Immunol* **177**: 333–40.

Clutterbuck A., Kenworthy J., Liddell J. (2007). Endotoxin reduction using disposable membrane adsorption technology in cGMP manufacturing. *BioPharm Int Suppl* **20**: 11–4.

Cohn E. (1947). The separation of blood into fractions of therapeutic value. *Ann Intern Med* **26**: 341–52.

Copmann T., Davis G., Garnick R., Landis J., Lubiniecki A., Massa T., Murano G., Kenneth S., Zezza D. (2001). One product, one process, one set of specifications: a proven quality paradigm for the safety and efficacy of biologic drugs. *BioPharm Int* **14**: 14–24.

Curling J., Gottschalk U. (2007). Process chromatography: five decades of innovation. *BioPharm Int* **20**: 70–93.

Desai M.A. (2000). Methods in biotechnology (Vol. 9). In Walker J.M. (Ed.), *Downstream Processing of Protein: Methods and Protocols.* Humana Press, Totowa, NJ.

Devy L., Huang L., Naa L., Yanamandra N., Pieters H., Frans N., Chang E., Tao Q., Vanhove M., Lejeune A., van Gool R., Sexton DJ., Kuang G., Rank D., Hogan S., Pazmany C., Ma Y.L., Schoonbroodt S., Nixon A.E., Ladner R.C., Hoet R., Henderikx P., TenHoor C., Rabbani S.A., Valentino M.L., Wood C.R., Dransfield D.T. (2009). Selective inhibition of matrix metalloproteinase-14 blocks tumor growth, invasion, and angiogenesis. *Cancer Res* **69**: 1517–26.

Diesenhofer J. (1981). Crystallographic refinement and atomic models of a human Fc fragment and its complex with fragment B of protein A from *Staphylococcus aureus* at 2.9- and 2.8-Angstrom resolution. *Biochemistry* **20**: 2361–70.

DiLeo M., Haimes E., Ley A., Chen J. (2007). Interaction of resin-buffer-antibody product influences the development of an optimal cation exchange purification process for specific monoclonal antibody product. Wilbio Waterside Antibody Conference Poster, San Juan, Puerto Rico, April.

Dixit M. (2008). Membranes and filtration: membrane filtration in the biopharm industry. *Filtra Separat* **45**: 18–21.

Doyle H., Mamula M. (2001). Post-translational protein modifications in antigen recognition and autoimmunity. *Trends Immunol* **22**: 443–9.

Dudgeon K., Famm K., Christ D. (2009). Sequence determinants of protein aggregation in human VH domains. *Prot Eng Des Sel* **22**: 217–20.

Dutton G. (2009). Strategies for removing viruses from cell lines. *Gen Eng News* **29**: 1.

Dyson H.J., Wright P.E., Scheraga H.A. (2006). The role of hydrophobic interactions in initiation and propagation of protein folding. *Proc Natl Acad Sci USA* **103**: 13057–61.

Eriksson K., Ljunglof A., Rodrigo G., Brekkan E. (2009). Mab contaminant removal with a multi-modal anion exchanger: a platform step to follow Protein A. *BioProcessing Int* **7**: 52–6.

Farid S.S. (2007). Process economics of industrial monoclonal antibody manufacture. *J Chromatogr B* **848**: 8–18.

Farid S.S. (2009). Economic drivers and trade-offs in antibody purification processes. *BioPharm Int Suppl* **22**: 38–42.

Faulkner J. (2000). The evolving role of product characterization during development. *BioPharm Int* **13**: 26–34.

FDA (2003a). Guidance for industry, drug product — chemistry, manufacture, and controls. http://www.fda.gov/cder/guidance/1215dft.pdf

FDA (2003b). Guidance for industry, drug substance — chemistry, manufacture, and controls. http://www.fda.gov/cder/guidance/3969dft.pdf

FDA (2006). Guidance for industry: quality systems approach to pharmaceutical CGMP regulations. http://www.fda.gov/cvm/guidance/published.html

Ferreira G.M., Dembecki J., Patel K., Arunakumari A. (2007). A two-column process to purify antibodies without Protein A. *BioPharm Int* **20**: 32–44.

Flatman S., Alam I., Gerard J., Mussa N. (2007). Process analytics for purification of monoclonal antibodies. *J Chromatogr B* **848**: 79–87.

Follman D.K., Fahrner R.L. (2004). Factorial screening of antibody purification processes using three chromatography steps without Protein A. *J Chromatogr A* **1024**: 79–85.

Francis R. (2007). Development of a MEP HyperCel IgG purification process for a commercial polyclonal antibody product. Presentation on SCI (The Society of Chemical Industry) Mixed Mode Chromatography-Separation Science and Technology Conference, June, London.

Franklin S.G. (2003). Removal of aggregate from an IgG4 product using CHT ceramic hydroxyapatite. *BioProcessing Int* **1**: 50–1.

Fraud N., Faber R., Kiss C., Wolfgang D., Hoerl H.H. (2008). A novel membrane adsorber for hydrophobic interaction chromatography. Bioprocess International Conference & Exhibition Poster, Anaheim, CA, USA.

Fraud N., Kuczewski M., Zarbis-Papastoitsis G., Hirai M. (2009). Hydrophobic membrane adsorbers for large-scale downstream processing. *BioPharm Int Suppl* **22**: 24–7.

Gagnon P. (1996). *Purification Tools for Monoclonal Antibodies*. Validated Biosystems Press, Tucson, AZ.

Gagnon P., Ng P., Aberrin C., Zhen J., He J., Mekosh H., Cummings L., Richieri R., Zaidi S. (2006). A ceramic hydroxyapatite based purification platform: simultaneous removal of leached protein A, aggregates, DNA, and endotoxins. *BioProcessing Int* **4**: 50–60.

Gagnon P. (2008). Purification of monoclonal antibodies by mixed mode chromatography. In U. Gottschalk (Ed.), *Process Scale Purification of Antibodies*, pp. 125–144. John Wiley and Sons, New York.

Gagnon P. (2009). Monoclonal antibody purification with hydroxyapatite. *New Biotechnol* **25**: 287–93.

Garber E., Demarest S.J. (2007). A broad range of Fab stabilities within a host of therapeutic IgGs. *Biochem Biophys Res Commun* **355**: 751–7.

GE Healthcare, 11001165AB. MabSelect SuRe, GE Data File.

GE Healthcare, 11001164AA. MabSelect SuRe studies on ligand toxicity, leakage, removal of leached ligand, and sanitization, GE Application Note.

GE Healthcare, 11000421AA. Ion exchange chromatography and chromatofocusing: Principle and methods.

GE Healthcare, 11001269AA. Hydrophobic interaction and reversed phase chromatography: Principle and methods.

GE Healthcare, 28907893AA. Selective removal of aggregates with Captoadhere, GE Application Note.

GE Healthcare, 28907889AA. Optimization of loading conditions on Captoadhere using design of experiments, GE Application Note.

GE Healthcare, 18102218AJ. Gel Filtration: Principles and methods.

Ghose S., Allen M., Hubbard B., Brooks C., Cramer S.M. (2005). Antibody variable region interactions with Protein A: implications for the development of generic purification processes. *Biotechnol Bioeng* **92**: 665–73.

Ghose S., Hubbard B., Cramer S. M. (2006). Evaluation and comparison of alternatives to Protein A chromatography mimetic and hydrophobic charge induction chromatographic stationary phases. *J Chromatogr A* **1122**: 144–52.

Ghosh R. (2002). Protein separation using membrane chromatography: opportunities and challenges. *J Chromatogr A* **952**: 13–27.

Giovannoni L., Ventani M., Gottschalk U. (2008). Antibody purification using membrane adsorbers: a membrane adsorber in retention mode can be used efficiently in a commercial antibody manufacturing process. *BioPharm Int* **21**: 26–30.

Glynn J. (2008). Process-scale precipitation of impurities in mammalian cell culture broth. *BioPharm Int Suppl* **21**.

Glynn J., Boyle D. (2010). Advances in monoclonal antibody purification. *BioPharm Int* **23**.

Gomez M.I., O'Seaghdha M., Magarge M., Foster T.J., Prince A.S. (2006). *Staphylococcus aureus* Protein A activates TNFR1 signaling through conserved IgG binding domains. *J Bio Chem* **281**: 20190–6.

Gottschalk U. (2006). The renaissance of protein purification: progress in downstream processes have not kept pace with increases in upstream yields. It is time for protein purification to make a comeback. *BioPharm Int Suppl* **19**: 8–19.

Gottschalk U. (2008). Bioseparation in antibody manufacturing: the good, the bad and the ugly. *Biotechnol Prog* **24**: 496–503.

Gottschalk U. (2009a). Generic platform processes — the Deus ex Machina of antibody purification? *BioPharm Int* **22**: 7.

Gottschalk U. (2009b). Tech Review: Downstream processing tomorrow — Great expectations? *European Biotechnology Science & Industry News* 111609.

Graille M., Stura E.A., Corper A.L., Sutton B.J., Taussig M.J., Charbonnier JB., Silverman G.J. (2000). Crystal structure of a *Staphylococcus aureus* protein A domain complexed with the Fab fragment of a human IgM antibody: structural basis for recognition of B-cell receptors and superantigen activity. *Proc Natl Acad Sci USA* **97**: 5399–404.

Grodzki AC., Berenstein E. (2010). Antibody purification: ammonium sulfate fractionation or gel filtration. *Methods Mol Biol* **588**: 15–26.

Guerrier L., Girot P., Schwartz W., Boschetti E. (2000). New method for the selective capture of antibodies under physiological conditions. *Bioseparation* **9**: 211–21.

Guerrier L., Flayeux I., Boschetti E. (2001). A dual-mode approach to the selective separation of antibodies and their fragments. *J Chromatogr B* **755**: 37–46.

Haake D.A., Franklin E.C., Frangione B. (1982). The modification of human immunoglobulin binding to staphylococcal protein A using diethylpyrocarbonate. *J Immunol* **129**: 190–192.

Habeeb A., Francis R. (1976). Preparation of human immunoglobulin by ammonium sulfate precipitation. *Vox Sang* **31**: 423–34.

Hahn R., Schlegel R., Jungbauer A. (2003). Comparison of protein A affinity sorbents. *J Chromatogr B* **790**: 35–51.

Hahn R., Bauerhansl P., Shimahara K., Wizniewski C., Tscheliessnig A., Jungbauer A. (2005). Comparison of protein A affinity sorbents: II. Mass transfer properties. *J Chromatogr A* **1093**: 98–110.

Hahn R., Shimahara K., Steindl F., Jungbauer A. (2006). Comparison of protein A affinity sorbents III. Life time study. *J Chromatogr A* **1102**: 224–31.

Halle B. (2004). Protein hydration dynamics in solution: a critical survey. *Phil Trans R Soc Lond B* **359**: 1207–24.

Hardin A.M., Harinarayan C., Malmquist G., Axen A., van Reis R. (2009). Ion exchange chromatography of monoclonal antibodies: effect of resin ligand density on dynamic binding capacity. *J Chromatogr A* **1216**: 4366–71.

Haslam H.C. (1913). Separation of proteins. Part III: Globulins. *Biochem J* **7**: 492–516.

Hermeling S., Crommelin D.J.A., Schellekens H., Jiskoot W. (2004). Structure-Immunogenicity relationship of therapeutic proteins. *Pharmaceut Res* **21**: 897–903.

Hjerten S., Yao K., Eriksson K.O., Johansson B. (1986). Gradient and isocratic high-performance hydrophobic interaction chromatography of proteins on agarose columns. *J Chromatogr* **359**: 99–109.

Hober S., Nord K., Linhult M. (2007). Protein A chromatography for antibody purification. *J Chromatogr B* **848**: 40–7.

Hongo-Hirasaki T., Yamaguchi K., Yanagida K., Okuyama K. (2006). Removal of small viruses (parvovirus) from IgG solution by virus removal filter Planova 20N virus removal filter. *J Membr Sci* **78**: 3–9.

Ibere G., Schwinn H., Josic D., Jungbauer A., Buchacher A. (2001). Improvement performance of protein separation by continuous annular chromatography in the sixe-exclusion mode, *J Chromatogr A* **921**: 15–24.

ICH and FDA, Q5A (1998). Viral safety evaluation of biotechnology products derived from cell lines of human or animal origin (Geneva, Switzerland and Rockville, MD).

Ionescu R.M., Vlasak J., Price C., Kirchmeier, M. (2007). Contribution of variable domains to the stability of humanized IgG1 monoclonal antibodies. *J Pharmaceut Sci* **97**: 1414–26.

Jagschies G., Gronberg A., Bjorkman T., Lacki K., Johansson H.J. (2006). Technical and economical evaluation of downstream processing options for monoclonal antibody (mAb) production. *BioPharm Int Suppl* **19**: 10–9.

Jansson B., Uhlen M., Nygren PA. (1998). All individual domains of staphylococcal protein A show Fab binding. *FEMS Immunol Med Microbiol* **20**: 69–78.

Jefferis R. (2001). Glycosylation of human IgG antibodies. *BioPharm* **14**: 19–27.

Jefferis R. (2006). Criteria for selection of IgG isotype and glycoform of antibody therapeutics. *BioProcessing Int* **4**: 40–4.

Jefferis R., Lefranc MP. (2009). Human Immunoglobulin allotypes: possible implications for immunogenicity. *mAbs* **1**: 332–8.

Johansson H.J., Ljunglof A., Eriksson K., Brekkan E., Rodrigo G., Bjorkman T. (2007). A novel multi-modal anion exchanger designed for monoclonal antibody purification. Presentation at IBC's 18th International antibody development & production, Feb. CA USA.

Jones S.D., Castillo F.J., Levine H.L. (2007). Advances in the development of therapeutic monoclonal antibodies. *BioPharm Int* **20**: 96–114.

Jossang T., Feder J., Rosenqvist E. (1988). Photon correlation spectroscopy of human IgG. *J Protein Chem* **7**: 165–71.

Kaleas K.A., Schmelzer C.H., Pizarro S.A. (2010). Industrial case study: evaluation of a mixed-mode resin for selective capture of a human growth factor recombinantly expressed in E. Coli. *J Chromatogr A* **1217**: 235–42.

Kats M. (2005). Forced degradation studies: regulatory considerations and implementation. *BioPharm Int* **18**: 32–42.

Kelley B. (2007). Very large scale monoclonal antibody purification: the case for conventional unit operation. *Biotechnol Prog* **23**: 995–1008.

Klein E. (2000). Review: affinity membranes: a 10-year review. *J Membr Sci* **179**: 1–27.

Kim J.S., Akeprathumchai S., Wickramasinghe S.R. (2001). Flocculation to enhance microfiltration. *J Membr Sci* **182**: 161–72.

Klyushnichenko V.E., Wulfson A.N. (1993). Recombinant human insulin-II. Size-exclusion HPLC of biotechnological precursors: factors influencing on retention and selectivity. *Pure Appl Chem* **65**: 2265–72.

Knudsen H.L., Fahrner R.L., Xu Y., Norling L.A., Blank G.S. (2001). Membrane ion-exchange chromatography for process-scale antibody purification. *J Chromatogr A* **907**: 145–54.

Kopaciewicz W., Regnier F.E. (1982). Non-ideal size-exclusion chromatography of protein: effects of pH at low ionic strength. *Anal Biochem* **126**: 8–16.

Kozlowski S. (2007). Protein Therapeutics and the regulation of quality: a brief history from an OBP perspective. *BioPharm Int* **20**: 37–55.

Kuczewski M., Zarbis-Papastoitsis G. (2009). Development of a bind/elute polishing step using a hydrophobic interaction membrane adsorber. ACS National Meeting, Washington, DC.

Kuczewski M., Schirmer E., Lain B., Zarbis-Papastoitsis G. (2010). PEG precipitation development of a bind/elute polishing step using a hydrophobic interaction membrane adsorber, ACS National Meeting, Washington, DC.

Kunz W., Henle J., Ninham B.W. (2004). "Zur Lehre von der Wirkung der Salze" (on the science of the effect of salt): Franz Hofmeister's historical papers. *Curr Opin Colloid and Interface Sci* **9**: 19–37.

Kurz, M. (2007). Regulatory considerations regarding quality aspects of monoclonal antibodies. *Biopharm Int* **20**: 44–52.

Lain B., Cacciuttolo M.A., Zarbis-Papastoitsis G. (2009). Development of a high-capacity Mab capture step based on cation-exchange chromatography. *BioProcessing Int* **7**: 26–34.

Langer E.S. (2010). Downstream processing: promising technologies are on the horizon. *BioPharm Int* **23**: 24–7.

Laurent T.C., Killander J. (1964). A theory of gel filtration and its experimental verification. *J Chromatogr* **14**: 317–30.

Laurent T.C. (1993). History of a theory. *J Chromatogr* **633**: 1–8.

Lebreton B.A., O'Connor D.A., Safta A., Sharma M. (2009). US Patent 0148435.

Lee A., Topping A., Razzaq A., Reiter K., Acosta A. (2009). Purifying a recalcitrant therapeutic recombinant protein with a mixed- mode chromatography sorbent. *Bioprocessing Int* **7**: 42–8.

Li J., Yang C., Xia Y., Bertino A., Glaspy J., Roberts M., Kuter D.J. (2001). Thrombocytopenia caused by the development of antibodies to thrombopoietin. *Blood* **98**: 3241–8.

Liedert B., Bassus S., Schneider C.K., Kalinke, U., Lower J. (2007). Safety of phase I clinical trials with monoclonal antibodies in Germany: the regulatory requirements viewed in the aftermath of the TGN1412 disaster. *Int J Clin Pharmacol Ther* **45**: 1–9.

Lim J.A.C., Sinclair A., Kim D.S., Gottschalk U. (2007a). Economic benefits of single-use membrane chromatography in polishing. *BioProcessing Int* **5**: 48–56.

Lim J.A.C., Sinclair A., Hirai M., Gottschalk U. (2007b). Disposable chromatography-counting the cost. *BioPharm Int Suppl* **20**: 34–40.

Liu J., Cheung A., Hickey J.L., Ghose S. (2009). Comparison of camelid antibody ligand to Protein A for monoclonal antibody purification, A stable alternative to Protein A chromatography. *BioPharm Int* **22**: 35–43.

Low D., O'Leary R., Pujar N.S. (2007). Future of antibody purification. *J Chromatogr B* **848**: 48–63.

Lundblad R.L., Bradshaw R.A. (2006). Addressing product improvement using chemical modification in biopharmaceutical manufacture. *BioProcessing Int* **4**: 30–6.

Luo R., Waghmare R., Krishnan M., Adams C., Poon E., Kahn D. (2006). High-concentration UFDF of a monoclonal antibody. *BioProcessing Int* **4**: 44–8.

Mahn A., Lienqueo M.E., Asenjo J.A. (2007). Optimal operation conditions for protein separation in hydrophobic interaction chromatography. *J Chromatogr B* **849**: 236–42.

Majors R.E, Lees A, Burkhardt M. (2009). Improving protein separations with mixed-mode chromatography. LCGC North America, Jan.

Matheus S., Friess W., Schwartz D., Mahler HC. (2009). Liquid high concentration IgG1 antibody formulations by precipitation. *J Pharm Sci* **98**: 3043–57.

Mhatre R., Nashabeh W., Schmalzing D., Yao X., Fuchs M., Whitney D., Regnier F. (1995). Purification of antibody Fab fragments by cation-exchange chromatography and pH gradient elution. *J Chromatogr A* **707**: 225–31.

Moks T., Abrahmsen L., Nilsson B., Hellman U., Sjoquist J., Uhlen M. (1986). Staphylococcal protein A consists of five IgG-binding domains. *Eur J Biochem* **156**: 637–43.

Mora J., Dolan S., Fraud N. (2006). Scaling up disposable membrane chromatography. *BioPharm Int* **19**.

Moreira A.R. (2007). The evolution of protein expression and cell culture. *BioPharm Int* **20**: 56–68.

Morrow J.K., Jr. (2007a). Optimizing antibody biomanufacture: better processes can consistently yield both quality and quantity. *Genet Eng News* **27**: 34.

Morrow J.K., Jr. (2007b). Glycosylation and the demands of antibody engineering. *BioPharm Int* **20**: 126–8.

Morrow J.K., Jr. (2008). Antibody processing platforms evolve: development and evaluation of new downstream methodologies is leveling the playing field. *Genet Eng News* **28**: 62.

Ng P., Cohen A., Gagnon P. (2006). Monoclonal antibody purification with CHT. *Genet Eng News* **26**: 60.

Norling L., Lute S., Emery R., Khuu W., Voisard M., Xu Y., Chen Q., Blank G., Brorson K. (2005). Impact of multiple re-use of anion-exchange chromatography media on viral removal. *J Chromatogr A* **1069**: 79–89.

Nunes V.S., Bonifacio M.J., Queiroz J.A., Passarinha L.A. (2009). Assessment of COMT isolation by HIC using a dual salt system and low temperature. *Biomed Chromatogr* **24**: 858–62.

O'Donnell K.J., Gershenow E., Sakima E., Hashimoto Y., Mitoma Y. (2006). Purification of monoclonal antibodies using new pore size optimized hydrophobic interaction resins. Bioprocess International Show & Exhibition Poster.

Oehme F., Peters J. (2010). Mixed-mode chromatography in downstream process development. *Biopharm Int Suppl* **23**.

Paliwal A., Asthagiri D., Abras D., Lenhoff A.M., Paulaitis M.E. (2005). Light-scattering studies of protein solutions: role of hydration in weak protein–protein interactions. *Biophys J* **89**: 1564–73.

Pall Life Sciences, Mercapto-Ethyl-Pyridine (MEP) Resin Instruction Manual.

Pavlaki M., Zucker S. (2003). Matrix metalloproteinase inhibitors (MMPIs): The beginning of phase I or the termination of phase III clinical trials. *Cancer Metast Rev* **22**: 177–203.

Pepinsky R.B., Silvian L., Berkowitz S.A., Farrington G., Lugovskoy A., Walus L., Eldredge J., Capili A., Mi S., Graff C., Garber E. (2010). Improving the solubility of anti-LINGO-1 monoclonal antibody Li33 by isotype switching and targeted mutagenesis. *Protein Sci* **19**: 954–66.

Phillips M., Cormier J., Ferrence J., Dowd C., Kiss R., Lutz H., Carter J. (2005). Performance of a membrane adsorber for trace impurity removal in biotechnology manufacturing. *J Chromatogr A* **1078**: 74–82.

Piaggio M.V., Peirotti M.B., Deiber J.A. (2005). Effect of background electrolyte on the estimation of protein hydrodynamic radius and net charge through capillary zone electrophoresis. *Electrophoresis* **26**: 3232–46.

Porath J, Flodin P. (1959). Gel Filtration: a method for desalting and group separation. *Nature* **183**: 1657–9.

Porath J. (1960). Gel filtration of proteins, peptides and amino acids. *Biochimica et Biophysica Acta* **39**: 193–207.

Pratt L.R., Chandler D. (1980). Effects of solute-solvent attractive forces on hydrophobic correlations. *J Chem Phys* **73**: 3434–41.

Przybycien T., Narahari S., Steele L. (2004). Alternative bioseparation operations: life beyond packed-bed chromatography. *Curr Opin Biotechnol* **15**: 469–78.

Queiroz J.A., Tomaz C.T., Carbal J.M.S. (2001). Hydrophobic interaction chromatography of proteins. *J Biotechnol* **87**: 143–59.

Ray S., Tarrach K. (2008). Viral clearance strategy using a three-tier orthogonal technology platform. *BioPharm Int* **21**: 50–9.

Rathore A.S., Saleki-Gerhardt A., Montgomery S.H., Tyler S.M. (2008–2009). Quality by design: industrial case studies on defining and implementing design space for pharmaceutical processes Parts 1–2. *BioPharm Int* **21**:15–9, **22**: 36–44.

Regenmortel M.H.V., Boven K., Bade, F. (2005). Immunogenicity of biopharmaceuticals: an example from Erythropoietin: protein structure, contaminants, formulation, container, and closure all can affect the immunogenicity of the product. *BioPharm Int* **18**: 36–52.

Reichert J.M., Pavlou A. (2004). Monoclonal antibodies market. *Nat Rev Drug Discov* **3**: 383–4.

Reichert J.M. (2009). Global antibody development trends. *mAbs* **1**: 1–2.

Rinderknecht E.H., Zapata G.A. (2006). US Patent 7038017.

Rios M. (2007). Eluting possibilities with mixed-mode chromatography. *Pharmaceut Technol* **31**: 40–8.

Rosenberg A.S., Worobec A. (2004a). A risk-based approach to immunogenicity concerns of therapeutic protein products: Part 1 — considering consequences of the immune response to a protein. *BioPharm Int* **17**: 22–6.

Rosenberg A.S., Worobec A. (2004b). A risk-based approach to immunogenicity concerns of therapeutic protein products: Part 2 — considering host-specific and product-specific factors impacting immunogenicity. *BioPharm Int* **17**: 34–42.

Rosenberg A.S., Worobec, A. (2005). A risk-based approach to immunogenicity concerns of therapeutic protein products: Part 3 — effects of manufacturing changes on immunogenicity and the utility of animal immunogenicity studies. *BioPharm Int* **18**: 32–6.

Roush D.J., Lu Y. (2008). Advances in primary recovery: centrifugation and membrane technology. *Biotechnol Prog* **24**: 488–95.

Sakamoto K., Ito Y., Hakaaki H., Soni P.B., Mori T., Sugimura K. (2009). Discovery and characterization of a peptide motif that specifically recognizes a non-native conformation of human IgG induced by acidic pH conditions. *J Biol Chem* **284**: 9986–93.

Sartorius, Virisart CPV Virus Removal Filters Technical Information.

Sasso E.H., Silverman G.J., Mannik M. (1989). Human IgM molecules that bind staphylococcal protein A contain VHIII H chains. *J Immunol* **142**: 2778–83.

Sasso E.H., Silverman G.J., Mannik M. (1991). Human IgA and IgG F(ab')2 that bind to staphylococcal protein A belong to the VHIII subgroup. *J Immunol* **147**: 1877–83.

Saxena A., Tripathi B.P., Kumar M., Shahi V.K. (2009). Membrane-based techniques for the separation and purification of proteins: an overview. *Adv Colloid Interface Sci* **145**: 1–22.

Schenerman M.A., Sunday B.R., Kozlowski S., Webber K., Gazzano-Santoro H., Mire-Sluis A. (2004). CMC strategy forum report. *BioProcessing Int* **2**: 42–53.

Schwartz L. (2003). Diafiltration for desalting or buffer exchange. *BioProcessing Int* **1**: 43–9.

Schwartz W., Jiao J.A., Ford J., Conrad D., Hamel J.F., Santambien P., Bradbury L., Robin T. (2004). Application of chemically-stable immunoglobulin-selective sorbents: capture and purification of antibodies with resolution of aggregate. *BioProcessing J* **3**: 53–62.

Schwartz W., Judd D., Wysocki M., Guerrier L., Birck-Wilson E., Boschetti E. (2001). Comparison of hydrophobic charge induction chromatography with affinity chromatography on Protein A for harvest and purification of antibodies. *J Chromatogr A* **908**: 251–63.

Senczuk A.M., Klinke R., Arakawa T., Vedantham G., Yigzaw Y. (2009). Hydrophobic interaction chromatography in dual salt system increases protein binding capacity. *Biotechnol Bioeng* **103**: 930–5.

Shadle P.J., Erickson J.C., Robert G.S., Smith T.M. (1995). US Patent 5429746.

Shah V. (2005). CMC system based approach for pharmaceutical quality. www.fda.gov.

Shankar G., Shores E., Wagner C., Mire-Sluis A. (2006). Scientific and regulatory considerations on the immunogenicity of biologics. *Trends Biotechnol* **24**: 274–80.

Shukla A.A., Hinckley P.J., Gupta P., Yinges Y., Hubbard B. (2005). Strategies to address aggregation during Protein A chromatography. *BioProcess Int* **3**: 36–45.

Shukla A.A., Gupta P., Han X. (2007a). Protein aggregation kinetics during Protein A chromatography: case study for an Fc fusion protein. *J Chromatogr A* **1171**: 22–8.

Shukla A.A., Hubbard B., Tressel T., Guhan S., Low D. (2007b). Downstream processing of monoclonal antibodies-application of platform approaches. *J Chromatogr B* **848**: 28–39.

Shukla A.A., Kandula J.R. (2008). Harvest and recovery of monoclonal antibodies from large-scale mammalian cell culture: comparing primary harvest techniques adopted in commercial-scale operations for monoclonal antibody products. *BioPharm Int* **21**: 18–26.

Shukla AA., Thommes J. (2010). Recent advances in large-scale production of monoclonal antibodies and related proteins. *Trends Biotechnol* **28**: 253–61.

Silverman G.J., Goodyear C.S., Siegel D.L. (2005). On the mechanism of staphylococcal protein A immunomodulation. *Transfusion* **45**: 274–80.

Singhvi R., Schorr C., O'Hara C., Xie L., Wang D. (1996). Clarification of animal cell culture process fluids using depth microfiltration. *Biopharm* **4**: 35–41.

Sofer G., Chirica L.C. (2006). Improving productivity in downstream processing. *BioPharm Int* **19**: 30–5.

Spencer J., Boschetti E., Bengio S. (2003). Replacing protein A sorbents for the large-scale manufacturing of recombinant antibodies: hydrophobic charge induction chromatography. *BioProcessing J* **2**: 67–72.

Staby A., Jensen I.H., Mollerup I. (2000). Comparison of chromatographic ion-exchange resins I. Strong anion-exchange resins. *J Chromatogr A* **897**: 99–111.

Staby A., Jensen I.H. (2001). Comparison of chromatographic ion-exchange resins II. More strong anion-exchange resins. *J Chromatogr A* **908**: 149–61.

Staby A., Sand M.B., Hansen R.G., Jacobsen J.H., Andersen L.A., Gerstenberg M., Bruus U.K., Jensen I.H. (2004). Comparison of chromatographic ion-exchange resins III. Strong cation-exchange resins. *J Chromatogr A* **1034**: 85–97.

Staby A., Sand M.B., Hansen R.G., Jacobsen J.H., Andersen L.A., Gerstenberg M., Bruus U.K., Jensen I.H. (2005). Comparison of chromatographic ion-exchange resins IV. strong and weak cation-exchange resins and heparin resins. *J Chromatogr A* **1069**: 65–77.

Staby A., Jacobsen J.H., Hansen R.G., Bruus U.K., Jensen I.H. (2006). Comparison of chromatographic ion-exchange resins V. strong and weak cation-exchange resins. *J Chromatogr A* **1118**: 168–79.

Staby A., Jense R.H., Bensch M., Hubbuch J., Dunweber D.L., Krarup J., Nielsen J., Lund M., Kida, S., Hansen T.B., Jensen I.H. (2007). Comparison of chromatographic ion-exchange resins VI. Weak anion-exchange resins. *J Chromatogr A* **1164**: 82–94.

Stein A., Kiesewetter A. (2007). Cation exchange chromatography in antibody purification: pH screening for optimized binding and HCP removal. *J Chromatogr B* **848**: 151–8.

Stockwin L.H., Holmes S. (2003). Antibodies as therapeutic agents: vive la renaissance! *Expert Opin Biol Ther* **3**: 1133–52.

Sun S., Gallo C., Kelley B. (2004). Removal of high molecular weight aggregates from an antibody preparation using ceramic hydroxyapatite chromatography. Patent application 032883.

Suzanne A., Bonnerjea J. (2007). Scale-up of monoclonal antibody purification processes. *J Chromatogr B* **848**: 64–78.

Svergun D.J., Richard S., Koch M.H.J., Sayers Z., Kuprin S., Zaccai G. (1998). Protein hydration in solution: experimental observation by X-ray and neutron scattering. *Proc Natl Acad Sci USA* **95**: 2267–72.

Swinnen K., Krul A., Van Goidsenhoven I., Van Tichelt N., Roosen A., Van Houdt K. (2007). Performance comparison of Protein A affinity resins for the purification of monoclonal antibodies. *J Chromatogr B* **848**: 97–107.

Tabrizi M.A., Bornstein G.G., Klakamp S.L., Drake A., Knight R., Roskos L. (2009). Translational strategies for development of monoclonal antibodies from discovery to the clinic. *Drug Discovery Today* **14**: 298–305.

Tarvers R.C., Church F.C. (1985). Use of high-performance size-exclusion chromatography to measure protein molecular weight and hydrodynamic radius: an investigation of the properties of the TSK3000SW column. *Int J Pept Prot Res* **26**: 539–49.

TeGenero (2005). Investigator's Brochure: TGN1412 humanized agonistic anti-CD28 monoclonal antibody. Press release.

Thommes J., Etzel M. (2007). Alternatives to chromatographic separations. *Biotechnol Prog* **23**: 42–5.

Thorpe R., Wadhwa M. (2003). Unwanted immunogenicity pf therapeutic biological products. *BioProcess Int* **1**: 60–9.

Tishchenko G.A., Bleha M., Skvor J., Bostik T. (1998). Effect of salt concentration gradient on separation of different types of specific immunoglobulins by ion-exchange chromatography on DEAE cellulose. *J Chromatogr B* **706**: 157.

Tosoh Bioscience, Toyopearl Gigacap Ion Exchange Resins Brochure.

Trexler-Schmidt M., Sze-Khoo S., Cothran A.R., Thai B.Q., Sargis S., Lebreton B., Kelley B., Blank G. S. (2009). Purification strategies to process 5g/L titers of monoclonal antibodies. *BioPharm Int Suppl* **22**: 8–15.

Valdes R., Ibarra N., Ruibal I., Beldarrain A., Noa E., Herrera N., Aleman R., Padilla S., Garcia J., Perez M., Morales R., Chong E., Reyes B., Quinones Y., Agraz A., Herrera L. (2002). Chromatographic removal combined with heat, acid and chaotropic inactivation of four model viruses. *J Biotechnol* **96**: 251–8.

van Holten R.W., Ciavarella D., Harmon F., Riester S. (2002). Incorporation of an additional viral-clearance step into a human immunoglobulin manufacturing process. *Vox Sang* **83**: 227–33.

van Reis R., Leonard L.C., Hsu C.C., Builder S. (1991). Industrial scale harvest of proteins from mammalian cell culture by tangential flow filtration. *Biotechnol Bioeng* **38**: 413–22.

van Reis R., Zydney A. (2001). Membrane separations in biotechnology. *Curr Opin Biotechnol* **12**: 208–11.

Vaughan T.J., Osbourn J.K., Tempest P.R. (1998). Human antibodies by design. *Nat Biotechnol* **16**: 535–9.

Voynov V., Chennamsetty N., Kayser V., Helk B., Trout B.L. (2009). Predictive tools for stabilization of therapeutic proteins. *mAbs* **1**: 580–2.

Walsh G. (2004). Modern Antibody-based therapeutics. *BioPharm Int* **17**: 18–25.

Walsh G., Jefferis R. (2006). Post-translational modifications in the context of therapeutic proteins. *Nat Biotechnol* **24**: 1241–52.

Walsh G. (2007). Engineering biopharmaceuticals. *BioPharm Int* **20**: 64–8.

Wang L., Kanani D.M., Ghosh R. (2006). Purification of humanized monoclonal antibodies by membrane-based hybrid bioseparation technique. *J Immunol Methods* **314**: 1–8.

Wang W., Singh S., Zeng D.L., King K., Nema S. (2007). Antibody structure, instability, and formulation. *J Pharmaceut Sci* **96**: 1–26.

Wang J., Diehl T., Watkins-Fischl M., Perkins D., Aguiar D., Arunakumari A. (2008). Optimizing the primary recovery step in non-affinity purification schemes for HumAbs. *BioPharm Int Suppl* **19**: 4–10.

Wang M., Diehl T., Aguiar D., Dai X.P., Arunakumari, A. (2009a). Precipitation of Process-derived impurities in non-Protein a purification schemes for antibodies. *BioPharm Int Suppl* **22**: 4–10.

Wang X.L., Das T.K., Singh S.K., Kumar S. (2009b). Potential aggregation prone regions in biotherapeutics. *mAbs* **1**: 254–67.

Weatherly G.T., Bouvier A., Lydiard D.D., Chapline J., Henderson I., Schrimsher J.L., Shepard S.R. (2002). Initial purification of recombinant botulinum neurotoxin fragments for pharmaceutical production using hydrophobic charge induction chromatography. *J Chromatogr A* **952**: 99–110.

Wen J., Jiang Y., Nahri L. (2008). Effect of carbohydrate on thermal stability of antibodies. *American Pharmaceutical Review* **11**: 98–104.

Westoby M., Chrostowski J., de Vilmorin P., Smelko JP., Romero JK. (2010). Effects of solution environment on mammalian cell fermentation broth properties: enhanced impurity removal and clarification performance. *Biotechnol Bioeng* **108**: 50–8.

Winter G., Milstein C. (1991). Man-made antibodies. *Nature* **349**: 293–9.

Wu D., Walter R.R. (1992). Effects of stationary phase ligand density on high-performance ion-exchange chromatography of protein. *J Chromatogr* **598**:7 –13.

Xia F., Nagrath D., Garde S., Cramer S.M. (2004). Evaluation of selectivity changes in HIC systems using a preferential interaction based analysis. *Biotechnol Bioeng* **87**: 354–63.

Xia F., Nagrath D., Cramer S.M. (2005). Effect of pH changes on water release values in hydrophobic interaction chromatographic systems. *J Chromatogr A* **1079**: 229–35.

Yigzaw Y., Piper R., Tran M., Shukla A.A. (2006). Exploitation of the adsorptive properties of depth filters for host cell protein removal during monoclonal antibody purification. *Biotechnol Prog* **22**: 288–96.

Yokoyama T., Murai K., Murozuka T., Wakisaka A., Tanifuji N., Tomono T. (2004). Removal of small non-enveloped viruses by nanofiltration. *Vox Sang* **86**: 225–9.

Zabrecky J. (2008). Why do so many biopharmaceuticals fail? *BioProcess Int* **6**: 26–34.

Zhao G., Dong X.Y., Sun Y. (2009). Ligands for mixed-mode protein chromatography: principles, characteristics and design. *J Biotechnol* **144**: 3–11.

Zhou J., Tressel T. (2008). Basic concepts in Q membrane chromatography for large-scale antibody production. *Biotechnol Prog* **22**: 341–9.

Zhou J.X., Dermawan S., Felix S., Flynn G., Stenson R., Tresse, T., Guhan S. (2007a). pH-Conductivity hybrid gradient cation-exchange chromatography for process scale monoclonal antibody purification. *J Chromatogr A* **1175**: 69–80.

Zhou J.X., Tressel T., Guhan S. (2007b). Disposable chromatography: single-use membrane chromatography as a polishing option during antibody production is gaining momentum. *BioPharm Int Suppl* **20**: 25–33.

Zhou J.X., Solamo F., Hong T., Shearer M. (2008). Viral clearance using disposable systems in monoclonal antibody commercial downstream processing. *Biotechnol Bioeng* **100**: 488–96.

Zhou D. (2009). Sartobind membrane adsorbers. Sartorius Stedim, Presentation at Dyax.

Zydney A.L. (2009). Membrane technology for purification of therapeutic proteins. *Biotechnol Bioeng* **103**: 227–30.

8

Design and Application of Immunoconjugates for Cancer Therapy

Sherif El Sheikh, Hans-Georg Lerchen*, Beate Müller-Tiemann* and Jörg Willuda**

8.1 Introduction

Antibody therapy has successfully been introduced into the clinic as a validated approach for treating malignancies. The advent of recombinant antibody technology to deliver tailor-made fully human antibodies with high selectivity and affinity against selective disease associated targets make them attractive agents for cancer immunotherapy and has led to the approval of leading products such as Rituxan (rituximab, 1997), Herceptin (trastuzumab, 1998), Erbitux (cetuximab, 2004) and Avastin (bevacizumab, 2004). However, efficacy of naked antibodies can be limited, especially in the treatment of solid tumors. This can be explained in part by redundancy and by-pass mechanisms of signal transduction pathways in cancer cells. Although one of the pathways may be specifically targeted by a monoclonal antibody (mAb), it may not be sufficient for complete elimination of the tumor cells. Furthermore, targeting of a singular transduction pathway by one mAb would also be inefficient in a tumor mass with heterogeneous cell populations as other molecular pathways may compensate and take a lead role in driving cell proliferation. Furthermore, factors such as slow extravasation from the blood supply and poor tumor vascularization and lymphatic drainage (Jain, 1990), *in situ* binding site accessibility (Saga *et al.*, 1995; Fujimori *et al.*, 1990) and antigen heterogeneity (Ruiz-Cabello *et al.*, 2002) are relevant in impeding efficient

* Bayer HealthCare AG, Germany.

and prolonged homogeneous intratumoral mAb uptake. Immune activation mechanisms such as antibody-dependent cellular cytotoxicity (ADCC) or complement-dependent mechanism (CDC), which are mediated by mAbs via their native Fc region, may also not be very effective in the immunosuppressive environment of solid tumors. In fact, the contributions of ADCC and CDC to clinical efficacy of approved antibodies have been documented (Klepfish *et al.*, 2007; Clynes *et al.*, 2000; Cartron *et al.*, 2002; Weng and Levy, 2003; Di Gaetano *et al.*, 2003; Zhang *et al.*, 2007; Musolino *et al.*, 2008), but the relevance of this mechanism of action to long-lasting clinical efficacy has not been proven. It appears that clinical efficacy of the approved antibodies is more a combination of several contributing activities, such as the intrinsic neutralizing or function-blocking activity of the antibody and of the above described immune-mediated mechanisms (Spector and Blackwell, 2009; Adams and Weiner, 2005).

Current efforts are aimed at overcoming the limitations and even potentiating conventional antibody therapeutics to further enhance clinical efficacy. Multiple strategies are evaluated preclinically and clinically (Carter, 2001; Wu and Senter, 2005; Brumlik *et al.*, 2008; Senter, 2009), and one of the most promising approaches is the antibody-mediated delivery of immunoconjugates as highly potent effector molecules. These molecules are referred to as antibody drug conjugates (ADCs) when a cytotoxic small molecule is attached or as radioimmunoconjugates (RICs) if a radionuclide is attached (Carter, 2001; Garnett, 2001; Wu and Senter, 2005; Ricart and Tolcher, 2007).

Within the field of immunoconjugate development, there have been attempts to combine the specificity of mAb therapy with effector molecules that exert potent cytotoxic activity to tumor cells by inducing either direct or indirect cell death and that ideally limit off-target toxicity to normal tissues. Coupling of the payload to the delivery vehicle is usually achieved via a linker or a chelator molecule, and stability of the linkage has a critical impact on clinical toxicology, pharmacology, and efficacy of the therapeutic (Brumlik *et al.*, 2008). In addition, immunoconjugates can be used for diagnostic applications with an appropriate imaging effector molecule (Kenanova and Wu, 2006).

Currently, three immunoconjugates have been approved by the Food and Drug Administration (FDA). Gemtuzumab ozagamycin is an anti-CD33 ADC targeting calicheamicin for the treatment of relapsed acute myeloid leukemia (AML) (Mylotarg, approved in 2000 but was voluntarily withdrawn by Pfizer in July 2010 due to negative results from post-marketing trials). Ibritumomab tuixetan (Zevalin, Biogen-Idec/Bayer HealthCare AG, approved 2002) is a murine anti-CD20 IgG1 RIC targeting ^{90}Y for the treatment of low grade or follicular, relapsed, or refractory CD20-positive B-cell non-Hodgkin's lymphoma and rituximab-refractory follicular non-Hodgkin's lymphoma. Tositumomab (Bexxar, GlaxoSmithKline, approved 2003) is a second murine anti-CD20 IgG2a RIC targeting ^{131}I for the treatment of

Table 8.1 ADC pipeline

Phase I	Phase II	Phase III	Market
SAR 3419 (anti-CD19) Lymphoma (Sanofi-Aventis w. IMGN*)	huC242-DM4 (CanAG) Gastric Cancer (IMGN)	WAY-207294 (anti-CD22) Lymphoma (Wyeth)	Mylotarg®** (anti-CD33) Leukemia (Wyeth/UCB)
BT-062 (DM4) Myeloma, Biotest with IMGN	huN901-DM1 (anti-CD56) Multiple Myeloma, SCLC (IMGN)	Herceptin-DM1 (anti-Her2) Breast Cancer Her2+ (Genentech/ Roche w. IMGN)	
IMGN388 (anti-integrin DM4) Melanoma, Sarcoma, Carcinoma (IMGN with Centocor)	CR011-vcMMAE (anti-GPNMB) Melanoma, Breast Cancer Curagen w. SGN)		
PSMA ADC Prostate Cancer (Progenics with SGN)	SGN35 (vcMMAE) (anti-CD30) Lymphoma (SGN)		
BIIB-015 (DM4) Cripto-positive solid tumors Biogen Idec with IMGN			
CA9-ADC Bayer HealthCare (with SGN)			

* IMGN = Immunogen; SGN = Seattle Genetics.
** Mylotarg, approved in 2000 but was voluntarily withdrawn by Pfizer in July 2010 due to negative results from post-marketing trials.

CD20-positive non-Hodgkin's lymphoma, with and without transformation, which is refractory to rituximab following chemotherapy. Several others are being tested with trastuzumab-DM1 being the most advanced in clinical development (Phase III; Genentech, Immunogen), and many others to follow (Tables 8.1 and 8.2). Together, they constitute the next wave of antibody therapeutics, some of which are likely to enter the market and ultimately improve patient outcomes. In this chapter we will discuss current approaches with a special focus on the design and application of immunoconjugates for drug and isotope delivery.

Table 8.2 Selected RICs in clinical development or on the market

Antibody	Type	Isotope	Target	Clinical Development	Indication	Reference
Ibritumomab tiuxetan (Zevalin)	mIgG1	^{90}Y	CD20	FDA approv	NHL	Ricart and Tolcher, 2007
Tositumomab (Bexxar)	mIgG2a	^{131}I	CD20	FDA approv	NHL	Ricart and Tolcher, 2007
chTNT-1/B (Cotara)	chIgG1	^{131}I	DNA	Phase II	Glioblastoma multiforma, anaplastic astrocytoma	Shapiro *et al.*, 2006; Jensen *et al.*, 2010
Labetuzumab	huIgG1	^{90}Y	CEA	Pending phase III	Colorectal, breast, lung, pancreatic, stomach	Liersch *et al.*, 2005, 2007
Epratuzumab	huIgG1	^{90}Y	CD22	Phase III	NHL	Milenic *et al.*, 2004
J591	Deimmunized mIgG	^{177}Lu	PSMA	Phase I	Prostate cancer	Tagawa *et al.*, 2010
HuM195	huIgG	^{213}Bi	CD33	Phase III	AML	Gibson *et al.*, 2002
81C6	chIgG	^{211}AT, ^{131}I	Tenascin	Phase I, II	AML, malignant brain tumors and metastases	Zalutsky *et al.*, 2000; Reardon *et al.*, 2006
L19SIP	Rec.antibody format	^{131}I	EDB-FN	Phase II	Haemotologic tumors, Solid tumors (brain mets)	Philogen SpA

CEA, carcinoembryonic antigen; NHL, non-Hodgkin's lymphoma; AML, acute myeloid leukaemia.

8.2 Immunoconjugates

8.2.1 *Antibody drug conjugates (ADCs)*

8.2.1.1 *The concept and composition of ADCs*

ADC technology was triggered by the development and availability of mAbs as well as highly potent cytotoxic agents from different classes and by the development of sophisticated linker chemistry during the last two decades (Wu and Senter, 2005; Chen *et al.*, 2005; Ricart and Tolcher, 2007; Chari, 2008).

Particular challenges in the design and synthesis of ADCs include: (1) identification of appropriate antigens which are selectively overexpressed in tumor tissue and which are efficiently internalized after binding their mAb ligands; (2) specificity of ADC binding to the target tumor antigen; (3) development of an appropriate linker chemistry ensuring efficient linkage of toxophores, high stability of ADCs during circulation and specific toxophore release in tumor tissue; and (4) physicochemical properties of ADCs and tissue penetration. To meet these challenges, ADCs are composed in a modular manner with three structural components (Fig. 8.1).

The targeting moiety is typically a monoclonal antibody which binds to a tumor-specific antigen. All ADCs currently on the market and in the clinic are based on a complete IgG format carrying the toxophore moiety. Currently available PK data (Schmidt and Wittrup, 2009) imply that the IgG molecule is the preferred antibody format for tumor targeting of ADC. Additionally, the IgG format allows a drug load ≥ 4 toxophore per antibody without compromising antibody binding activity. In principle other recombinant antibody formats are possible, but a greater understanding of the therapeutic potential and toxicity profile of ADCs with altered PK profile and metabolism is required. Most important are the target specificity of the antibody and the selection criteria for appropriate antigen targets, as discussed in Section 8.2.1.2 below. Aspects referring to different antibody formats and PK-properties are also discussed in 8.2.2.3.

For the most advanced ADCs in clinical trials their antigen target is specifically expressed on the surface of cancer cells. After antigen recognition and binding, the ADC-antigen complex is internalized into endosomes. A key factor in targeted drug delivery is the extent and rate of ADC endocytosis. Subsequent fusion with lysosomes finally leads to the complete degradation of the antibody and potential recycling of the antigen to the cell surface.

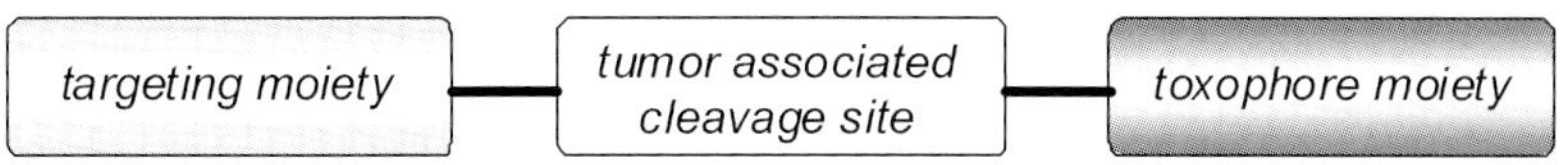

Figure 8.1 Modular composition of ADCs

To achieve a tumor-specific cytotoxic effect exceeding the effect of the mAb alone, a specific effector chemistry consisting of a potent toxophore moiety linked to the antibody via a tumor-selective cleavage site has to be developed. After modification of the mAb with respective effector moieties, an efficient target binding of the ADC needs to be retained. Recent effector chemistry strategies are reviewed in Section 8.2.1.3.

Most of the concepts for tumor-associated cleavage of ADCs leverage the cascade of tumor cell recognition of the mAb component, internalization and subsequent intracellular trafficking of the ADC-antigen complex. Appropriate cleavage mechanisms mediating toxophore release include the reductive cleavage of disulfide bonds, a hydrolytic cleavage of hydrazones, acetals and cis-aconitate-like amides in the acidic pH of the lysosomes, peptide cleavage by lysosomal enzymes (e.g. cathepsins) or a release of the effector moieties subsequently to complete mAb degradation in the lysosomes (Dyba *et al.*, 2004). The latter effector moieties may consist of mAb derived lysine or cysteine moieties which are still carrying the toxophore load (Erickson *et al.*, 2009). This cascade of ADC binding to the antigen, internalization of the complex and subsequent release of the toxophore is illustrated in Fig. 8.2.

Toxophore moieties from various classes and with different modes of action have been employed in ADCs (Hamann *et al.*, 2005). They include DNA-interactive agents (endiynes, duocarmycins, anthracyclins) tubulin binders (maytansinoids, auristatins, taxanes, vinca alkaloids) and topoisomersase inhibitors (camptothecins). After intracellular release from the ADC, the free toxophore may penetrate the cell membrane, either by diffusion or mediated by active transporters. If the physicochemical properties of the toxophores released from cancer cells allow a membrane penetration of cells in the vicinity, these cells may also be hit by the cytotoxic agent. While the initial cytotoxic effect of the ADC clearly depends on antigen-mediated uptake and intracellular toxophore release, the secondary effect of the released toxophore on cells in the vicinity is independent of any antigen expression. This non-antigen-mediated effect on cells in the vicinity is called a "bystander effect", which may be beneficial in the tumor tissue expressing heterogeneous levels of antigen (Kovtun *et al.*, 2006). However, redistribution of the released free toxophore into the circulation may also be associated with increased side effects.

8.2.1.2 *Target selection*

One of the advantages of harnessing mAbs for cytotoxic drug delivery is the high specificity of these biomolecules. Such drug-loaded antibodies will specifically direct their payload to their molecular targets, ideally overexpressed and accessible in the tumor tissue. This will lead to accumulation of the mAb in the tumor tissue and subsequent tumor-specific drug delivery. Methods for identification

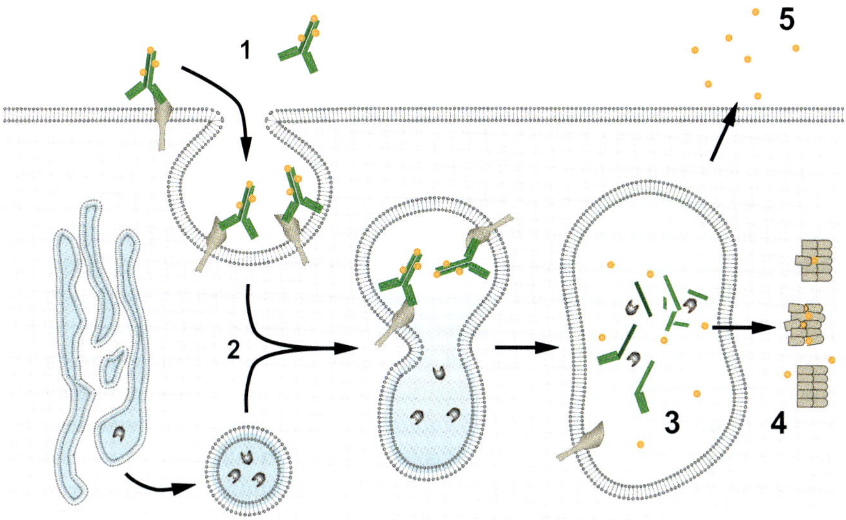

Figure 8.2 Cellular binding, uptake and intracellular trafficking of ADCs (1) After ADC binding to its target antigen, the ADC–antigen complex moves into a clathrin-coated pit (2), which fuses with endosomes, lysosomes and vesicles formed by Golgi. (3) Proteases and hydrolases, activated at the acidic pH of the late endosomes and lysosomes degrade the ADC and cleave the active drug. (4) The drug molecules then cross vesicle membranes and enter the cytoplasm, where they may interact with their intracellular target. (5) Some of the released active drug molecules may also leave the cell and kill neighboring cells if they are able to penetrate the cellular membranes (bystander effect).

and validation of cell surface targets for antibody targeting in oncology were discussed previously (Schultze and Vonderheide, 2001; Carter and Banwell, 2004), and examples of ADC targets were reviewed (Teicher, 2009).

Conceptually, it is well understood that expression of the specific antigen elsewhere in the body might lead to accumulation of the ADC in healthy tissues and that it has the potential to cause unwanted toxicities upon toxophore release (Deonarain, 2008). Thus, tumor specificity in terms of antigen expression is a hallmark of antigens addressed in ADC approaches. Furthermore, these antigens need to be accessible to the ADC via the circulation. Accessibility of the antigen is essential and can influence efficacy and toxicity but is challenging to model without clinical data. Accessibility can vary from tissue to tissue and takes vascularization and organ pressure into account (Rudnick and Adams, 2009). Additionally, for most of the toxophore/linker technologies currently in the clinic, internalization of the antibody plays a significant role (Kovtun and Goldmacher, 2007). Many of the current ADCs rely on the release

Current ADCs in clinical development are based on linker chemistries that release the active toxophore within the lysosome. Thus, the propensity of an antigen-ADC complex on the cell surface to internalize into the cell is a driver of efficacy (Kovtun and Goldmacher, 2007; Chari, 2008; Sun *et al.*, 2009). Three classes of internalization routes have been described: (a) rapid internalization via clathrin-coated pit mediated endocytosis, (b) internalization via caveolae-mediated endocytosis, and (c) internalization via pinocytosis (Grant and Donaldson, 2009; Sun *et al.*, 2009).

The internalization via clathrin-coated pits distinguishes between cell surface antigens (i) continuously co-localized with clathrin-coated pits and ii) those which localize to clathrin-coated pits in a ligand-induced or receptor-mediated manner. The transferrin receptor belongs to the former class and is constitutively and rapidly internalized. However, due to its wide expression, it is not suitable for ADCs (Daniels *et al.*, 2006b; Daniels *et al.*, 2006a; Grant and Donaldson, 2009). The latter class of receptors is triggered by binding of their respective ligands. A prominent example is the EGF-receptor family (Madshus and Stang, 2009; Sorkin and Goh, 2008), among which HER2 (as mentioned above) is the target for the clinically most advanced ADC for treatment of solid tumors, trastuzumab-DM1. This shows that members of this class can be overexpressed in tumors (Rossi *et al.*, 2003; Nahta and Esteva, 2006) and can be associated with tumor biology and tumor progression. Internalization of cell surface antigens which internalize via the caveola, or non-clathrin-dependent pathways, are less well understood and GPI-anchored proteins are among these (Hansen and Nichols, 2009). The endosomal fate of such cell surface antigens is reported to be the endoplasmic reticulum and the Golgi, compartments not characterized by proteolytic activity (Razani *et al.*, 2002). Lack of proteolytic activity within the endosome has direct implications for the design of linker chemistry and efficacy of ADCs targeting such cell surface molecules. Lastly, internalization via pinocytosis represents the background activity of any cytotoxic molecule in the extracellular space not specifically binding to the cell membrane (Jedema *et al.*, 2004). This activity can be measured as cytotoxic background activity of control ADCs *in vitro*. Internalization rates between the classes of target molecules can differ by orders of magnitude and must greatly influence efficacy. Nevertheless, a conclusive dataset describing the correlation between potency and internalization rates across various target classes has not yet been published.

Therefore, selecting the target antigen of an ADC is critical, but there are no firm rules. The decision must be based on a careful balance of the specificity of an antigen for tumor expression, its heterogeneity in tumor tissue, its accessibility from the circulation and its rate of internalization. Similar criteria apply for the selection of RIC targets (discussed in detail in Section 8.2.2). However, the targeted antigen in RIC approaches does not necessarily need to be internalized. Again, tumor selectivity, accessibility and a critical antigen density are crucial to allow for high total dose delivery. In most approaches, molecules on the tumor cell surface are targeted, some of which are internalized (Carter and Banwell, 2004). Alternative approaches to those described above employ highly selective and abundantly expressed antigens present in the tumor extracellular matrix such as Tenascin-C (Ventimiglia *et al.*, 1992) or EDB-fibronectin (Menrad and Menssen, 2005), or DNA, which becomes accessible in necrotic areas of the tumor (Shapiro *et al.*, 2006).

8.2.1.3 *Effector chemistry*

8.2.1.3.1 Toxophore moieties

Despite the availability of a large number of drugs for traditional cancer therapy (i.e. non-targeted application of cytotoxic agents based on the principle that rapidly proliferating tumor cells are more susceptible to poisoning), few are suitable as "warheads" in ADCs due to several special requirements.

One of the foremost criteria for a suitable toxophore is high potency. This is due to two distinct peculiarities of ADCs. First, only a limited number of cytotoxic molecules can be loaded onto an antibody without compromising its antigen binding affinity or impairing its physicochemical properties, which could potentially lead to aggregation and precipitation. Second, as ADCs are internalized into the cell only after binding to antigens on the cell surface, the limited number of antigens per cell (typically $<1 \times 10^5$/cell) (Chari, 2008) restricts the attainable drug concentration within the cell to about 10^{-7} M (Chen *et al.*, 2005; Chari, 1998). This necessitates the warhead to have subnanomolar potency in order to be effective. The majority of current clinically used cytotoxic agents do not meet this requirement. For example, the commonly used chemotherapeutics vincristine and doxorubicin have IC_{50} values typically in the one- to three-digit nanomolar range (Ferlin *et al.*, 2005; Stefanska *et al.*, 1999), and other drugs from the alkylating agent and antimetabolite classes such as chlorambucil or methotrexate need even higher concentrations to be effective (Endo *et al.*, 1987). An issue with potency is that cancer cells may develop resistance by upregulation of drug efflux pumps, and thus a toxophore non-susceptible to multiple drug resistance would be highly desirable. Another important criterion is the chemical stability of the cytotoxic agent. It must be stable not only in the bloodstream but, more importantly, also under the acidic and proteolytic conditions of the lysosomes where the antibody is decomposed and the toxophore is released (Chen *et al.*, 2005). A suitable toxophore should also be free of immunogenicity. Some cytotoxic agents of large molecular size, such as bacterial or plant toxins such as diphtheria toxin (Collier, 1975) or ricin (Lord *et al.*, 1994) can potentially be recognized and destroyed by the host immune system, rendering them ineffective. The fourth requisite is the presence of adequate attachment sites such as hydroxyl, amino, carboxyl, or other groups that allow the chemical ligation of the toxophore to the antibody via a suitable linker. As can be seen from the above list of criteria, only a small proportion of the known cytotoxic agents are eligible for conjugation to antibodies; a selection of those under current scrutiny will be discussed below.

Calicheamicins In 1987, the isolation and structural characterization of a series of bacterial metabolites derived from *Micromonospora echinospora* ssp. *calichensis*, found in a soil sample from Texas, was described by scientists from the Lederle Laboratories (now a subsidiary of Pfizer). Apart from their unusual structures featuring an enediyne subunit, they were particularly noteworthy for their exceedingly high cytotoxicity, outclassing doxorubicin by more than three orders of magnitude in murine tumor models. They were named calicheamicins, with Calicheamicin γ_1^I being the most prominent representative (Lee *et al.*, 1987b; Lee *et al.*, 1987a; Lee *et al.*, 1989).

The calicheamicins feature a unique mode of action, inducing double-strand breaks after binding to the minor groove of DNA (Dedon and Goldberg, 1992; Lee *et al.*, 1987b; Lee *et al.*, 1987a; Lee *et al.*, 1991; Salzberg and Dedon, 2000). The mechanism by which double-stranded DNA is cleaved by the calicheamicins starts with the attack of a nucleophile (usually glutathione) at the trisulfide group, setting free a thiolate that undergoes an intramolecular Michael addition with the reactive α,β-unsaturated ketone. The resulting tricyclic intermediate undergoes a Bergman cyclization (Bergman, 1973), leading to a benzene diradical which can in turn interact with DNA to produce a DNA diradical. The DNA diradical is unstable and readily cleaved by O_2, which leads to scission of the chain (Fig. 8.3).

The DNA damage brought about by the calicheamicins leads to apoptosis in most cases, since double-strand DNA breaks are usually irreparable. The extraordinarily high potency of calicheamicin and calicheamicin analogs has led to their evaluation as new cancer therapeutics, but unfortunately they suffer from a very narrow therapeutic index and delayed toxicity, which prohibits further development as independent compounds. However, calicheamicin γ_1^I meets most of the above-mentioned requirements for conjugation with an antibody. It is extremely potent, requiring only sub-picomolar concentrations to be effective. It is also sufficiently stable, not immunogenic and contains a trisulfide group which not only represents the "match cord" whose degradation triggers the Bergman cyclization, but also constitutes an excellent attachment point for linkage with an antibody. It therefore clearly offers the potential for enlargement of the therapeutic window by targeted delivery to tumor cells. Because of these favorable properties, calicheamicin γ_1^I-antibody conjugates have been under intense investigation for several years, culminating in FDA approval of gemtuzumab ozagamicin (Mylotarg®).

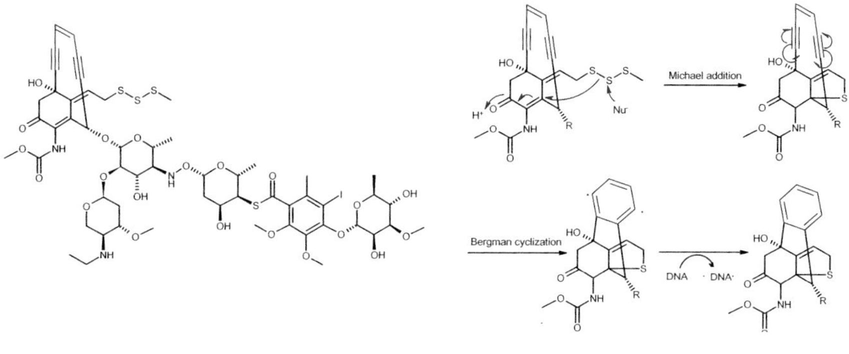

Figure 8.3 Calicheamicin γ_1^I and its mechanism of DNA cleavage

Auristatins The first description of a highly toxic sea hare of the genus *Dolabella* dates back to the year 60 A.D. (Pliny, Historia Naturalis, Lib.9, Lib.32, ca. 60 A.D.); extracts from this animal (later named *Dolabella auricularia*) were already renowned for their toxic properties during the reign of Nero (A.D. 37–68) (Pettit, 1997), when they were successfully employed for poisoning of competitors. For the following nearly two millennia, the potential of the Indian Ocean sea hare with respect to medical application remained unrecognized, until in 1972 a research group led by George R. Pettit began to isolate and characterize extremely active cytotoxic substances from *D. auricularia* (Pettit, *et al.*, 1976; Pettit, 1997). The peptidic compounds were called dolastatins; the most potent constituent was dolastatin 10 (Fig. 8.4; Pettit *et al.*, 1987), a very strong tubulin polymerization inhibitor. At the time of its discovery, dolastatin 10 was the most potent antineoplastic substance known ($ED_{50} = 4.6 \times 10^{-5}$ µg/mL against PS leukemia cells). *D. auricularia* contains only minute amounts of dolastatin 10 (~1 mg from 100 kg of sea hare) (Pettit *et al.*, 1981), and therefore extraction does not constitute a reasonable means of accessing considerable quantities of the natural product. Fortunately, total synthesis allowed access to larger amounts of dolastatin 10 (Pettit *et al.*, 1989), permitting its evaluation *in vivo*. However, none of the clinical trials pursued to date have supported the hope for effective cancer treatment with dolastatin 10 (Vaishampayan *et al.*, 2000; Hoffman *et al.*, 2003; Kindler *et al.*, 2005).

The peptidic nature of the dolastatins allows facile access to derivatives with altered pharmacodynamic, PK, and physicochemical profiles by *de novo* synthesis. Efforts directed to the development of such derivatives have resulted in the discovery of the auristatins, with monomethyl auristatin E (MMAE) and monomethyl auristatin F (MMAF) as two prominent examples of this class of compounds (Doronina *et al.*, 2003; Doronina *et al.*, 2006; Pettit and Barkoczy,

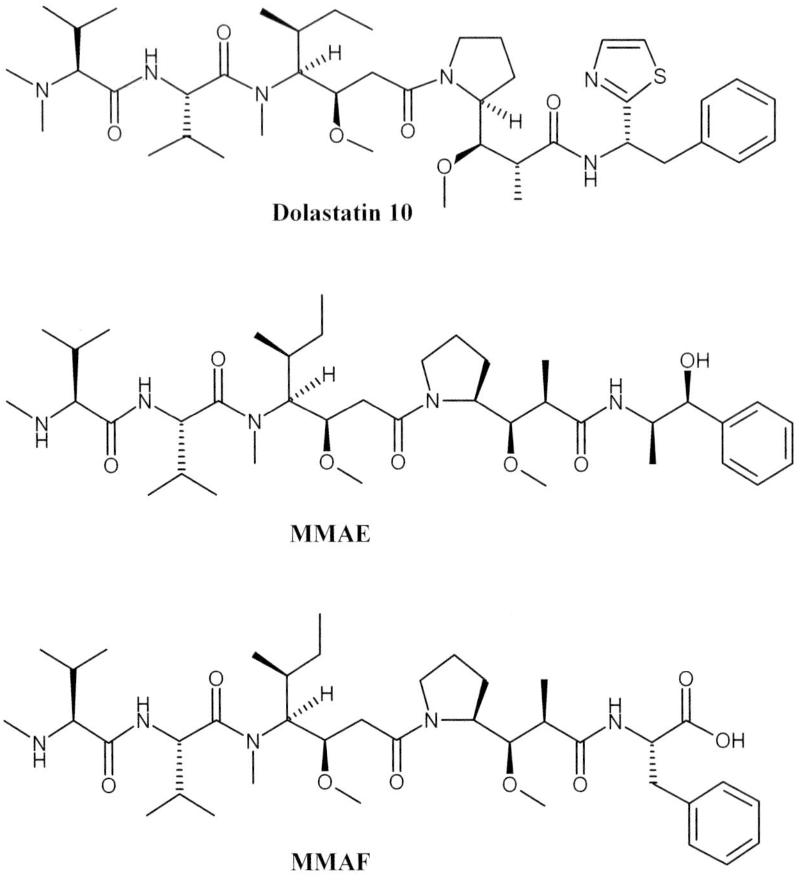

Figure 8.4 Dolastatin 10 and the synthetic auristatins MMAE and MMAF

1997), which offer advantages over dolastatin 10 in terms of solubility and synthetic accessibility (Fig. 8.4).

Special attention is being directed to the incorporation of auristatins into antibody-conjugates, since the synthetic compounds are not only highly stable (much more than the calicheamicins) and meet the potency criteria for toxophores in ADCs, but they also contain several attachment sites: MMAE can either be conjugated via the C-terminal hydroxyl group or the N-terminal secondary amine. MMAF shares the N-terminal secondary amine and can potentially also be conjugated via the C-terminal carboxy group.

Numerous conjugates of MMAE and MMAF have been prepared and will be discussed in Section 8.2.1.3.2.

Maytansinoids In 1972, a paper by Kupchan appeared describing the isolation and structural elucidation of a highly potent cytotoxic constituent of the fruits, roots and stem bark of the *Maytenus serrata* shrub (Kupchan *et al.*, 1972). The compound, called maytansine by its discoverers, is a chlorine-containing macrolactam with an epoxide moiety (Fig. 8.5). Initial *in vitro* cytotoxicity studies showed it has median ED_{50} values in the range of 0.6–2 nM (Wolpert-Defilippes *et al.*, 1975).

It was found that maytansine is, like the dolastatins and auristatins, a very potent inhibitor of tubulin polymerization (Remillard *et al.*, 1975). This prompted investigation of maytansine as a new cancer treatment option, but similar to the aforementioned toxophores, maytansine failed to live up to the high expectations. Clinical trials with maytansine showed only marginal benefit at tolerated doses (Issell and Crooke, 1978; Blum and Kahlert, 1978), and therefore further development of maytansine was not pursued. Numerous semi-synthetic derivatives of maytansine, named maytansinoids, have been prepared, establishing clear structure–activity relations (Cassady *et al.*, 2004). These identify the C3 *N*-acyl-*N*-methyl-L-alanyl ester side chain, the C4-C5 epoxide, the C9 carbinol, and the C11 and C13 double bonds as crucial for potency; however, the nature of the *N*-acyl group can be varied without significantly compromising biological activity (Kupchan *et al.*, 1978). This makes the *N*-acyl group a suitable attachment point for conjugation with an antibody. Several approaches have been undertaken in incorporating maytansinoids into ADCs, the most advanced of which are currently in clinical Phases II and III and will be discussed in Section 8.2.1.3.2.

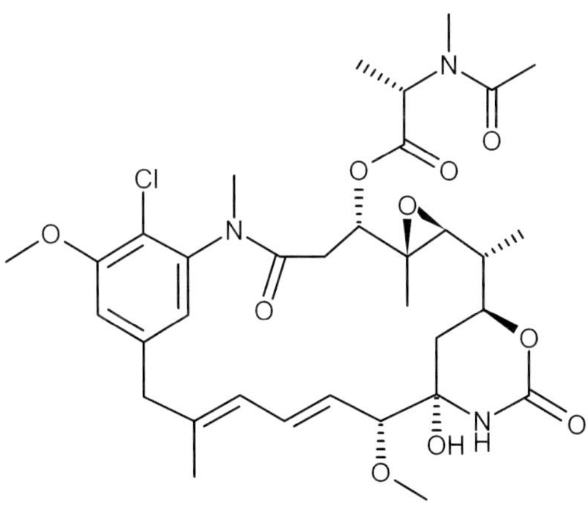

Figure 8.5 Maytansine

Further toxophores Apart from the above-mentioned major toxophore classes used in ADCs, several other compounds have been or are currently being investigated in conjunction with targeted cancer therapy. To date none of them have been successful in clinical trials, which can be attributed to insufficient potency in most of the cases. However, as more potent derivatives are currently being investigated, a final candidate cannot yet be named. Table 8.3 gives a concise overview of these toxophores and their application in ADCs.

Apart from these toxophore classes, a large number of further cytotoxic natural products and synthetic derivatives thereof are being investigated with respect to their usefulness in targeted cancer therapy. As a "one size fits all" toxophore can never exist due to the heterogeneity of cancer, the identification of further cytotoxic compounds with different mechanisms of action is highly desirable, in particular in the context of ADCs.

8.2.1.3.2 Choice of linker and case studies

Toxophore attachment to ADCs and impact of drug load on the ADC profile The major goal of ADC delivery strategies is to combine the specificity of a monoclonal antibody for binding and uptake into tumor cells with the high potential of cytotoxic compounds to eradicate respective target cells. To achieve this goal, those cytotoxic agents have to be attached covalently to the antibody molecules. In most cases this conjugation is hampered by the dramatically different physical properties of the antibody, preferring the buffered aqueous environment and the often highly lipophilic drug molecules with sparing solubility under these conditions. Typically the protein side chains of solvent accessible lysine or cysteine residues are employed for toxophore attachment.

The toxophore attachment to lysine side chains is possible without prior processing of the antibody molecule. Appropriate lysine side chain directed coupling chemistry strategies preferably include activated ester derivatives of a linker-modified drug. About 80 lysine residues are present on an IgG, but only about ten lysine side chains are preferentially accessible for chemical modification. This lysine attachment results in a random loading of the mAb with average drug loads (toxophores per mAb), typically in the range between two and eight. Drug loading of the mAb may have a major impact on binding, aggregation state, PK and biodistribution of the resulting ADC. The optimal drug load to obtain an appropriate physicochemical profile and a good efficacy/safety balance differs with each individual ADC, but once determined the optimal load can be adjusted in a reproducible manner depending on the chemistry employed and the reaction conditions (Oflazoglu *et al.*, 2008; Hamblett *et al.*, 2004). Another factor that limits the drug load is the tendency to self-aggregation of the linked toxophores as found for calicheamicin

Table 8.3 Other toxophores used in ADCs

Compound Class (Example)	Structure	Mechanism of Action	References	Development Status
Anthracyclines (Doxorubicin)	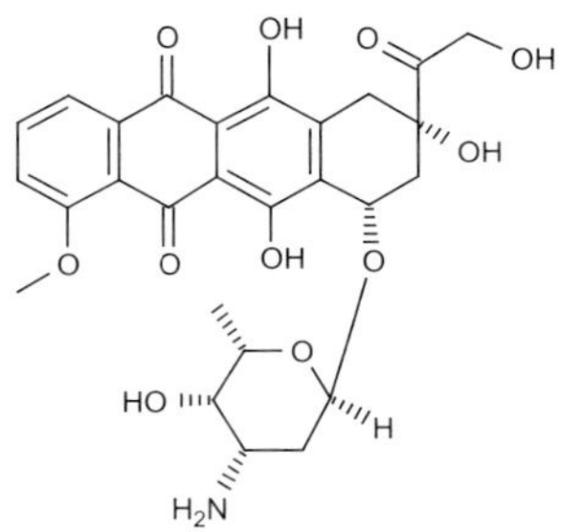	DNA intercalation	Smith *et al.*, 2001; Trail *et al.*, 1993; King *et al.*, 2003	Discontinued/no development reported
Taxanes (paclitaxel)	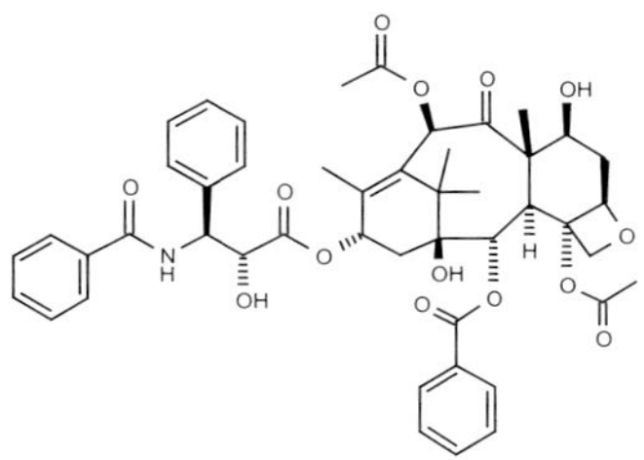	Tubulin stabilization	Guillemard *et al.*, 2001; Safavy *et al.*, 2003; Chen *et al.*, 2005; Ojima *et al.*, 1996, 1997a, 1997b, 1998, 1999; Miller *et al.*, 2004a, 2004b; Baloglu *et al.*, 2004	No clinical development reported
Cyclopropaindoles (duocarmycin A)	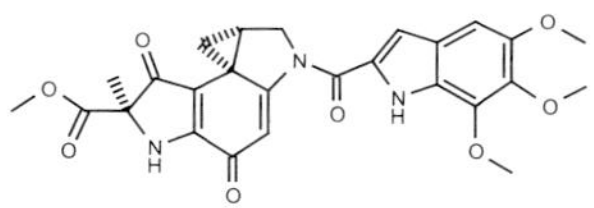	DNA alkylation	Ducry and Stump, 2010	Phase I

(*Continued*)

Table 8.3 (*Continued*)

Compound Class (Example)	Structure	Mechanism of Action	References	Development Status
Quinoline alkaloids (camptothecin)	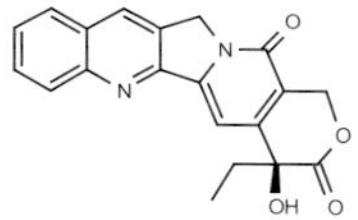	Topoisomerase I inhibition	Walker *et al.*, 2002; Lerchen *et al.*, 2001; Burke *et al.*, 2009	No clinical development reported
Vinca alkaloids (vincristine)	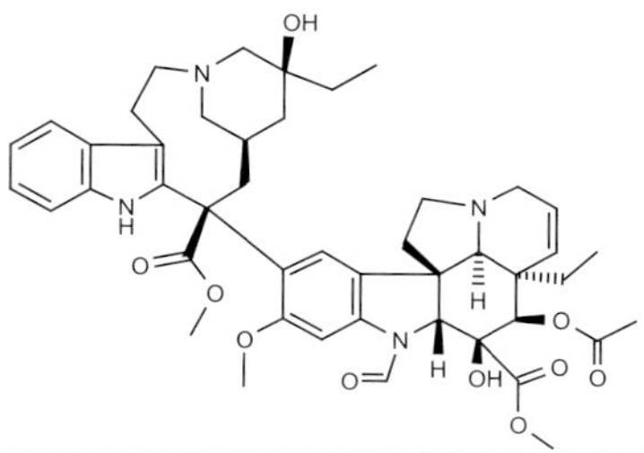	Tubulin inhibition	Patents US4845200 and S5094849	No clinical development reported

conjugates (Hamann *et al.*, 2002). A laborious optimization of the conjugation process in the presence of additives as performed with Mylotarg (Hollander *et al.*, 2008) may be required to provide monomeric ADCs and may differ in each individual case depending on the antibody, linker and drug.

Toxophore attachment to cysteine side chains typically requires a reduction of the interchain disulfides. IgG1 molecules contain 12 intrachain and four interchain disulfide bridges. The latter are more susceptible to reduction by agents such as dithothreitol (DTT), and a consistent number of reactive thiol groups can be generated (Willner *et al.*, 1993). Therefore, cysteine-linked immunoconjugates are often more homogeneous compared to lysine modified ADCs. Cysteine-directed coupling chemistry strategies leverage the nucleophilicity of the thiol functional groups that readily react with haloacyl moieties, or are more preferred in a Michael addition with maleinimide functionalized linker moieties attached to the drug. After reduction of the interchain disulfides, a maximum drug load of eight can be achieved without a negative impact on antigen binding. Despite the fact that binding is not affected, drug load was shown to be a key design parameter for ADCs. This is shown in a systematic investigation of the impact of different drug loads on the profile of anti-CD30-MMAE ADCs (Hamblett *et al.*, 2004). Although the *in vitro* potency of ADCs increases with higher drug load, this is not reflected *in vivo*. A drug load of four was found to be optimal with respect to efficacy and therapeutic window, and led to the development of SGN-35 (Seattle Genetics/Millenium). ADCs with a higher drug load of eight were found to be more toxic and displayed a three- and five-fold higher clearance compared to ADCs with loads of four and two, respectively (Hamblett *et al.*, 2004).

Some work has been directed towards recombinant strategies for the production of ADCs with predetermined sites for drug attachment, either by replacing intrachain cysteines of anti-CD30 cAC10 antibody by serines (McDonagh *et al.*, 2006), or by incorporation of cysteine residues into anti-Muc16 antibody at defined sites for drug conjugation (Junutula *et al.*, 2008). Junutula *et al.*, with the anti-Muc16 "Thiomab" approach, obtained ADCs with a drug load of 1.6, which were shown to be equipotent but better tolerated than conventional ADCs with a drug load of 3.1.

Strategies to enable tumor-associated toxophore release and requirements for linker stability To achieve a beneficial impact on both efficacy and tolerability of respective ADCs, the toxophore moiety and the particular linker have to meet several requirements. Due to the long half-life of IgGs and IgG-derived ADCs in circulation, a high stability of toxophore and linker has to be ensured. To specifically unfold the cytotoxic potential solely at the site of action in tumor tissue, toxophore release is preferably achieved inside the tumor cells. Typically, antigen binding of ADCs is associated with an efficient internalization into the tumor cells which ends up in the lysosomal compartments (Fig. 8.2). These intracellular compartments are highly equipped with an enzyme machinery working at a decreased pH of approximately 4.5–5.5. This internalization route and subsequent intracellular trafficking provides the rationale for the development of several release mechanisms of free drugs from ADCs.

Acid sensitive linkers and case studies Several ADCs were designed with an acid-labile cleavage site to enable toxophore release under the acidic pH environment inside the lysosomal compartments (Garnett, 2001). However, acid-labile cleavage is a rate-dependent phenomenon proportional to pH with a ten-fold increase of cleavage at a decrease of pH of one unit; as a consequence, a high rate of lysosomal cleavage is always accompanied with a certain amount of toxophore released in the circulation. Cis-aconityl functionalities (Shen and Ryser, 1981), ortho esters, acetals, ketals (Srinivasachar and Neville, 1989), acid labile esters and hydrazones have been investigated as potential cleavage sites. However, only the latter have been used in clinically tested ADCs.

One example is *BR96-Dox* (Trail *et al.*, 1993), a conjugate of doxorubicin and the mAb BR96. It targets a Lewis-Y-related antigen on the surface of tumor cells and has been developed for the potential treatment of lung cancer. Due to the limited potency of the drug, ADC doses in pre-clinical models exceeded 100 mg/kg, but it did achieve tumor eradication. The drug was investigated in Phase II studies against colon and lung cancer and in Phase III in against breast cancer. Only limited anti-tumor activity could be demonstrated in the clinic, which has been attributed to the modest potency of the drug, instability of the linker, and the presence of the LeY antigen in the gut, leading to gastrointestinal toxicities (Fig. 8.6).

Another example is gemtuzumab ozogamicin (Mylotarg®), which contains an acid-labile hydrazone linker between the CDR-grafted humanized mAb P67.6 targeting the CD33 antigen on myeloid cells and N-acetyl calicheamicin γ_1^I

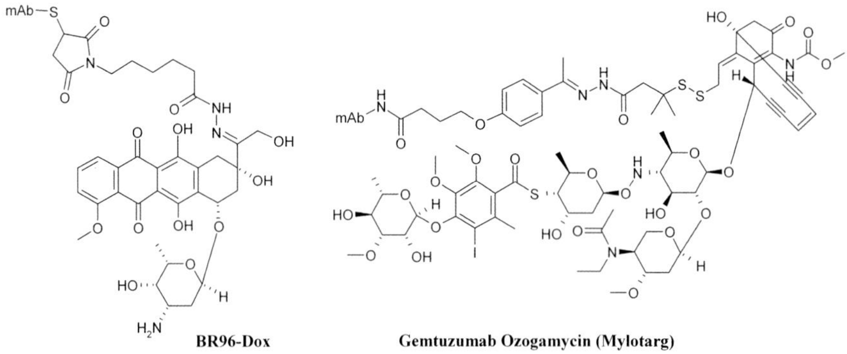

BR96-Dox **Gemtuzumab Ozogamycin (Mylotarg)**

Figure 8.6 Hydrazone linker in immunoconjugates BR96-Dox and Mylotarg

(Hamann *et al.*, 2002). The conjugation technology is rather sophisticated. The mAb is coupled to 4-(4-acetyl-phenoxy) butanoic acid via an amide bond. The acetophenone carbonyl is then utilized for conjugation with calicheamicin via an alkanoylhydrazone linker, which is in turn connected to a disulfide moiety. Although both the hydrazone and the disulfide represent potential cleavage sites, it was shown that the hydrazone bond is hydrolyzed before the disulfide is reduced (Hamann *et al.*, 2002).

The approved drug consists of a 1:1 mixture of unconjugated P67.6 and P67.6 conjugated to 4–6 mol of N-acetyl calicheamicin γ_1^I, resulting in an average drug loading of two to three calicheamicin molecules per mAb (Bross *et al.*, 2001). (A higher drug loading is prohibited due to unfavorable alteration of the conjugate's physicochemical properties, leading to aggregation and precipitation.) It is approved for the treatment of acute myeloid leukemia (AML) at doses of 9 mg/m^2 given twice at two-week intervals (Larson *et al.*, 2005; Bross *et al.*, 2001; Giles *et al.*, 2001). However, in July 2010, Pfizer announced a voluntary withdrawal of the product from the market, as post-marketing trials had shown an increased death rate among AML patients receiving Mylotarg® in addition to chemotherapy. It is conceivable that non-specific drug release due to the instability of the linker may have contributed to this unfavorable outcome.

A new calicheamicin conjugate that binds to the CD22 antigen on normal B cells and on B cell lymphomas, inotuzumab ozogamicin, employs the same conjugation technology as in Mylotarg and is currently in Phase III clinical trials (DiJoseph *et al.*, 2004; DiJoseph *et al.*, 2005).

Enzymatically cleavable linker and case studies To increase ADC stability during systemic circulation, peptide-based linker technologies have been developed and utilized for attaching various drugs to antibodies. These peptide spacers are designed with a cleavage site susceptible for proteolytic cleavage within the lysosomes. Cathepsin B is one of the abundant lysosomal enzymes identified as an appropriate protease to cleave ADCs. Simple dipeptide sequences such as Phenylalanine–Lysine and Valine–Citrulline have been identified, initially in doxorubicine conjugates with BR96, as appropriate cleavage sites with high plasma stability and good cathepsin cleavage (Dubowchik *et al.*, 2002). To further enhance cleavage efficiency and rate of toxophore release, self-immolative spacer groups such as p-aminobenzyloxycarbonyl have been introduced, which facilitate toxophore release via fragmentation subsequent to enzymatic cleavage (Carl *et al.*, 1981). This linker chemistry has been adapted to the auristatin family of toxophores, a highly potent and stable class of tubulin inhibitors that interact with the vinca-alkaloid binding site of tubulin. A methyl group from the amino terminus of auristatin E could be removed without loss of potency, and the resulting monomethyl auristatin E derivative now allows for attachment of peptide linker.

Conjugation with different antibodies has been performed, and the anti-CD30 immunoconjugate (cAc10-Val-Cit-PABC-MMAE; SGN-35) has gained high interest in clinical Phase I studies in patients with Hodgkin's lymphoma (Doronina *et al.*, 2003). The SGN-35 ADC and the mechanism of drug release are illustrated in Fig. 8.7. The effector moiety is attached to cysteine residues of the anti-CD30 cA10 antibody, and an average drug load of four was established to achieve the optimal ADC profile (Hamblett *et al.*, 2004; Doronina *et al.*, 2003). Monomethyl auristatin E is released after ADC internalization into the lysosomes upon cathepsin B cleavage and subsequent fragmentation of the p-aminobenzyloxycarbonyl moiety. In Phase I clinical trials, among 22 patients treated every three weeks with SGN-35 at doses between 1.2 and 3.6 mg/kg that could be evaluated, 23% complete regressions and 45% objective responses were observed (Senter, 2009). SGN-35 (brentuximab vedotin) is currently being evaluated in Phase II trials for Hodgkin's lymphoma and anaplastic large cell lymphoma (Seattle Genetics and Millenium).

Reductive cleavable linker and case studies Alternatively to hydrolytic and proteolytic linker cleavage and toxophore release from ADCs, the reductive cleavage of disulfide bonds has been explored as a potential cleavage site in linker moieties. This strategy follows the rationale that disulfide-containing linkers are stable at physiological pH and are susceptible to scission mediated by reduced glutathione, which is reported to be in millimolar concentrations inside of cancer cells, but only in micromolar concentrations in the circulation (Meister and Anderson, 1983). Such linkers with disulfide bridges are particularly preferred by research groups at Immunogen Inc. and were employed in

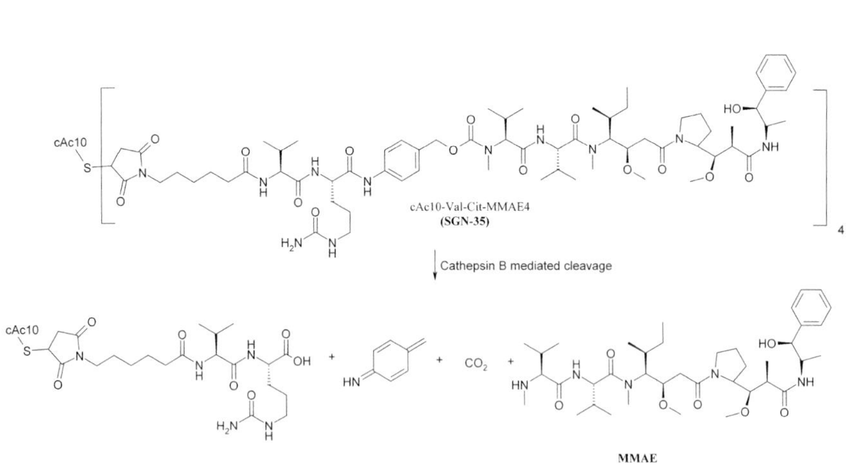

Figure 8.7 Composition and enzyme mediated drug release of SGN-35

ADCs with maytansine-derived toxophore moieties (Chari, 2008). The semi-synthetic preparation of such ADCs starts from a mixture of ansamitocins obtained by fermentation of the antibacterial actinomycetes strain *Actinosynnema pretiosum*. Subsequent reductive cleavage of the resulting product mixture of C3 esters provides maytansinol. Starting from maytansinol ADCs are synthesized in a multi-step procedure with formation of the disulfide bridge as the final step (Fig. 8.8; Widdison *et al.*, 2006).

The exact composition of the disulfide linker with respect to length and methyl group substitution pattern of the alkyl chains has a significant impact on stability and activity of respective ADCs (Widdison *et al.*, 2006).

Two ADCs with different cleavable disulfide linkers are in Phase I and II clinical trials (Fig. 8.9). The SPP linker with a methyl group on the antibody side of the disulfide is employed to link DM1 with the anti-CD56 antibody huN901 to target CD56-expressing solid tumors such as small-cell lung cancer (SCLC), Merkel cell carcinoma (MCC), ovarian cancer, carcinoid, and other neuroendocrine tumors as well as multiple myeloma (MM) and certain other types of hematological cancers (Woll *et al.*, 2009). The SPDB linker with two geminal methyl groups on the drug side of the disulfide is employed to link DM4 with the humanized version of the murine monoclonal antibody C242 that binds the CanAg antigen, which is highly expressed on the surface of colon, pancreatic, gastric and some lung cancers.

In a side-by-side comparison of ADCs with the mAb huC242 via different linkers, the SPDP-DM4 linker shows a superior profile. Greater anti-tumor activity of huC242-SPDB-DM4 is attributed to higher stability of the disulfide

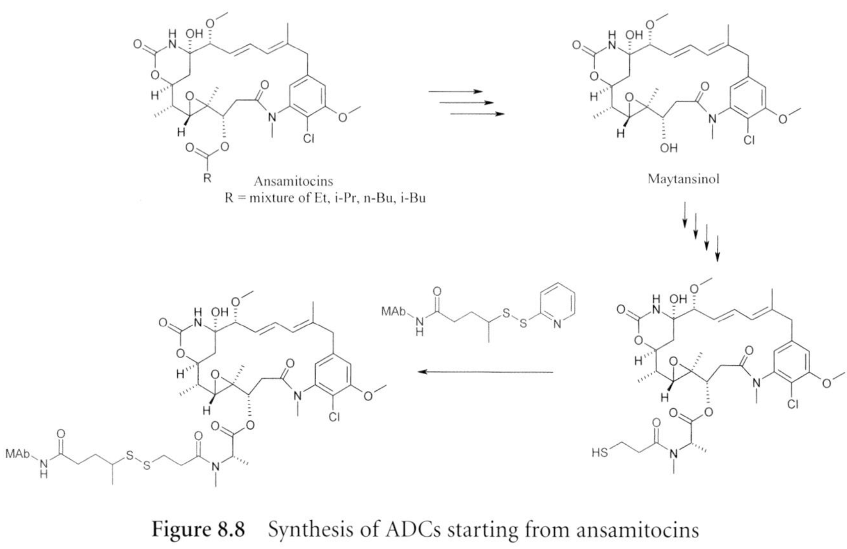

Figure 8.8 Synthesis of ADCs starting from ansamitocins

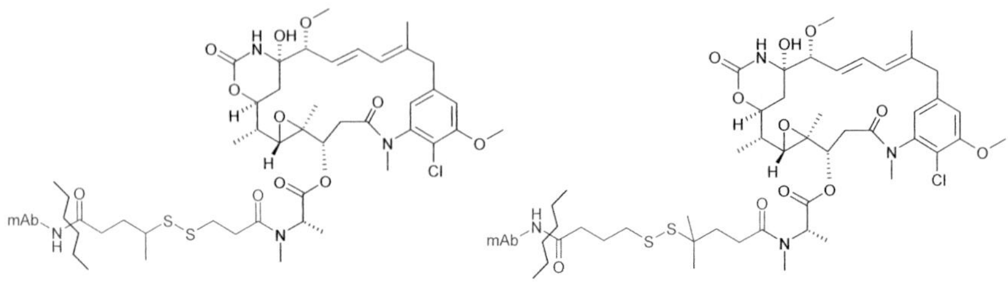

mAb-SPP-DM1
utilized in huN901-SPP-DM1 (IMGN901)

mAb-SPDB-DM4
utilized in huC242-SPDB-DM4

Figure 8.9 Maytansinoid-derived ADCs with reductively cleavable disulfide linker moieties

bond in circulation and higher stability of the DM4 drug with the thiol at a tertiary carbon center (Widdison *et al.*, 2006). Further investigation of the intracellular metabolism suggested S-Methyl DM4 as the highly active metabolite, which also exhibits a pronounced bystander effect (Erickson *et al.*, 2006).

Linker without predetermined breaking points and case studies Various ADCs with different antibodies and toxophores from different drug classes have been synthesized without a predetermined breaking point. Consequently, those ADCs are characterized by high stability in systemic circulation. After antigen binding and internalization the ADCs are exposed to the efficient enzyme machinery within the lysosomes. The degradation of the ADCs in these cellular compartments may result in the formation of active intracellular metabolites and subsequently in tumor-specific cell killing. The most prominent example is trastuzumab-DM1 (Immunogen and Genentech), where the maytansinoid drug DM1 is linked to the HER2-binding antibody trastuzumab via a non-cleavable thioether linker (Burke *et al.*, 2009; Lewis Phillips *et al.*, 2008). The active intracellular metabolite is a toxophore-linker-modified lysine derivative, which may be formed after complete antibody degradation within the lysosomes (Fig. 8.10; Erickson *et al.*, 2006). Metabolite is only weakly active when added to the culture medium of tumor cells, probably due to its inability to cross the cellular membrane. Trastuzumab-DM1 ADC is currently in Phase III clinical trials as second-line treatment for HER2-positive metastatic breast cancer.

ADCs with non-cleavable linkers have been further investigated in combination with the auristatin class of toxophores. Interestingly, in contrast to the MMAE mentioned above, MMAF has been shown to be compatible with non-cleavable maleidocaproyl linkers (Fig. 8.11; Doronina *et al.*, 2006; Alley *et al.*, 2009; Oflazoglu *et al.*, 2008). MMAF with a charged carboxy group at the C-terminal end is more than 100-fold less cytotoxic compared to MMAE against a panel of tumor cells in cytotoxicity assays *in vitro*. This may be due to its low ability to penetrate cellular

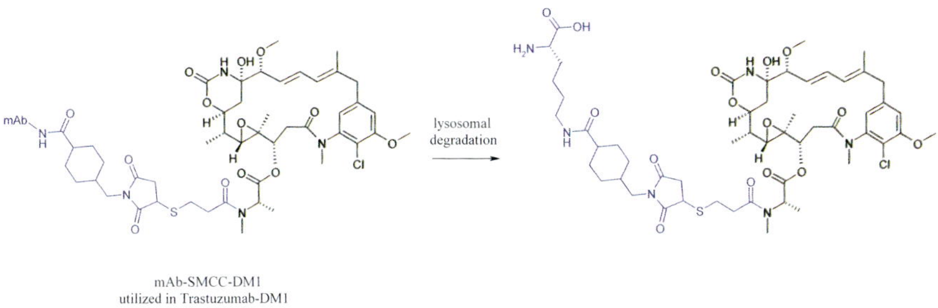

Figure 8.10 Trastuzumab-DM1 and bioactive intracellular degradation product

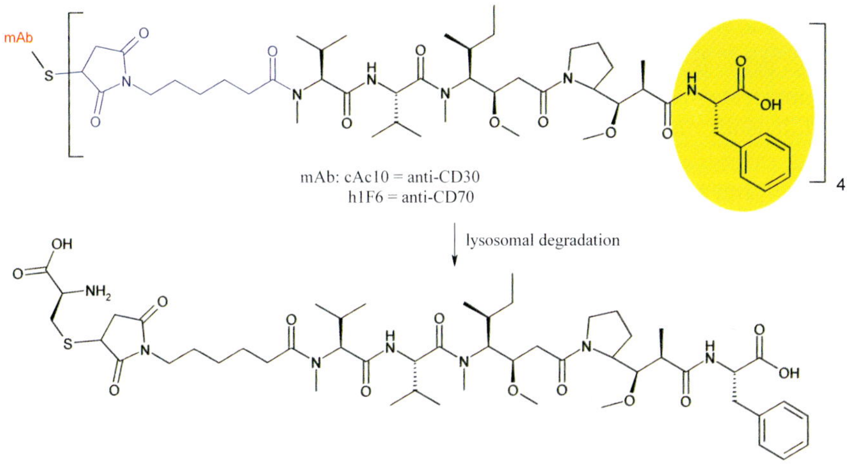

Figure 8.11 Monomethyl auristatin F linked to cysteine side chains of mAbs via non-cleavable maleidocaproyl linker and bioactive cellular degradation product

membranes of tumor cells (Doronina *et al.*, 2006). However, when this MMAF toxophore is linked to internalized antibodies such as cAC10 (anti-CD30) or h1F6 (anti-CD70), immunoconjugates are obtained with more than 1,000-fold higher anti-proliferative activity. Interestingly, this effect is achieved with both a cathepsin-cleavable linker (vc-linker) composed as described above for SGN-35 with MMAE (Fig. 8.12) and a non-cleavable maleimidocaproyl (mc) linker. As detected by LC–MS (liquid chromatography coupled with mass spectrometry) the cellular degradation product of the ADC with the non-cleavable mc linker is a toxophore-linker-modified cysteine derivative, which may be formed after complete ADC breakdown in the lysosomes (Fig. 8.11; Doronina *et al.*, 2006). This lysosomal ADC degradation leading to the release of an active metabolite only inside of tumor cells may translate to an improved safety profile of respective ADCs.

This ADC design with a stable linker in combination with MMAF as effector moiety has been utilized in an anti-CD70-directed ADC (h1F16-mc-MMAF), designated as SGN-75. SGN-75 effects tumor regressions in a panel of CD-70-expressing patient-derived renal cell carcinoma starting at dose levels of 0.5 mg/kg (Oflazoglu *et al.*, 2008). Biodistribution studies with dual-labeled anti-CD70 h1F6 MMAF immunoconjugate (^{14}C label in drug; ^{3}H label in Ab) in nude mice with CD-70-positive 786-O tumors confirmed specific tumor targeting, tumor retention of cleaved low molecular weight MMAF metabolite, and ADC stability in circulation (Alley *et al.*, 2009).

A Phase I study to evaluate the safety and tolerability of SGN-75 in patients with CD70-positive non-Hodgkin's lymphoma or metastatic renal cell carcinoma was started in September 2009.

Target dependency of linker design The individual case studies mentioned above demonstrate that the question of the optimal linker design in ADCs is still open and depends on numerous factors. ADCs are in clinical development utilizing either cleavable or non-cleavable linkers. Cleavable linkers may have a benefit in efficacy and bystander effect, whereas non-cleavable linkers may be advantageous with respect to an enhanced safety margin (Doronina *et al.*, 2006). Recent systematic studies (Polson *et al.*, 2009) examined the impact of linker and drug on efficacy, pharmacokinetics and tolerability in various ADCs directed against seven different targets for the treatment of non-Hodgkin's lymphoma (CD19, CD20, CD21, CD22, CD72, CD79b and CD180). In general, cleavable linkers tend to be more robust, and respective ADCs are active against a broad panel of targets. In contrast, the efficacy of ADCs with non-cleavable linkers is more selective for the respective targeted cell population. Target-associated properties such as a high internalization rate, high copy number and homogeneous expression in tumor tissue may be beneficial for such non-cleavable ADCs, which require efficient cellular uptake and subsequent lysosomal breakdown for release of their effector moiety. Finally, it has been shown that ADCs directed to the targets CD22 CD79b can cause tumor regressions independently of the constitution of the linker and the toxophore employed.

8.2.2 *Radioimmunoconjugates (RICs)*

8.2.2.1 *The concept and development of RICs*

RICs are ligand-targeted (protein) molecules for the systemic delivery of radionuclides to sites of disease. They consist of a targeting moiety, which is usually a mAb, to which a radionuclide is attached either covalently or indirectly via a protein-bound chelator group. Targeting of radionuclides is used for diagnostic purposes or for therapeutic treatment of malignancies such as cancer and is

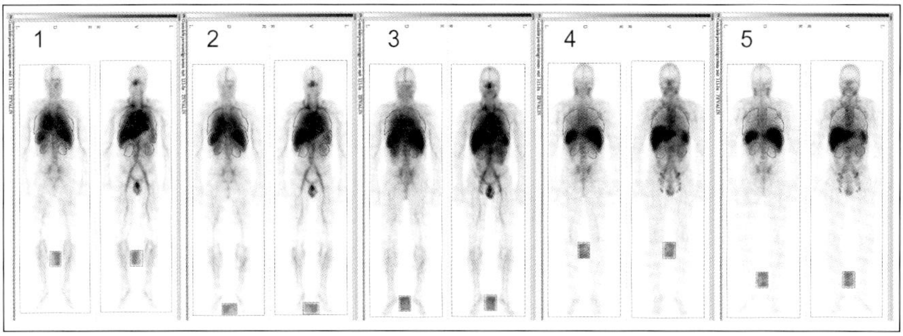

Figure 8.12 Sequence of planar whole body scans (anterior, V; and posterior, D) captured at hour 1, 4, 21, 93, and 140 (1–5) after intravenous injection of ^{111}In-ibritumomab tiuxetan in a male patient (35 years) suffering from relapsed follicular lymphoma (grade II, Ann Arbor stage IVa) with prominent inguinal lymph node involvement. (Zevalin scans where kindly provided by Dr. Sandrock, Hospital Chemnitz, Dept. of Nuclear Medicine.)

referred to as radioimmunotherapy (RIT). Antibody targeting leads to a selective accumulation of the radionuclide at its target site while the non-targeted tissues are spared. The resulting contrast to non-target tissues is used for visualization in diagnostic applications. In the treatment of disease a therapeutic window is achieved that is not obtained with a non-targeted radiation therapy and that allows a more selective irradiation of target tissue with lower side effects (Milenic *et al.*, 2004).

RIT using RICs has been developing since the 1950s when it was recognized that IgGs are able to mediate selective delivery of radiation to a target tissue (Pressman and Korngold, 1953; Korngold and Pressman, 1954; Bale and Spar, 1957; Bale *et al.*, 1960; McCardle *et al.*, 1966). In immunoscintigraphy studies, immune sera raised against α-fetoprotein and carcinoembryonic antigen (CEA) were used first in pre-clinical and then in clinical studies, leading to the localization of different cancers (Goldenberg *et al.*, 1974; Goldenberg *et al.*, 1978; Mach *et al.*, 1980; Kim *et al.*, 1980; van Nagell *et al.*, 1980). Rodent antibody fragments, which became available with the development of the mAb technology in the early 1980s, allowed early cancer detection (Mach *et al.*, 1983; Goldenberg *et al.*, 1990a; Goldenberg *et al.*, 1993) but suffered from poor sensitivity and image resolution (Goldenberg and Larson, 1992; Goldenberg *et al.*, 1997; Sharkey and Goldenberg, 2006).

With the demonstration of selective antibody targeting of cancers, interest in therapeutic application was generated, but due to the murine origin of the antibodies it soon became apparent that repeated administration was limited by human anti-mouse mAbs (HAMA) (Larson, 1990; Murray *et al.*, 1994). Luckily, this issue could be overcome in the 1990s by introduction of chimerization and humanization technologies (Goldenberg *et al.*, 1990b; Goldenberg and Larson, 1992).

The next wave of research focused on understanding the requirements for efficient tumor targeting and its improvement. Identified parameters for optimization of RIT were the choice of the optimal target antigen with the most selective tissue expression pattern, the PK-optimized antibody format, and the most suitable radionuclide and radiochemistry (Goldenberg and Sharkey, 2007). As a result of these efforts, clinical efficacy of RICs could be demonstrated in locoregional applications (Alvarez *et al.*, 1997; Mahe *et al.*, 1999; Riva *et al.*, 1999; Meredith *et al.*, 2001; Goetz *et al.*, 2003; Paganelli *et al.*, 2006; Reardon *et al.*, 2006), in adjuvant settings (Liersch *et al.*, 2005; Zalutsky, 2006), and in patients with radiosensitive lymphomas (Meredith and Knox, 2006; Goldenberg, 2001; Meredith, 2006; Witzig, 2006; Dreyling *et al.*, 2007). Most impressive results were obtained with partial and complete responses in treatment of lymphoma, and they are related to the radiosensitivity and accessibility of these tumors (Goldenberg and Sharkey, 2007; Sharkey and Goldenberg, 2008).

Achieving efficacy with RICs in solid tumor applications, however, is more challenging. Only minor successes with some partial responses have been reported, and this may be due to insufficient dose delivery to the less radiosensitive solid tumor tissue mass. Efficient delivery is limited by the PK properties of the employed antibody molecules and by hematotoxicity, which is usually the dose-limiting organ toxicity. Other limiting factors correspond to biological barriers such as poor extravasation from the vasculature, slow diffusion into the tissue, tumor heterogeneity and heterogeneous antigen expression (Sharkey and Goldenberg, 2008).

Current research focuses on the optimal design of RICs and the development of appropriate treatment strategies to overcome these hurdles and to improve the clinical efficacy of drug candidates. In the following section, we discuss the options for radionuclides as effector molecules and the selection of the appropriate antibody format.

8.2.2.2 *Radioisotopes, radiolabeling and case studies*

RIT selection of the appropriate radionuclide as effector molecule is key and should be performed based on the therapeutic strategy, the disease indication and the lesion size. Radionuclides differ in their chemistry, energy emission(s), including range and tissue penetration, half-life and availability. To allow optimal dose delivery, radioactive decay of the radionuclide and pharmacokinetic half-life of the RIC should match. Available isotopes fall into two main categories of emissions: β-emitters such as ^{131}I, ^{90}Y, ^{177}Lu, ^{67}Cu; and α-emitters such as ^{213}B and ^{211}At (Srivastava and Dadachova, 2001).

RIT with β-emitters has been used successfully in hematological malignancies (Cheson, 2006) where antigen accessibility is relatively high. It has also been proposed to be used in cancers that do not respond to conventional therapies, in particular small solid tumors with heterogeneous target antigen expression where a bystander effect is required (Sharkey and Goldenberg, 2008).

β-emitters have a relatively long emission path length with low linear energy transfer (LET). Energy deposition occurs at some distance from the actual decay event, resulting in a bystander effect in which not only the antigen-positive cell is hit by the radiation but also the surrounding tissue. For ADCs this is advantageous when the targeted antigen has a heterogeneous expression pattern, and not all tumor areas will be targeted homogeneously by the RIC. Uniform radiation of a target lesion is possible when the emission range exceeds the radius of the targeted lesion (O'Donoghue *et al.*, 1995). A limitation of β-emitters is their use in settings where only single tumor cells need to be targeted such as in micrometastatic disease, leukemias or disseminated disease (Humm, 1986). In such cases, β-emitters may not be an appropriate choice.

Current research is focused on the selection of the most suitable β-emitting isotope. Iodine isotopes, for example ^{131}I, have been widely used for reasons of convenience, availability and easy radiolabeling chemistry. Other clinically relevant β-emitters include ^{90}Y, ^{67}Cu, ^{186}Re, and more recently ^{177}Lu. Direct radioiodination occurring with tyrosine moieties of the protein has dominated this field. The convenience of the method, however, is compromised by potentially rapid de-iodination of the protein post-internalization, an event that does not occur when radiometals are used (Mattes *et al.*, 1997). De-iodination leads to a loss of the delivered dose, as the released iodine is rapidly excreted from the cell and circulation. This needs to be taken into account when interpreting biodistribution data where the "real" antibody delivery can be underestimated, while for therapeutic studies suboptimal dose delivery and deposition is the consequence. To minimize this effect, residualizing iodination methods have been used (Qu *et al.*, 2005; Vaidyanathan *et al.*, 2009), or antigens can be targeted that are not internalized.

Alternatively β-emitting radiometals can be used. ^{90}Y is a pure β-emitter and delivers 4.5 times more radiation per mCi to a tumor than does ^{131}I. The greater emission range of ^{90}Y means that most of the decay energy is deposited in tumors only if their diameter is 1 cm or more, and it might be disadvantageous as its longer emission range can cause irradiation of normal tissue surrounding the lesion. Other radiometal isotopes such as ^{177}Lu and ^{67}Cu are investigated as they offer the advantage of a decreased β-emission energy and range that may result in better tolerability, which has already been demonstrated in clinical trials (Mulligan *et al.*, 1995; Alvarez *et al.*, 1997; O'Donnell *et al.*, 1999). A consistent limitation of β-emitters is that irradiation occurs during circulation of the antibody, which subsequently causes myelosuppression (Knox and Meredith, 2000).

Another limitation of radiometal isotopes is that they are trapped intracellularly. This is advantageous in a targeted tumor; however, it also occurs during excretion and degradation and results in high radiation exposure of normal organs such as liver, spleen and kidney (Milenic *et al.*, 2004). Several strategies have been pursued in pre-clinical studies to reduce kidney uptake: pre-administration of D-Lysine (Behr *et al.*, 1998), use of metabolizable peptide linkers which are cleaved at renal or hepatic surface before uptake (Meares *et al.*, 1988; Peterson and Meares, 1999), linkers allowing excretion of radiometal, suitable chelation chemistry (Li *et al.*, 2002) and charge modifications (Arano, 1998).

The most prominent examples for β-emitting RICs are Zevalin (ibritumomab, the murine parent of rituximab labeled with ^{90}Y) and Bexxar (tositumomab labeled with ^{131}I) (Cheson, 2006). Zevalin is given as a maximum dose of 32 mCi (0.3–0.4 mg/kg) after an initial clearing dose of rituximab and an imaging step with ^{111}In ibritumomab if biodistribution is favorable (Fig. 8.12). Initial Phase I/II studies demonstrated anti-tumor efficacy and safety, and durable responses were obtained with minimal non-hematological toxicity as side effects. Comparing Zevalin and Bexxar, both appear comparable for efficacy measures. Yet, one or the other may be preferable for a specific indication. For example, Zevalin may work better for a bulky tumor masses because of its longer path length, while it is expected for Bexxar to induce less myelosuppression, although the frequency of serious infections or other complications is similar (Cheson, 2006). In some cases the decision between the agents may be based on convenience and cost-effectiveness. For a detailed review on Zevalin and Bexxar please see Cheson (2006).

To summarize, β-emitting epitopes constitute the most important class of therapeutic nuclides, and selection of the right β-emitting isotope with its pros and cons needs to be carefully considered. However, their advantages outweigh their limitations, and they will continue to dominate pre-clinical and clinical trials.

The development of novel RICs with α-emitters has created new opportunities and theoretical advantages. RIT with α-emitters appears particularly suitable for the elimination of circulating single cells, cell clusters, or for the treatment of early micrometastases (Humm, 1986; Couturier *et al.*, 2005). Additionally, bone-marrow purging for transplant conditioning, as well as selective tumor vasculature targeting with the goal of tumor eradication seem to be obvious avenues for future work (Bethge *et al.*, 2003). Adjuvant therapies for the treatment of residual disease may also be worth pursuing (Hartmann *et al.*, 1994; Hassfjell and Brechbiel, 2001).

α-emitting nuclides have a high LET of > 20 to several hundred of kEV per micrometer and a short path length of 50 to 100 μm in biological tissues. This results in a high ionization potential per unit path length and a high absorbed dose in a range corresponding to single cell diameters. The high energy radiation typically causes complex DNA double-strand breaks that are difficult to repair, thus leading to effective cell killing close to the decay site (Humm, 1986). The main obstacle in using α-emitters is the limited availability and complex logistics. Thus, only a limited number of studies using ^{212}Bi, ^{213}Bi, ^{211}At, and ^{225}Ac have been performed (McDevitt *et al.*, 1998; McDevitt *et al.*, 2001; Hassfjell and Brechbiel, 2001). One example is the anti-CD33 mAb M195, which has been administered with ^{213}Bi to patients with relapsed or refractory acute myelocytic leukemia, marking the first antibody-targeted radiation therapy trial in humans using an α-emitter in a setting addressing minimal residual tumor cells (Jurcic *et al.*, 2002; Milenic *et al.*, 2004).

Protocols for coupling of radionuclides to biomolecules are well established and differ for halogen, metal and transition elements. Basically, conditions have to be chosen that allow formation of a stable linkage to the mAb within the constraints imposed by isotope chemistry and half-life. Direct radioiodination is well established and will not be discussed here. Coupling of radiometal isotopes require chelation via bifunctional chelators that allow stable complex formation with the radiometal isotope and binding to the protein carrier. Development of bifunctional chelators (BCAs) is an active area of research, as radiometal isotopes are diverse in their properties and coordination chemistry, and no single chelator is suitable for all metals (Milenic *et al.*, 2004). Notable examples are 1,4,7,10-tetra-azacyclododecance-N,N,N,N-tetraacetic acid (DOTA), Diethylenetriaminepentaacetic anhydride (DTPA) and its analog 2-(p-isothiocyanatobenzyl)-6-methyl-DTPA (MX-DTPA) (Tiuxetan, which is used in Zevalin) (Fig. 8.13). DOTA is widely applied as it allows complexation of β-emitters such as ^{90}Y and ^{177}Lu (Mulligan *et al.*, 1995; Alvarez *et al.*, 1997; Milenic *et al.*, 2002; Forrer *et al.*, 2009). However, radiolabeling kinetics to DOTA are normally very slow and require an elevated temperature, which may cause a significant loss of antibody immunoreactivity. The advantage of using DOTA and its analogs as chelators is the kinetic inertness of their radiometal chelates. A major advantage of using DTPA analogs as chelators is their extremely high radiolabeling efficacy under mild conditions, but the kinetic lability of their metal chelates often results in dissociation of radiometal from the metal chelate. For a detailed discussion on coupling of radiometal isotopes and chelator chemistry, please refer to the review by Liu (2008).

8.2.2.3 *Antibody formats for RIT and case studies*

In contrast to an ADC, which ideally requires antigen binding and internalization for activation of the toxophore, RICs contain "active" effector molecules at the time of administration of the drug and thus deliver unwanted irradiation to normal tissues as a side effect. Therefore, the doses administered and delivered to critical organs are limiting. Most RIT studies and advanced clinical programs have been performed with full IgG antibodies or enzymatically

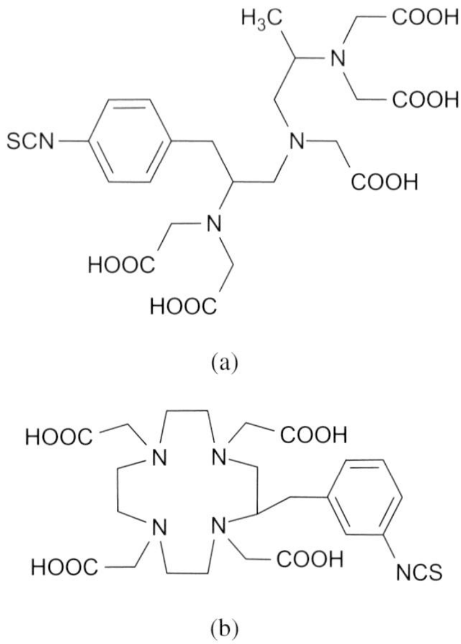

Figure 8.13 Selected examples of bifunctional chelators used in clinical trials. Conjugation occurs through the isothiocyanate. Radiometal isotopes interact with the chelate via the amine and carboxylate groups. a: MX-DTPA, tiuxetan; b: c-DOTA.

prepared derivatives. Meanwhile, recombinant antibody technology has been introduced, and a number of different antibody formats with different binding and PK properties are available (Fig. 8.14). The intensive pre-clinical analyses of these antibody formats in tumor-targeting and efficacy studies have enabled the identification of antibody formats with the best tumor-to-noise ratios and therapeutic windows. They have also allowed the optimization of treatment strategies. In the following section, we will discuss the properties of different antibodies and their relevance for RIT.

The targeting properties of a full IgG or a recombinant antibody for a given biologic system is determined by their PK and biophysical properties. The PK behavior is dependent on the size, molecular composition, isotype and antigen specificity. Biophysical properties comprise the thermal and proteolytic stability in the serum, valency and affinity to the antigen. Avidity (effective affinity from more than one binding site) rather than affinity has been concluded to be an important factor for tumor retention at high total doses (Rudnick and Adams, 2009). Taken together, these factors determine how fast and at which dose an antibody accumulates at its target site, how well it penetrates and is retained, how fast it clears from the circulation and non-target tissue, and to what extent it is catabolized via the liver or excreted via the kidney.

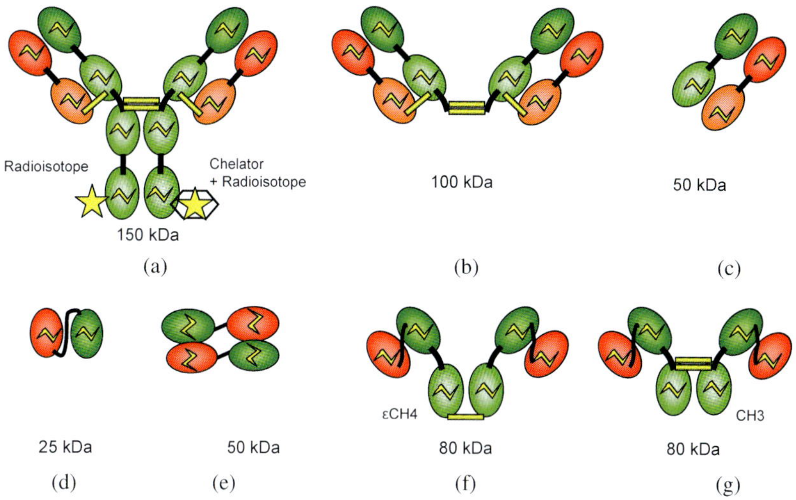

Figure 8.14 Antibody formats for radionuclide delivery prepared by enzymatic cleavage or molecular engineering. Full-size IgG antibodies (a) are either directly radiolabeled by radioiodination via tyrosine residues or indirectly by radiolabeling of a chelator. The chelator was coupled either to carbohydrates or directly to the protein. Other formats are $F(ab')_2$ (b) or Fab fragments (c). Single chain Fv are created by linking the variable domain of the heavy (V_H) and light chain (V_L) with an amino acid linker (d). A dimeric diabody forms by employing a shorter 5-amino acid linker (e). Recombinant mini-antibody formats are homodimers of scFv or Fab molecules fused to a dimerizing domain which can be an Fc part or a domain of it yielding a dimeric antibody molecule with intermediate molecular size. The constant domain C_H4 of an IgE molecule is used in the small immune protein format (f) and a C_H3 domain in the Flex minibody format (g).

Full IgG antibodies are large molecules of approximately 150 kDa in size, and as such clear very slowly from the circulation with a half-life of days to weeks. The long circulation time is due to the poor extravasation of the large antibody molecules from the vasculature into the tissue, and to recycling of IgGs by a neonatal Fc receptor (FcRn)-mediated mechanism (Zalevsky *et al.*, 2009). The high serum concentrations of IgG act as a reservoir for antibodies to enter the tumor tissue slowly over days as the tumor vasculature is "leakier" than that of most normal tissues (Jain, 1990; Folkman *et al.*, 1971), and the maximum concentration at the target site is reached after days at a relatively high dose level (Table 8.4). However, IgGs also enter other tissues, and it takes several days until clearance of the antibody from non-target tissues and blood result in a difference which can be visualized in imaging applications. For therapeutic applications with IgG-based RICs, the total dose delivery to the tumor is limited by the simultaneous delivery of radionuclide to the radiosensitive bone marrow, causing hematoxicity. In

Table 8.4 Radionuclides for targeted immunotherapy

Radionuclide	Type	Half-Life	Emax (MeV)	Mean Range (mm)	Imaging Possible
^{90}Y	β	2.7 d	2.3	2.76	No
^{131}I	β, γ	8.0 d	0.81	0.4	Yes
^{177}Lu	β, γ	6.7 d	0.5	0.28	Yes
^{153}Sm	β, γ	2.0 d	0.8	0.53	Yes
^{186}Re	β, γ	3.8 d	1.1	0.92	Yes
^{188}Re	β, γ	0.7 d	2.1	2.43	Yes
^{67}Cu	β, γ	2.6 d	0.6	0.6	Yes
^{225}Ac	α, β	10 d	5.8	0.04–0.1	Yes
^{213}Bi	α	46 min	5.9	0.04–0.1	Yes
^{212}Bi	α	1.0 h	6.1	0.04–0.1	Yes
^{211}At	α	7.2 h	5.9	0.04–0.1	Yes
^{212}Pb	α	10.6 h	0.6	0.6	Yes
^{125}I	Auger	60.1 d	0.2	0.001–0.02	Yes

(based on Milenic *et al.*, 2004).

mouse xenograft models, tumor uptake is typically between 10 and 30% of the injected activity per gram of tumor, but in humans, with a larger vascular and extravascular volume of distribution, this accretion is reduced to less than 0.1% per gram (Siegel *et al.*, 1990; Buchsbaum, 1995), which is only a small percentage of the total injected dose.

Once the IgG antibody has left the circulation and enters the tumor, penetration into the tissue is poor, and a perivascular and heterogeneous IgG distribution can be found (Yokota *et al.*, 1992). It has been described that diffusion of a large molecule against a high interstitial pressure gradient is an extremely slow and inefficient process (Jain and Baxter, 1988). The antibody, upon entering a tissue, may also be captured by the early accessible antigens close to the vessels, preventing a deeper entry into the tissue and a more homogeneous penetration. This effect has been referred to as antibody-binding barrier and is dependent on the antibody dose and affinity (Fujimori *et al.*, 1990; Baxter and Jain, 1991; Thurber *et al.*, 2008). Possible approaches to optimizing delivery include reducing the size, increasing the dose, lowering affinity and maintaining avidity (Baxter *et al.*, 1994).

Size reduction of an antibody from an IgG to divalent $F(ab')_2$, (~100 kDa) to a monovalent Fab (~50 kDa) or a single-chain Fv (scFv) fragment (25 kDa) results in progressively faster clearance from the blood. These molecules retain the binding specificity of the IgG but binding to the FcRn is abolished as the Fc part is removed. By reducing the antibody size, the tumor maximum is reached more quickly and a higher selectivity (tumor-to-organ ratio) and earlier visualization are achieved in imaging experiments (Batra *et al.*, 2002; Wittel *et al.*, 2005). This is at the cost of a proportionally lower delivered dose and a shorter residence time, as the blood reservoir of circulating antibody fragments is smaller. For imaging purposes this may be beneficial, but delivering a critical dose becomes more difficult and must be compensated by more frequent dosing in therapeutic applications. Another effect of reducing antibody size is greater and more homogeneous tissue penetration (Yokota *et al.*, 1992). The change in antibody molecular size and format also results in an altered tissue distribution, since larger molecules (>60 kDa) are mainly metabolized by the liver while smaller ones (<60 kDa) are excreted through the kidney.

Based on these observations, new formats have been engineered by recombinant antibody technology to identify the right combination of PK and antigen-binding properties. Constructs have been described in which the C_H2 domain is deleted, yielding a bivalent IgG-like molecule of ~ 100 kDa that is not recognized by the FcRn (Mueller *et al.*, 1990; Slavin-Chiorini *et al.*, 1993; Slavin-Chiorini *et al.*, 1995; Slavin-Chiorini *et al.*, 1997; Leung *et al.*, 1999; Rogers *et al.*, 2005). Other bi- and multivalent constructs have been built from scFv fragments, which have the tendency to di- and multimerize when the peptide linker between V_H and V_L is shortened (Lawrence *et al.*, 1998; Hudson and Kortt, 1999; Wu and Yazaki, 2000; Todorovska *et al.*, 2001; Power *et al.*, 2003; Holliger and Hudson, 2005). In another group of constructs, scFv fragments are linked to an IgG Fc-domain, which may be mutated to lack FcRn binding and to accelerate blood clearance (Olafsen *et al.*, 2006; Roopenian and Akilesh, 2007; Presta, 2008). The pre-clinical analysis of these constructs in tumor-targeting studies led to the identification of formats with an improved selectivity for imaging and therapy where blood clearance was increased and tumor uptake was maintained. These formats have a molecular weight above the renal threshold of 60 kDa and a valency allowing bivalent binding and high avidity to enhance tumor retention and penetration (Kenanova and Wu, 2006).

A recent example of a new RIC in clinical development with optimized PK features is ^{131}I-L19SIP (Philogen SpA), which is a human recombinant antibody directed against the extradomain B type III (EDB-FN) of the extracellular matrix protein fibronectin (Menrad and Menssen, 2005). EDB-FN is an angiogenesis-related and tumor stroma-associated antigen that is tumor-selective and abundantly expressed in many tumors. ^{131}I-L19SIP is a homodimer of the anti-EDB scFv L19 (Tarli *et al.*, 1999) fused to the ε-C_H4 sequence of an IgE antibody, which results in a disulfide-linked dimeric 80 kDa molecule referred to as a small immune protein (SIP) format. Targeting of EDB with ^{131}I-L19SIP has been demonstrated pre-clinically (Berndorff *et al.*, 2005) and clinically (Sauer *et al.*, 2009). A high tumor uptake of ^{131}I with long residual time within the tumor was found, while the molecule was rapidly cleared from the circulation. This results in a favorable therapeutic window, as effective radiation dose delivery is possible at doses tolerable for the dose-limiting organ (red blood marrow). The *in vivo* accessibility of EDB-FN was confirmed in three lymphoma patients in whom the lymphoma lesions were visualized by scintigraphy with ^{131}I-L19SIP (Fig. 8.15) and in two relapsed Hodgkin's lymphoma patients. ^{131}I-L19SIP RIT induced a sustained partial response (Menrad and Menssen, 2005; Berndorff *et al.*, 2005; Sauer *et al.*, 2009).

It has also been shown that the PK behavior of targeting molecules can be modified by other strategies. For example, enhanced tumor targeting has been achieved by co-administration of large amounts of human IgG, which blocked binding to the FcRn (Jaggi *et al.*, 2007). Smaller formats with rapid plasma clearance were optimized for increased circulation by addition of polyethylene glycol chains (PEG), which increased the molecular weight (Kubetzko *et al.*, 2006), or by fusion of an antibody fragment specific for serum albumin (Stork *et al.*, 2007; Muller *et al.*, 2007; Yazaki *et al.*, 2008).

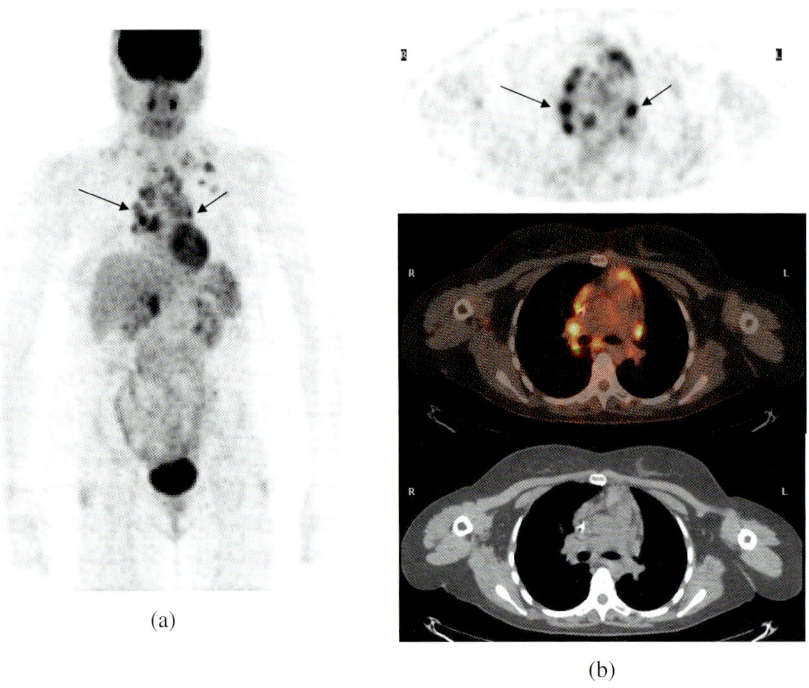

Figure 8.15 ^{131}I-L19SIP dosimetry used to identify individual patients for subsequent radioimmunotherapy (RIT): FDG-PET/CT scan of a patient diagnosed with relapsed advanced Hodgkin's lymphoma, showing active lymphoma lesions (a). Transversal ^{131}I-L19SIP SPECT (top), CT scan (bottom), and overlay (center) of the same patient indicating co-localization of paracardial lymph node enlargements with ^{131}I-L19SIP targeted tissues. The paracardia lymph node enlargements were identified as active lymphoma lesions in FDG-PET/CT (see arrows in a and b). (This figure was provided as a courtesy of Paola Erba, Pisa Medical School, Pisa, Italy, and Hans-Dietrich Menssen, Philogen SpA, Sovicille, Italy.)

Non-immunoglobulin molecules have also been used for tumor targeting studies. Examples are anti-HER2 Affibodies (7,5 kDa), which are binders derived from a library based on the *Staphylococcus aureus* protein A framework (Orlova *et al.*, 2006; Namavari *et al.*, 2008) and anti-HER2 binding molecules isolated from a library of randomized ankyrin-repeat molecules (Zahnd *et al.*, 2010). Interestingly, improved selectivity and tumor targeting have been achieved with high affinity variants (up to picomolar affinity) of these small monovalent molecules, which was not expected from earlier studies with anti-HER2 scFv molecules (Zahnd *et al.*, 2010).

A model that was developed based on the available data in the literature predicts the complex interplay between molecular size, affinity and tumor uptake (Schmidt and Wittrup, 2009). In the typical size range for proteins, the model uncovers a complex trend in which intermediate-sized targeting agents (~ 25 kDa) have the lowest tumor uptake, whereas higher tumor uptake levels are achieved by smaller and larger agents. Small peptides accumulate rapidly in the tumor but require extremely high affinity to be retained, whereas larger proteins can achieve highest retention with > 100-fold weaker binding compared to the small peptides and intermediate-sized targeting agents (Schmidt and Wittrup, 2009). This evaluation indicates that the conventional IgG is the preferred format. However, for new applications in imaging and RIT, smaller peptide compounds and non-immunoglobulin scaffold binders deserve further study.

Table 8.5 Comparison of targeting properties of antibody formats in preclinical animal models*

Antibody Format	Molecular Weight (kDa)	Half-Life Range**	Time to Optimum Tumor Targeting***	Relative Total Dose Delivery****
IgG	150	>24 h	48–168 h	++++
scFv-C_H4 (SIP), disulfide linked	80	6–12 h	12–48 h	++–++++
F(ab')$_2$	100	6–12 h	12–48 h	++–+++
Fab	50	>1 h	6–24 h	++
Diabody	50	>1 h	6–24 h	++
scFv	25	0.5–2 h	6–24 h	+

*based on work published in Kenanova and Wu, 2006 and references cited there.
**Estimated time for 50% clearance from blood
***Suitable tumor-to-blood ratio for imaging
****Total dose delivery resulting from high uptake and long retention; low to highest dose delivery indicated by + – ++++

A further development in tumor targeting is a concept referred to as "pre-targeting." The idea is to separate targeting and radionuclide delivery into two distinct steps and thus achieve higher selectivity and improved tumor-targeted dose as compared to targeting of directly radiolabeled immunoconjugates. In a first step the unlabeled bispecific mAb, consisting of an anti-tumor and an anti-chelate or anti-hapten Fab, is administered and after several days, when the antibody has localized to the tumor and the unbound circulating fraction has cleared from the body, the radiolabeled chelate or hapten is given (Reardan *et al.*, 1985; Goodwin *et al.*, 1986; Goodwin *et al.*, 1988; Goodwin and Meares, 1999; Chang *et al.*, 2002). Once injected the radiolabeled chelate or hapten rapidly extravasates into the tissues and is retained at high doses by the anti-chelate Fab, while the remaining material is rapidly excreted via the kidneys. Pre-targeting has been used for imaging (Stickney *et al.*, 1991) and a limited number of Phase I clinical applications in advanced solid tumors such as colorectal, medulary, thyroid, ovarian and brain cancers (Paganelli *et al.*, 1999; Knox *et al.*, 2000; Paganelli *et al.*, 2001; Grana *et al.*, 2002; Bartolomei *et al.*, 2004; Boiardi *et al.*, 2005; Forero-Torres *et al.*, 2005; Shen *et al.*, 2005). The method has been optimized by evaluation of different bispecific formats (Atwell *et al.*, 1996; Asano *et al.*, 2002; Lu *et al.*, 2003; Kipriyanov, 2003; Rossi *et al.*, 2003; Hayashi *et al.*, 2004; Kontermann, 2005; Das and Suresh, 2005; Rossi *et al.*, 2005; Wu *et al.*, 2007; Zhang *et al.*, 2007; Shen *et al.*, 2007), and the best dose for targeting and retention has been found in animal studies with bispecific antibodies with two binding sites each for the tumor associated antigen, and the radiolabeled hapten is given in the second step (Karacay *et al.*, 2002). Another variation makes use of the high affinity biotin/streptavidin system, where streptavidin has an exceptionally high affinity for biotin (10^{-15}M) compared to the usual antibody affinity to haptens in the low nanomolar range (Hnatowich, 1987). This extraordinary affinity assures an almost covalent binding strength between the pre-targeted antibody–streptavidin conjugate and the radiolabeled biotin. Using this strategy, a very rapid and efficient tumor uptake and an optimized therapeutic window with reduced hematologic toxicity has been obtained, and the advantage has been reported in several model systems (Gautherot *et al.*, 1997; Kraeber-Bodere *et al.*, 1999a; Kraeber-Bodere *et al.*, 1999b; Gautherot *et al.*, 2000; Axworthy *et al.*, 2000; Press *et al.*, 2001; Pagel *et al.*, 2003; Subbiah *et al.*, 2003; Sharkey *et al.*, 2005; Karacay *et al.*, 2005; Sharkey and Goldenberg, 2008).

Compared to traditional RIC strategies, "pre-targeting" has great potential. However, it is a highly complex strategy that requires fine-tuning of several components in preclinical and clinical studies. Therefore progress is slow and strategies employing directly radiolabeled antibody formats have a higher mid-term realization potential with regard to clinical development.

8.3 Conclusion

After more than 20 years of research in the area of toxin and radio-conjugated antibodies, this field has undergone many learning cycles, but it is now gaining significant momentum in clinical trials, particularly with the successes of trastuzumab-DM1 in Phase II clinical trials, being the first ADC to show promise in the treatment of solid tumors (Alley *et al.*, 2010). Trastuzumab-DM1 has successfully overcome many of the shortcomings of first generation ADCs and incorporates several important selection criteria for target antigens, including high overexpression compared to normal tissues, rapid internalization via receptor-mediated endocytosis and high antigen density reported for the eligible patient population. Additionally, trastuzumab-DM1 includes a maytansinoid derivative, one of the most potent toxophores currently used in the field, outperforming first generation toxophores such as doxorubicin. Furthermore, trastuzumab-DM1 harbors a stable linker, which contributes to low systemic toxicity compared to that observed with many alternative linkers with built-in cleavage sites that allow non-specific release of the toxophore. Considering the deep understanding that has been gained with this model therapeutic, it can thus be expected that additional ADCs beyond trastuzumab-DM1 will be clinically successful for other solid tumor indications.

Similar principles hold true for RIT approaches. Clinical evidences increasingly contribute to our knowledge of the prerequisites for effective RICs by selection of the right target and clinical settings, isotope conjugation chemistry and antibody format. The development of Zevalin and Bexxar demonstrated that RIT as a therapeutic principle is an option for patients when other modalities have failed and that RICs effectively complement the anticancer armory when used in the right setting. RICs can be used in monotherapy, but there is a strong rationale to combine RIT with chemotherapy to sensitize tumors to radiation, or with external beam radiation as a boost to achieve very high radiation doses. Another advantage is that patients who are likely to benefit from RIT can be selected by a simple dosimetry step using the same molecule as for radiation therapy. The recent progress is likely to further improve the treatment of hematologic cancers as well, and there is good hope that the success of RIT can be extended to the treatment of less radiosensitive solid tumors. Optimized RICs such as ^{131}I-L19SIP constitute a new generation of therapeutics that will hopefully deliver higher and thus more effective doses to tumors.

ADCs and RICs will likely be used in different clinical settings and stages of disease due to their different modes of action. It is also likely that ADCs will soon be introduced into existing chemotherapy protocols, and RICs can potentially be used in indications where radiotherapy is already part of the treatment

Burke P.J., Senter P.D., Meyer D.W., Miyamoto J.B., Anderson M., Toki B.E., Manikumar G., Wani M.C., Kroll D.J., Jeffrey S.C. (2009). Design, synthesis, and biological evaluation of antibody-drug conjugates comprised of potent camptothecin analogues. *Bioconjug Chem* **20**: 1242–50.

Carl P.L., Chakravarty P.K., Katzenellenbogen J.A. (1981). A novel connector linkage applicable in prodrug design. *J Med Chem* **24**: 479–80.

Carter P. (2001). Improving the efficacy of antibody-based cancer therapies. *Nat Rev Cancer* **1**: 118–29.

Carter P.S., Banwell P.E. (2004). Necrotising fasciitis: a new management algorithm based on clinical classification. *Int Wound J* **1**: 189–98.

Cartron G., Dacheux L., Salles G., Solal-Celigny P., Bardos P., Colombat P., Watier H. (2002). Therapeutic activity of humanized anti-CD20 monoclonal antibody and polymorphism in IgG Fc receptor FcgammaRIIIa gene. *Blood* **99**: 754–8.

Cassady J.M., Chan K.K., Floss H.G., Leistner E. (2004). Recent developments in the maytansinoid antitumor agents. *Chem Pharm Bull (Tokyo)* **52**: 1–26.

Chang C.H., Sharkey R.M., Rossi E.A., Karacay H., McBride W., Hansen H.J., Chatal J.F., Barbet J., Goldenberg D.M. (2002). Molecular advances in pretargeting radioimunotherapy with bispecific antibodies. *Mol Cancer Ther* **1**: 553–63.

Chari R.V. (1998). Targeted delivery of chemotherapeutics: tumor-activated prodrug therapy. *Adv Drug Deliv Rev* **31**: 89–104.

Chari R.V. (2008). Targeted cancer therapy: conferring specificity to cytotoxic drugs. *Acc Chem Res* **41**: 98–107.

Chen J., Jaracz S., Zhao X., Chen S., Ojima I. (2005). Antibody-cytotoxic agent conjugates for cancer therapy. *Expert Opin Drug Deliv* **2**: 873–90.

Cheson B.D. (2006). Radioimmunotherapy of non-Hodgkin's lymphomas. *Curr Drug Targets* **7**: 1293–300.

Clynes R.A., Towers T.L., Presta L.G., Ravetch J.V. (2000). Inhibitory Fc receptors modulate *in vivo* cytoxicity against tumor targets. *Nat Med* **6**: 443–6.

Collier R.J. (1975). Diphtheria toxin: mode of action and structure. *Bacteriol Rev* **39**: 54–85.

Couturier O., Supiot S., Degraef-Mougin M., Faivre-Chauvet A., Carlier T., Chatal J.F., Davodeau F., Cherel M. (2005). Cancer radioimmunotherapy with alpha-emitting nuclides. *Eur J Nucl Med Mol Imaging* **32**: 601–14.

Daniels T.R., Delgado T., Helguera G., Penichet M.L. (2006a). The transferrin receptor part II: targeted delivery of therapeutic agents into cancer cells. *Clin Immunol* **121**: 159–76.

Daniels T.R., Delgado T., Rodriguez J.A., Helguera G., Penichet M.L. (2006b). The transferrin receptor part I: Biology and targeting with cytotoxic antibodies for the treatment of cancer. *Clin Immunol* **121**: 144–58.

Das D., Suresh M.R. (2005). Producing bispecific and bifunctional antibodies. *Methods Mol Med* **109**: 329–46.

Dedon P.C., Goldberg I.H. (1992). Free-radical mechanisms involved in the formation of sequence-dependent bistranded DNA lesions by the antitumor antibiotics bleomycin, neocarzinostatin, and calicheamicin. *Chem Res Toxicol* **5**: 311–32.

Deonarain M.P. (2008). Recombinant antibodies for cancer therapy. *Expert Opin Biol Ther* **8**: 1123–41.

Di Gaetano N., Cittera E., Nota R., Vecchi A., Grieco V., Scanziani E., Botto M., Introna M., Golay J. (2003). Complement activation determines the therapeutic activity of rituximab *in vivo*. *J Immunol* **171**: 1581–7.

DiJoseph J.F., Armellino D.C., Boghaert E.R., Khandke K., Dougher M.M., Sridharan L., Kunz A., Hamann P.R., Gorovits B., Udata C., *et al.* (2004). Antibody-targeted chemotherapy with CMC-544: a CD22-targeted immunoconjugate of calicheamicin for the treatment of B-lymphoid malignancies. *Blood* **103**: 1807–14.

DiJoseph J.F., Popplewell A., Tickle S., Ladyman H., Lawson A., Kunz A., Khandke K., Armellino D.C., Boghaert E.R., Hamann P., *et al.* (2005). Antibody-targeted chemotherapy of B-cell lymphoma using calicheamicin conjugated to murine or humanized antibody against CD22. *Cancer Immunol Immunother* **54**: 11–24.

Doronina S.O., Mendelsohn B.A., Bovee T.D., Cerveny C.G., Alley S.C., Meyer D.L., Oflazoglu E., Toki B.E., Sanderson R.J., Zabinski R.F., *et al.* (2006). Enhanced activity of monomethylauristatin F through monoclonal antibody delivery: effects of linker technology on efficacy and toxicity. *Bioconjug Chem* **17**: 114–24.

Doronina S.O., Toki B.E., Torgov M.Y., Mendelsohn B.A., Cerveny C.G., Chace D.F., DeBlanc R.L., Gearing R.P., Bovee T.D., Siegall C.B., *et al.* (2003). Development of potent monoclonal antibody auristatin conjugates for cancer therapy. *Nat Biotechnol* **21**: 778–84.

Dreyling M., Trumper L., von Schilling C., Rummel M., Holtkamp U., Waldmann A., Wehmeyer J., Freund M. (2007). Results of a national consensus workshop: therapeutic algorithm in patients with follicular lymphoma — role of radioimmunotherapy. *Ann Hematol* **86**: 81–7.

Dubowchik G.M., Firestone R.A., Padilla L., Willner D., Hofstead S.J., Mosure K., Knipe J.O., Lasch S.J., Trail P.A. (2002). Cathepsin B-labile dipeptide linkers for lysosomal release of doxorubicin from internalizing immunoconjugates: model studies of enzymatic drug release and antigen-specific *in vitro* anticancer activity. *Bioconjug Chem* **13**: 855–69.

Ducry L., Stump B. (2010). Antibody-drug conjugates: linking cytotoxic payloads to monoclonal antibodies. *Bioconjug Chem* **21**: 5–13.

Dyba M., Tarasova N.I., Michejda C.J. (2004). Small molecule toxins targeting tumor receptors. *Curr Pharm Des* **10**: 2311–34.

Endo N., Takeda Y., Kishida K., Kato Y., Saito M., Umemoto N., Hara T. (1987). Target-selective cytotoxicity of methotrexate conjugated with monoclonal anti-MM46 antibody. *Cancer Immunol Immunother* **25**: 1–6.

Erickson H.K., Park P.U., Widdison W.C., Kovtun Y.V., Garrett L.M., Hoffman K., Lutz R.J., Goldmacher V.S., Blattler W.A. (2006). Antibody-maytansinoid conjugates are activated in targeted cancer cells by lysosomal degradation and linker-dependent intracellular processing. *Cancer Res* **66**: 4426–33.

Erickson H.K., Phillips G.L., Provenzano C.A., Leipold D.D., Pinkas J., Mai E., Audette C., Gupta M., Johnson H., Tibbitts J. (2009). The effect of linker on target cell catabolism and PK/PD of trastuzumab-maytasinoid conjugates. Paper presented at: AACR — NCI — EORTC.

Ferlin M.G., Chiarelotto G., Gasparotto V., Dalla Via L., Pezzi V., Barzon L., Palu G., Castagliuolo I. (2005). Synthesis and in vitro and *in vivo* antitumor activity of 2-phenylpyrroloquinolin-4-ones. *J Med Chem* **48**: 3417–27.

Folkman J., Merler E., Abernathy C., Williams G. (1971). Isolation of a tumor factor responsible for angiogenesis. *J Exp Med* **133**: 275–88.

Forero-Torres A., Shen S., Breitz H., Sims R.B., Axworthy D.B., Khazaeli M.B., Chen K.H., Percent I., Besh S., LoBuglio A.F., *et al.* (2005). Pretargeted radioimmunotherapy (RIT) with a novel anti-TAG-72 fusion protein. *Cancer Biother Radiopharm* **20**: 379–90.

Forrer F., Chen J., Fani M., Powell P., Lohri A., Muller-Brand J., Moldenhauer G., Maecke H.R. (2009). *In vitro* characterization of (177)Lu-radiolabelled chimeric anti-CD20 monoclonal antibody and a preliminary dosimetry study. *Eur J Nucl Med Mol Imaging* **36**: 1443–52.

Fujimori K., Covell D.G., Fletcher J.E., Weinstein J.N. (1990). A modeling analysis of monoclonal antibody percolation through tumors: a binding-site barrier. *J Nucl Med* **31**: 1191–8.

Garnett M.C. (2001). Targeted drug conjugates: principles and progress. *Adv Drug Deliv Rev* **53**: 171–216.

Gautherot, E., Bouhou, J., Le Doussal, J.M., Manetti, C., Martin, M., Rouvier, E., Barbet, J. (1997). Therapy for colon carcinoma xenografts with bispecific antibody-targeted, iodine-131-labeled bivalent hapten. *Cancer* **80**: 2618–23.

Gautherot E., Rouvier E., Daniel L., Loucif E., Bouhou J., Manetti C., Martin M., Le Doussal J.M., Barbet J. (2000). Pretargeted radioimmunotherapy of human colorectal xenografts with bispecific antibody and 131I-labeled bivalent hapten. *J Nucl Med* **41**: 480–7.

Gibson A.D. (2002). Phase III trial of a humanized anti-CD33 antibody (HuM195) in patients with relapsed or refractory acute myeloid leukemia. *Clin Lymphoma* **3**: 18–9.

Giles F.J., Kantarjian H.M., Kornblau S.M., Thomas D.A., Garcia-Manero G., Waddelow T.A., David C.L., Phan A.T., Colburn D.E., Rashid A., *et al.* (2001). Mylotarg (gemtuzumab ozogamicin) therapy is associated with hepatic venoocclusive disease in patients who have not received stem cell transplantation. *Cancer* **92**: 406–13.

Goetz C., Riva P., Poepperl G., Gildehaus F.J., Hischa A., Tatsch K., Reulen H.J. (2003). Locoregional radioimmunotherapy in selected patients with malignant glioma: experiences, side effects and survival times. *J Neurooncol* **62**: 321–8.

Goldenberg D.M. (2001). The role of radiolabeled antibodies in the treatment of non-Hodgkin's lymphoma: the coming of age of radioimmunotherapy. *Crit Rev Oncol Hematol* **39**: 195–201.

Goldenberg D.M., DeLand F., Kim E., Bennett S., Primus F.J., van Nagell J.R., Jr., Estes N., DeSimone P., Rayburn P. (1978). Use of radiolabeled antibodies to carcinoembryonic antigen for the detection and localization of diverse cancers by external photoscanning. *N Engl J Med* **298**: 1384–6.

Goldenberg D.M., Goldenberg H., Sharkey R.M., Higginbotham-Ford E., Lee R.E., Swayne L.C., Burger K.A., Tsai D., Horowitz J.A., Hall T.C., *et al.* (1990a). Clinical studies of cancer radioimmunodetection with carcinoembryonic antigen monoclonal antibody fragments labeled with 123I or 99mTc. *Cancer Res* **50**: 909s–21s.

Goldenberg D.M., Goldenberg H., Sharkey R.M., Lee R.E., Horowitz J.A., Hall T.C., Hansen H.J. (1990b). *In-vivo* antibody imaging for the detection of human tumors. *Cancer Treat Res* **51**: 273–92.

Goldenberg D.M., Juweid M., Dunn R.M., Sharkey R.M. (1997). Cancer imaging with radiolabeled antibodies: new advances with technetium-99m-labeled monoclonal antibody Fab' fragments, especially CEA-Scan and prospects for therapy. *J Nucl Med Technol* **25**: 18–23; quiz 34.

Goldenberg D.M., Larson S.M. (1992). Radioimmunodetection in cancer identification. *J Nucl Med* **33**: 803–14.

Goldenberg D.M., Preston D.F., Primus F.J., Hansen H.J. (1974). Photoscan localization of GW-39 tumors in hamsters using radiolabeled anticarcinoembryonic antigen immunoglobulin G. *Cancer Res* **34**: 1–9.

Goldenberg D.M., Sharkey R.M. (2007). Novel radiolabeled antibody conjugates. *Oncogene* **26**: 3734–44.

Goldenberg D.M., Wlodkowski T.J., Sharkey R.M., Silberstein E.B., Serafini A.N., Garty II, Van Heertum R.L., Higginbotham-Ford E.A., Kotler J.A., Balasubramanian N., *et al.* (1993). Colorectal cancer imaging with iodine-123-labeled CEA monoclonal antibody fragments. *J Nucl Med* **34**: 61–70.

Goodwin D.A., Meares C.F. (1999). Pretargeted peptide imaging and therapy. *Cancer Biother Radiopharm* **14**: 145–52.

Goodwin D.A., Meares C.F., David G.F., McTigue M., McCall M.J., Frincke J.M., Stone M.R., Bartholomew R.M., Leung J.P. (1986). Monoclonal antibodies as reversible equilibrium carriers of radiopharmaceuticals. *Int J Rad Appl Instrum B* **13**: 383–91.

Goodwin D.A., Meares C.F., McCall M.J., McTigue M., Chaovapong W. (1988). Pretargeted immunoscintigraphy of murine tumors with indium-111-labeled bifunctional haptens. *J Nucl Med* **29**: 226–34.

Grana C., Chinol M., Robertson C., Mazzetta C., Bartolomei M., De Cicco C., Fiorenza M., Gatti M., Caliceti P., Paganelli G. (2002). Pretargeted adjuvant radioimmunotherapy with yttrium-90-biotin in malignant glioma patients: a pilot study. *Br J Cancer* **86**: 207–12.

Grant B.D., Donaldson J.G. (2009). Pathways and mechanisms of endocytic recycling. *Nat Rev Mol Cell Biol* **10**: 597–608.

Guillemard V., Saragovi H.U. (2001). Taxane-antibody conjugates afford potent cytotoxicity, enhanced solubility, and tumor target selectivity. *Cancer Res* **61**, 694–9.

Hamann P.R., Hinman L.M., Beyer C.F., Greenberger L.M., Lin C., Lindh D., Menendez A.T., Wallace R., Durr F.E., Upeslacis J. (2005). An anti-MUC1 antibody-calicheamicin conjugate for treatment of solid tumors. Choice of linker and overcoming drug resistance. *Bioconjug Chem* **16**: 346–53.

Hamann P.R., Hinman L.M., Hollander I., Beyer C.F., Lindh D., Holcomb R., Hallett W., Tsou H.R., Upeslacis J., Shochat D., *et al.* (2002). Gemtuzumab ozogamicin, a potent and selective anti-CD33 antibody-calicheamicin conjugate for treatment of acute myeloid leukemia. *Bioconjug Chem* **13**: 47–58.

Hamblett K.J., Senter P.D., Chace D.F., Sun M.M., Lenox J., Cerveny C.G., Kissler K.M., Bernhardt S.X., Kopcha A.K., Zabinski R.F., *et al.* (2004). Effects of drug loading on the antitumor activity of a monoclonal antibody drug conjugate. *Clin Cancer Res* **10**: 7063–70.

Hansen C.G., Nichols B.J. (2009). Molecular mechanisms of clathrin-independent endocytosis. *J Cell Sci* **122**: 1713–21.

Hartmann F., Horak E.M., Garmestani K., Wu C., Brechbiel M.W., Kozak R.W., Tso J., Kosteiny S.A., Gansow O.A., Nelson D.L., *et al.* (1994). Radioimmunotherapy of nude mice bearing a human interleukin 2 receptor alpha-expressing lymphoma utilizing the alpha-emitting radionuclide-conjugated monoclonal antibody 212Bi-anti-Tac. *Cancer Res* **54**: 4362–70.

Hassfjell S., Brechbiel M.W. (2001). The development of the alpha-particle emitting radionuclides 212Bi and 213Bi, and their decay chain related radionuclides, for therapeutic applications. *Chem Rev* **101**: 2019–36.

Hayashi H., Asano R., Tsumoto K., Katayose Y., Suzuki M., Unno M., Kodama H., Takemura S., Yoshida H., Makabe K., *et al.* (2004). A highly effective and stable bispecific diabody for cancer immunotherapy: cure of xenografted tumors by bispecific diabody and T-LAK cells. *Cancer Immunol Immunother* **53**: 497–509.

Hnatowich D.J. (1987). Biodistribution of indium-111-labeled monoclonal antibodies. *J Nucl Med* **28**: 1924–5.

Hoffman M.A., Blessing J.A., Lentz S.S. (2003). A phase II trial of dolastatin-10 in recurrent platinum-sensitive ovarian carcinoma: a Gynecologic Oncology Group study. *Gynecol Oncol* **89**: 95–8.

Hollander I., Kunz A., Hamann P.R. (2008). Selection of reaction additives used in the preparation of monomeric antibody-calicheamicin conjugates. *Bioconjug Chem* **19**: 358–61.

Holliger P., Hudson P.J. (2005). Engineered antibody fragments and the rise of single domains. *Nat Biotechnol* **23**: 1126–36.

Hudson P.J., Kortt A.A. (1999). High avidity scFv multimers; diabodies and triabodies. *J Immunol Methods* **231**: 177–89.

Humm J.L. (1986). Dosimetric aspects of radiolabeled antibodies for tumor therapy. *J Nucl Med* **27**: 1490–7.

Issell B.F., Crooke S.T. (1978). Maytansine. *Cancer Treat Rev* **5**: 199–207.

Iyer A.K., Khaled G., Fang J., Maeda H. (2006). Exploiting the enhanced permeability and retention effect for tumor targeting. *Drug Discov Today* **17–18**: 812–8.

Jaggi J.S., Carrasquillo J.A., Seshan S.V., Zanzonico P., Henke E., Nagel A., Schwartz J., Beattie B., Kappel B.J., Chattopadhyay D., *et al.* (2007). Improved tumor imaging and therapy via i.v. IgG-mediated time-sequential modulation of neonatal Fc receptor. *J Clin Invest* **117**: 2422–30.

Jain R.K. (1990). Physiological barriers to delivery of monoclonal antibodies and other macromolecules in tumors. *Cancer Res* **50**: 814s–9s.

Jain R.K., Baxter L.T. (1988). Mechanisms of heterogeneous distribution of monoclonal antibodies and other macromolecules in tumors: significance of elevated interstitial pressure. *Cancer Res* **48**: 7022–32.

Jedema I., Barge R.M., van der Velden V.H., Nijmeijer B.A., van Dongen J.J., Willemze R., Falkenburg J.H. (2004). Internalization and cell cycle-dependent killing of leukemic cells by Gemtuzumab Ozogamicin: rationale for efficacy in CD33-negative malignancies with endocytic capacity. *Leukemia* **18**: 316–25.

Jensen R.L., Shan J.S., Freimark B.D., Harris D.A., King S.W., Lai J., Parseghian M.H. (2010). Clinical update: treatment of glioblastoma multiforme with radiolabeled antibodies that target tumor necrosis. *Curr Can Ther Rev* **6**: 13–8.

Junutula J.R., Raab H., Clark S., Bhakta S., Leipold D.D., Weir S., Chen Y., Simpson M., Tsai S.P., Dennis M.S., *et al.* (2008). Site-specific conjugation of a cytotoxic drug to an antibody improves the therapeutic index. *Nat Biotechnol* **26**: 925–32.

Jurcic J.G., Larson S.M., Sgouros G., McDevitt M.R., Finn R.D., Divgi C.R., Ballangrud A.M., Hamacher K.A., Ma D., Humm J.L., *et al.* (2002). Targeted alpha particle immunotherapy for myeloid leukemia. *Blood* **100**: 1233–9.

Karacay H., Brard P.Y., Sharkey R.M., Chang C.H., Rossi E.A., McBride W.J., Ragland D.R., Horak I.D., Goldenberg D.M. (2005). Therapeutic advantage of pretargeted radioimmunotherapy using a recombinant bispecific antibody in a human colon cancer xenograft. *Clin Cancer Res* **11**: 7879–85.

Karacay H., Sharkey R.M., McBride W.J., Griffiths G.L., Qu Z., Chang K., Hansen H.J., Goldenberg D.M. (2002). Pretargeting for cancer radioimmunotherapy with bispecific antibodies: role of the bispecific antibody's valency for the tumor target antigen. *Bioconjug Chem* **13**: 1054–70.

Kenanova V., Wu A.M. (2006). Tailoring antibodies for radionuclide delivery. *Expert Opin Drug Deliv* **3**: 53–70.

Kim E.E., DeLand F.H., Nelson M.O., Bennett S., Simmons G., Alpert E., Goldenberg D.M. (1980). Radioimmunodetection of cancer with radiolabeled antibodies to alpha-fetoprotein. *Cancer Res* **40**: 3008–12.

Kindler H.L., Tothy P.K., Wolff R., McCormack R.A., Abbruzzese J.L., Mani S., Wade-Oliver K.T., Vokes E.E. (2005). Phase II trials of dolastatin-10 in advanced pancreaticobiliary cancers. *Invest New Drugs* **23**: 489–93.

King H.D., Staab A.J., Pham-Kaplita K., Yurgaitis D., Firestone R.A., Lasch S.J., Trail P.A. (2003). BR96 conjugates of highly potent anthracyclines. *Bioorg Med Chem Lett* **13**, 2119–22.

Kipriyanov S.M. (2003). Generation and characterization of bispecific tandem diabodies for tumor therapy. *Methods Mol Biol* **207**: 323–33.

Klepfish A., Schattner A., Ghoti H., Rachmilewitz E.A. (2007). Addition of fresh frozen plasma as a source of complement to rituximab in advanced chronic lymphocytic leukaemia. *Lancet Oncol* **8**: 361–2.

Knox S.J., Goris M.L., Tempero M., Weiden P.L., Gentner L., Breitz H., Adams G.P., Axworthy D., Gaffigan S., Bryan K., *et al.* (2000). Phase II trial of yttrium-90-DOTA-biotin pretargeted by NR-LU-10 antibody/streptavidin in patients with metastatic colon cancer. *Clin Cancer Res* **6**: 406–14.

Knox S.J., Meredith R.F. (2000). Clinical radioimmunotherapy. *Semin Radiat Oncol* **10**: 73–93.

Kontermann R.E. (2005). Recombinant bispecific antibodies for cancer therapy. *Acta Pharmacol Sin* **26**: 1–9.

Korngold L., Pressman D. (1954). The localization of antilymphosarcoma antibodies in the Murphy lymphosarcoma of the rat. *Cancer Res* **14**: 96–9.

Kovtun Y.V., Audette C.A., Ye Y., Xie H., Ruberti M.F., Phinney S.J., Leece B.A., Chittenden T., Blattler W.A., Goldmacher V.S. (2006). Antibody-drug conjugates designed to eradicate tumors with homogeneous and heterogeneous expression of the target antigen. *Cancer Res* **66**: 3214–21.

Kovtun Y.V., Goldmacher V.S. (2007). Cell killing by antibody-drug conjugates. *Cancer Lett* **255**: 232–40.

Kraeber-Bodere F., Faibre-Chauvet A., Sai-Maurel C., Gautherot E., Fiche M., Campion L., Le Boterff J., Barbet J., Chatal J.F., Thedrez P. (1999a). Bispecific antibody and bivalent hapten radioimmunotherapy in CEA-producing medullary thyroid cancer xenograft. *J Nucl Med* **40**: 198–204.

Kraeber-Bodere F., Faivre-Chauvet A., Sai-Maurel C., Campion L., Fiche M., Gautherot E., Le Boterff J., Barbet J., Chatal J.F., Thedrez P. (1999b). Toxicity and efficacy of radioimmunotherapy in carcinoembryonic antigen-producing medullary thyroid cancer xenograft: comparison of iodine 131-labeled F(ab′)2 and pretargeted bivalent hapten and evaluation of repeated injections. *Clin Cancer Res* **5**: 3183s–9s.

Kubetzko S., Balic E., Waibel R., Zangemeister-Wittke U., Pluckthun A. (2006). PEGylation and multimerization of the anti-p185HER-2 single chain Fv fragment 4D5: effects on tumor targeting. *J Biol Chem* **281**: 35186–201.

Kupchan S.M., Ashmore J.W., Sneden A.T. (1978). Structure-activity relationships among *in vivo* active germacranolides. *J Pharm Sci* **67**: 865–7.

Kupchan S.M., Komoda Y., Court W.A., Thomas G.J., Smith R.M., Karim A., Gilmore C.J., Haltiwanger R.C., Bryan R.F. (1972). Maytansine, a novel antileukemic ansa macrolide from Maytenus ovatus. *J Am Chem Soc* **94**: 1354–6.

Lambert J.M. (2005). Drug-conjugated monoclonal antibodies for the treatment of cancer. *Curr Opin Pharmacol* **5**: 543–9.

Larson R.A., Sievers E.L., Stadtmauer E.A., Lowenberg B., Estey E.H., Dombret H., Theobald M., Voliotis D., Bennett J.M., Richie M., *et al.* (2005). Final report of the efficacy and safety of gemtuzumab ozogamicin (Mylotarg) in patients with CD33-positive acute myeloid leukemia in first recurrence. *Cancer* **104**: 1442–52.

Larson S.M. (1990). Clinical radioimmunodetection, 1978–1988: overview and suggestions for standardization of clinical trials. *Cancer Res* **50**: 892s–8s.

Lawrence L.J., Kortt A.A., Iliades P., Tulloch P.A., Hudson P.J. (1998). Orientation of antigen binding sites in dimeric and trimeric single chain Fv antibody fragments. *FEBS Lett* **425**: 479–84.

Lee M.D., Dunne T.S., Chang C.C., Ellestad G.A., Siegel M.M., Morton G.O., McGahren W.J., Borders D.B. (1987a). Calichemicins, a novel family of antitumor antibiotics. 2. Chemistry and structure of calichemicin Ψ_1^I. *J Am Chem Soc* 3466–8.

Lee M.D., Dunne T.S., Siegel M.M., Chang C.C., Morton G.O., Borders D.B. (1987b). Calichemicins, a novel family of Antitumor Antibiotics. 1. Chemistry and partial structure of Calichemicin Ψ_1^I. *J Am Chem Soc*: 3464–6.

Lee M.D., Ellestad G.A., Borders D.B. (1991). Chalicheamicins: Discovery, structure, chemistry, and interaction with DNA. *Acc Chem Res* 235–43.

Lee M.D., Manning J.K., Williams D.R., Kuck N.A., Testa R.T., Borders D.B. (1989). Calicheamicins, a novel family of antitumor antibiotics. 3. Isolation, purification and characterization of calicheamicins beta 1Br, gamma 1Br, alpha 2I, alpha 3I, beta 1I, gamma 1I and delta 1I. *J Antibiot (Tokyo)* **42**: 1070–87.

Lerchen H.G., Baumgarten J., von dem Bruch K., Lehmann T.E., Sperzel M., Kempka G., Fiebig H.H. (2001). Design and optimization of 20-O-linked camptothecin glycoconjugates as anticancer agents. *J Med Chem* **44**: 4186–95.

Leung S.O., Qu Z., Hansen H.J., Shih L.B., Wang J., Losman M.J., Goldenberg D.M., Sharkey R.M. (1999). The effects of domain deletion, glycosylation, and long IgG3 hinge on the biodistribution and serum stability properties of a humanized IgG1 immunoglobulin, hLL2, and its fragments. *Clin Cancer Res* **5**: 3106s–17s.

Lewis Phillips G.D., Li G., Dugger D.L., Crocker L.M., Parsons K.L., Mai E., Blattler W.A., Lambert J.M., Chari R.V., Lutz R.J., *et al.* (2008). Targeting HER2-positive breast cancer with trastuzumab-DM1, an antibody-cytotoxic drug conjugate. *Cancer Res* **68**: 9280–90.

Li L., Olafsen T., Anderson A.L., Wu A., Raubitschek A.A., Shively J.E. (2002). Reduction of kidney uptake in radiometal labeled peptide linkers conjugated to recombinant antibody fragments. Site-specific conjugation of DOTA-peptides to a Cys-diabody. *Bioconjug Chem* **13**: 985–95.

Liersch T., Meller J., Kulle B., Behr T.M., Markus P., Langer C., Ghadimi B.M., Wegener W.A., Kovacs J., Horak I.D., *et al.* (2005). Phase II trial of carcinoembryonic antigen radioimmunotherapy with 131I-labetuzumab after salvage resection of colorectal metastases in the liver: five-year safety and efficacy results. *J Clin Oncol* **23**: 6763–70.

Liersch T., Meller J., Bittrich M., Kulle B., Becker H., Goldenberg D.M. (2007). Update of carcinoembryonic antigen radioimmunotherapy with (131)I-labetuzumab after salvage resection of colorectal liver metastases: comparison of outcome to a contemporaneous control group. *Ann Surg Oncol* **14**: 2577–90.

Liu S. (2008). Bifunctional coupling agents for radiolabeling of biomolecules and target-specific delivery of metallic radionuclides. *Advanced Drug Delivery Reviews* **60**: 1347–70.

Lord J.M., Roberts L.M., Robertus J.D. (1994). Ricin: structure, mode of action, and some current applications. *FASEB J* **8**: 201–8.

Lu D., Jimenez X., Zhang H., Atkins A., Brennan L., Balderes P., Bohlen P., Witte L., Zhu Z. (2003). Di-diabody: a novel tetravalent bispecific antibody molecule by design. *J Immunol Methods* **279**: 219–32.

Mach J.P., Chatal J.F., Lumbroso J.D., Buchegger F., Forni M., Ritschard J., Berche C., Douillard J.Y., Carrel S., Herlyn M., *et al.* (1983). Tumor localization in patients by radiolabeled monoclonal antibodies against colon carcinoma. *Cancer Res* **43**: 5593–600.

Mach J.P., Forni M., Ritschard J., Buchegger F., Carrel S., Widgren S., Donath A., Alberto P. (1980). Use of limitations of radiolabeled anti-CEA antibodies and their fragments for photoscanning detection of human colorectal carcinomas. *Oncodev Biol Med* **1**: 49–69.

Madshus I.H., Stang E. (2009). Internalization and intracellular sorting of the EGF receptor: a model for understanding the mechanisms of receptor trafficking. *J Cell Sci* **122**: 3433–9.

Mahe M.A., Fumoleau P., Fabbro M., Guastalla J.P., Faurous P., Chauvot P., Chetanoud L., Classe J.M., Rouanet P., Chatal J.F. (1999). A phase II study of intraperitoneal radioimmunotherapy with iodine-131-labeled monoclonal antibody OC-125 in patients with residual ovarian carcinoma. *Clin Cancer Res* **5**: 3249s–53s.

Mattes M.J., Shih L.B., Govindan S.V., Sharkey R.M., Ong G.L., Xuan H., Goldenberg D.M. (1997). The advantage of residualizing radiolabels for targeting B-cell lymphomas with a radiolabeled anti-CD22 monoclonal antibody. *Int J Cancer* **71**: 429–35.

McCardle R.J., Harper P.V., Spar I.L., Bale W.F., Andros G., Jiminez F. (1966). Studies with iodine-131-labeled antibody to human fibrinogen for diagnosis and therapy of tumors. *J Nucl Med* **7**: 837–47.

McDevitt M.R., Ma D., Lai L.T., Simon J., Borchardt P., Frank R.K., Wu K., Pellegrini V., Curcio M.J., Miederer M., *et al.* (2001). Tumor therapy with targeted atomic nanogenerators. *Science* **294**: 1537–40.

McDevitt M.R., Sgouros G., Finn R.D., Humm J.L., Jurcic J.G., Larson S.M., Scheinberg D.A. (1998). Radioimmunotherapy with alpha-emitting nuclides. *Eur J Nucl Med* **25**: 1341–51.

McDonagh C.F., Turcott E., Westendorf L., Webster J.B., Alley S.C., Kim K., Andreyka J., Stone I., Hamblett K.J., Francisco J.A., *et al.* (2006). Engineered antibody-drug conjugates with defined sites and stoichiometries of drug attachment. *Protein Eng Des Sel* **19**: 299–307.

Meares C.F., McCall M.J., Deshpande S.V., DeNardo S.J., Goodwin D.A. (1988). Chelate radiochemistry: cleavable linkers lead to altered levels of radioactivity in the liver. *Int J Cancer Suppl* **2**: 99–102.

Meister A., Anderson M.E. (1983). Glutathione. *Annu Rev Biochem* **52**: 711–60.

Menrad A., Menssen H.D. (2005). ED-B fibronectin as a target for antibody-based cancer treatments. *Expert Opin Ther Targets* **9**: 491–500.

Meredith R.F. (2006). Ongoing investigations and new uses of radioimmunotherapy in the treatment of non-Hodgkin's lymphoma. *Int J Radiat Oncol Biol Phys* **66**: S23–9.

Meredith R.F., Alvarez R.D., Partridge E.E., Khazaeli M.B., Lin C.Y., Macey D.J., Austin J.M., Jr., Kilgore L.C., Grizzle W.E., Schlom J., *et al.* (2001). Intraperitoneal radioimmunochemotherapy of ovarian cancer: a phase I study. *Cancer Biother Radiopharm* **16**: 305–15.

Meredith R.F., Knox S.J. (2006). Clinical development of radioimmunotherapy for B-cell non-Hodgkin's lymphoma. *Int J Radiat Oncol Biol Phys* **66**: S15–22.

Milenic D.E., Brady E.D., Brechbiel M.W. (2004). Antibody-targeted radiation cancer therapy. *Nat Rev Drug Discov* **3**: 488–99.

Milenic D.E., Garmestani K., Chappell L.L., Dadachova E., Yordanov A., Ma D., Schlom J., Brechbiel M.W. (2002). *In vivo* comparison of macrocyclic and acyclic ligands for radiolabeling of monoclonal antibodies with 177Lu for radioimmunotherapeutic applications. *Nucl Med Biol* **29**: 431–42.

Miller M.L., Roller E.E., Wu X., Leece B.A., Goldmacher V.S., Chari R.V., Ojima I. (2004). Synthesis of potent taxoids for tumor-specific delivery using monoclonal antibodies. *Bioorg Med Chem Lett* **14**: 4079–82.

Miller M.L., Roller E.E., Zhao R.Y., Leece B.A., Ab O., Baloglu E., Goldmacher V.S., Chari R.V. (2004). Synthesis of taxoids with improved cytotoxicity and solubility for use in tumor-specific delivery. *J Med Chem* **47**: 4802–5.

Mueller B.M., Reisfeld R.A., Gillies S.D. (1990). Serum half-life and tumor localization of a chimeric antibody deleted of the CH2 domain and directed against the disialoganglioside GD2. *Proc Natl Acad Sci USA* **87**: 5702–5.

Muller D., Karle A., Meissburger B., Hofig I., Stork R., Kontermann R.E. (2007). Improved pharmacokinetics of recombinant bispecific antibody molecules by fusion to human serum albumin. *J Biol Chem* **282**: 12650–60.

Mulligan T., Carrasquillo J.A., Chung Y., Milenic D.E., Schlom J., Feuerstein I., Paik C., Perentesis P., Reynolds J., Curt G., *et al.* (1995). Phase I study of intravenous Lu-labeled CC49 murine monoclonal antibody in patients with advanced adenocarcinoma. *Clin Cancer* Res **1**: 1447–54.

Murray J.L., Macey D.J., Kasi L.P., Rieger P., Cunningham J., Bhadkamkar V., Zhang H.Z., Schlom J., Rosenblum M.G., Podoloff D.A. (1994). Phase II radioimmunotherapy trial with 131I-CC49 in colorectal cancer. *Cancer* **73**: 1057–66.

Musolino A., Naldi N., Bortesi B., Pezzuolo D., Capelletti M., Missale G., Laccabue D., Zerbini A., Camisa R., Bisagni G., *et al.* (2008). Immunoglobulin G fragment C receptor polymorphisms and clinical efficacy of trastuzumab-based therapy in patients with HER-2/neu-positive metastatic breast cancer. *J Clin Oncol* **26**: 1789–96.

Nahta R., Esteva F.J. (2006). Herceptin: mechanisms of action and resistance. *Cancer Lett* **232**: 123–38.

Namavari M., Padilla De Jesus O., Cheng Z., De A., Kovacs E., Levi J., Zhang R., Hoerner J.K., Grade H., Syud F.A., *et al.* (2008). Direct site-specific radiolabeling of an Affibody protein with 4-[18F]fluorobenzaldehyde via oxime chemistry. *Mol Imaging Biol* **10**: 177–81.

O'Donnell R.T., DeNardo G.L., Kukis D.L., Lamborn K.R., Shen S., Yuan A., Goldstein D.S., Carr C.E., Mirick G.R., DeNardo S.J. (1999). A clinical trial of radioimmunotherapy with 67Cu-2IT-BAT-Lym-1 for non-Hodgkin's lymphoma. *J Nucl Med* **40**: 2014–20.

O'Donoghue J.A., Bardies M., Wheldon T.E. (1995). Relationships between tumor size and curability for uniformly targeted therapy with beta-emitting radionuclides. *J Nucl Med* **36**: 1902–9.

Oflazoglu E., Stone I.J., Gordon K., Wood C.G., Repasky E.A., Grewal I.S., Law C.L., Gerber H.P. (2008). Potent anticarcinoma activity of the humanized anti-CD70 antibody h1F6 conjugated to the tubulin inhibitor auristatin via an uncleavable linker. *Clin Cancer Res* **14**: 6171–80.

Ojima I. (1997a). Preparation of taxoids as antitumor agents. Patent written in English. Application: WO 97-US2971 19970304. Priority: US 96-608003 19960304. CAN 127:248260 AN 1997:617970 CAPLUS.

Ojima I. (2008). Guided molecular missiles for tumor-targeting chemotherapy — case studies using the second-generation taxoids as warheads. *Acc Chem Res* **41**: 108–19.

Ojima I, Bounaud P-Y, Bernacki R.J. (1999). Designing taxanes to treat multidrug-resistant tumors. *Mod Drug Discov* **2**: 51–2.

Ojima I., Bounaud P.Y., Takeuchi C., Pera P., Bernacki R.J. (1998). New taxanes as highly efficient reversal agents for multidrug resistance in cancer cells. *Bioorg Med Chem Lett* **8**: 189–94.

Ojima I., Slater J.C., Michaud E., Kuduk S.D., Bounaud P.Y., Vrignaud P., Bissery M.C., Veith J.M., Pera P., Bernacki R.J. (1996). Syntheses and structure-activity relationships of the second-generation antitumor taxoids: exceptional activity against drug-resistant cancer cells. *J Med Chem* **39**: 3889–96.

Ojima I, Slater J.C., Pera P. Veith J.M., Abouabdellah A, Begue, J-P, Bernacki R.J. (1997b). Synthesis and biological activity of novel 3′-trifluoromethyl taxoids. *Bioorg Med Chem Lett* **7**: 133–8.

Olafsen T., Kenanova V.E., Wu A.M. (2006). Tunable pharmacokinetics: modifying the *in vivo* half-life of antibodies by directed mutagenesis of the Fc fragment. *Nat Protoc* **1**: 2048–60.

Orlova A., Nilsson F.Y., Wikman M., Widstrom C., Stahl S., Carlsson J., Tolmachev V. (2006). Comparative in vivo evaluation of technetium and iodine labels on an anti-HER2 affibody for single-photon imaging of HER2 expression in tumors. *J Nucl Med* **47**: 512–19.

Paganelli G., Bartolomei M., Ferrari M., Cremonesi M., Broggi G., Maira G., Sturiale C., Grana C., Prisco G., Gatti M., *et al.* (2001). Pre-targeted locoregional radioimmunotherapy with 90Y-biotin in glioma patients: phase I study and preliminary therapeutic results. *Cancer Biother Radiopharm* **16**: 227–35.

Paganelli G., Bartolomei M., Grana C., Ferrari M., Rocca P., Chinol M. (2006). Radioimmunotherapy of brain tumor. *Neurol Res* **28**: 518–22.

Paganelli G., Grana C., Chinol M., Cremonesi M., De Cicco C., De Braud F., Robertson C., Zurrida S., Casadio C., Zoboli S., *et al.* (1999). Antibody-guided three-step therapy for high grade glioma with yttrium-90 biotin. *Eur J Nucl Med* **26**: 348–57.

Pagel J.M., Hedin N., Subbiah K., Meyer D., Mallet R., Axworthy D., Theodore L.J., Wilbur D.S., Matthews D.C., Press O.W. (2003). Comparison of anti-CD20 and anti-CD45 antibodies for conventional and pretargeted radioimmunotherapy of B-cell lymphomas. *Blood* **101**: 2340–8.

Peterson J.J., Meares C.F. (1999). Enzymatic cleavage of peptide-linked radiolabels from immunoconjugates. *Bioconjug Chem* **10**: 553–7.

Pettit G.R. Ode R.H., Herald C.L., Von Dreele R.B., Michel C. (1976). The isolation and structure of dolatriol. *J Am Chem Soc* **98**: 4677–8.

Pettit G.R. (1997) 'The Dolastatins' in Herz W., Kirby G.W., Moore R.E., Steglich W., Tamm C. (Eds.), *Progress in the chemistry of organic natural products — 70th ed.* New York, NY: Springer-Verlag; pp. 1–79.

Pettit G.R., Barkoczy J. (1997). Tumor inhibiting tetrapeptide bearing modified phenethylamides, U.P. Dept., ed. (US), pp. 514–517.

Pettit G.R., Kamano Y., Fujii Y., Herald C.L., Inoue M., Brown P., Gust D., Kitahara K., Schmidt J.M., Doubek D.L., *et al.* (1981). Marine animal biosynthetic constituents for cancer chemotherapy. *J Nat Prod* **44**: 482–5.

Pettit G.R., Kamano Y., Herald C.L., Tuinman A.A., Boettner F.E., Kizu H., Schmidt J.M., Baczynskyj L., Tomer K.B., Bontems R.J. (1987). The isolation and structure of a remarkable marine animal antineoplastic constituent: Dolastatin 10[1a]. *J Am Chem Soc* **109**: 6883–5.

Pettit G.R., Singh S.B., Hamel E., Lin C.M., Alberts D.S., Garcia-Kendall D. (1989). Isolation and structure of the strong cell growth and tubulin inhibitor combretastatin A-4. *Experientia* **45**: 209–11.

Polson A.G., Calemine-Fenaux J., Chan P., Chang W., Christensen E., Clark S., de Sauvage F.J., Eaton D., Elkins K., Elliott J.M., *et al.* (2009). Antibody-drug conjugates for the treatment of non-Hodgkin's lymphoma: target and linker-drug selection. *Cancer Res* **69**: 2358–64.

Power B.E., Kortt A.A., Hudson P.J. (2003). Generation of recombinant multimeric antibody fragments for tumor diagnosis and therapy. *Methods Mol Biol* **207**: 335–50.

Press O.W., Corcoran M., Subbiah K., Hamlin D.K., Wilbur D.S., Johnson T., Theodore L., Yau E., Mallett R., Meyer D.L., *et al.* (2001). A comparative evaluation of conventional and pretargeted radioimmunotherapy of CD20-expressing lymphoma xenografts. *Blood* **98**: 2535–43.

Pressman D., Korngold L. (1953). The *in vivo* localization of anti-Wagner-osteogenic-sarcoma antibodies. *Cancer* **6**: 619–23.

Presta L.G. (2008). Molecular engineering and design of therapeutic antibodies. *Curr Opin Immunol* **20**: 460–70.

Qu Z., Griffiths G.L., Wegener W.A., Chang C.H., Govindan S.V., Horak I.D., Hansen H.J., Goldenberg D.M. (2005). Development of humanized antibodies as cancer therapeutics. *Methods* **36**: 84–95.

Razani B., Woodman S.E., Lisanti M.P. (2002). Caveolae: from cell biology to animal physiology. *Pharmacol Rev* **54**: 431–67.

Reardan D.T., Meares C.F., Goodwin D.A., McTigue M., David G.S., Stone M.R., Leung J.P., Bartholomew R.M., Frincke J.M. (1985). Antibodies against metal chelates. *Nature* **316**: 265–8.

Reardon D.A., Akabani G., Coleman R.E., Friedman A.H., Friedman H.S., Herndon J.E., 2nd, McLendon R.E., Pegram C.N., Provenzale J.M., Quinn J.A., *et al.* (2006). Salvage radioimmunotherapy with murine iodine-131-labeled antitenascin monoclonal antibody 81C6 for patients with recurrent primary and metastatic malignant brain tumors: phase II study results. *J Clin Oncol* **24**: 115–22.

Remillard S., Rebhun L.I., Howie G.A., Kupchan S.M. (1975). Antimitotic activity of the potent tumor inhibitor maytansine. *Science* **189**: 1002–5.

Ricart A.D., Tolcher A.W. (2007). Technology insight: cytotoxic drug immunoconjugates for cancer therapy. *Nat Clin Pract Oncol* **4**: 245–55.

Riva P., Franceschi G., Frattarelli M., Riva N., Guiducci G., Cremonini A.M., Giuliani G., Casi M., Gentile R., Jekunen A.A., *et al.* (1999). 131I radioconjugated antibodies for the locoregional radioimmunotherapy of high-grade malignant glioma — phase I and II study. *Acta Oncol* **38**: 351–9.

Rogers B.E., Roberson P.L., Shen S., Khazaeli M.B., Carpenter M., Yokoyama S., Brechbiel M.W., LoBuglio A.F., Buchsbaum D.J. (2005). Intraperitoneal radioimmunotherapy with a humanized anti-TAG-72 (CC49) antibody with a deleted CH2 region. *Cancer Biother Radiopharm* **20**: 502–13.

Roopenian D.C., Akilesh S. (2007). FcRn: the neonatal Fc receptor comes of age. *Nat Rev Immunol* **7**: 715–25.

Rossi E.A., Chang C.H., Losman M.J., Sharkey R.M., Karacay H., McBride W., Cardillo T.M., Hansen H.J., Qu Z., Horak I.D., *et al.* (2005). Pretargeting of carcinoembryonic antigen-expressing cancers with a trivalent bispecific fusion protein produced in myeloma cells. *Clin Cancer Res* **11**: 7122s–9s.

Rossi E.A., Sharkey R.M., McBride W., Karacay H., Zeng L., Hansen H.J., Goldenberg D.M., Chang C.H. (2003). Development of new multivalent-bispecific agents for pretargeting tumor localization and therapy. *Clin Cancer Res* **9**: 3886S–96S.

Rudnick S.I., Adams G.P. (2009). Affinity and avidity in antibody-based tumor targeting. *Cancer Biother Radiopharm* **24**: 155–61.

Ruiz-Cabello F., Cabrera T., Lopez-Nevot M.A., Garrido F. (2002). Impaired surface antigen presentation in tumors: implications for T cell-based immunotherapy. *Semin Cancer Biol* **12**: 15–24.

Safavy A., Bonner J.A., Waksal H.W., Buchsbaum D.J., Gillespie G.Y., Khazaeli M.B., Arani R., Chen D.T., Carpenter M., Raisch K.P. (2003). Synthesis and biological evaluation of paclitaxel-C225 conjugate as a model for targeted drug delivery. *Bioconjug Chem* **14**, 302–10.

Saga T., Neumann R.D., Heya T., Sato J., Kinuya S., Le N., Paik C.H., Weinstein J.N. (1995). Targeting cancer micrometastases with monoclonal antibodies: a binding-site barrier. *Proc Natl Acad Sci USA* **92**: 8999–9003.

Salzberg A.A., Dedon P.C. (2000). DNA bending is a determinant of calicheamicin target recognition. *Biochemistry* **39**: 7605–12.

Sauer S., Erba P.A., Petrini M., Menrad A., Giovannoni L., Grana C., Hirsch B., Zardi L., Paganelli G., Mariani G., *et al.* (2009). Expression of the oncofetal ED-B-containing fibronectin isoform in hematologic tumors enables ED-B-targeted 131I-L19SIP radioimmunotherapy in Hodgkin lymphoma patients. *Blood* **113**: 2265–74.

Schmidt M.M., Wittrup K.D. (2009). A modeling analysis of the effects of molecular size and binding affinity on tumor targeting. *Mol Cancer Ther* **8**: 2861–71.

Schultze J.L., Vonderheide R.H. (2001). From cancer genomics to cancer immunotherapy: toward second-generation tumor antigens. *Trends Immunol* **22**: 516–23.

Senter P.D. (2009). Potent antibody drug conjugates for cancer therapy. *Curr Opin Chem Biol* **13**: 235–44.

Shapiro W.R., Carpenter S.P., Roberts K., Shan J.S. (2006). (131)I-chTNT-1/B mAb: tumour necrosis therapy for malignant astrocytic glioma. *Expert Opin Biol Ther* **6**: 539–45.

Sharkey R.M., Goldenberg D.M. (2006). Advances in radioimmunotherapy in the age of molecular engineering and pretargeting. *Cancer Invest* **24**: 82–97.

Sharkey R.M., Goldenberg D.M. (2008). Use of antibodies and immunoconjugates for the therapy of more accessible cancers. *Adv Drug Deliv Rev* **60**: 1407–20.

Sharkey R.M., Karacay H., Cardillo T.M., Chang C.H., McBride W.J., Rossi E.A., Horak I.D., Goldenberg D.M. (2005). Improving the delivery of radionuclides for imaging and therapy of cancer using pretargeting methods. *Clin Cancer Res* **11**: 7109s–21s.

Shen J., Vil M.D., Jimenez X., Zhang H., Iacolina M., Mangalampalli V., Balderes P., Ludwig D.L., Zhu Z. (2007). Single variable domain antibody as a versatile building block for the construction of IgG-like bispecific antibodies. *J Immunol Methods* **318**: 65–74.

Shen S., Forero A., LoBuglio A.F., Breitz H., Khazaeli M.B., Fisher D.R., Wang W., Meredith R.F. (2005). Patient-specific dosimetry of pretargeted radioimmunotherapy using CC49 fusion protein in patients with gastrointestinal malignancies. *J Nucl Med* **46**: 642–51.

Shen W.C., Ryser H.J. (1981). cis-Aconityl spacer between daunomycin and macromolecular carriers: a model of pH-sensitive linkage releasing drug from a lysosomotropic conjugate. *Biochem Biophys Res Commun* **102**: 1048–54.

Siegel J.A., Pawlyk D.A., Lee R.E., Sasso N.L., Horowitz J.A., Sharkey R.M., Goldenberg D.M. (1990). Tumor, red marrow, and organ dosimetry for 131I-labeled anti-carcinoembryonic antigen monoclonal antibody. *Cancer Res* **50**: 1039s–42s.

Slavin-Chiorini D.C., Horan Hand P.H., Kashmiri S.V., Calvo B., Zaremba S., Schlom J. (1993). Biologic properties of a CH2 domain-deleted recombinant immunoglobulin. *Int J Cancer* **53**: 97–103.

Slavin-Chiorini D.C., Kashmiri S.V., Lee H.S., Milenic D.E., Poole D.J., Bernon E., Schlom J., Hand P.H. (1997). A CDR-grafted (humanized) domain-deleted antitumor antibody. *Cancer Biother Radiopharm* **12**: 305–16.

Slavin-Chiorini D.C., Kashmiri S.V., Schlom J., Calvo B., Shu L.M., Schott M.E., Milenic D.E., Snoy P., Carrasquillo J., Anderson K., *et al.* (1995). Biological properties of chimeric domain-deleted anticarcinoma immunoglobulins. *Cancer Res* **55**: 5957s–67s.

Smith S. (2001) Technology evaluation: SGN-15. *Curr Opin Mol Ther* **3**: 295s–303s.

Sorkin A., Goh L.K. (2008). Endocytosis and intracellular trafficking of ErbBs. *Exp Cell Res* **314**: 3093–106.

Spector N.L., Blackwell K.L. (2009). Understanding the mechanisms behind trastuzumab therapy for human epidermal growth factor receptor 2-positive breast cancer. *J Clin Oncol* **27**: 5838–47.

Srinivasachar K., Neville D.M., Jr. (1989). New protein cross-linking reagents that are cleaved by mild acid. *Biochemistry* **28**: 2501–9.

Srivastava S., Dadachova E. (2001). Recent advances in radionuclide therapy. *Semin Nucl Med* **31**: 330–41.

Stefanska B., Dzieduszycka M., Bontemps-Gracz M.M., Borowski E., Martelli S., Supino R., Pratesi G., De Cesare M., Zunino F., Kusnierczyk H., *et al.* (1999). 8,11-dihydroxy-6-[(aminoalkyl)amino]-7H-benzo[e]perimidin-7-ones with activity in multidrug-resistant cell lines: synthesis and antitumor evaluation. *J Med Chem* **42**: 3494–501.

Stickney D.R., Anderson L.D., Slater J.B., Ahlem C.N., Kirk G.A., Schweighardt S.A., Frincke J.M. (1991). Bifunctional antibody: a binary radiopharmaceutical delivery system for imaging colorectal carcinoma. *Cancer Res* **51**: 6650–5.

Stork R., Muller D., Kontermann R.E. (2007). A novel tri-functional antibody fusion protein with improved pharmacokinetic properties generated by fusing a bispecific single-chain diabody with an albumin-binding domain from streptococcal protein G. *Protein Eng Des Sel* **20**: 569–76.

Subbiah K., Hamlin D.K., Pagel J.M., Wilbur D.S., Meyer D.L., Axworthy D.B., Mallett R.W., Theodore L.J., Stayton P.S., Press O.W. (2003). Comparison of immunoscintigraphy, efficacy, and toxicity of conventional and pretargeted radioimmunotherapy in CD20-expressing human lymphoma xenografts. *J Nucl Med* **44**: 437–45.

Sun Y., Yu F., Sun B.W. (2009). Antibody-drug conjugates as targeted cancer therapeutics. *Yao Xue Xue Bao* **44**: 943–52.

Tagawa S.T., Beltran H., Vallabhajosula S., Goldsmith S.J., Osborne J., Matulich D., Petrillo K., Parmar S., Nanus D.M., Bander N.H. (2010). Anti-prostate-specific membrane antigen-based radioimmunotherapy for prostate cancer. *Cancer* **116**: 1075–83.

Tarli L., Balza E., Viti F., Borsi L., Castellani P., Berndorff D., Dinkelborg L., Neri D., Zardi L. (1999). A high-affinity human antibody that targets tumoral blood vessels. *Blood* **94**: 192–8.

Teicher B.A. (2009). Antibody-drug conjugate targets. *Curr Cancer Drug Targets* **9**: 982–1004.

Thurber G.M., Schmidt M.M., Wittrup K.D. (2008). Antibody tumor penetration: transport opposed by systemic and antigen-mediated clearance. *Adv Drug Deliv Rev* **60**: 1421–34.

Todorovska A., Roovers R.C., Dolezal O., Kortt A.A., Hoogenboom H.R., Hudson P.J. (2001). Design and application of diabodies, triabodies and tetrabodies for cancer targeting. *J Immunol Methods* **248**: 47–66.

Trail P.A., Willner D., Lasch S.J., Henderson A.J., Hofstead S., Casazza A.M., Firestone R.A., Hellstrom I., Hellstrom K.E. (1993). Cure of xenografted human carcinomas by BR96-doxorubicin immunoconjugates. *Science* **261**: 212–15.

Vaidyanathan G., Jestin E., Olafsen T., Wu A.M., Zalutsky M.R. (2009). Evaluation of an anti-p185(HER2) (scFv-C(H)2-C(H)3)2 fragment following radioiodination using two different residualizing labels: SGMIB and IB-Mal-D-GEEEK. *Nucl Med Biol* **36**: 671–80.

Vaishampayan U., Glode M., Du W., Kraft A., Hudes G., Wright J., Hussain M. (2000). Phase II study of dolastatin-10 in patients with hormone-refractory metastatic prostate adenocarcinoma. *Clin Cancer Res* **6**: 4205–8.

van Nagell J.R., Jr., Kim E., Casper S., Primus F.J., Bennett S., DeLand F.H., Goldenberg D.M. (1980). Radioimmunodetection of primary and metastatic ovarian cancer using radiolabeled antibodies to carcinoembryonic antigen. *Cancer Res* **40**: 502–6.

Ventimiglia J.B., Wikstrand C.J., Ostrowski L.E., Bourdon M.A., Lightner V.A., Bigner D.D. (1992). Tenascin expression in human glioma cell lines and normal tissues. *J Neuroimmunol* **36**: 41–55.

Walker M.A., Dubowchik G.M., Hofstead S.J., Trail P.A., Firestone R.A. (2002). Synthesis of an immunoconjugate of camptothecin. *Bioorg Med Chem Lett* **12**: 217–9.

Weng W.K., Levy R. (2003). Two immunoglobulin G fragment C receptor polymorphisms independently predict response to rituximab in patients with follicular lymphoma. *J Clin Oncol* **21**: 3940–7.

Widdison W.C., Wilhelm S.D., Cavanagh E.E., Whiteman K.R., Leece B.A., Kovtun Y., Goldmacher V.S., Xie H., Steeves R.M., Lutz R.J., *et al.* (2006). Semisynthetic maytansine analogues for the targeted treatment of cancer. *J Med Chem* **49**: 4392–408.

Willner D., Trail P.A., Hofstead S.J., King H.D., Lasch S.J., Braslawsky G.R., Greenfield R.S., Kaneko T., Firestone R.A. (1993). (6-Maleimidocaproyl)hydrazone of doxorubicin — a new derivative for the preparation of immunoconjugates of doxorubicin. *Bioconjug Chem* **4**: 521–7.

Wittel U.A., Jain M., Goel A., Chauhan S.C., Colcher D., Batra S.K. (2005). The *in vivo* characteristics of genetically engineered divalent and tetravalent single-chain antibody constructs. *Nucl Med Biol* **32**: 157–64.

Witzig T.E. (2006). Radioimmunotherapy for B-cell non-Hodgkin lymphoma. *Best Pract Res Clin Haematol* **19**: 655–68.

Woll P.J., Lorigan P., O'Brien M., Villalona-Calero M., Mastico R.A., Lutz R.J., Carrigan C., O'Keeffe J., Qin A., O'Leary J., *et al.* (2009). Clinical experience of IMGN901 (BB-10901, huN901-DM1) in patients with Merkel Cell Carcinoma (MCC). Paper presented at: AACR — NCI — EORTC.

Wolpert-Defilippes M.K., Adamson R.H., Cysyk R.L., Johns D.G. (1975). Cytotoxic action of maytansine, a novel ansa macrolide. *Biochemical Pharmacology* **24**: 751–4.

Wu A.M., Senter P.D. (2005). Arming antibodies: prospects and challenges for immunoconjugates. *Nat Biotechnol* **23**: 1137–46.

Wu A.M., Yazaki P.J. (2000). Designer genes: recombinant antibody fragments for biological imaging. *Q J Nucl Med* **44**: 268–83.

Wu C., Ying H., Grinnell C., Bryant S., Miller R., Clabbers A., Bose S., McCarthy D., Zhu R.R., Santora L., *et al.* (2007). Simultaneous targeting of multiple disease mediators by a dual-variable-domain immunoglobulin. *Nat Biotechnol* **25**: 1290–7.

Yazaki P.J., Kassa T., Cheung C.W., Crow D.M., Sherman M.A., Bading J.R., Anderson A.L., Colcher D., Raubitschek A. (2008). Biodistribution and tumor imaging of an anti-CEA single-chain antibody-albumin fusion protein. *Nucl Med Biol* **35**: 151–8.

Yokota T., Milenic D.E., Whitlow M., Schlom J. (1992). Rapid tumor penetration of a single-chain Fv and comparison with other immunoglobulin forms. *Cancer Res* **52**: 3402–8.

Zahnd C., Kawe M., Stumpp M.T., de Pasquale C., Tamaskovic R., Nagy-Davidescu G., Dreier B., Schibli R., Binz H.K., Waibel R., *et al.* (2008). Efficient tumor targeting with high-affinity designed ankyrin repeat proteins: effects of affinity and molecular size. *Cancer Res* **70**: 1595–605.

Zalevsky J., Leung I.W., Karki S., Chu S.Y., Zhukovsky E.A., Desjarlais J.R., Carmichael D.F., Lawrence C.E. (2009). The impact of Fc engineering on an anti-CD19 antibody: increased Fcgamma receptor affinity enhances B-cell clearing in nonhuman primates. *Blood* **113**: 3735–43.

Zalutsky M.R. (2006). Targeted alpha-particle therapy of microscopic disease: Providing a further rationale for clinical investigation. *J Nucl Med* **47**: 1238–40.

Zalutsky M.R., Vaidyanathan G. (2000). Astatine-211-labeled radiotherapeutics: an emerging approach to targeted alpha-particle radiotherapy. *Curr Pharm Des* **6**: 1433–55.

Zhang W., Gordon M., Schultheis A.M., Yang D.Y., Nagashima F., Azuma M., Chang H.M., Borucka E., Lurje G., Sherrod A.E., *et al.* (2007). FCGR2A and FCGR3A polymorphisms associated with clinical outcome of epidermal growth factor receptor expressing metastatic colorectal cancer patients treated with single-agent cetuximab. *J Clin Oncol* **25**: 3712–18.

9

Dual-Targeting Bispecific Antibodies as New Therapeutic Modalities for Cancer

*Zhenping Zhu**

9.1 Introduction

Antibodies quickly evolve by recombination, somatic mutation, and clonal selection *in vivo* to acquire the exquisite specificity and high affinity needed for an effective immune response. The modular structure of antibodies that permits *in vivo* reorganization also makes them exceptionally amenable to molecular manipulation and facilitates the development of desirable properties. As a result, the field of antibody engineering has developed rapidly in recent years, as researchers exploit the ease with which antibodies can be genetically modified to develop ever more powerful therapeutics. So far, 28 antibody-based products have been approved by the Food and Drug Administration (FDA) for therapeutic applications, including ten for oncology indications. While the inoculation of animals with antigen and high-throughput screening of hybridoma clones is still common practice, *in vitro* selection is rapidly becoming the norm. Through the use of synthetic or semi-synthetic libraries in microscopic selection techniques, such as via phage display (Sidhu, 2005), yeast surface display (Colby *et al.*, 2004), and ribosomal display (Lipovsek and Pluckthun, 2004), identifying an antibody to a given target has become a routine, although by no means trivial, matter.

Once an antibody specific to a particular antigen is identified, numerous mechanisms of action can be exploited for therapy. These mechanisms include: (1) As an

* Kadmon Pharmaceuticals, New York, NY, USA.

irinotecan (Cunningham *et al.*, 2004). In head and neck cancer, the addition of Erbitux® to radiation treatment increases overall survival from 29.3 month to 49.0 months and locoregional control of the disease from 14.9 to 24.4 months, without significantly increasing toxicity (Bonner *et al.*, 2006). Herceptin®, an antibody to the extracellular domain (ECD) of HER2, increases the response, time to progression, and overall survival of breast cancer patients on paclitaxel therapy from 32% to 50%, 4.6 months to 7.4 months, and 20.3 months to 25.1 months, respectively, without substantial additional side effects (Slamon *et al.*, 2001). Further, the addition of Avastin®, an antibody directed against VEGF, to an irinotecan/5-fluorouracil/leucovorin regimen for first-line treatment of CRC increases the median survival from 15.6 months to 20.3 months, progression-free survival from 6.2 months to 10.6 months, and the response rate from ~ 35% to ~ 45% (Hurwitz *et al.*, 2004).

9.2.2 *Combinations of antibodies in cancer therapy*

While combination treatment with antibodies and chemo-/radiotherapy is clearly effective, improvements must still be made. Since the toxicity of these regimens is attributable primarily to the cytotoxic agents, fine tuning of the dosage of these components is unlikely to yield dramatic results. To this end, accumulating evidence suggest that a better strategy for increasing efficacy while limiting systemic toxicity is through combinations of mAb directed against multiple proteins associated with tumor growth and survival in order to inhibit specific mechanisms of tumor escape.

The most obvious potential antibody-based combination therapy is simultaneous administration of two or more approved antibodies. For example, the combination of Erbitux® and Avastin®, which have both been approved for CRC, is particularly noteworthy. A Phase II study (81 patients) has been performed to test the effect of the combination of both antibodies with or without irinotecan for irinotecan-refractory CRC (Saltz *et al.*, 2007). Of those receiving Erbitux®, Avastin®, and irinotecan, 37% achieved a partial response and had a median time to progression of 7.9 months, while 20% of those receiving just Erbitux® and Avastin® had a partial response, with a median time to progression of 5.6 months. Although not all the drug permutations were included, and inter-study comparisons were not statistically valid, the relatively high response rate of Erbitux®/Avastin®/irinotecan combination (37% response) relative to Erbitux®/irinotecan alone (23% response, historical rate) suggests that antibody combination therapy warrants further study. Phase III trials with this combination for CRC are underway. Similarly, multiple clinical trials using a combination of Herceptin® and Avastin® are being conducted in various breast cancer patients. Since this field is still in its infancy, most of the evidence supporting the use of antibody-based combination therapies comes from pre-clinical studies.

9.2.2.1 *Targeting EGFR and VEGF receptor 2 (VEGFR2)*

Analogous to the Erbitux®/Avastin® combination, preclinical studies using antibodies against EGFR and VEGFR2 have been performed in gastric (Jung *et al.*, 2002), colon (Shaheen *et al.*, 2001; Tonra *et al.*, 2009), and glioblastoma cancer models (Lamszus *et al.*, 2005). Erbitux® augments the tumor growth inhibition of an anti-VEGFR2 antibody, DC101, in only the gastric model, but it increases endothelial cell apoptosis in all three models, and significantly reduces the number of satellite microtumors in the glioblastoma model. This peculiar therapeutic effect of Erbitux® stems from the fact that EGFR expression in gliomas is associated with migration and invasion (Penar *et al.*, 1997), and anti-VEGFR2 antibody therapy, while significantly decreasing tumor volume, also increases tumor cell invasion and the formation of satellite tumors (Kunkel *et al.*, 2001). Together, these results indicate that inhibition of EGFR signaling may be a beneficial supplement to anti-angiogenic therapies for various tumor types.

9.2.2.2 *Targeting EGFR and insulin-like growth factor receptor (IGFR)*

While inhibition of EGFR activity clearly slows the growth of EGFR-expressing cell lines, and has demonstrated clinical efficacy, it is naive to assume that all EGFR-expressing tumors are equally sensitive to EGFR-inhibitory drugs. In one particular glioblastoma cell line, resistance to anti-EGFR therapy is achieved through upregulation of the insulin-like growth factor receptor (IGFR) (Chakravarti *et al.*, 2002). There is also evidence of synergy between EGF and IGF-I signaling in tumor cells (Krane *et al.*, 1991; Kobari *et al.*, 1998). To this end, an antibody to the ECD of IGFR, IMC-A12, has since been developed (Burtrum *et al.*, 2003). This antibody has anti-tumor activity on its own in multiple tumor xenograft models and is currently in Phase II trials in various cancer indications. In pre-clinical studies, combination of IMC-A12 with IMC-11F8, an antibody to EGFR, inhibited tumor growth to a greater extent than either monotherapy (Lu *et al.*, 2005), demonstrating the effectiveness of antibody-based combination therapy. The fact that similar therapeutic effects are obtained in the same study from a BsAb constructed from the parental Fv of IMC-A12 and IMC-11F8 underscores the significance of an antibody-based dual-receptor targeting approach (see below for further discussion).

9.2.2.3 *Targeting EGFR and HER2*

It is well known that misregulation of the ErbB family of receptor tyrosine kinases is a cause of many cancers. Overexpression of HER2 can be particularly disruptive to homeostasis because of its ability to heterodimerize with the other family members (Leahy, 2004). To test the effect of interfering with both EGFR and

HER2, the effect of both Erbitux® and Herceptin® on the growth of human ovarian cancer cells has been tested (Ye *et al.*, 1999). Each antibody as a single agent inhibits cell growth by 55% and 35%, respectively, with increased populations of cells in G1 arrest. The effect of combining both antibodies is additive, resulting in cell growth inhibition of 80%. These data clearly suggest that combination therapy with Erbitux® and Herceptin® could be useful in cancers caused by misregulation of the ErbB family of receptor tyrosine kinases.

9.2.2.4 *Targeting different epitopes of HER2 and EGFR*

The use of multiple antibodies can also be therapeutic if they target different epitopes of the same protein. Both Herceptin® and Omnitarg® bind HER2. Structures of each antibody in complex with the ECD of the receptor show that they bind distinct epitopes. Herceptin® binds the juxtamembrane domain IV (Cho *et al.*, 2003) and inhibits ECD cleavage, which has been implicated as a mechanism for activation of the cytoplasmic domain (Molina *et al.*, 2001), but does not directly inhibit activation by heterodimerization with the other ErbB family members (Fendly *et al.*, 1990). Omnitarg® binds the dimerization domain II (Franklin *et al.*, 2004), and directly inhibits dimerization and phosphorylation (Agus *et al.*, 2002), but does not inhibit ECD cleavage. Combination therapy with these two antibodies would thus provide complementary mechanisms of action for disabling HER2 activation. In fact, Herceptin® and Omnitarg® have been shown to synergistically inhibit the survival of breast cancer cells in an *in vitro* study (Nahta *et al.*, 2004). Similarly, the same additive and/or synergistic antitumor activity has been observed with the combination of two anti-EGFR antibodies that bind to separate, non-overlapping, epitopes on the ECD of the receptor (Dechant *et al.*, 2008).

9.2.3 *Alternative approaches to antibody combinations*

9.2.3.1 *Oligoclonal antibodies*

If treatment with multiple antibodies provides opportunities to disable cancer cells by multiple mechanisms, then theoretically, more is better. The bulk of the evidence in favor of oligoclonal antibodies (defined mixtures of mAb) comes not from trials in cancer, but from other diseases in which escape from therapy can also be life threatening: neurotoxin poisoning. Relative to monoclonal treatment, oligoclonal antibody cocktails are 200-fold more potent at neutralizing tetanus toxin *in vivo* (Volk *et al.*, 1984), and a cocktail of three antibodies is 20,000-fold more potent against botulinum toxin (Nowakowski *et al.*, 2002). The increased potency is likely due to three factors. First and foremost, the functional affinity of the oligoclonal cocktail is increased because each antibody species in the cocktail binds a different epitope of the toxin. Since simultaneous dissociation of multiple antibodies is rare, free toxin is at extremely low concentration. Second, obstruction of a greater surface area of the toxin on its own may result in better inhibition. Third, large immunocomplexes may be cleared faster by the reticulo-endothelial system. Unfortunately, the development and characterization of each of the components in the cocktail may be prohibitively time-consuming and expensive. The development, manufacturing and commercialization of an antibody cocktail may also face significant regulatory challenges.

9.2.3.2 *Polyclonal antibodies*

An alternative to defined mixtures of mAb is the use of polyclonal antibodies. Often regarded as primitive reagents best suited for molecular biology applications, polyclonal antibodies are regaining favor as potential therapeutic agents. Perhaps less *en vogue* than mAb, purified polyclonal IgGs from human serum are approved drugs and account for ~ $3 billion of the IgG market (Haurum and Bregenholt, 2005). Recently, instead of relying on serum donors for IgGs, recombinant methods for generating polyclonal antibodies have been developed (Sharon *et al.*, 2005). Basically, phage display of a Fab library is used in conjunction with negative selection to isolate many antibodies that specifically bind cancer cells. The DNA encoding these Fab fragments is recloned *en masse* into a mammalian IgG expression vector, and polyclonal IgG harvested (Williams and Sharon, 2002a; Williams *et al.*, 2002b). In one experiment, the polyclonal IgG was demonstrated to be more effective than a monoclonal antibody at inhibiting the growth of the cancer cell line against which it was raised (Liebman *et al.*, 2004). Alternatively, cell lines expressing the mAb of interest can be banked and maintained individually, and then combined just before the final production process.

While the concept of polyclonal Abs is appealing, and could well be efficacious in animal models, they are likely to be met with skepticism by the FDA. First, the argument in favor of polyclonal therapy is more of an extension of oligoclonal therapy than an evidence-based argument. Moreover, unlike polyclonal IgG derived from serum donors, which have obviously been proven "safe" in humans, an undefined mixture of recombinant IgGs is both a safety and an efficacy concern, particularly as batch consistency in the manufacturing process would be very difficult to test and control.

9.3 Bispecific Antibodies (BsAb) in Cancer Therapy

9.3.1 *BsAb in re-directing effector cells and other cytotoxic agents*

BsAb are molecules that can simultaneously bind two separate and distinct antigens (or different epitopes of the same antigen). The primary use of BsAb has been to re-direct cytotoxic immune effector cells for enhanced killing of tumor cells by ADCC and other cytotoxic mechanisms mediated by the effector cells. In this context, one arm of the BsAb binds an antigen on tumor cells, and the other binds a determinant expressed on effector cells, such as CD3, CD16, or CD64, which are expressed on T lymphocytes, natural killer (NK) cells, or other mononuclear cells (Xiong *et al.*, 2002; Kipriyanov *et al.*, 2002; James *et al.*, 2001; Baeuerle *et al.*, 2003; Zhu *et al.*, 1996). By cross-linking tumor and effector cells, the BsAb not only brings the effector cells within the proximity of the tumor cells but also simultaneously triggers their activation, leading to effective tumor cell killing. Among the binding and activation molecules on effector cells, CD3 on T cells is clearly the one that has been utilized the most by investigators in the field. CD3 is expressed on the surface of all T cells, including cytotoxic T cells, which are not only abundant but can be stimulated to proliferate and kill target cells efficiently. Thus a BsAb that is directed against both an antigen expressed on a tumor cell surface and CD3 has a capacity to activate and re-direct cytotoxic T cells to kill the tumor cells. Effective tumor cell killing *in vitro* and tumor inhibition *in vivo* have been demonstrated with a variety of anti-CD3-based BsAb molecules (Xiong *et al.*, 2002; Kipriyanov *et al.*, 2002; James *et al.*, 2001; Baeuerle *et al.*, 2003). However, results from a limited number of early stage clinical studies with these BsAb were far less impressive, albeit promising (Cao and Lam, 2003). In addition to CD3, T cells require a second co-stimulatory signal, e.g. CD28/B7 pathway, for full activation. Without the second signal, the T cells may even become anergic. Based on these findings, anti-CD3-based BsAb were combined with an anti-CD28 antibody or an anti-CD28-based BsAb. These combinations led to enhanced tumor-killing efficacy *in vitro*, and improved anti-tumor activity *in vivo* (Bohlen *et al.*, 1997; Grosse-Hovest *et al.*, 1999; Cochlovius *et al.*, 2000). As an alternative, *ex vivo* T cell pre-activation, for example, with anti-CD3, anti-CD28, and/or IL2, has been shown to enhance the anti-tumor activity of anti-CD3-based BsAb in animal models (Peipp and Valerius, 2002; Cao and Lum, 2003; Lum and Davol, 2005; Müller and Kontermann, 2007).

Significant clinical benefits were recently reported with the use of an anti-CD19 x CD3 BsAb, in BiTE, or "bispecific T-cell-engager," format (Bargou *et al.*, 2008). BiTE are recombinant BsAb consisting of two scFv fragments that are linked together via a short peptide linker (Wolf *et al.*, 2005). One scFv of BiTE is directed against a tumor antigen on target cell surface, and the other is specific for CD3 on T cells. BiTE is capable of transiently engaging both T cells and tumor cells, while potently activating T cells' cytolytic mechanism, leading to efficient tumor cell lysis, without the need of co-stimulating signals or pre-activation of the T effector cells. Potent tumor killing activity *in vitro* and tumor inhibitory efficacy *in vivo* have been demonstrated with a number of BiTE directed against different tumor antigens, including CD19, EpCAM, CEA, CD33, EphA2, EGFR, HER2, and MCSP (Baeuerle *et al.*, 2009a). The most advanced candidate, blinatumomab, an anti-CD19 x anti-CD3 BiTE, demonstrated potent anti-tumor activity in a variety of *in vitro*, *ex vivo*, and *in vivo* models (Cheadle, 2006). Blinatumomab is currently in Phase I trials in patients with relapsed NHL and in a Phase II trial in patients with acute B cell lymphoblastic leukemia. In a Phase I trial published recently (Bargou *et al.*, 2008), blinatumomab induced partial and complete regression in NHL patients at very low doses (15 $\mu g/m^2$/day). All seven patients receiving 60 $\mu g/m^2$/day of blinatumomab showed tumor regression. The toxicity profile of blinatumomab appeared to be manageable, mostly lymphopenia, leukopenia, fever, and enzyme abnormality. No evidence of global T cell activation such as cytokine release syndrome was noted. A second BiTE, an anti-EpCAM x anti-CD3 BsAb, is undergoing Phase I trials in patients with late-stage lung and gastrointestinal cancers (Schlereth *et al.*, 2005; Amann *et al.*, 2009; Kufer *et al.*, 2009).

On April 20, 2009, a "quadroma," or hybrid hybridoma-derived rat/mouse chimeric BsAb, Catumaxomab (Removab®), received European market approval for the intraperitoneal treatment of malignant ascites in patients with EpCAM-positive carcinomas. Catumaxomab is a product of the so-called Triomab class, a tri-functional BsAb comprising a Fab fragment of a mouse IgG2a antibody specific to EpCAM, a Fab fragment of a rat IgG2b antibody specific to CD3, and a hybrid mouse/rat Fc domain (Ruf *et al.*, 2007; Shen and Zhu, 2008). The hybrid mouse/rat Fc domain facilitates a one-step purification of catumaxomab from the monospecific parental antibodies in large-scale productions though Protein A columns, based on the difference in binding strength to Protein A between the BsAb and its parental monospecific antibodies (mouse IgG> mouse/rat chimeric IgG>> rat IgG). Triomab® antibodies bind to cancer-specific surface antigens (via the anti-tumor antigen arm) and recruit both T cells (via the anti-CD3 arm) as well as accessory cells, such as macrophages, dendritic cells, and natural killer cells (via the Fc domain) to the tumor site. As a result, they provide high potency in cancer cell killing, activating both arms of the immune system — the adaptive one with cytotoxic T cells as effectors and the innate one including accessory effector cells (Ruf *et al.*, 2007; Shen and Zhu, 2008).

Complete data on the Phase II/III trial were reported recently and confirmed the clinical benefit of catumaxomab treatment in cancer patients with malignant ascites (Fresenius, 2007; Shen and Zhu, 2008). A total of 258 patients were enrolled in the trial, including 129 ovarian cancer patients and an additional 129 patients with malignant ascites caused by other forms of epithelial tumors (gastric, 51%; breast, 10%; pancreatic, 7%; colorectal, 6%; others, 26%). Patients were randomized in a 2:1 ratio, with 170 patients to receive catumaxomab following paracentesis on days 0, 3, 7, and 10 by intraperitoneal (i.p.) infusions of 10, 20, 50, and 150 μg, respectively, and 88 patients to be treated by paracentesis therapy alone. In the catumaxomab treatment arm, 131 patients received all four infusions. Catumaxomab significantly prolonged the puncture-free survival (the primary endpoint) with a median of 46 days, compared to that of 11 days in patients who were treated by paracentesis only ($P < 0.0001$). The median time to the first paracentesis therapy improved to 77 days in the catumaxomab group vs. 13 days in the control group ($P < 0.0001$). Statistical analysis showed that the median time to progression of the 170 patients who received catumaxomab therapy, either as a whole group, as subgroups categorized by patients with ovarian cancer (85 patients), or other primary cancers (85 patients), was significantly prolonged compared to their corresponding patient populations in the control group (111, 111, and 110 days in the antibody group vs. 35, 35, and 34 days in the control group, respectively). The median survival of the 131 patients who had received all four catumaxomab infusions was 86 days vs. that of 68 days for patients in the control group ($P = 0.0085$) (Fresenius, 2007; Shen and Zhu, 2008). In addition to catumaxomab, a second trifunctional BsAb constructed based on the same Triomab technique, ertumaxomab, directed against both HER2 and CD3, is currently under Phase I clinical development (Kiewe *et al.*, 2006; Kiewe and Thiel, 2008; Jäger *et al.*, 2009).

In addition to effector cell targeting, BsAb have also been used to enrich the tumor/normal tissue localization ratio of chemo- or radiotherapeutic agents. In this setting, one arm of the BsAb binds an antigen expressed on the cell targeted for destruction, and the other binds either directly to a cytotoxic agent, or a carrier coupled with the cytotoxic agent, such as a chemotherapeutic drug, radioisotope, or toxin. The naked BsAb is administered first, and after sufficient time has passed for the BsAb to bind tumor cells and clear from normal tissue, the cytotoxic molecule is delivered, with rapid accumulation in the tumor, due to its affinity for the tumor-bound BsAb (Corvalan *et al.*, 1988; Ford *et al.*, 2001; Ferrini *et al.*, 2001; Chang *et al.*, 2002).

9.3.2 *BsAb that simultaneously target two tumor-associated molecules (dual-targeting BsAb)*

Although there are many potential benefits to using approved mAb in combination therapy, there are also many drawbacks. The broad clinical applications of antibody combination therapy have not been pursued significantly thus far, because only ten antibodies have been approved for cancer, and because their approved indications are mostly non-overlapping. The characterization and development of a mAb-based therapeutic can be a very time-consuming process. Repeating the process of research, development, manufacturing, and pursuing clinical trials and regulatory pathways for multiple antibodies as single agents, only to combine them in one therapy, would present an unacceptable cost burden and development delay to most pharmaceutical and biotech companies. Furthermore, developing a mAb for an indication already met by one made by another company poses a serious investment risk. From the perspective of the consumer (and insurance companies), the cost of mAb therapy is staggering — a typical mAb regimen costs tens of thousands of dollars; the addition of a second mAb to a treatment regimen could make such treatments prohibitively expensive.

An attractive alternative to mAb combinations is the use of BsAb that simultaneously target two tumor-associated antigens (e.g. growth factor receptors), leading to downregulation/blockade of two or more signal transduction pathways critical to tumor proliferation/survival, which provides enhanced anti-tumor activity (Lu *et al.*, 2001; 2004b; 2005; Jimenez *et al.*, 2005; Marvin and Zhu, 2006). BsAb have the potential to alleviate the aforementioned issues associated with mAb combinations by combining two treatment modalities in one molecule and providing unique biophysical properties that can be exploited for novel treatments. If BsAb could be produced in similar quantities and quality as the normal mAb IgG, then significant cost savings could also be achieved.

9.3.2.1 *BsAb to extracellular targets*

As mentioned previously, it has recently been determined that in some cases, upregulation of IGFR compensates for inhibition of EGFR, and renders the cells insensitive to anti-EGFR therapy (Chakravarti *et al.*, 2002). It is therefore plausible that a BsAb inhibiting both receptors would have increased efficacy over either mAb. With this in mind, two bispecific IgG-like molecules targeting both EGFR and IGFR have been constructed from the variable domains of IMC-11F8 and IMC-A12.

In one format, the scFvs are fused to the constant domains of an IgG ($scFv_{EGFR}$-C_L and $scFv_{IGFR}$-C_H1-C_H2-C_H3), and bispecificity is achieved through C_L-C_H1 dimerization (Zuo *et al.*, 2000; Lu *et al.*, 2004b). This construct, called $(scFv)_4$-IgG, blocks ligand binding to both receptors, and inhibits EGF- and IGF-I-induced activation of EGFR, IGFR, and their respective downstream signaling pathways. In a pancreatic tumor cell line that expresses high levels of both EGFR and IGFR, treatment with anti-EGFR or anti-IGFR antibody alone only inhibited growth by 25–35%, while treatment with both monospecific antibodies or the bispecific construct inhibited growth by ~ 80% and ~ 60%, respectively. These results are promising and validate the approach of using a BsAb as a dual-targeting signal transduction inhibitor.

In the another format, a bispecific IgG-like antibody that targets EGFR and IGFR is made by substitution of the C_L and C_H1 domains with V_L and V_H domains conferring secondary specificity, resulting in a diabody–Fc fusion, called a "di-diabody" (Lu *et al.*, 2005). The two chains (V_LIGFR-V_HEGFR and V_LEGFR-V_HIGFR-CH2-CH3) associate by the natural affinity of V_L and V_H domains. This 150 kDa protein binds both receptors simultaneously and blocks the signaling pathways stimulated by EGF and IGF-I. Furthermore, the di-diabody triggers internalization and degradation of IGFR and also mediates ADCC activity on tumor cells that express EGFR and/or IGFR. Most importantly, the di-diabody effectively inhibited the growth of two tumor xenografts *in vivo*. Unfortunately, this construct showed significantly reduced stability in nude mouse models. Unlike a normal IgG, in which the light chains (LCs) and heavy chains (HCs) are covalently linked by inter-chain disulfide bonds, the di-diabody relies on a non-covalent association between the V_L and V_H domains to remain intact. Dissociation *in vivo* between the two polypeptide chains leads to inactivation of the protein, as the lighter fragment (~ 25 kDa) is rapidly cleared from serum by renal filtration. This instability can probably be surmounted by further engineering. For example, disulfide bonds can be introduced between the V_L and V_H (Zhu *et al.*, 1997; Jung *et al.*, 1994; FitzGerald *et al.*, 1997), and the V_L–V_H interface can be improved by evolution of better packing interactions (Worn and Pluckthun, 2001).

Recently, anti-angiogenic bispecific diabodies have been constructed by combining the scFvs that bind the ECD of VEGFR1 and VEGFR2 (Lu *et al.*, 2001), or VEGFR2 and VEGFR3 (Jimenez *et al.*, 2005). The dual-receptor inhibitors offer significant advantages over single-receptor inhibitors. For example, blocking both VEGFR1 and VEGFR2 could neutralize endothelial cell activation by not only placenta growth factor and VEGF-B (both are VEGFR1-specific ligands), but also by VEGF (which binds VEGFR1 and VEGFR2), VEGF-C, VEGF-D, and VEGF-E (the latter three all bind both VEGFR2 and VEGFR3). Therefore, cancerous escape from inhibition of one signaling cascade by upregulation of others is thwarted by the inhibition of the other receptor.

We believe that BsAb dual-targeting the ECD of signal transduction proteins offer a valuable approach for cancer treatment. In addition to simultaneous inhibition of two signaling pathways, these constructs, when built with an intact Fc domain, may also support targeted effector mechanisms such as ADCC and CMC, and are expected to have long serum half lives. It is also possible to create bispecific constructs that target distinct epitopes on the same receptor to provide different mechanisms of inhibition (similar to Herceptin® and Omnitarg®, described above) or to convert non-inhibitory monospecific antibodies into inhibitory bispecific ones (see below for further discussion) (Lu *et al.*, 1999).

9.3.2.2 *Intrabodies*

BsAb in the form of "intrabodies," i.e. antibodies that expressed as intracellular proteins, can be used as gene therapy agents to inhibit signal transduction pathways intracellularly. A bispecific intrabody that inhibits two growth factor receptors involved in angiogenesis has been expressed as an endoplasmic reticulum (ER)-retained protein (Jendreyko *et al.*, 2003; 2005). In this report, an adenovirus encodes a bispecific IgG-like protein comprising an N-terminal fusion of an anti-Tie2 scFv and a C-terminal fusion of an anti-VEGFR2 scFv to the Fc of an IgG, along with a C-terminal KDEL ER-retention tag (i.e. $scFv_{Tie2}$-Fc-$scFv_{VEGFR2}$-KDEL). The authors demonstrated that when the BsAb is expressed, both Tie2 and VEGFR2 do not express at the cell surface, but rather co-localize with the BsAb in the ER. Remarkably, the surface depletion of the target proteins by the BsAb persists for 15 days. When a monospecific tetravalent version of the anti-VEGFR2 scFv is used ($scFv_{VEGFR2}$-Fc-$scFv_{VEGFR2}$-KDEL), only surface depletion of VEGFR2 is achieved (Jendreyko *et al.*, 2005). The induced "knockout" of both receptors results in 90% inhibition of capillary formation in human endothelial cells, whereas the monospecific scFvs for VEGFR2 and Tie2 inhibit capillary formation by only 44% and 11%, respectively. Significantly, peripheral injection of the adenovirus encoding this bispecific construct greatly inhibited tumor growth *in vivo*, and the inhibition by the bispecific intrabody was statistically more significant than that of the monospecific tetravalent intrabody, indicating that the combination therapy does in fact yield better results. As anti-angiogenic therapy has been shown to be an effective treatment option in humans, the ability to knock out two receptors in the angiogenesis cascade with a bispecific antibody would thus be of great value. Unfortunately, similar to other forms of gene therapy, selective delivery of the bispecific intrabodies to the desired cells is technically challenging (Cao and Heng, 2005).

9.3.3 *Novel applications of BsAb*

9.3.3.1 *Binding avidity enhancement*

One very valuable benefit of BsAb is the enhanced avidity they pose for their antigen (Lu *et al.*, 1999). High binding affinity (or avidity) is usually beneficial and, in some cases, is required for the biological activity of a therapeutic antibody. In addition to having intrinsic high affinity on a binding unit (i.e. a Fab or scFv) to antigen basis, normal IgG antibodies also exploit the avidity effect to increase their association with antigens due to their bivalent binding towards the targets. To this end, a BsAb directed against two separate (non-overlapping) epitopes on the same target molecule may most likely possess an increased binding avidity due to its bivalent paratopic binding. Similarly, a tetravalent IgG-like BsAb, with bivalent binding for each antigen, are expected, once bound to cell-surface antigens, to dissociate at a very slow rate; two dissociation events must occur simultaneously for the BsAb to be free from the cell. In addition, bivalent binding to a target, particularly a cell-surface receptor, under many circumstances is a prerequisite for antibody function, e.g. cross-linking the receptors in order to stimulate activation, induce apoptosis or promote receptor internalization.

9.3.3.2 *Epitope cross-linking for acquired antagonistic activity*

In addition to avidity enhancement, a BsAb directed against two non-overlapping epitopes on the same antigen molecule, or so-called "bi-paratopic targeting," may also acquire novel properties that are not associated with either of the parent antibodies. For example, we have demonstrated that a bispecific diabody that binds two distinct epitopes on VEGFR2 cross-links the epitopes and effectively blocks the binding of VEGF to the receptor, whereas the parent scFvs from which the bispecific diabody is derived do not, on their own or in combination, block VEGF binding (Lu *et al.*, 1999). Cross-linking two separate epitopes within the same receptor molecule may introduce new steric hindrance for ligand binding, and/or induce conformation changes in the receptor, preventing it from binding ligand. It will be interesting to see if this principle is applicable to other receptors for which ligand-blocking antibodies are difficult to identify.

9.3.3.3 *Fine-tuning antibody specificity towards tumor cells*

An emerging and intriguing concept is the use of BsAb to further fine-tune the specificity of anti-tumor antibodies to target cells. In this context, a BsAb is constructed from two antibodies directed against different tumor antigens, each of low to moderate binding affinity. Only simultaneous-bivalent binding (or cross-linking) of the two target antigens on the same tumor cell surface by the BsAb would result in the strong association that is required to trigger biological processes, such as downregulation of signaling pathways, internalization of BsAb-conjugated toxins, ADCC and CMC. By contrast, monovalent binding of the BsAb to cells that only express one of the target antigens would result in weak association, which may not be sufficient to induce any meaningful cellular activity and thus would preserve the cells from the side effects of antibody-mediated activities. This is significant since most of the targets currently being used as tumor cell identifiers are, in fact, not truly "tumor-specific," but rather "tumor-associated." i.e. they are also expressed in certain normal tissues/cells, albeit at lower densities than in tumor cells. By identifying and constructing BsAb to pairs of targets simultaneously expressed on a given type of tumor, one could expect enhanced antibody specificity towards the targeted cells while sparing normal cells of unwanted side effects. This concept was elegantly demonstrated in a recent report using an anti-HER2 × anti-HER3 BsAb (Robinson *et al.*, 2008). In this report, the authors constructed an anti-ErbB3 × anti-ErbB2 bispecific scFv and demonstrated that the BsAb selectively targeted tumor cells *in vitro* and *in vivo* that co-express the two target antigens over tumor cells that express only one target antigen, or normal cells that express low levels of both antigens. Further, the bispecific scFvs also exhibited significantly greater *in vivo* targeting of ErbB2/ErbB3 double positive tumors than derivative molecules that contain only one functional arm targeting ErbB2 or ErbB3. Taken together, these results clearly suggest that BsAb selected to co-target critical functional pairs of tumor-associated antigens could enhance the targeting specificity of antibody-based cancer therapeutics.

9.4 Engineering and Production of Recombinant IGG-Like BsAb

While the concept of BsAb as a treatment for cancer has been around for twenty years (Barr *et al.*, 1987), no recombinant BsAb has actually been approved for commercialization in the clinics. A major obstacle in the general development of BsAb has been the difficulty of producing materials of sufficient quality and quantity for both preclinical and clinical studies. Initially, the main route to the production of BsAbs was by co-expression of both the LCs and both the HCs of two parent antibodies of different specificities (antibody A and antibody B) in a single cell via either the hybrid hybridoma technique (Milstein and Cuello, 1983) or DNA co-transfection. Unfortunately, assuming that all the four polypeptide chains are equally expressed and there is no pairing preference between any particular LC and HC, in addition to the desired heterodimeric BsAb product (LC_A-HC_A plus LC_B-HC_B), there is also a large number of undesired products formed from the ten molecules that result from the 16 permutations of LC and HC pairings. Consequently, the desired binding-competent BsAb are a minor product (in theory, 1/8th of the total), and purification from the other products is very difficult. Another traditional method for BsAb production is chemical conjugation of two antibodies (or their fragments) of different specificities (Brennan *et al.*, 1985), although this method is by no means simple. Furthermore, the chemical modification process may inactivate the antibody or promote aggregation. Since purification from undesired products remains difficult, the resulting low yield and poor quality BsAb makes this process, like the hybrid hybridoma and DNA co-transfection, unsuitable for the large-scale production required for clinical development.

Ideally, a BsAb would be homogenous, stable, and well expressed, like normal IgG. Thus the major requirements of efficient BsAb production are: (1) a novel structural format that promotes or obligates the formation of homogenous, bispecific proteins; and (2) an efficient expression system in prokaryotic or eukaryotic cells that leads to high level production. In early years, most BsAb research had focused on the creation of bispecific fragments, and significant progress has been made in the past decade in this field (Carter *et al.*, 1995; Todorovska *et al.*, 2001). BsAb fragments are smaller than full-length IgGs, so they have better solid tumor penetration rates, but their small size and lack of an intact Fc also results in their being cleared rapidly from circulation, leading to a short *in vivo* half-life. Further, the BsAb fragments do not require glycosylation and therefore can be produced with high yield in bacteria. Recently there has been an increased interest in IgG-like BsAb. These molecules contain an intact Fc, which endows them with the effector functions, such as ADCC and CMC and half-life of normal IgG, but permuted variable domain organization to endow them with bispecific, and in many cases tetravalent, binding (Marvin and Zhu, 2005). The engineering and application of various BsAb fragments have been reviewed extensively (Carter *et al.*,

1995; Todorovska *et al.*, 2001; Kipriyanov and Le Gall, 2004; Kontermann, 2005). Thus, I will focus here on the latest advancements and ongoing developments in recombinant production of full-length, IgG-like, i.e. Fc domain-containing BsAb.

9.4.1 *Increasing heterodimer:homodimer ratio*

9.4.1.1 *"Knobs-into-holes" BsAb IgG*

Since one of the major barriers to the production of BsAb in a single-cell system, e.g. hybrid hybridoma, is the undesired formation of HC homodimers, a first and obvious solution is to re-engineer the C_H3 domain of the Fc region so as to favor HC heterodimerization over homodimerization. This concept, first developed by Carter and colleagues at Genentech (Ridgway *et al.*, 1996), is based on the idea that simple mutations can be introduced into one half of the C_H3 dimer such that the steric complementarity required for C_H3/C_H3 association obligates the mutated C_H3 domain to pair with a C_H3 domain that has different, accommodating mutations (Fig. 9.1). Specifically, a "knob" mutation (T366W in the C_H3 domain, chosen by inspection of the crystal structure of the Fc) (Deisenhofer, 1981) is made into one HC to introduce a larger residue at the C_H3 dimer interface and create a steric barrier to homodimerization. To promote heterodimerization, an accommodating "hole" (Y407A) mutation is engineered into the CH3 domain of the other HC. Co-expression of these two HCs results in the formation of a mostly (92%) heterodimeric product, but with a stability that is significantly lower than that of the wild-type (knob:T366W + hole:Y407A mutant, Tm = 65.4°C; wild-type, Tm = 80.4°C). To optimize and stabilize the heterodimerized molecules, variants with enhanced stability were selected from a phage display library in which residues near the "hole" (T366, L368, Y407) were randomized (Atwell *et al.*, 1997). The resulting variant (knob: T366W; hole: T366S, L368A, Y407V) has an increased Tm (69.4°C) and formed predominantly heterodimers when co-transfected in 293 cells. This example clearly demonstrates the utility of the knob-into-holes approach. It is possible there are other sets of residues within the CH3 domains that could be mutated/optimized to further increase BsAb stability and the heterodimer:homodimer ratio.

While this approach solves the HC homodimerization problem, and reduces the total number of potential LC/HC pairings in a co-transfected single host from ten to four, thus in theory increasing the production of the functional BsAb from 12.5% to 25%, it does not address the equally problematic mispairing of LCs and HCs from two different antibodies (e.g. LC_A-HC_B and LC_B-HC_A mispairing, instead of the correct LC_A-HC_A and LC_B-HC_B pairing). This can be resolved by similarly redesigning the LC–HC interaction to incorporate knobs and holes at the V_L–V_H interface (Zhu *et al.*, 1997). Based on inspection of the V_L–V_H interface of the anti-HER2 Fab fragment, 4D5 Fab (Eigenbrot *et al.*, 1993), knob mutations and the complementary hole mutations were designed for a diabody fragment

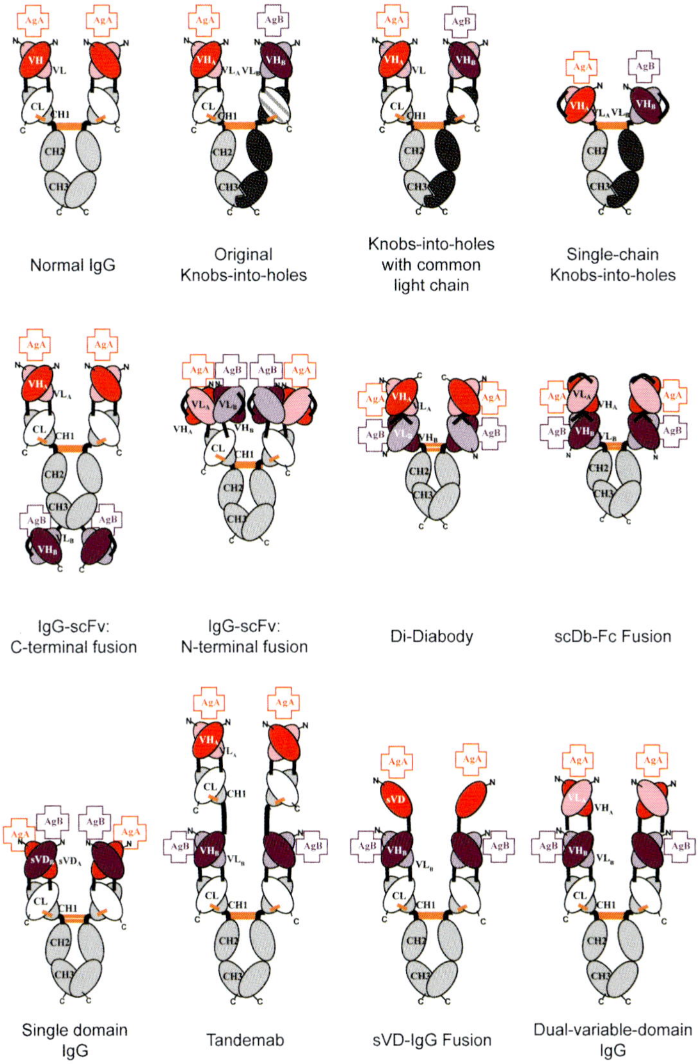

Figure 9.1 Schematic of various Fc-containing, IgG-like, recombinant BsAb formats

Details of each construct are described in the text. HC constant domains = gray (and dark gray for knobs-into-holes); LC constant domain = white (and diagonal hashed for knobs-into-holes IgG); V_H binding to antigen A = red; V_L binding to antigen A = light red; Common LC for knobs-into-holes = patched light red and light purple; V_H binding to antigen B = purple; V_L binding to antigen B = light purple; Covalent polypeptide linkers = black; Interchain disulfide bonds = orange; Antigens A and B are indicated by colored crosses, red and purple, respectively. sVD = single variable domain; N- and C- termini are indicated for each polypeptide chain.

containing the Fvs of 4D5 and an anti-CD3 antibody (Zhu *et al.*, 2006). For one of the variants (termed "v5"; knob: VH-V37F, L45W; hole: VL-Y87A, F98M), 92% of the products were heterodimers, and near wild-type affinity for antigens was retained (Zhu *et al.*, 1997). Although these mutations were used to promote heterodimerization in a bispecific diabody, the principle should translate easily to a full-length IgG format.

9.4.1.2 *Common light chain "knobs-into-holes" BsAb IgG*

An alternative approach to solving the LC–HC mispairing problem is to construct BsAb using two antibodies of different specificities that share an identical LC (Merchant *et al.*, 1998; Fig. 9.1). In a proof-of-concept study, a large single-chain Fv (scFv) phage display library was used to screen for antibodies to a variety of antigens. The V_L fragments of selected scFvs were compared and those with identical amino acid sequences, but paired with different V_H fragments for different antigen-binding specificities, were selected to create BsAb. Functional IgG-like BsAb were formed with high yield (>95% of total IgG products) when a single LC was co-transfected in a host with two different HCs that were engineered to incorporate the original knobs and holes mutations (knob: T366W, hole: T366S, L368A, Y407V) and a heterodimer promoting/stabilizing disulfide bond (S354C, Y349'C).

This highly engineered system demonstrates that the knobs-into-holes principle, when combined with a common light chain, can be used to effectively produce a near homogenous population of BsAb IgG. It has also been suggested that, in addition to the knob-into-holes approach, there are other alternative ways in which Fc heterodimerization can be achieved, for example, by charge engineering within the C_H3 domain, and that BsAb with even higher ratios of heterodimer formation may be accomplished. However, an obvious drawback of these methods is that the inclusion of multiple mutations in the C_H3 domains might pose an immunogenic risk in a therapeutic setting. Furthermore, it requires the identification of antibodies with common LC, which is rare, particularly for high affinity antibodies.

9.4.1.3 *ScFv–Fc "knobs-into-holes" BsAb*

The LC–HC mispairing problem may also be circumvented by fusing the V_L and V_H in a scFv format (Xie *et al.*, 2005). In this construct, an anti-HER2 scFv is fused to an Fc with the "knob" mutation T366Y, and an anti-CD16 scFv is fused to an Fc with the "hole" mutation Y407T (Fig. 9.1). Efficient pairing of the two chains, via the knobs-into-holes mechanism, was demonstrated (but not quantitated), and specificity for both antigens was confirmed by cell surface-binding analysis. The scFv–Fc "knobs-into-holes" molecule is also able to cross-link the two target antigens, as shown by its ability to induce higher NK cell-mediated cytotoxicity than a monospecific anti-HER2 antibody alone. One could hypothesize that this format, which is slightly smaller (120 kDa) than a normal IgG (150 kDa), might provide slight gains in tumor penetration, while maintaining the longer half-life and effector functions provided by the intact Fc region.

9.4.2 *Ig–scFv fusions*

To circumvent the LC–HC mispairing and HC homodimerization issues completely, a number of formats have been developed to construct and produce BsAb that achieve their bispecificity and product homogeneity not by molecular alteration of the component domains, but by direct addition of a new antigen-binding specificity to a fully functional antibody or an antibody-like molecule, e.g. an IgG antibody or a scFv–Fc fusion protein (Fig. 9.1).

9.4.2.1 *IgG C-terminal scFv fusion*

One approach is to fuse a scFv specific for one antigen to the carboxy terminus of a full-length IgG antibody specific for another antigen, creating a tetravalent bispecific IgG-like antibody, IgG–scFv fusion (Coloma and Morrison, 1997; Fig. 9.1). The specificity at the N-terminal end (in this case, for dextran) is defined by a normal Fab. The specificity at the C-terminal end (in this case, for dansyl) is conferred by an anti-dansyl scFv. The fusion construct has an affinity approximately ten-fold lower for dansyl than expected (as compared to its parent IgG), primarily due to a slower on-rate. This may be due to steric hindrance from the rest of the IgG, an inability of both scFv fragments to bind antigen simultaneously (whether all four binding sites were occupied was not tested), or an artifact resulting from conversion of the original anti-dansyl IgG to scFv format. Regardless, the constructs in both a full-length IgG-like format (V_{LA}-C_L plus V_{HA}-C_H1-C_H2-C_H3-$scFv_B)_2$ and a F(ab′)2-like format (V_{LA}-C_L plus V_{HA}-C_H1-$scFv_B)_2$ were capable of binding both target antigens. Furthermore, the IgG-like version is capable of binding C1q, presumably through the intact Fc fragment, although it is unable to trigger a complete complement cascade. This format has the potential to allow the creation of tetravalent molecules with some effector function, but the full potential of the latter remains to be investigated.

9.4.2.2 *IgG N-terminal scFv fusion*

In an alternative approach, the two scFvs of different specificities are fused, respectively, to the N-termini of the constant light (C_L) and the first constant heavy (C_H1) domains of an IgG: ($scFv_A$)-C_L and ($scFv_B$)-C_H1-C_H2-C_H3. Co-expression of the two polypeptide chains in a single host results in the formation of a tetravalent bispecific molecule, Bs(scFv)$_4$-IgG (Zuo *et al.*, 2000; Fig. 9.1). Only homogenous BsAb are generated, due to the natural heterodimerization between the C_L and the C_H1 domains. Similar to the C-terminal fusion, this molecule is also amenable to truncation: a bivalent bispecific Fab-like molecule can be created without the Fc fragment. The ability of this construct to cross-link antigens (in this case, two distinct epitopes on VEGFR2) was demonstrated, but it is not clear whether all four binding sites are or can be occupied simultaneously.

Two other Bs(scFv)$_4$-IgG molecules were recently constructed using the same two scFvs directed against EGFR (IMC-11F8) and IGFR (IMC-A12), but in different fusion orientations (Lu *et al.*, 2004b). Both constructs, $scFv_{anti\text{-}EGFR}$-C_L plus $scFv_{anti\text{-}IGFR}$-C_H1-C_H2-C_H3, or $scFv_{anti\text{-}IGFR}$-C_L plus $scFv_{anti\text{-}EGFR}$-C_H1-C_H2-C_H3, blocked EGF and IGF from binding their respective receptors and inhibited signal transduction pathways activated by both EGF and IGF, whereas a monospecific antibody only inhibited the pathway stimulated by a single ligand. In addition, the BsAbs inhibited tumor cell proliferation *in vitro* at a level that is on a par with the combination of the two parental IgG antibodies. Furthermore, both Bs(scFv)$_4$-IgG demonstrated very good stability when incubated *in vitro* in mouse serum at 37°C for up to seven days. We believe that this Bs(scFv)$_4$-IgG format should be applicable to BsAb construction from two antibodies directed against any pair of antigens. An alternative approach to this Bs(scFv)$_4$-IgG format is to directly fuse a scFv to the N-terminus of an IgG LC, HC, or both, creating a scFv–IgG fusion (as opposed to the IgG–scFv fusion described in Section 4.2.1).

9.4.2.3 *Tandem scFv–Fc fusion*

An alternative to N-terminal scFv–IgG fusion, tandem scFv [scFv(A)-scFv(B)], was fused to the Fc domain of an IgG to form a tetravalent BsAb (Natsume *et al.*, 2006). In one report, a scFv to TAG-72 and a scFv to MUC1 were fused in tandem and then to Fc. The BsAb was expressed effectively in CHO cells, recognized both target antigens, and importantly, was able to mediate ADCC to tumor cells expressing either TAG-72 or MUC-1.

9.4.2.4 *Single domain binder N-terminal or C-terminal fusions*

In addition to scFvs, single variable domain (sVD) binders, from a single antibody variable domain (V_L or V_H), and novel scaffolds such as fibronectin, ankyrin repeats, lipocalin, and A protein, also represent excellent building blocks for the construction of IgG-like BsAb molecules (Holliger and Hudson, 2005; Saerens *et al.*, 2008; Wesolowski *et al.*, 2009). In one report, a single V_H domain binder to PDGFRα was first isolated from a phage library. The single V_H domain was then fused to the N-terminus of V_L of an IgG antibody, directed against PDGFRβ (Shen *et al.*, 2006), or either the N-terminus of V_L or the C-terminus of C_H3 of an

IgG antibody directed against mouse VEGFR2 (Shen *et al.*, 2007). All three BsAb constructs were expressed well in mammalian cells, and demonstrated comparable binding capacity to both targets to their monospecific parents. Further, the BsAb constructs were able to cross-link both target antigens in solution, and inhibited the function of both targets in cell-based assays.

9.4.3 *Diabody–Fc fusions*

Another method by which homogenous populations of IgG-like BsAb can be constructed is by replacement of the Fab fragment with a bispecific diabody. Diabodies are a derivative of the scFv construct (Holliger *et al.*, 1993). An scFv is composed of a V_H and a V_L domain connected by a flexible linker of approximately 15 amino acids, e.g. $(Gly_4Ser)_3$, that permits self-assembly into an antigen-binding competent form. If the linker is shortened to five amino acids, e.g. Gly_4Ser, self-assembly is impossible, and two scFvs interact with each other to form a bivalent molecule of two interlinked polypeptides, the V_L of one chain associating with the V_H of the other. If V_L and V_H with specificities for different antigens comprise the diabody, i.e. V_{HA}–V_{LB} and V_{HB}–V_{LA} (the two so-called "cross-over" scFvs), bispecific bivalent diabodies are formed, with one binding site for each antigen. In addition to assembling as functional heterodimers, the cross-over scFvs can also assemble as non-functional homodimers. Fortunately, purification of properly heterodimerized molecules can easily be achieved by one round of affinity chromatography. Further, as noted earlier, the knobs-into-holes technique can be used to re-engineer the Fv interface to promote the correct heterodimeric V_H–V_L pairing (Zhu *et al.*, 1997). Diabodies have shown to be useful for antigen cross-linking (Holliger *et al.*, 1993; Zhu *et al.*, 1996), and their small size is valuable for tumor penetration (Colcher *et al.*, 1990). However, like many other smaller bispecific fragments, diabodies lack functional Fc domains and the corresponding effector function (Kipriyanov *et al.*, 2002). Recently, they were fused to the Fc domain of an IgG to create tetravalent IgG-like BsAb.

9.4.3.1 *Single-chain diabody–Fc fusion*

In this approach, a single-chain diabody (scDb) is first constructed by fusing both "cross-over" scFvs of a bispecific diabody with a flexible linker (Kipriyanov *et al.*, 1999; Alt *et al.*, 1999). This construct is fused to an Fc fragment (or just a C_H3 domain) to create a tetravalent bispecific IgG-like molecule (Fig. 9.1). In this format, one polypeptide with six domains is produced: V_{HA}-V_{LB}-V_{HB}-V_{LA}-C_H2-C_H3, which then assembles into IgG-like dimers via the Fc domains. The scDb-Fc is bispecific and bivalent for both antigens and has a full Fc (although effector function activity was not tested). The stability of the scDb–Fc fusion was not reported, but the multiple exposed non-human polypeptide linkers within the molecule

insulin-like growth factor I receptor blocks ligand-dependent signaling and inhibits human tumor growth in vivo. *Cancer Res* **63**: 8912–21.

Cao T., Heng B.C. (2005). Intracellular antibodies (intrabodies) versus RNA interference for therapeutic applications. *Ann Clin Lab Sci* **35**: 227–9.

Cao Y., Lam L. (2003). Bispecific antibody conjugates in therapeutics. *Adv Drug Deliv Rev* **55**: 171–97.

Carter P., Ridgway J., Zhu Z. (1995). Toward the production of bispecific antibody fragments for clinical applications. *J Hematother* **4**: 463–70.

Chakravarti A., Loeffler J.S., Dyson N.J. (2002). Insulin-like growth factor receptor I mediates resistance to anti-epidermal growth factor receptor therapy in primary human glioblastoma cells through continued activation of phosphoinositide 3-kinase signaling. *Cancer Res* **62**: 200–7.

Chames P., Baty D. (2009). Bispecific antibodies for cancer therapy. *Curr Opin Drug Discov Devel* **12**: 276–83.

Chang C.H., Sharkey R.M., Rossi E.A., Karacay H., McBride W., Hansen H.J., Chatal J.F., Barbet J., Goldenberg D.M. (2002). Molecular advances in pretargeting radioimunotherapy with bispecific antibodies. *Mol Cancer Ther* **1**: 553–63.

Chang C.H., Rossi E.A., Goldenberg D.M. (2007). The dock and lock method: a novel platform technology for building multivalent, multifunctional structures of defined composition with retained bioactivity. *Clin Cancer Res* **13**: 5586s–91s.

Cheadle E.J. (2006). MT-103 Micromet/MedImmune. *Curr Opin Mol Ther* **8**: 62–8.

Cho H.S., Mason K., Ramyar K.X., Stanley A.M., Gabelli S.B., Denney D.W., Jr., Leahy D.J. (2003). Structure of the extracellular region of HER2 alone and in complex with the Herceptin Fab. *Nature* **421**: 756–60.

Clynes R.A., Towers T.L., Presta L.G., Ravetch J.V. (2000). Inhibitory Fc receptors modulate in vivo cytoxicity against tumor targets. *Nat Med* **6**: 443–6.

Cobleigh M.A., Vogel C.L., Tripathy D., Robert N.J., Scholl S., Fehrenbacher L., Wolter J.M., Paton V., Shak S., Lieberman G., Slamon D.J. (1999). Multinational study of the efficacy and safety of humanized anti-HER2 monoclonal antibody in women who have HER2-overexpressing metastatic breast cancer that has progressed after chemotherapy for metastatic disease. *J Clin Oncol* **17**: 2639–48.

Cochlovius B., Kipriyanov S.M., Stassar M.J., Schuhmacher J., Benner A., Moldenhauer G., Little M. (2000). Cure of Burkitt's lymphoma in severe combined immunodeficiency mice by T cells, tetravalent CD3 x CD19 tandem diabody, and CD28 costimulation. *Cancer Res* 60: 4336–4341.

Colby D.W., Kellogg B.A., Graff C.P., Yeung Y.A., Swers J.S., Wittrup K.D. (2004). Engineering antibody affinity by yeast surface display. *Methods Enzymol* **388**: 348–358.

Colcher D., Bird R., Roselli M., Hardman K.D., Johnson S., Pope S., Dodd S.W., Pantoliano M.W., Milenic D.E., Schlom J. (1990). In vivo tumor targeting of a recombinant single-chain antigen-binding protein. *J Natl Cancer Inst* **82**: 1191–7.

Coloma M.J., Morrison S.L. (1997). Design and production of novel tetravalent bispecific antibodies. *Nat Biotechnol* **15**: 159–63.

Corvalan J.R., Smith W., Gore V.A. (1998). Tumour therapy with Vinca alkaloids targeted by a hybrid-hybrid monoclonal antibody recognising both CEA and Vinca alkaloids. *Int J Cancer Suppl* 2: 22–5.

Cunningham D., Humblet Y., Siena S., Khayat D., Bleiberg H., Santoro A., Bets D., Mueser M., Harstrick A., Verslype C., Chau I., Van Cutsem E. (2004). Cetuximab monotherapy and cetuximab plus irinotecan in irinotecan-refractory metastatic colorectal cancer. *N Engl J Med* **351**: 337–45.

Dechant M., Weisner W., Berger S., Peipp M., Beyer T., Schneider-Merck T., Lammerts van Bueren J.J., Bleeker W.K., Parren P.W., van de Winkel J.G., Valerius T. (2008). Complement-dependent tumor cell lysis triggered by combinations of epidermal growth factor receptor antibodies. *Cancer Res* **68**: 4998–5003.

Deisenhofer J. (1981). Crystallographic refinement and atomic models of a human Fc fragment and its complex with fragment B of protein A from Staphylococcus aureus at 2.9- and 2.8-A resolution. *Biochemistry* **20**: 2361–70.

Eigenbrot C., Randal M., Presta L., Carter P., Kossiakoff A.A. (1993). X-ray structures of the antigen-binding domains from three variants of humanized anti-p185HER2 antibody 4D5 and comparison with molecular modeling. *J Mol Biol* **229**: 969–95.

Els Conrath K., Lauwereys M., Wyns L., Muyldermans S. (2001). Camel single-domain antibodies as modular building units in bispecific and bivalent antibody constructs. *J Biol Chem* **276**: 7346–50.

Ewert S., Honegger A., Plückthun A. (2004). Stability improvement of antibodies for extracellular and intracellular applications: CDR grafting to stable frameworks and structure-based framework engineering. *Methods* **34**: 184–99.

Fendly B.M., Winget M., Hudziak R.M., Lipari M.T., Napier M.A., Ullrich A. (1990). Characterization of murine monoclonal antibodies reactive to either the human epidermal growth factor receptor or HER2/neu gene product. *Cancer Res* **50**: 1550–8.

Ferrara N., Hillan K.J., Gerber H.P., Novotny W. (2004). Discovery and development of bevacizumab, an anti-VEGF antibody for treating cancer. *Nat Rev Drug Discov* **3**: 391–400.

Ferrini S., Sforzini S., Canevari S. (2001). Bispecific monoclonal antibodies for the targeting of type I ribosome-inactivating proteins against hematological malignancies. *Methods Mol Biol* **166**: 177–92.

FitzGerald K., Holliger P., Winter G. (1997). Improved tumour targeting by disulphide stabilized diabodies expressed in Pichia pastoris. *Protein Eng* **10**: 1221–5.

Ford C.H., Osborne P.A., Rego B.G., Mathew A. (2001). Bispecific antibody targeting of doxorubicin to carcinoembryonic antigen-expressing colon cancer cell lines in vitro and in vivo. *Int J Cancer* **92**: 851–5.

Franklin M.C., Carey K.D., Vajdos F.F., Leahy D.J., de Vos A.M., Sliwkowski M.X. (2004). Insights into ErbB signaling from the structure of the ErbB2-pertuzumab complex. *Cancer Cell* **5**: 317–28.

Fresenius A.G. (2007). Secondary endpoints of phase II/III study confirm clear benefits from treatment with Removab for patients with malignant ascites. *Fresenius AG PRESS RELEASE* 2007 July 17.

Govindan S.V., Griffiths G.L., Hansen H.J., Horak I.D., Goldenberg D.M. (2005). Cancer therapy with radiolabeled and drug/toxin-conjugated antibodies. *Technol Cancer Res Treat* **4**: 375–91.

Grosse-Hovest L., Brandl M., Dohlsten M., Kalland T., Wilmanns W., Jung G. (1999). Tumor-growth inhibition with bispecific antibody fragments in a syngeneic mouse melanoma model: the role of targeted T-cell co-stimulation via CD28. *Int J Cancer* **80**: 138–44.

Haurum J., Bregenholt S. (2005). Recombinant polyclonal antibodies: therapeutic antibody technologies come full circle. *IDrugs* **8**: 404–9.

Holliger P., Hudson P.J. (2005). Engineered antibody fragments and the rise of single domains. *Nat Biotechnol* **23**: 1126–36.

Holliger P., Prospero T., Winter G. (1993). "Diabodies": small bivalent and bispecific antibody fragments. *Proc Natl Acad Sci U S A* **90**: 6444–8.

Holt L.J., Herring C., Jespers L.S., Woolven B.P., Tomlinson I.M. (2003). Domain antibodies: proteins for therapy. *Trends Biotechnol* **21**: 484–90.

Hurwitz H., Fehrenbacher L., Novotny W., Cartwright T., Hainsworth J., Heim W., Berlin J., Baron A., Griffing S., Holmgren E., Ferrara N., Fyfe G., Rogers B., Ross R., Kabbinavar F. (2004). Bevacizumab plus irinotecan, fluorouracil, and leucovorin for metastatic colorectal cancer. *N Engl J Med* **350**: 2335–42.

Ichikawa K., Liu W., Zhao L., Wang Z., Liu D., Ohtsuka T., Zhang H., Mountz J.D., Koopman W.J., Kimberly R.P., Zhou T. (2001). Tumoricidal activity of a novel anti-human DR5 monoclonal antibody without hepatocyte cytotoxicity. *Nat Med* **7**: 954–60.

Jäger M., Schoberth A., Ruf P., Hess J., Lindhofer H. (2009). The trifunctional antibody ertumaxomab destroys tumor cells that express low levels of human epidermal growth factor receptor 2. *Cancer Res* **69**: 4270–6.

James N.D., Atherton P.J., Jones J., Howie A.J., Tchekmedyian S., Curnow R.T. (2001). A phase II study of the bispecific antibody MDX-H210 (anti-HER2 x CD64) with GM-CSF in HER2 + advanced prostate cancer. *Br J Cancer* **85**: 152–6.

Jendreyko N., Popkov M., Beerli R.R., Chung J., McGavern D.B., Rader C., Barbas C.F. (2003). Intradiabodies, bispecific, tetravalent antibodies for the simultaneous functional knockout of two cell surface receptors. *J Biol Chem* **278**: 47812–19.

Jendreyko N., Popkov M., Rader C., Barbas C.F. (2005). Phenotypic knockout of VEGF-R2 and Tie-2 with an intradiabody reduces tumor growth and angiogenesis *in vivo*. *Proc Natl Acad Sci USA* **102**: 8293–8.

Jimenez X., Lu D., Brennan L., Persaud K., Liu M., Miao H., Witte L., Zhu Z. (2005). A recombinant, fully human, bispecific antibody neutralizes the biological activities mediated by both vascular endothelial growth factor receptors 2 and 3. *Mol Cancer Ther* **4**: 427–34.

Jordan J.L., Arndt J.W., Hanf K., Li G., Hall J., Demarest S., Huang F., Wu X., Miller B., Glaser S., Fernandez E.J., Wang D., Lugovskoy A. (2009). Structural understanding of stabilization patterns in engineered bispecific Ig-like antibody molecules. *Proteins* **77**: 832–41.

Jung S.H., Pastan I., Lee B. (1994). Design of interchain disulfide bonds in the framework region of the Fv fragment of the monoclonal antibody B3. *Proteins* **19**: 35–47.

Jung Y.D., Mansfield P.F., Akagi M., Takeda A., Liu W., Bucana C.D., Hicklin D.J., Ellis L.M. (2002). Effects of combination anti-vascular endothelial growth factor receptor and anti-epidermal growth factor receptor therapies on the growth of gastric cancer in a nude mouse model. *Eur J Cancer* **38**: 1133–40.

Kiewe P., Hasmüller S., Kahlert S., Heinrigs M., Rack B., Marmé A., Korfel A., Jäger M., Lindhofer H., Sommer H., Thiel E., Untch M. (2006). Phase I trial of the trifunctional anti-HER2 x anti-CD3 antibody ertumaxomab in metastatic breast cancer. *Clin Cancer Res* **12**: 3085–91.

Kiewe P., Thiel E. (2008). Ertumaxomab: a trifunctional antibody for breast cancer treatment. *Expert Opin Investig Drugs* **17**: 1553–8.

Kipriyanov S.M., Cochlovius B., Schafer H.J., Moldenhauer G., Bahre A., Le Gall F., Knackmuss S., Little M. (2002). Synergistic antitumor effect of bispecific CD19 × CD3 and CD19 × CD16 diabodies in a preclinical model of non-Hodgkin's lymphoma. *J Immunol* **169**: 137–44.

Kipriyanov S.M., Le Gall F. (2004). Recent advances in the generation of bispecific antibodies for tumor immunotherapy. *Curr Opin Drug Discov Devel* **7**: 233–42.

Kipriyanov S.M., Moldenhauer G., Schuhmacher J., Cochlovius B., Von der Lieth C.W., Matys E.R., Little M. (1999). Bispecific tandem diabody for tumor therapy with improved antigen binding and pharmacokinetics. *J Mol Biol* **293**: 41–56.

Knight D.M., Trinh H., Le J., Siegel S., Shealy D., McDonough M., Scallon B., Moore M.A., Vilcek J., Daddona P. (1993). Construction and initial characterization of a mouse-human chimeric anti-TNF antibody. *Mol Immunol* **30**: 1443–53.

Kobari M., Kullenberg B., Bjorkman A., Matsuno S., Ihse I., Axelson J. (1998). The inhibitory effect of an EGF receptor-specific tyrosine kinase inhibitor on pancreatic cancer cell lines was more potent than inhibitory antibodies against the receptors for EGF and IGF I. *Int J Pancreatol* **24**: 85–95.

Kontermann R.E. (2005). Recombinant bispecific antibodies for cancer therapy. *Acta Pharmacol Sin* **26**: 1–9.

Krane J.F., Murphy D.P., Carter D.M., Krueger J.G. (1991). Synergistic effects of epidermal growth factor (EGF) and insulin-like growth factor I/somatomedin C (IGF-I) on keratinocyte proliferation may be mediated by IGF-I transmodulation of the EGF receptor. *J Invest Dermatol* **96**: 419–24.

Kufer P., Schlereth B., Baeuerle P.A., Friedrich M. (2009). Antitumor activity of an EpCAM/CD3-bispecific BiTE antibody during long-term treatment of mice in the absence of T-cell anergy and sustained cytokine release. *J Immunother* **32**: 452–64.

Kunkel P., Ulbricht U., Bohlen P., Brockmann M.A., Fillbrandt R., Stavrou D., Westphal M., Lamszus K. (2001). Inhibition of glioma angiogenesis and growth in vivo by systemic treatment with a monoclonal antibody against vascular endothelial growth factor receptor-2. *Cancer Res* **61**: 6624–8.

Lamszus K., Brockmann M.A., Eckerich C., Bohlen P., May C., Mangold U., Fillbrandt R., Westphal M. (2005). Inhibition of glioblastoma angiogenesis and invasion by combined treatments directed against vascular endothelial growth factor receptor-2, epidermal growth factor receptor, and vascular endothelial-cadherin. *Clin Cancer Res* **11**: 4934–40.

Leahy D.J. (2004). Structure and function of the epidermal growth factor (EGF/ErbB) family of receptors. *Adv Protein Chem* **68**: 1–27.

Lenz H.J., Van Cutsem E., Khambata-Ford S., Mayer R.J., Gold P., Stella P., Mirtsching B., Cohn A.L., Pippas A.W., Azarnia N., Tsuchihashi Z., Mauro D.J., Rowinsky E.K. (2006). Multicenter phase II and translational study of cetuximab in metastatic colorectal carcinoma refractory to irinotecan, oxaliplatin, and fluoropyrimidines. *J Clin Oncol* **24**: 4914–21.

Li S., Schmitz K., Jeffery P.D., Wiltzius J., Kussie P., Ferguson K.M. (2005). Structural basis for inhibition of the Epidermal Growth Factor Receptor by cetuximab/Erbitux. *Cancer Cell* **7**: 301–11.

Liebman M.A., Williams B.R., Daley K.M., Sharon J. (2004). Generation and preliminary characterization of an antibody library with preferential reactivity to human colorectal cancer cells as compared to normal human blood cells. *Immunol Lett* **91**: 179–88.

Lipovsek D., Pluckthun A. (2004). In-vitro protein evolution by ribosome display and mRNA display. *J Immunol Methods* **290**: 51–67.

Lu D., Kotanides H., Jimenez X., Zhou Q., Persaud K., Bohlen P., Witte L., Zhu Z. (1999). Acquired antagonistic activity of a bispecific diabody directed against two different epitopes on vascular endothelial growth factor receptor 2. *J Immunol Methods* **230**: 159–71.

Lu D., Jimenez X., Zhang H., Wu Y., Bohlen P., Witte L., Zhu Z. (2001). Complete inhibition of vascular endothelial growth factor (VEGF) activities with a bifunctional diabody directed against both VEGF kinase receptors, fms-like tyrosine kinase receptor and kinase insert domain-containing receptor. *Cancer Res* **61**: 7002–8.

Lu D., Jimenez X., Zhang H., Atkins A., Brennan L., Balderes P., Bohlen P., Witte L., Zhu Z. (2003). Di-diabody: a novel tetravalent bispecific antibody molecule by design. *J Immunol Methods* **279**: 219–32.

Lu D., Jimenez X., Witte L., Zhu Z. (2004a). The effect of variable domain orientation and arrangement on the antigen-binding activity of a recombinant human bispecific diabody. *Biochem Biophys Res Commun* **318**: 507–13.

Lu D., Zhang H., Ludwig D., Persaud A., Jimenez X., Burtrum D., Balderes P., Liu M., Bohlen P., Witte L., Zhu Z. (2004b). Simultaneous blockade of both the epidermal growth factor receptor and the insulin-like growth factor receptor signaling pathways in cancer cells with a fully human recombinant bispecific antibody. *J Biol Chem* **279**: 2856–65.

Lu D., Zhang H., Koo H., Tonra J., Balderes P., Prewett M., Corcoran E., Mangalampalli V., Bassi R., Anselma D., Patel D., Kang X., Ludwig D.L., Hicklin D.J., Bohlen P., Witte L., Zhu Z. (2005). A fully human recombinant IgG-like bispecific antibody to both the epidermal growth factor receptor and the insulin-like growth factor receptor for enhanced antitumor activity. *J Biol Chem* **280**: 19665–72.

Lum L.G., Davol P.A. (2005). Retargeting T cells and immune effector cells with bispecific antibodies. *Cancer Chemother Biol Response Modif* **22**: 273–91.

Marvin J.S., Zhu Z. (2005). Recombinant approaches to IgG-like bispecific antibodies. *Acta Pharmacol Sin* **26**: 649–58.

Marvin J.S., Zhu Z. (2006). Bispecific antibodies for dual-modality cancer therapy: killing two signaling cascades with one stone. *Curr Opin Drug Discov Devel* **9**: 184–93.

McLaughlin P., Grillo-Lopez A.J., Link B.K., Levy R., Czuczman M.S., Williams M.E., Heyman M.R., Bence-Bruckler I., White C.A., Cabanillas F., Jain V., Ho A.D., Lister J., Wey K., Shen D., Dallaire B.K. (1998). Rituximab chimeric anti-CD20 monoclonal antibody therapy for relapsed indolent lymphoma: half of patients respond to a four-dose treatment program. *J Clin Oncol* **16**: 2825–33.

Merchant A.M., Zhu Z., Yuan J.Q., Goddard A., Adams C.W., Presta L.G., Carter P. (1998). An efficient route to human bispecific IgG. *Nat Biotechnol* **16**: 677–81.

Michaelson J.S., Demarest S.J., Miller B., Amatucci A., Hughes I., Snyder W.B., Wu X., Huang F., Phan S., Gao S., Doern A., Lugovskoy A., Joseph I., Bailly V., Wang X, Garber E., Browning J., Glaser S.M. (2009). Anti-tumor activity of stability-engineered IgG like bispecific antibodies targeting TRAIL-R2 AND LTBetaR. *Monocolonal Antibodies* **1**: 128–41.

Miller K., Meng G., Liu J., Hurst A., Hsei V., Wong W.L., Ekert R., Lawrence D., Sherwood S., DeForge L., Gaudreault J., Keller G., Sliwkowski M., Ashkenazi A., Presta L. (2003). Design, construction, and in vitro analyses of multivalent antibodies. *J Immunol* **170**: 4854–61.

Milstein C., Cuello A.C. (1983). Hybrid hybridomas and their use in immunohistochemistry. *Nature* **305**: 537–40.

Molina M.A., Codony-Servat J., Albanell J., Rojo F., Arribas J., Baselga J. (2001). Trastuzumab (herceptin), a humanized anti-Her2 receptor monoclonal antibody, inhibits basal and activated Her2 ectodomain cleavage in breast cancer cells. *Cancer Res* **61**: 4744–9.

Müller D., Kontermann R.E. (2007). Recombinant bispecific antibodies for cellular cancer immunotherapy. *Curr Opin Mol Ther* **9**: 319–26.

Muyldermans S., Atarhouch T., Saldanha J., Barbosa J.A., Hamers R. (1994). Sequence and structure of VH domain from naturally occurring camel heavy chain immunoglobulins lacking light chains. *Protein Eng* **7**: 1129–35.

Nahta R., Hung M.C., Esteva F.J. (2004). The HER-2-targeting antibodies trastuzumab and pertuzumab synergistically inhibit the survival of breast cancer cells. *Cancer Res* **64**: 2343–6.

Natsume A., Wakitani M., Yamane-Ohnuki N., Shoji-Hosaka E., Niwa R., Uchida K., Satoh M., Shitara K. (2006). Fucose removal from complex-type oligosaccharide enhances the antibody-dependent cellular cytotoxicity of single-gene-encoded bispecific antibody comprising of two single-chain antibodies linked to the antibody constant region. *J Biochem* **140**: 359–68.

Nowakowski A., Wang C., Powers D.B., Amersdorfer P., Smith T.J., Montgomery V.A., Sheridan R., Blake R., Smith L.A., Marks J.D. (2002). Potent neutralization of botulinum neurotoxin by recombinant oligoclonal antibody. *Proc Natl Acad Sci USA* **99**: 11346–50.

Peipp M., Valerius T. (2002). Bispecific antibodies targeting cancer cells. *Biochem Soc Trans* **30**: 507–11.

Penar P.L., Khoshyomn S., Bhushan A., Tritton T.R. (1997). Inhibition of epidermal growth factor receptor-associated tyrosine kinase blocks glioblastoma invasion of the brain. *Neurosurgery* **40**: 141–51.

Ridgway J.B., Presta L.G., Carter P. (1996). 'Knobs-into-holes' engineering of antibody CH3 domains for heavy chain heterodimerization. *Protein Eng* **9**: 617–21.

Riechmann L., Muyldermans S. (1999). Single domain antibodies: comparison of camel VH and camelised human VH domains. *J Immunol Methods* **231**: 25–38.

Robinson M.K., Hodge K.M., Horak E., Sundberg A.L., Russeva M., Shaller C.C., von Mehren M., Shchaveleva I., Simmons H.H., Marks J.D., Adams G.P. (2008). Targeting ErbB2 and ErbB3 with a Bispecific-Single Chain Fv Enhances Targeting Selectivity and Induces a Therapeutic Effect *in vitro*. *Br J Cancer* **99**: 1415–25.

Rossi E.A., Goldenberg D.M., Cardillo T.M., McBride W.J., Sharkey R.M., Chang C.H. (2006). Stably tethered multifunctional structures of defined composition made by the dock and lock method for use in cancer targeting. *Proc Natl Acad Sci USA* **103**: 6841–6.

Rossi E.A., Goldenberg D.M., Cardillo T.M., Stein R., Chang C.H. (2009). Hexavalent bispecific antibodies represent a new class of anticancer therapeutics: 1. Properties of anti-CD20/CD22 antibodies in lymphoma. *Blood* **113**: 6161–71.

Ruf P., Gires O., Jäger M., Fellinger K., Atz J., Lindhofer H. (2007). Characterisation of the new EpCAM–specific antibody HO-3: implications for trifunctional antibody immunotherapy of cancer. *Br J Cancer* **97**: 315–21.

Saltz L.B., Lenz H.J., Kindler H.L., Hochster H.S., Wadler S., Hoff P.M., Kemeny N.E., Hollywood E.M., Gonen M., Quinones M., Morse M., Chen H.X. (2007). Randomized phase II trial of cetuximab, bevacizumab, and irinotecan compared with cetuximab and bevacizumab alone in irinotecan-refractory colorectal cancer: the BOND-2 study. *J Clin Oncol* **25**: 4557–61.

Saerens D., Ghassabeh G.H., Muyldermans S. (2008). Single-domain antibodies as building blocks for novel therapeutics. *Curr Opin Pharmacol* **8**: 600–8.

Schlereth B., Fichtner I., Lorenczewski G., Kleindienst P., Brischwein K., da Silva A., Kufer P., Lutterbuese R., Junghahn I., Kasimir-Bauer S., Wimberger P., Kimmig R., Baeuerle P.A. (2005). Eradication of tumors from a human colon cancer cell line and from ovarian cancer metastases in immunodeficient mice by a single-chain Ep-CAM-/CD3-bispecific antibody construct. *Cancer Res* **65**: 2882–9.

Shaheen R.M., Ahmad S.A., Liu W., Reinmuth N., Jung Y.D., Tseng W.W., Drazan K.E., Bucana C.D., Hicklin D.J., Ellis L.M. (2001). Inhibited growth of colon cancer carcinomatosis by antibodies to vascular endothelial and epidermal growth factor receptors. *Br J Cancer* **85**: 584–9.

Sharkey R.M., Karacay H., Cardillo T.M., Chang C.H., McBride W.J., Rossi E.A., Horak I.D., Goldenberg D.M. (2005). Improving the delivery of radionuclides for imaging and therapy of cancer using pretargeting methods. *Clin Cancer Res* **11**: 7109s–21s.

Sharon J., Liebman M.A., Williams B.R. (2005). Recombinant polyclonal antibodies for cancer therapy. *J Cell Biochem* **96**: 305–13.

Shen J., Vil M.D., Jimenez X., Iacolina M., Zhang H., Zhu Z. (2006). Single variable domain-IgG fusion. A novel recombinant approach to Fc domain-containing bispecific antibodies. *J Biol Chem* **281**: 10706–14.

Shen J., Vil M.D., Jimenez X., Zhang H., Iacolina M., Mangalampalli V., Balderes P., Ludwig D.L., Zhu Z. (2007). Single variable domain antibody as a versatile building block for the construction of IgG-like bispecific antibodies. *J Immunol Methods* **318**: 65–74.

Shen J., Zhu Z. (2008). Catumaxomab, a rat/murine hybrid trifunctional bispecific monoclonal antibody for the treatment of cancer. *Curr Opin Mol Ther* **10**: 273–84.

Sidhu S.S. (2005). *Phage Display In Biotechnology And Drug Discovery*, Marcel Dekker, New York, NY.

Silverman D.H., Delpassand E.S., Torabi F., Goy A., McLaughlin P., Murray J.L. (2004). Radiolabeled antibody therapy in non-Hodgkins lymphoma: radiation protection, isotope comparisons and quality of life issues. *Cancer Treat Rev* **30**: 165–72.

Slamon D.J., Leyland-Jones B., Shak S., Fuchs H., Paton V., Bajamonde A., Fleming T., Eiermann W., Wolter J., Pegram M., Baselga J., Norton L. (2001). Use of chemotherapy plus a monoclonal antibody against HER2 for metastatic breast cancer that overexpresses HER2. *N Engl J Med* **344**: 783–92.

Stasi R. (2008). Gemtuzumab ozogamicin: an anti-CD33 immunoconjugate for the treatment of acute myeloid leukaemia. *Expert Opin Biol Ther* **8**: 527–40.

Todorovska A., Roovers R.C., Dolezal O., Kortt A.A., Hoogenboom H.R., Hudson P.J. (2001). Design and application of diabodies, triabodies and tetrabodies for cancer targeting. *J Immunol Methods* **248**: 47–66.

Tonra J.R., Corcoran E., Deevi D.S., Steiner P., Kearney J., Li H., Ludwig D.L., Zhu Z., Witte L., Surguladze D., Hicklin D.J. (2009). Prioritization of EGFR/IGF-IR/VEGFR2 combination targeted therapies utilizing cancer models. *Anticancer Res* **29**: 1999–2007.

Volk W.A., Bizzini B., Snyder R.M., Bernhard E., Wagner R.R. (1984). Neutralization of tetanus toxin by distinct monoclonal antibodies binding to multiple epitopes on the toxin molecule. *Infect Immun* **45**: 604–9.

Wesolowski J., Alzogaray V., Reyelt J., Unger M., Juarez K., Urrutia M., Cauerhff A., Danquah W., Rissiek B., Scheuplein F., Schwarz N., Adriouch S., Boyer O., Seman M., Licea A., Serreze D.V., Goldbaum F.A., Haag F., Koch-Nolte F. (2009). Single domain antibodies: promising experimental and therapeutic tools in infection and immunity. *Med Microbiol Immunol* **198**: 157–74.

Williams B.R., Sharon J. (2002a). Polyclonal anti-colorectal cancer Fab phage display library selected in one round using density gradient centrifugation to separate antigen-bound and free phage. *Immunol Lett* **81**: 141–8.

Williams B.R., Sompuram S.R., Sharon J. (2002b). Generation of anti-colorectal cancer fab phage display libraries with a high percentage of diverse antigen-reactive clones. *Comb Chem High Throughput Screen* **5**: 489–99.

Wolf E., Hofmeister R., Kufer P., Schlereth B., Baeuerle P.A. (2005). BiTEs: bispecific antibody constructs with unique anti-tumor activity. *Drug Discov Today* **10**: 1237–44.

Worn A., Pluckthun A. (2001). Stability engineering of antibody single-chain Fv fragments. *J Mol Biol* **305**: 989–1010.

Wu C., Ying H., Grinnell C., Bryant S., Miller R., Clabbers A., Bose S., McCarthy D., Zhu R.R., Santora L., Davis-Taber R., Kunes Y., Fung E., Schwartz A., Sakorafas P., Gu J., Tarcsa E., Murtaza A., Ghayur T. (2007). Simultaneous targeting of multiple disease mediators by a dual-variable-domain immunoglobulin. *Nat Biotechnol* **25**: 1290–7.

Xie Z., Guo N., Yu M., Hu M., Shen B. (2005). A new format of bispecific antibody: highly efficient heterodimerization, expression and tumor cell lysis. *J Immunol Methods* **296**: 95–101.

Xiong D., Xu Y., Liu H., Peng H., Shao X., Lai Z., Fan D., Yang M., Han J., Xie Y., Yang C., Zhu Z. (2002). Efficient inhibition of human B-cell lymphoma xenografts with an anti-CD20 x anti-CD3 bispecific diabody. *Cancer Lett* **177**: 29–39.

Yagita H., Takeda K., Hayakawa Y., Smyth M.J., Okumura K. (2004). TRAIL and its receptors as targets for cancer therapy. *Cancer Sci* **95**: 777–83.

Yan, L., Zhu, Z. (2006). Development of antibody-based therapeutics for oncology indications. *Drug Dev Res* **67**: 699–728.

Ye D., Mendelsohn J., Fan Z. (1999). Augmentation of a humanized anti-HER2 mAb 4D5 induced growth inhibition by a human-mouse chimeric anti-EGF receptor mAb C225. *Oncogene* **18**: 731–8.

Zapata G., Ridgway J.B., Mordenti J., Osaka G., Wong W.L., Bennett G.L., Carter P. (1995). Engineering linear F(ab′)2 fragments for efficient production in Escherichia coli and enhanced antiproliferative activity. *Protein Eng* **8**: 1057–62.

Zhu Z., Presta L.G., Zapata G., Carter P. (1997). Remodeling domain interfaces to enhance heterodimer formation. *Protein Sci* **6**: 781–8.

Zhu Z., Zapata G., Shalaby R., Snedecor B., Chen H., Carter P. (1996). High level secretion of a humanized bispecific diabody from Escherichia coli. *Biotechnology (NY)* **14**: 192–96.

Zuo Z., Jimenez X., Witte L., Zhu Z. (2000). An efficient route to the production of an IgG-like bispecific antibody. *Protein Eng* **13**: 361–7.

10

Antibody Fragments and Alternate Protein Scaffolds

Lioudmila Tchistiakova, William J. J. Finlay†, Stephane Olland*, Helen Dooley‡ and Davinder Gill**

10.1 Introduction

Considerable knowledge of antibody:antigen recognition based on X-ray crystal structures reveals that the six CDRs of antibodies often form planar-binding interfaces, making binding to grooves and the catalytic sites of antigens rather difficult. Therefore it becomes very important to diversify the binding potential of novel therapeutics to achieve the specific recognitions of unique structural features on each target. Over the last 20 years, molecular engineering technologies have made it possible to create many protein constructs based on antibody fragments. These consist of essentially the antibody variable regions in a variety of different presentation modes, with *in vivo* half-life extension achieved through use of Fc regions or alternative approaches. This approach, derived from initial monoclonal antibody studies, provides the possibility of exploring several antigen recognition options, which alter the geometry of the interaction with the target as well as molecular size. The smallest single polypeptide-binding fragment that can be derived from the initial antibody consists of V_H and V_L domains joined through a flexible peptide linker and has been named "single-chain variable fragment" (scFv). For the past ten years, scFvs have been used as building blocks in various scaffolds such as diabodies (scFv-scFv, BiTE), multi-bodies, minibodies scFv-CH3, SMIPs (scFv–Fc fusions), and Scorpions (Fc-based fusions for dual targeting) (Kontermann, 2010).

*Pfizer, Inc., USA, † Pfizer, Inc., Ireland, ‡ Pfizer, Inc., UK.

In this review we will mostly focus on the Fc-containing novel scaffolds for targeting one antigen (SMIPs) and for dual targeting (Scorpions). We also address the emerging opportunity afforded by avian-derived antibodies to generate human therapeutics. We conclude with the next frontier of alternate protein scaffolds that include camelid-derived V_HH antibodies as well as the shark-derived IgNAR proteins.

10.2 Small Modular Immunopharmaceuticals

SMIPs represent a class of scFv-based scaffolds, pioneered by Seattle-based Trubion Pharmaceuticals. SMIPs are mono-specific therapeutics — drugs that recognize and attach to a single antigen to initiate biological activity. They consist of two identical polypeptide chains dimerized through the Fc region or through additional cysteine bonds in the hinge region. These hinges connect target recognition binding domains (BDs) with the Fc region (Fig. 10.1). SMIPs use human antibody Fc regions for both half-life extension and effector function capabilities.

The advantage of the SMIP approach is a modular design that allows one to customize the properties of the molecule by independently optimizing the function and activity of each structural component of SMIP (binding domain, hinge and Fc region). This results in a biotherapeutic drug with unique structure/function properties as described below. Since the binding domains of SMIPs are in a different configuration than in antibodies, and because they are connected to the Fc only through a short hinge region, the overall geometry of the molecule is quite

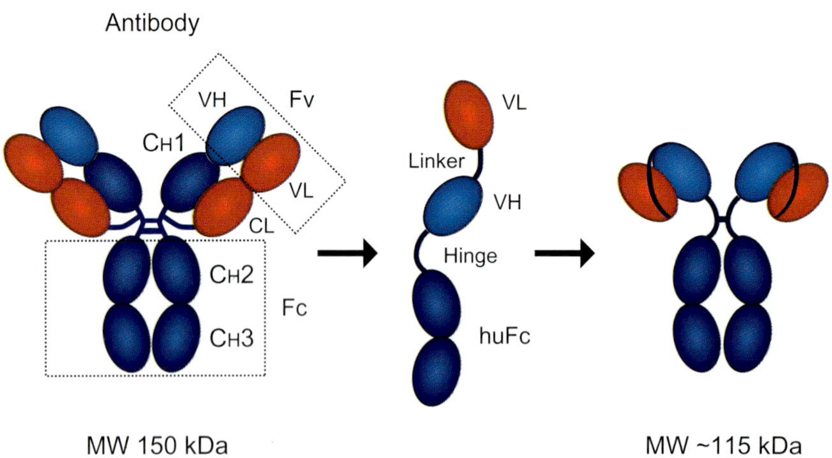

Figure 10.1 Design of SMIP molecules from classical antibodies

different. For example, the distance between two binding domains in SMIPs is approximately 60–70 Å, while in conventional mAbs it could be 95–180 Å (Sandin *et al.*, 2004). This difference was exemplified by converting an anti-CD3 monoclonal antibody into a SMIP. Trubion has developed an anti-CD3 SMIP based on OKT3 antibody and shown that it retains immunosuppressive activity while demonstrating significant reductions in cytokine release.

The structural format of SMIP proteins permits the engineering of a range of distances between the binding domains to find the best format for engaging cell surface receptors and to achieve bivalency if needed. Plus, SMIP proteins can be engineered to balance target signal induction, complement-dependent cytotoxicity (CDC) and antibody-dependent cellular cytotoxicity (ADCC).

10.2.1 *Design and optimization of SMIP properties*

The binding domains (BD) of SMIPs can include in addition to scFv, soluble proteins such as cytokines and growth factors, or extracellular domains of cell-surface receptors. They can be optimized for high affinity, fine specificity and good expression and solubility properties based on well-validated methods of scFv optimization (phage display, ribosome display, and rational optimization). The scFv domains can be constructed using human or humanized antibodies. The optimization of the orientation of V_H and V_L domains (V_H-V_L or V_L-V_H) and the length of the flexible linker connecting these domains is often required to preserve binding properties. If the linker between V_L and V_H domains is too short, or the interaction between V_H and V_L domains is weak, the binding domains can fold in a diabody configuration when the V_H of one polypeptide chain interacts with the V_L of the second chain (Wu *et al.*, 2001). That creates a possibility of aggregate formation and loss of binding activity and requires specific optimization of SMIP folding properties. As such, the optimization of SMIPs can benefit from the accumulated knowledge of scFv stability and solubility (Ewert *et al.*, 2004).

The hinge domain in a SMIP provides the flexible linkage between BDs and the Fc and can significantly influence overall SMIP properties such as the distance between BDs, stability, and effector functions. Several sources of hinge regions have been explored including flexible Gly-Ser linkers and hinge regions derived from various isotypes of human antibodies. The introduction of cysteine residues in this region allows covalent cross-linking of the two polypeptide chains and provides additional stabilization to the entire molecule. Interestingly, similar to monoclonal antibodies, mutations in the hinge region can also affect the C1q binding and CDC activity of the molecule (Dall'Acqua *et al.*, 2006).

Since the half-life extending domain of SMIPs is designed based on the Fc region of immunoglobulins, it provides all of the advantages of effector function capacity found in mAbs, such as enhanced ADCC through glycosylation modification or introducing mutations improving Fcγ receptor binding. From a

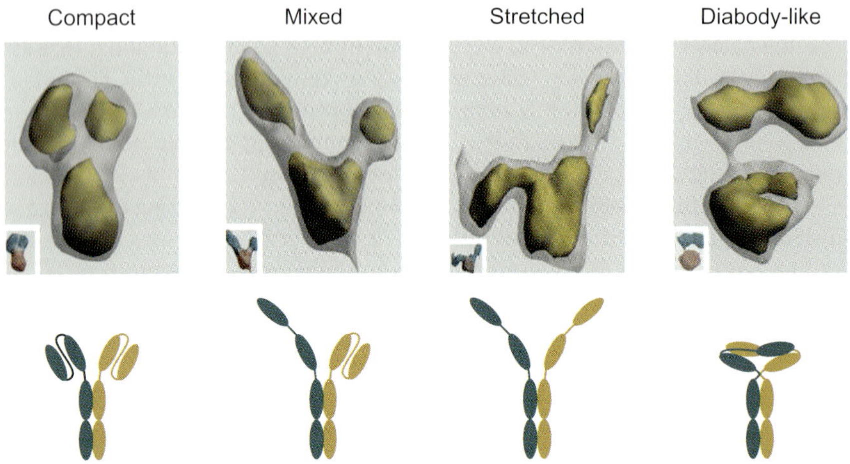

Figure 10.2 Conformations of SMIP molecules in solution determined by Protein Tomography

different perspective, the attenuation of the effector function of SMIPs can also be achieved by introducing mutations in Fcγ receptor and C1q binding regions. For example, a humanized anti-CD20 SMIP generated with wild-type Fc or Fc with a single amino acid mutation, P331S, known to reduce *in vitro* CDC activity in the mAb format, showed comparable *in vitro* binding to normal B cells and ADCC, but the P331S mutant had no detectable CDC activity (Dunussi-Joannopoulos *et al.*, 2009).

Despite the fact that detailed crystal structure information is available for the fragments used in SMIP design (Fc, scFvs and diabody), understanding of the overall folding of SMIP molecules remains unclear. Attempts at generating crystal structures of SMIP proteins have not been successful to date. However, intriguing insights into the folding of SMIP molecules in solution was achieved by applying Protein Tomography technology (Fig. 10.2).

For three different SMIP molecules analyzed, Protein Tomography showed that each molecule could form four different conformations: (1) compact with each binding domain in scFv conformation, (2) mixed when one of the binding domains lost V_H–V_L interaction and exposed both domains in the tandem fashion, (3) stretched when both binding domains are in the open conformation, and (4) diabody-like when the V_H and V_L of one polypeptide chain interact with V_L and V_H of the second chain. Although it was unclear from this study if these conformational changes in SMIPs are interchangeable and reversible, the majority of analyzed molecules appeared to accommodate compact conformation. This structural flexibility could potentially provide advantages in the modular design of SMIP molecules. It could also pose challenges in manufacturing by affecting the overall stability of this scaffold. However, that has not been reported so far. For example, several clinical candidates were successfully generated based on SMIP scaffold: anti-CD20 SMIP TRU-015 (Hayden-Ledbetter *et al.*, 2009), humanized anti-CD20 SMIP SBI-089 (Dunussi-Joannopoulos *et al.*, 2009), and anti-CD37 SMIP TRU-016 (Robak *et al.*, 2009). All clinical candidates have shown very good expression in recombinant mammalian expression systems, reliable manufacturing properties and demonstrated high potency in preclinical and clinical settings.

10.2.2 *Preclinical and clinical validation of SMIP-based biotherapeutics*

A chimeric SMIP (TRU-015) based on an anti-CD20 mouse monoclonal antibody has been reported (Hayden-Ledbetter *et al.*, 2009). Preclinical development of this SMIP demonstrated the absence of immunogenicity in cynomolgus monkeys and superior B cell depletion compared with rituximab. TRU-015 has been tested in Phase I/IIa trials in RA patients and has demonstrated robust B cell depletion. Preliminary results of the first Phase I study also showed that TRU-015 is clinically well tolerated and has an extended serum half-life of 12–19 days (Burge *et al.*, 2008).

Subsequently, a humanized version of TRU-015 (SBI-087) was developed with very interesting findings. In addition to binding domain humanization, modifications in the hinge regions were introduced in SBI-087 to improve molecular stability. These modifications resulted in changes in functional properties. SBI-087 binds CD20 on the surface of primary B cells with an affinity similar to TRU-015. However, despite the fact that its maximal binding capacity is only 50% of that of TRU-015, the CDC activity of SBI-087 was significantly higher compared not only to TRU-015, but even to rituximab (Dunussi-Joannopoulos *et al.*, 2009). It is reasonable to hypothesize that the binding properties of SBI-087 together with its SMIP configuration bring the C1q binding site proximal to the cell membrane, favor capturing of C1q and provide more efficient CDC lysis of B cells. Interestingly, structural differences of SBI-087 versus TRU-015 had no effect on its ADCC activity (Dunussi-Joannopoulos *et al.*, 2009). To evaluate efficacy and PK properties a single infusion of 1 or 10 mg/kg of SBI-087 or rituximab was administered intravenously to cynomolgus monkeys, and B cell depletion and recovery were monitored for six months. SBI-087 demonstrated a favorable PK profile with a mean elimination half-life of 5.4 ± 1.1 and 7.8 ± 1.5 days in the 1 and 10 mg/kg dose groups, respectively. However, SBI-087 demonstrated efficient B cell depletion and had more sustained activity in the peripheral blood and bone marrow compared to rituximab (Dunussi-Joannopoulos *et al.*, 2009; Nickerson-Nutter, submitted). Most importantly, SBI-087 had significantly higher potency in depleting B cells in the lymph nodes and significantly prolonged activity compared to rituximab, suggesting that a different scaffold format and higher CDC activity may be responsible for effective B cell depletion in some tissues of cynomolgus monkeys. Currently, SBI-087 is being evaluated in Phase I clinical trials for systemic lupus erythematosus (SLE) and has entered Phase II clinical trial for the treatment of RA.

10.3 Fc-Based Scaffolds for Dual Targeting

Scorpion therapeutics are single-chain proteins comprised of an N-terminal binding domain, an effector domain based on immunoglobulin Fc regions and a C-terminal binding domain that are produced as disulfide-linked dimers. As in the case of SMIPs, Scorpions are also based on a modular design that includes a choice of binding domains and Fc functionality. The resulting proteins exhibit appropriate effector function and multivalent (mono- or multi-specific) binding capacities as shown in Fig. 10.3.

The Scorpion format provides the basis for development of single therapies that simultaneously inhibit multiple ligand/receptor interactions or that manipulate cellular signaling pathways by cross-linking multiple cell surface receptors. The binding domains of a Scorpion can be a scFv, single binding domains (V_HH, IgNAR), functional parts of cell surface receptors, ligands, or other naturally occurring proteins that bind to the target of interest. Since this format allows two binding domains of each specificity to be combined in one molecular entity, it permits a wide range of functions such as simultaneous multivalent engagement of different cell-surface targets (Blankenship *et al.*, 2009). It also provides efficient blockade of two different cytokines such as tumor necrosis factor and IL-6, or blockade of synergistic cell activation pathways to achieve better efficacy such as dual targeting of CTLA-4 ligand and IL-10 cytokine. In addition, Scorpion molecules retain immunoglobulin effector functions such as long *in vivo* half-life and Fc-dependent cellular cytotoxicity (FcDCC) activity, if desired.

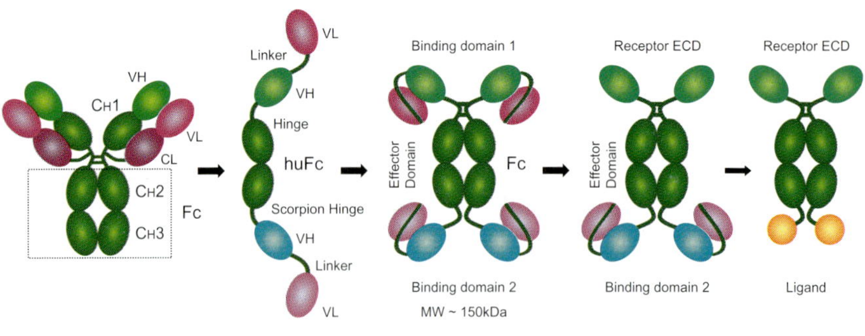

Figure 10.3 Structure and design of Scorpion molecules

Data presented by Trubion at the 2009 American Association of Cancer Research (AACR) meeting demonstrated the selectivity and high potency of the bi-specific CD79BxDR Scorpion molecule, its differentiation from mono-specific approaches and its potential as a therapeutic for B cell depletion, particularly in diseases refractory to CD20-targeted therapies (Blankenship *et al.*, 2009). This molecule can simultaneously bind with high affinity and specificity to two different targets: CD79B, a lineage-restricted component of the B cell antigen receptor and HLA-DR, an HLA Class II molecule. Preclinically, this molecule elicited strong ADCC as well as highly potent direct killing, superior to each of the parental antibodies. In addition, data also demonstrated that in assays using the NHL-derived DoHH2 cell line, the CD79BxDR Scorpion was 100× more potent than both rituximab and the combination of the individual mono-specific SMIP molecules against the respective targets. This enhanced potency was shown to be restricted to the depletion of B cells but not CD3+ T cells (Blankenship *et al.*, 2009).

Another dual-specific Scorpion therapeutic was generated to achieve blockade of two pro-inflammatory cytokines, TNFα and IL-6. TNFα is a well established target in autoimmune diseases and yet not all patients respond to existing TNFα targeted therapies (Tracey *et al.*, 2008). Interestingly, IL-6 inhibition (e.g. tocilizumab) has been shown to benefit RA patients, including non-TNF responders. Therefore, a dual agent that inhibits both IL-6 and TNFα might be a particularly potent combination. The IL-6xTNFα Scorpion was developed by combining the extracellular domain of the TNFα receptor at the N-terminus of human Fc with anti-IL-6 scFv at the C-terminus of the molecule. In preclinical models, this molecule also demonstrated the ability to bind both targets with high affinity and good potency in *in vitro* functional assays.

As mentioned above, the Scorpion design also allows for the combination of receptor extracellular domains and soluble ligands. This concept has been explored with a combination of CTLA-4 × IL-10 using the extracellular domain of CTLA-4 with IL-10 cytokine. Soluble CTLA-4 is a negative regulator of T cell signaling and IL-10 is a known immunosuppressant. Physical juxtaposition of the CTLA-4 ectodomain with IL-10 creates a dual-specific agent that brings more potent immune suppression of T cells than either agent alone. Preliminary results with the CTLA4 × IL-10 molecules demonstrate significantly enhanced activity in comparison to CTLA-4-Fc (unpublished results).

While the Scorpion strategy is appealing, the main challenges with such molecules are related to complete retention of the functional activity of all binding domains and product stability. Practical methods for producing high quality, bi-specific molecules that have good product stability, long serum half-life, straightforward purification, and scalable production have been challenging. While limited data are available on the expression properties and solubility of Scorpions, similar types of bi-functional antibodies (BsAb) have significant folding issues and are prone to aggregation (Michaelson *et al.*, 2009). In the Michaelson study, extensive protein engineering including introduction of specific stabilizing mutations or Cys-Cys covalent bonds were required to improve properties of anti-TRAIL-R2 × anti-Lymphotoxin β receptor BsAb. Despite these challenges, it has been possible to generate Scorpion molecules at the research scale with desired functional and pharmaceutical properties in multiple cases. It has also been possible to use effector function and half-life extension strategies based on the SMIP design. For instance, despite their different design and slightly larger size, Scorpion molecules have enhanced ADCC activity (Blankenship *et al.*, 2009).

10.4 Avians as a Source of Potent Antibody Therapeutics

The full exploitation of the SMIP and Scorpion structural formats described above requires a robust pipeline producing highly potent and stable scFv molecules. Despite the significant progress made with these multi-specific molecules, a major difficulty in their development is the rapid identification of potently neutralizing antibodies that are good drug candidates and are naturally functional in the scFv format. To accelerate the examination of a drug discovery program's therapeutic hypothesis, many investigators have historically made monoclonal antibodies (mAb) their first priority, due to their potentially high affinity and potency. Murine mAb technology (Kohler and Milstein, 1975) was a paradigm shift for biomedical research and allowed the development of truly uniform, highly specific biotherapeutics to many drug targets. In recent years however, it has become increasingly clear that the reliance on murine IgG leads to a number of significant practical difficulties in antibody drug discovery: (1) Tolerance — humans and rodents are closely phylogenetically related, leading to the problem of thymic tolerance restricting the murine IgG response after immunization. (2) Epitope restriction — when an antibody response to a human protein is raised in mice or rats, the high degree of sequence similarity between murine and human proteins (often >80% linear amino acid identity) may lead to a restricted possible number of immunogenic epitopes. mAbs generated from mice are also usually specific to the human target sequence and possibly that of a closely related test species such as rhesus macaque, but rarely show any reactivity to the mouse ortholog. This is a major consideration if an antibody is to pass through proof of concept pre-clinical analyses in mouse disease models (Lee *et al.*, 2006). (3) Inefficient sampling of diversity — mAbs are raised via the fusion of splenic B cells to a mouse myeloma line, followed by limiting dilution of the pooled cell population and the random identification of target-specific antibodies, often by a simple direct ELISA. This overall process is intrinsically inefficient as the screening process only allows a few thousand clones, at most, to be sampled. (4) Importantly, mAbs are isolated as full-length immunoglobulins. This means that the V_H–V_L family pairings in each molecule are uncontrollable and even after humanization into favorable frameworks, stable function in the scFv format is unpredictable.

Several species have been used as alternative sources of polyclonal antisera, but have never been used for mAb generation, partly due to the lack of good myeloma fusion partner cell lines. Nonetheless, the germline immunoglobulin genes of many of these animals have now been fully sequenced. This means that after immunization, their lymphoid tissues can be harvested and their variable region repertoire assembled into a display-friendly "antibody fragment" library format such as V_HH (camelids only), IgNAR (sharks) or scFv (Andris-Widhopf *et al.*, 2000). The resulting phage library is then selected on the protein used for immunization and any other related proteins of interest. This system has been used to derive novel antibodies from mice (Krebber *et al.*, 1997), rabbits (Li *et al.*, 2000), sheep (Charlton *et al.*, 2001), camelids (Arbabi-Ghahroudi *et al.*, 1997), cows (O'Brien *et al.*, 1999), sharks (Dooley *et al.*, 2003), and chickens (Yamanaka *et al.*, 1996).

10.4.1 *Reasons to consider chickens as an alternative immune source*

10.4.1.1 *Phylogenetic distance*

For therapeutic antibody discovery, the ideal immune host should have maximal evolutionary distance from mammals, V genes that are closely related to humans (and are therefore humanizable), proven ability to produce high affinity neutralizing antibodies, and a well-established platform to isolate the desired clones. The chicken is a logical choice as a host for immunization with mammalian proteins, as their evolutionary divergence from mammals is estimated to have occurred 310 million years ago in the early Mesozoic period (Hedges, 2002). Chickens are also the main laboratory model for the approximately 9,600 known avian species, and the full genome sequence has been derived (Wallis *et al.*, 2004). More importantly, the high degree of evolutionary divergence between chickens and humans means that the amino acid sequence homology between the mammalian and avian orthologs of most target proteins is typically much lower than between humans and rodents (Wallis *et al.*, 2004). This means that chickens immunized with proteins that are highly conserved amongst mammals can frequently exhibit very high titer, high affinity serum antibody responses with the capacity to target a broad spectrum of epitopes (Lemamy *et al.*, 1999; Ikemori *et al.*, 1993; Stuart *et al.*, 1988). This gives chickens an advantage over all mammalian immune hosts as they can target epitopes that are highly conserved across multiple mammalian orthologs of a target protein.

These beneficial aspects of the chicken immunization model have been known for many years, leading to the routine use of chickens as a host for production of polyclonal antibodies. In particular, the availability of the egg yolk antibody termed IgY has been heavily exploited as laying hens will export ~100 milligrams of IgY per yolk, which allows harvesting of the antibodies directly from the eggs without the need for serum sampling (Polson *et al.*, 1985). These polyclonal antibodies have been highly successful reagents in immunochemistry (Larsson *et al.*, 1991), diagnostics (Tsen *et al.*, 2003), proteomics (Qian *et al.*, 2008) and have even been used in experimental treatment of *Helicobacter pylori* gastritis (Attallah *et al.*, 2009). Attempts to generate chicken monoclonal antibodies via hybridoma technology have been less successful, however, due to the low antibody expression and instability associated with chicken myeloma cell lines (Nakamura, 2000; Michael *et al.*, 1998).

Technologies such as phage display have now overcome the limitations of chicken hybridoma technology, making recombinant chicken antibodies highly accessible. The potential of avian recombinant antibodies was first highlighted in a study where a scFv library was derived from the spleens of chickens immunized with both mouse and rat serum albumins (Yamanaka *et al.*, 1996). This study successfully generated highly specific scFv antibodies that targeted the mouse and rat serum albumin orthologs, a feat that could not be achieved with murine hybridoma technology due to tolerance issues. In another key study (Nakamura *et al.*, 2004), high affinity antibodies were generated from immunized chickens that strongly cross-react with multiple orthologs (murine, ovine, bovine) of the highly conserved mammalian prion protein (PrP). This was the first example of such mAbs being raised *in vivo*. Since these seminal studies, multiple groups have shown that recombinant antibodies derived from immunized chickens are excellent performers in diverse settings such as immunodetection of allergens and haptenic shellfish toxins (Finlay *et al.*, 2006; Finlay *et al.*, 2005).

10.4.1.2 *Ease of use*

Chickens are a very simple model animal to handle. They are highly accessible, very affordable, and easily housed in a generic animal house. As small animals, very little protein immunogen is required to raise a strong immunoglobulin response in chickens. Approximately 200 μg/bird of purified protein is sufficient to carry out a full immunization regime, which may be as short as 60 days in total (Finlay *et al.*, 2005; Gassmann *et al.*, 1990). Most importantly, chickens can be immunized simultaneously with multiple proteins of interest, with as many as eight proteins being used successfully in a single immunization scheme (Chiliza *et al.*, 2008; Hof *et al.*, 2008; Finlay *et al.*, 2005). The target proteins of interest are mixed in a single adjuvant preparation, and each immunized animal receives all proteins simultaneously. A small (10^7–10^8 diversity) individual phage display library can then be easily generated by a single investigator and specific antibodies derived via selection of the library separately on the individual proteins originally used for immunization (Finlay *et al.*, 2005). The resulting individual scFv antibody clones are fully specific by Western blot and ELISA, showing no reactivity to their co-immunogens (Hof *et al.*, 2008; Finlay *et al.*, 2005). The uniformity of the antibody V genes derived subsequently allows the rapid and simple reformatting of selected clones to scFv–Fc fusion or IgG formats (Nakamura *et al.*, 2003).

10.4.1.3 *High affinity*

The cloning of chicken antibodies via display technologies has allowed the detailed characterization of their specificity and affinity. In one definitive study (Leonard *et al.*, 2007), phage display was used to generate panels of immune chicken scFvs to the inflammatory biomarker C-reactive protein from an immune library of only 3×10^7 total clones. BIAcore technology was used to perform high throughput affinity measurements for 96 clones after phage selection, identifying clones that preferentially recognize the multimeric and/or monomeric forms of the protein with affinities ranging from 25 nM to 350 pM. Chicken anti-PrP scFvs have also been described with reported affinities reaching 15 pM, placing them among the highest affinity anti-protein scFvs reported to date from any source (Nishibori *et al.*, 2006).

10.4.1.4 *Variable gene uniformity*

In antibody drug discovery it is highly desirable to have as much uniformity and predictability as possible in therapeutic hits, allowing greater confidence in the expression, stability, and manufacturability characteristics of any lead compound. The chicken immunoglobulin repertoire is of predictable performance, as chickens generate their immunoglobulin repertoire from a single set of functional V_H, J_H, V_L, and J_L germline sequences (McCormack *et al.*, 1993). Chickens have an advantage as they do not exhibit the disruptive non-canonical cysteine usage that is common in the V_κ sequences of laboratory rabbit strains such as the New Zealand White (Popkov *et al.*, 2003). Importantly, the two germline V-gene sequences of chickens are homologous to the human V_λ and V_H3 germline families (Tsurushita *et al.*, 2004), which are associated with creating highly stable and soluble antibodies that generally express well in both prokaryotic and eukaryotic expression systems. Chicken immunoglobulins have also shown beneficial biophysical properties, exhibiting high stability to changes in pH and temperatures up to 70°C (Hatta *et al.*, 1993; Shimizu *et al.*, 1992). Indeed, chicken scFvs have been shown to be stable in bacterial culture supernatants for up to one month at room temperature (Chiliza *et al.*, 2008).

10.4.1.5. *Chicken antibodies can be humanized*

While the majority of studies addressing recombinant chicken antibody technology have focused on the generation of reagent antibodies for diagnostics, the therapeutic potential of these antibodies has not gone unrecognized. As with all non-human antibodies, the V genes of chickens are very likely to need humanization before they can be used in a clinical setting. Two key studies have now addressed this requirement. (Tsurushita *et al.*, 2004) generated a scFv phage display library from the lymphoid tissue of chickens immunized with both human and mouse interleukin-12 (IL-12) (a target in the treatment of inflammatory disorders). A scFv was then isolated that cross-reacts equally to both human and mouse orthologs of IL-12, and this clone was subjected to humanization. The V_H and V_L chains of the scFv were found to be most similar to the human germline DP54 (71% homology) and DPL16 (68% homology) respectively, and the chicken CDRs were first grafted into these frameworks and then expressed in full-length IgG format. Through a classical series of structurally-led "back mutations" (Kettleborough *et al.*, 1991), full activity and specificity of the antibody was returned, leaving a total of 11 chicken amino acids in the frameworks (six in the light chain and five in the heavy). Another antibody was also humanized with eight final back mutations. Despite the significance of the Tsurushita study, it did not report the neutralizing potential of the humanized anti-IL-12 antibody. In a later study (Akamatsu *et al.*, 2007), the same group again used immunized chickens to generate anti-IL-12 antibodies, but in this case used mammalian cell surface display (analogous to yeast display) instead of phage display. In this study they showed clearly that antibodies could be isolated that exhibited neutralization of both human and mouse IL-12 in cell-based assays.

Another chicken antibody with therapeutic potential, which binds PrP from multiple species, has also been humanized (Nishibori *et al.*, 2006). In this study the team decided to take a library approach to humanization, as opposed to the rational design method used by Tsurushita *et al.* (2004). Through oligo-mediated gene synthesis and PCR, the CDRs of the chicken antibody were grafted onto germline human V_H3 (DP47) and V_L (DPL16) frameworks. Several framework positions were then allowed amino acid permutations including both the human and chicken residues at each position: V_L positions 46, 66, 69, and 71, plus V_H 47, 49, 67, 75, 76, and 78. Through phage display and re-selection of functional clones, the authors showed that full activity could be restored with 7–8 back mutations and that positions L46, H47 and H78 were required to remain as chicken sequence.

While no published reports yet exist for the immunogenicity of humanized chicken antibodies in humans, one minor study has been performed in mice (Tateishi *et al.*, 2008). On conversion of their chicken antibody to a chimeric IgG with chicken V regions and mouse constant regions, no immunoglobulin response could be measured in mice dosed weekly over nine weeks, while IgY was strongly immunogenic. These early findings suggest that, as for murine antibodies, the main determinants of chicken antibody immunogenicity may be found in the constant regions rather than in the V regions. With antibodies containing CDRs from mice, rats, camels, and synthetic diversity all now showing success in the clinic, it is clear that humanized antibodies derived from chickens are poised to follow them.

10.5 V_HH Antibodies

Sections 1–4 of this chapter have dealt with the search for novel therapeutics based on optimized dual-domain antibodies. A logical alternative, however, would be modular therapeutics built from single-domain binding proteins that can simplify, stabilize, and reduce the size of the final molecule. It was therefore a significant development in the protein biotherapeutic field when, in 1993, the work of Raymond Hamers at the Free University in Brussels was published (Hamers-Casterman *et al.*, 1993). The Hamers group identified a previously unknown immunoglobulin form purified from camel serum. Instead of presenting the four classic polypeptide chains typically present in a conventional antibody with two heavy and two light chains, the newly described form was found to be missing a light chain, and following the loss of the splice consensus site, not to have the expected constant C_H1 domain in the heavy chain (Nguyen *et al.*, 1999). These unique antibodies, also called heavy chain-only antibodies (V_HH antibodies) were shown to represent over 50% of the total Ig population in camel sera (Nguyen *et al.*, 2001; Muyldermans and Lauwereys, 1999). The camel IgG_1 was shown to have the standard architecture of conventional antibodies, whereas the $\gamma2$ and $\gamma3$ isotypes were identified as V_HH antibodies. Here, the V_HH antibody is composed of the C_H3-C_H2 domains linked to an unpaired variable heavy chain domain (Fig. 10.4; Harmsen *et al.*, 2000).

10.5.1 *V_HH genetics*

Since the discovery of the camelid V_HH antibodies, their origin, evolution, and potential applications have been extensively studied (Harmsen *et al.*, 2007b; Conrath *et al.*, 2003). The earliest work comprehensively describing germline organization of the camelid genes of the conventional antibody and V_HH antibody types was published in 2003 (Conrath *et al.*, 2003). V_HH antibodies have evolved from conventional Abs and developed independently from them while maintaining mechanisms for class switching and affinity maturation.

The class switching and maturation in a typical immunoglobulin response is characterized by tight regulation allowing only fully functional heterotetrameric antibodies to be exported to the surface of B cells. Conventional mAbs build their diversity by combining both heavy and light chain recombination events together with somatic hypermutations and class switching. Homodimeric heavy chain antibodies rely on a

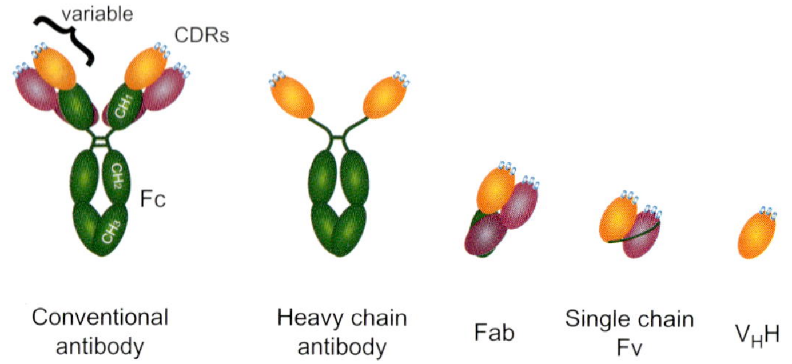

Figure 10.4 Schematic representation of conventional antibody, heavy chain antibody and their relevant antigen binding fragments

The single-chain Fv is a non-natural fusion of the variable heavy and light chains of conventional antibody through a peptide linker. The V_HH fragment is monomeric and highly homologous to the V_H domain but does not require the V_L domain for antigen recognition or good solubility.

single variable domain for antigen recognition but it has been shown that their diversity is no less than their conventional counterpart (De Genst *et al.*, 2006). In camelid V_HH, the V_H domains are encoded by distinct sets of V gene segments. Studies by Nguyen *et al.*, (1999), show that V_HH antibodies and conventional Abs share some D segments as well as an identical J_H region (Nguyen *et al.*, 1998).

A systematic analysis of V_HH germline sequences and comparison to the V_H counterparts sheds light on understanding how these single-chain antibodies compensate at the genomic level for their lack of light chain (Riechmann and Muyldermans, 1999). The frequency and position of hypermutation hotspots have been shown to be enhanced in V_HH compared to regular V_H, leading directly to a higher level of variability in CDR1 and CDR2 sequences. V_HH antibodies also contain a unique cysteine residue typically within the CDR3. This extra cysteine allows creation of an additional disulfide bridge that can alter the CDR1 or CDR3 fold, permitting access to neo-epitopes (Nguyen *et al.*, 2002). In addition, a region near the CDRs bears a higher frequency of insertions or deletions and is thought to have a significant impact on the domain loop structures (Nguyen *et al.*, 2000; Wilson and de Bouteiller, 1998).

10.5.2 *V_HH structural features*

Sequence analysis of camelid V_HH domains has shown a very close homology to conventional V_H domains. Similar to a V_H immunoglobulin fold, the V_HH is a beta-barrel formed by two anti-parallel beta sheets (Fig. 10.5).

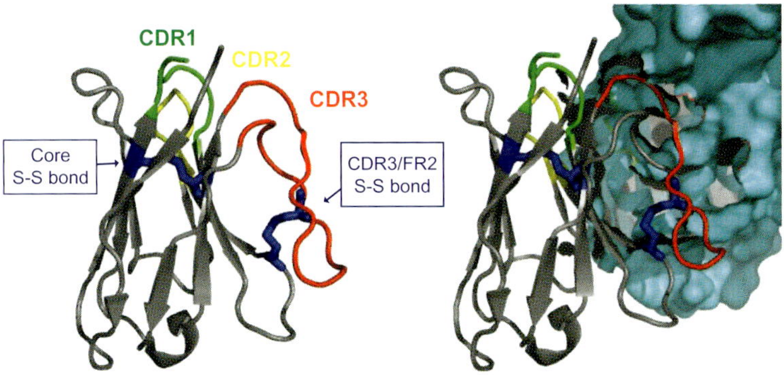

Figure 10.5 Camelid V_HH domain from a complex structure with porcine pancreatic alpha-amylase (pdb: 1KXT)

Un-liganded V_HH left panel; liganded V_HH right panel with amylase in surface representation in cyan. The long V_HH CDR3 (19 amino acids in red) extends along the classical hydrophobic face of the would-be V_L/V_H interface pairing and allows extensive contact with the antigen. The standard immunoglobulin core disulfide bridge is present and an additional non-canonical disulfide bridge linking CDR3 and framework 2 stabilizes the extended CDR3.

Compared to conventional mAbs, V_HH antibodies completely lack the light chain and the constant C_H1 domain of the heavy chain (Muyldermans *et al.*, 1994). Consequently the V_HH single variable domain does not have the need for the residues designed to interact with and stabilize the missing V_L domain. It is not a surprise to find hydrophobic to hydrophilic substitutions for V_HH hallmark residues in the framework 2 region where the V_H folds against the V_L in conventional antibodies. This region is responsible for the V_H–V_L interaction and is highly conserved throughout evolution. The substitutions V37F/Y, G44E, L45R, and W47G are the drivers for the ability of the V_HH to fold as an independent binding domain without the need for a cognate V_L pair (Harmsen and De Haard, 2007a).

The sum of structural data on antibody–antigen complexes reveals that the conventional immunoglobulin variable domains only rarely take advantage of their complete set of six CDRs to actually contact the antigen. As expected by the VDJ recombination and somatic hypermutation mechanisms, the CDR3 of the V_H and V_L chains present the highest diversity and therefore are key determinants of the antigen specificity. Due to their single-domain nature, it was anticipated at first that the V_HH antibody repertoire size would be smaller. It is clear today that camelids have evolved new maturation mechanisms to alleviate this risk (Nguyen *et al.*, 2002). The V_HH antibody variable loops have evolved to not only compensate for the lack of light chain but in some instances to develop

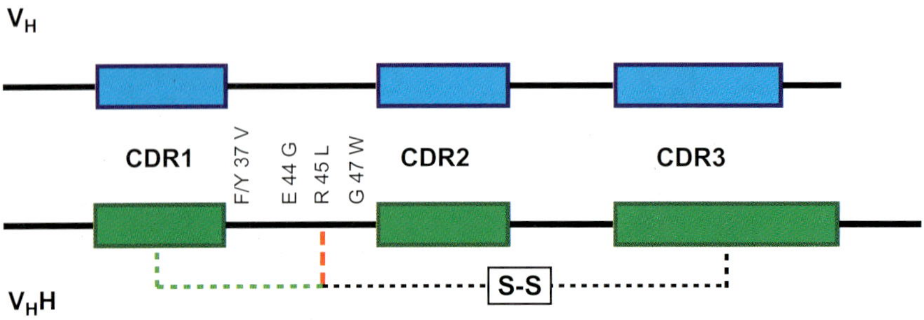

Figure 10.6 Schematic comparison of the sequence and structural elements specific to V_H and V_HH

All three V_HH CDRs exhibit a larger genomic and structural repertoire compared to V_H. CDR1 and CDR3 can be significantly larger in V_HH with CDR3 routinely reaching lengths of 16–18 amino acids compared to the 9–12 average of the V_H. The four substitutions in the framework 2 region are the hallmarks of the V_HH antibody, giving it its monomeric high solubility properties. The CDR3 often contains a cysteine residue responsible for an additional disulfide bridge with either CDR1 (green) or the position 45 (red) of the framework 2 region.

original structural features allowing the recognition of a diversity of epitopes (Fig. 10.6).

Even though a typical V_H can be expressed in the absence of its V_L pair, it becomes insoluble at concentrations above 1 mg/ml (Davies and Riechmann, 1994). Interestingly, removal of the hydrophobic patch and simple incorporation of the four V_HH antibody hallmark residues into a germlined V_H could lead to a "camelized" domain, well behaved in solution. Additionally, V_HH antibodies have a superior stability against denaturation by heat, extreme pH, as well as from proteases (Omidfar *et al.*, 2007). V_HH antibodies have been successfully expressed in a range of expression hosts from microbes to mammalian cells. As single genes with no requirement for post-translation modification, recombinant technologies applied to the V_HH antibodies have yielded large secreted amounts of very well-behaved proteins (Joosten *et al.*, 2005; Frenken *et al.*, 2000).

10.5.3 *V_HH antibody as therapeutics and perspectives*

Several factors have contributed to the recent success of the V_HH antibodies. Their monomeric nature, excellent biophysical properties, and their relative ease of generation have supported a variety of biotechnological development. Multiple research organizations are testing V_HH product candidates in human clinical trials to address their pharmacokinetic and immunogenicity profile.

V_HH antibodies are smaller than Fab and scFv antibody fragments. At 15 kDa in size, V_HH antibodies suffer from fast renal clearance and, without adequate protein engineering, they have short *in vivo* half-lives. Therefore, half-life extension is a key requirement for applications of V_HH antibodies in chronic settings. The good news is that as single-chain polypeptides, the V_HH antibodies are very simple to engineer. A modification at their N- or C-terminus is typically possible without loss of function. Fusing a V_HH antibody to serum albumin or to an immunoglobulin Fc fragment does efficiently extend its *in vivo* half life. Natural or non-natural polymers have also been validated as hydrodynamic volume enhancers to avoid kidney clearance. In addition, humanization of V_HH antibodies appears quite achievable, typically requiring only a very limited number (<3–10 typically) of back-mutations from the closest human V_H germline sequence. Together with their very high stability and low aggregation, it gives the humanized V_HH antibody lower immunogenicity risk.

V_HH antibody programs are now entering clinical trials. A V_HH antibody targeting von Willebrand Factor has been developed by Ablynx BV to reduce the risk of thrombosis in patients with acute coronary syndrome and thrombotic thrombocytopenic purpura. In a Phase I clinical study with healthy volunteers, this antibody was reported to be safe and well-tolerated without eliciting signs of immunogenicity. In addition, a V_HH antibody targeting TNFα is being clinically developed jointly by Ablynx and Pfizer. One exciting avenue being pursued with V_HH antibodies is the possibility of oral administration. Strains of *Lactobacillus* expressing a rotavirus-specific V_HH antibody (Pant *et al.*, 2006) or a TNF-specific V_HH antibody (Vandenbroucke *et al.*, 2010) were shown to be efficacious in mouse models when delivered orally. This opens up new vistas for non-parenteral delivery of V_HH antibody-based therapeutics.

10.6 Shark IgNAR-Derived Therapeutics

The camelid V_HH domain outlined above offers exciting new possibilities, but it is not the only naturally occurring single-domain binding protein with biopharmaceutical potential. In the 1990s a novel immunoglobulin isotype, IgNAR, was discovered in the serum of nurse sharks (*Ginglymostoma cirratum*). Similar to the V_HH, electron microscopy studies showed IgNAR to be a heavy-chain homodimer which naturally does not associate with light chains. Instead, IgNAR binds antigen with a single, highly soluble V domain which is tethered to C_H1 via a flexible, hinge-like region (Roux *et al.*, 1998; Greenberg *et al.*, 1995). The IgNAR scaffold is therefore an attractive alternative to the camelid V_HH and has the same potential to be combined into multi-valent and/or multi-specific biotherapeutic constructs.

IgNAR genes are arranged in clusters (Litman *et al.*, 1985), with each cluster containing a single variable (V) region, diversity (D) regions, a joining (J) segment, and the constant domains required to generate a functional molecule. As clusters have very high (>90%) identity, diversity is primarily produced via the use of three D regions, necessitating four rearrangement events to generate a functional V domain. Combined with unusually extensive N-region addition, the primary repertoire shows a wide range of CDR3 length (7–34 aa) and amino acid composition (Dooley and Flajnik, 2006a; Diaz *et al.*, 2002). The V region is then subject to extraordinary levels of antigen-driven somatic mutation (Diaz *et al.*, 1998) and can tolerate high mutational loads in non-binding regions of the molecule (Dooley *et al.*, 2006b).

10.6.1 *Structural features of IgNAR-derived single domains*

While the IgNAR V is a typical immunoglobulin superfamily (IgSF) domain it exhibits features of several types, making it difficult to classify. Sequence and structural homology are highest with T cell receptor V regions. It also has a deletion of CDR2, replacing it with a hypervariable "belt" (the HV2 loop) (Stanfield *et al.*, 2004). In contrast, camelid V_HH are derived from a *bona fide* IgG, which has lost the ability to interact with light chain. Thus, they exhibit high identity to mammalian V_Hs with the exception of a few key residues in framework 2 which stabilize the domain and increase solubility (Nguyen *et al.*, 1998; Desmyter *et al.*, 1996).

Three "types" of IgNAR V have been described and can be differentiated by the presence and position of non-canonical cysteine residues (Diaz *et al.*, 2002; Greenberg *et al.*, 1995). Type I have single non-canonical cysteine residues in framework 2 and framework 4 which disulfide bond with the two cysteines in CDR3. Thus, the CDR3 folds over the hydrophobic surface which would be the V_H/V_L interface in conventional antibodies (Stanfield *et al.*, 2004). In contrast, CDR3 protrudes from the top of type II V regions and is either tethered by a disulfide bond to CDR1 (type IIa) (Dooley *et al.*, 2006b) or is unbound (type IIb) and can drift away from CDR1 (Streltsov *et al.*, 2004). *In vivo* studies show positive selection of mutations in the HV2 of type I V regions except the CDR1 of type II, i.e. the hypervariable loop in closest proximity to CDR3. Finally, type III V regions are essentially type IIs, but encode an invariant tryptophan (W^{31}) within CDR1. In nurse shark, type III is expressed early and has a highly conserved CDR3 with low frequency of somatic hypermutation (Diaz *et al.*, 2002).

10.6.2 *Isolation of antigen-specific IgNAR-derived single domains*

Initial work showed that IgNAR V regions could be expressed in bacteria and displayed on bacteriophage (Dooley *et al.*, 2003; Nuttall *et al.*, 2001). Selection of libraries generated from nurse sharks immunized with hen egg white lysozyme (HEL) gave rise to high affinity, highly specific clones (Dooley *et al.*, 2003). Study of clones matured *in vivo* showed that the ancestral (unmutated) sequence had nanomolar affinity which was increased (~ 3-fold) by a single change in the HV4 loop. Interestingly, many of the somatic mutations outside the binding region increase the net positive charge, an attribute which appears important for the solubility of single-domain binding regions (Arbabi-Ghahroudi *et al.*, 2009; Dooley *et al.*, 2006b). Concurrently, Nuttall and colleagues constructed semi-synthetic libraries by grafting randomized CDR3s onto framework scaffolds derived from the natural IgNAR repertoire of wobbegong sharks (*Orectolobus maculatus*, a close relative of nurse shark). From these libraries, clones specific for a number of targets were isolated: a protease from *Porphyromonas gingivalis*, the human mitochondrial translocase receptor Tom70, and the malarial membrane protein AMA1 (Nuttall *et al.*, 2004; Nuttall *et al.*, 2003; Nuttall *et al.*, 2001). Although most of these clones were of moderate affinity, subsequent work has matured a number of the binders (see below). The Goldman group first reported the isolation of antigen-specific IgNAR V regions from more distant shark species, the spiny dogfish (*Squalus acanthias*) and smooth dogfish (*Mustelus canis*). Small naive and semi-synthetic libraries (randomized CDR3) yielded binders to various biodefence targets that were of low affinity but showed utility as biosensors in a multivalent format (Liu *et al.*, 2007a; Liu *et al.*, 2007b).

10.6.3 *Properties of IgNAR-derived single domains*

10.6.3.1 *Targeting novel epitopes*

Study of HEL-binding V domains showed the epitope bound is distinct from those recognized by conventional (V_H-V_L) antibodies (Dooley *et al.*, 2006b; Stanfield *et al.*, 2004). The vast majority of anti-HEL mouse mAbs bind one of three flat epitopes with planar binding sites (Davies *et al.*, 1988). However, the IgNAR-derived binding domains recognize an epitope nestled between those of the mouse mAbs, directly overlying the recessed active site; indeed, their CDR3 loops actually penetrate the active site (Fig. 10.7) and can inhibit carbohydrate binding (Stanfield *et al.*, 2004). Similarly, the semi-synthetic AMA1-binding clones have extended CDR3s that insert into a hydrophobic cleft on the protein surface, contacting highly conserved buried residues (Henderson *et al.*, 2007). In this way they avoid the highly variable loops on the surface of AMA1 and thus can inhibit erythrocyte invasion by multiple *Plasmodium* species (Henderson *et al.*, 2007; Nuttall *et al.*, 2004). The prevalence of such clones from IgNAR V (and camelid V_HH) libraries suggests this mode of binding is required to increase the surface area of interaction with antigen (De Genst *et al.*, 2006; Stanfield *et al.*, 2004). This trait has important implications for their future therapeutic development as such cavities and clefts have been shown to be integral to many enzyme-substrate and receptor-ligand interactions (Laskowski *et al.*, 1996).

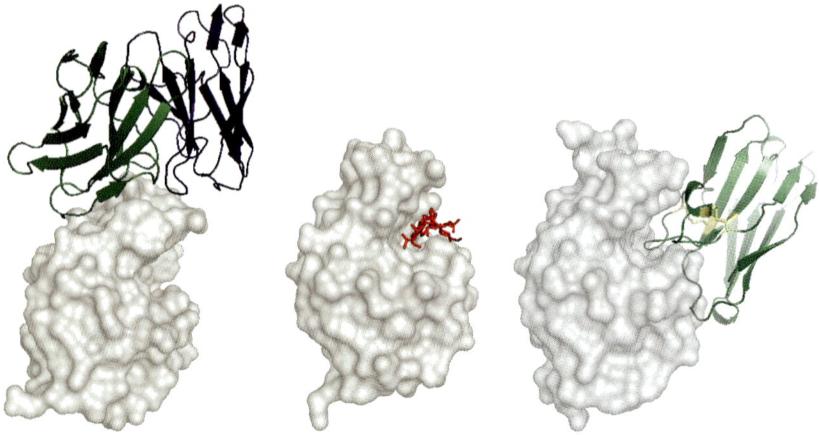
(a) (b) (c)

Figure 10.7 3D structures of anti-HEL antibody binding regions in complex with their antigen

(**a, b**) Mouse-derived Fv D44.1 (pbd code 1mlc. HEL surface in gray, D44.1 in blue and green) exemplifies the binding of conventional antibodies to sites distinct from the substrate binding site of HEL (pdb code 1lsz; oligosaccharide in red) via their (generally) planar combining sites. (c) In contrast, the shark HEL-5A7 domain (pdb code 1sq2; green ribbon with non-canonical cys in yellow) interacts directly with the active site and inhibits oligosaccharide binding. Although HEL-5A7 binds using only two CDR loops the interaction buries equivalent surface area as D44.1 (Stanfield *et al.*, 2004).

10.6.3.2 *Superior stability*

Binding domains derived from shark IgNAR show a number of advantageous properties when compared with those from conventional antibodies. These include their small size (~ 12 kDa) and single domain nature. In addition, their compact nature and presence of additional intra-domain disulfide bonds confer stability; isolated domains have a very high tolerance to thermal denaturation, considerably outperforming conventional mAbs at high temperatures (Liu *et al.*, 2007a; Dooley *et al.*, 2003). Circular dichroism (CD) studies have shown individual clones can withstand temperatures of 50–60°C before unfolding and, more surprisingly, spontaneously refolding back to their native (antigen binding) conformation upon cooling (Stanfield *et al.*, 2007).

10.6.3.3 *In vitro maturation*

Engineering lead clones to increase affinity and/or specificity has been conducted with many molecules, and IgNAR-derived binding domains are no exception.

Error-prone PCR was applied to a low-affinity AMA1-specific IgNAR V region and higher affinity variants selected. Mutations mapped to CDR1 and CDR3, with the best of the variants having a ~ ten-fold improved affinity through a single (P90L) change in CDR3 (Nuttall *et al.*, 2004). Henderson and colleagues also targeted the CDR3 of this clone, varying its length and introducing aromatic and hydrophobic residues. In this case a G92R mutation at the tip of CDR3 increased affinity for AMA-1 by ~ 50-fold. Comparison of variants showed that *in vitro* inhibition of erythrocyte invasion by parasites correlated with affinity, requiring a threshold of at least 600 nM to inhibit invasion (Henderson *et al.*, 2007). An alternative method employed by Kopsidas and colleagues involved iterative rounds of error-prone amplification using the RNA-dependent RNA polymerase of Qβ bacteriophage followed by ribosome display selection. The P90L change in CDR3 was again selected, this time in combination with a K61R change in the HV4 loop, giving a 23-fold enhancement in affinity (Kopsidas *et al.*, 2006). Mutations in CDR1 and HV4 were also noted during *in vivo* maturation; in this case a S61R change in HV4 gave a more modest three-fold increase in affinity (Dooley *et al.*, 2006b). Thus, where the characteristics of selected binders are considered sub-optimal, *in vitro* mutation can be used to improve affinity, or modify other features desirable for their intended application.

10.6.3.4 *Ease of formatting*

At ~ 12 kDa the IgNAR V domain is the smallest (antibody-derived, natural) binding domain discovered to date. In unmodified format this makes it ideal for imaging purposes, where reagents should be small enough to maximize tissue penetration while minimizing serum half-life, thus facilitating rapid clearance of payload (e.g. radioisotopes). In contrast, for use as therapeutic reagents, these domains will require some means to extend their serum half-life, thereby increasing their efficacy and extending the time between treatments. There is little doubt that IgNAR-derived binding domains can be linked to human Fc regions, as has been successfully done for camelid V_HH domains (Coppieters *et al.*, 2006). Alternate methods for prolonging serum half-life, such as PEGylation or the targeting of serum albumin (HSA) to facilitate *in vivo* recycling, could also be attempted to slow clearance rates (Harmsen *et al.*, 2007b). Single domain binders have a distinct advantage over those derived from conventional antibodies in the relative ease with which they can be engineered into multivalent formats. Depending on the binding domains selected, formatting can increase avidity and/or enable simultaneous binding to multiple targets. The only report thus far of reformatting showed that AMA1-binding domains could be dimerized via the addition of C-terminal helix-loop-helix motifs, increasing the functional affinity for immobilized antigen ~ ten-fold. In contrast, when two shark V domains were joined in tandem with a 15 residue linker, only the N-terminal domain could still bind antigen. Structural modeling suggests the linker enters the second V domain close to CDR3, and likely obstructs binding (Simmons *et al.*, 2006). However, detailed testing of different binding domains in a range of formats will be required to show the utility of such constructs.

10.6.3.5 *IgNAR-derived single domains as human therapeutics*

It is well known that the administration of therapeutic proteins derived from non-human species usually evokes some degree of humoral response, especially where the treatment necessitates repeated administration (reviewed in Aarden *et al.*, 2008). Although the IgNAR V domain is a bona fide IgSF protein it is also very obviously "non-human," raising questions regarding the potential immunogenicity of IgNAR-derived therapeutics. In preliminary studies the repeated immunization of mice with nurse shark V regions emulsified in Freund's adjuvant did not elicit a detectable antibody titer (Dooley and Flajnik, unpublished data). This mirrors previous reports that repeated administration of camelid V_HH domains does not induce B or T cell responses in mice (Cortez-Retamozo *et al.*, 2002). However, it should not be assumed that this apparent immunosilence is equally true for humans. Therefore, the ability to define and reduce the "foreignness" of IgNAR V regions needs further exploration. The IgSF nature of the IgNAR V region means we can test the same methods employed to humanize antibodies sourced from other species. The anti-malarial, AMA1-binding domain is one of the few IgNAR generated for human therapy. However, there are reports of IgNAR-derived binding domains that bind insulin-like growth factor receptor 1, a cell-surface tyrosine kinase receptor overexpressed in a number of cancers. The lead candidate is reported to antagonize receptor signaling and inhibits growth of the MCF-7 breast cancer cell line *in vitro* (www.adalta.com).

10.6.3.6 *Perspectives*

Although the shark IgNAR-derived single domain platform is still very much "in development," the molecules studied thus far demonstrate many intrinsic qualities that can be exploited for the development of future therapeutics. As mentioned, IgNAR V regions are naturally very small; therefore, in a monovalent format this would lead to rapid renal clearance of unbound molecules (Cortez-Retamozo *et al.*, 2002), a desirable characteristic for imaging reagents or specific delivery of some toxic payloads. Additionally, as has been found with similarly sized V_HH domains, they will likely have better tissue penetration into sites such as inflamed synovial joints, solid tumors, or perhaps across the blood–brain barrier. Alternately by increasing their physical size or introducing binding moieties to various human serum proteins, their *in vivo* half-life can be significantly extended (reviewed in Van Bockstaele *et al.*, 2009). The relative ease of engineering of multivalent and/or multi-specific formats can be used to improve their therapeutic characteristics (Coppieters *et al.*, 2006).

However, we perceive the main advantage of these molecules to be their ability to access sites that might be inaccessible using conventional antibodies; as detailed above, naturally derived single domains (whether from shark or camel) show a propensity to bind crevices and clefts on proteins to increase their binding affinity. Exploiting this binding mode may enable discrimination of members of closely-related protein families; such families generally have a highly similar gross structure but differ in the exact nature of their binding sites, a requirement if they are to recognize specific targets. The ability of IgNAR-derived binding domains to penetrate such sites could be exploited to push binding toward the family member of interest, while avoiding highly related family members; the result would be diminished off-target binding along with any associated side effects. While the immunogenicity of these molecules needs further study, problems of antigenicity for antibodies from other non-human species has been successfully overcome, leading to some of the current antibody-based "blockbuster drugs." We envision that IgNAR-derived binding domains will eventually lead to new biotherapeutic drugs with improved pharmacokinetics, increased target specificity and/or the capability to deliver toxic payloads more effectively, with fewer side effects than those currently employed.

10.7 Summary

As discussed in this chapter, mAbs have undoubtedly enjoyed tremendous success in the past decade and are now an established modality in the treatment of diseases such as cancer and autoimmune indications. This success however has also given rise to new questions regarding the durability of mAbs as mainstay biotherapeutic drugs particularly in the current environment of rising healthcare costs. Challenges facing mAb developers and beneficiaries within the United States have been discussed (Cohen and Wilson, 2009). In fact, aside from the cost issue itself, the very health benefit provided by mAb therapies has been brought into question (Neumann, 2009). This type of uncertainty has therefore fueled a need within the industry to explore alternate methods to both reduce costs as well as to enhance the performance of mAb-based drugs.

This chapter covers the frontiers of mAb development using fragments or alternate protein scaffolds. Clearly, this is a very broad area and it is not possible to do justice to the explosion that has been witnessed in this field over the past five years. However, the strategies we have discussed are ones that are being pursued in our organization. We believe that the fundamental opportunity with antibody fragments and alternate scaffolds is to build modular biotherapeutic drugs that deliver superior pharmacology at competitive costs with enhanced safety and patient convenience profiles. SMIPs and Fc-based bi-specific antibodies achieve this through enhanced ADCC/CDC or dual targeting of receptors and soluble targets. V_HH antibodies and IgNAR V regions represent the smallest protein folds that encapsulate key features of the antibody binding domains. The added advantages with these novel scaffolds are their high stability and superior pharmaceutical properties that lend themselves to more cost-effective formulation and delivery options. This could prove to be a competitive advantage, particularly in emerging markets where storage and delivery of protein pharmaceuticals require large infrastructure costs that those economies may not be able to bear.

There remain significant challenges in developing multi-functional proteins that mimic mAb function. First, it is not clear to what extent manufacturing costs alone could alleviate the overall cost of mAb-based therapies. While operational expenses related to the establishment of conventional biopharmaceutical manufacturing facilities could be obviated, the actual cost of goods are likely to be similar for antibody fragments and alternate scaffolds as compared to conventional mAbs. Second, the overall safety and tolerability profiles of newer protein modalities still needs to be addressed across a broad set of patient populations and across multiple dosing strategies. Third, related to the safety issue, immunogenicity continues to be an open question particularly for the non-human proteins such as the camelid-derived V_HH antibodies and the shark IgNAR scaffold, even though these proteins are humanized before entry into clinical studies. Last, principles of titrating a bi-specific drug against more than one target and therefore dosing of the bi-specific product are not well established. This is an issue that will need to be addressed on a product-by-product basis in the clinic. Despite all these challenges, our field has witnessed the regulatory approval of the first multi-functional mAb product last year (Catumaxomab for treatment of malignant ascites). There is no doubt that with the pressures around cost and performance of mAbs, the use of mAb fragments and alternate protein scaffolds will continue to grow and will one day deliver superior value to patients who desperately need them.

References

Aarden L., Ruuls S.R., Wolbink G. (2008). Immunogenicity of anti-tumor necrosis factor antibodies-toward improved methods of anti-antibody measurement. *Curr Opin Immunol* **20**: 431–5.

Akamatsu Y., Pakabunto K., Xu Z., Zhang Y., Tsurushita N. (2007). Whole IgG surface display on mammalian cells: application to isolation of neutralizing chicken monoclonal anti-IL-12 antibodies. *J Immunol Methods* **327**: 40–52.

Andris-Widhopf J., Rader C., Steinberger P., Fuller R., Barbas C.F., III. (2000). Methods for the generation of chicken monoclonal antibody fragments by phage display. *J Immunol Methods* **242**: 159–81.

Arbabi-Ghahroudi M., Desmyter A., Wyns L., Hamers R., Muyldermans S. (1997). Selection and identification of single domain antibody fragments from camel heavy-chain antibodies. *FEBS Lett* **414**: 521–6.

Arbabi-Ghahroudi M., Mackenzie R., Tanha J. (2009). Selection of non-aggregating VH binders from synthetic VH phage-display libraries. *Methods Mol Biol* **525**: 187–216.

Attallah A.M., Abbas A.T., Ismail H., Abdel-Raouf M., El-Dosoky I. (2009). Efficacy of passive immunization with IgY antibodies to a 58-kDa H. pylori antigen on severe gastritis in BALB/c mouse model. *J Immunoassay Immunochem* **30**: 359–77.

Blankenship J.W., Strobel S., Tan P., Grosmaire L., Simons A., Clark C., Bae G., Espling E., Tamman J., Wiens J., Whal A. (2009). Anti-CD79Bxanti-DR SCORPION™ Molecule: A single chain, bispecific immunotherapeutic with potent *in vitro* activity against B-cell lymphoma. 100th Annual Meeting of the AACR, April 18–22, Denver, CO.

Burge D.J., Bookbinder S.A., Kivitz A.J., Fleischmann R.M., Shu C., Bannink J. (2008). Pharmacokinetic and pharmacodynamic properties of TRU-015, a CD20-directed small modular immunopharmaceutical protein therapeutic, in patients with rheumatoid arthritis: a Phase I, open-label, dose-escalation clinical study. *Clin Ther* **30**: 1806–16.

Charlton K., Harris W.J., Porter A.J. (2001). The isolation of super-sensitive anti-hapten antibodies from combinatorial antibody libraries derived from sheep. *Biosens Bioelectron* **16**: 639–46.

Chiliza T.E., Van W.W., Du Plessis D.H. (2008). Single-chain antibody fragments from a display library derived from chickens immunized with a mixture of parasite and viral antigens. *Hybridoma (Larchmt)* **27**: 413–21.

Cohen J. Wilson A. (2009). New challenges to medicare beneficiary access to mAbs. *mAbs* **1**: 56–66.

Conrath K.E., Wernery U., Muyldermans S., Nguyen V.K. (2003). Emergence and evolution of functional heavy-chain antibodies in Camelidae. *Dev Comp Immunol* **27**: 87–103.

Coppieters K., Dreier T., Silence K., de Haard H., Lauwereys M., Casteels P., Beirnaert E., Jonckheere H., Van de Wiele C., Staelens L., Hostens J., Revets H., Remaut E., Elewaut D., Rottiers P. (2006). Formatted anti-tumor necrosis factor alpha VHH proteins derived from camelids show superior potency and targeting to inflamed joints in a murine model of collagen-induced arthritis. *Arthritis Rheum* **54**: 1856–66.

Cortez-Retamozo V., Lauwereys M., Hassanzadeh G.G., Gobert M., Conrath K., Muyldermans S., De Baetselier P., Revets H. (2002). Efficient tumor targeting by single-domain antibody fragments of camels. *Int J Cancer* **98**: 456–62.

Dall'Acqua W.F., Cook K.E., Damschroder M.M., Woods R.M., Wu H. (2006). Modulation of the effector functions of a human IgG1 through engineering of its hinge region. *J Immunol* **177**: 1129–38.

Davies D.R., Sheriff S., Padlan E.A. (1988). Antibody-antigen complexes. *J Biol Chem* **263**: 10541–4.

Davies J., Riechmann L. (1994). 'Camelising' human antibody fragments: NMR studies on VH domains. *FEBS Lett* **339**: 285–90.

De Genst E., Saerens D., Muyldermans S., Conrath K. (2006). Antibody repertoire development in camelids. *Dev Comp Immunol* **30**(1–2): 187–98.

Desmyter A., Transue T.R., Ghahroudi M.A., Thi M.H., Poortmans F., Hamers R., Muyldermans S., Wyns L. (1996). Crystal structure of a camel single-domain VH antibody fragment in complex with lysozyme. *Nat Struct Biol* **3**: 803–11.

Diaz M., Greenberg A.S., Flajnik M.F. (1998). Somatic hypermutation of the new antigen receptor gene (NAR) in the nurse shark does not generate the repertoire: possible role in antigen-driven reactions in the absence of germinal centers. *Proc Natl Acad Sci USA* **95**: 14343–8.

Diaz M., Stanfield R.L., Greenberg A.S., Flajnik M.F. (2002). Structural analysis, selection, and ontogeny of the shark new antigen receptor (IgNAR): identification of a new locus preferentially expressed in early development. *Immunogenetics* **54**: 501–12.

Dooley H., Flajnik M.F., Porter A.J. (2003). Selection and characterization of naturally occurring single-domain (IgNAR) antibody fragments from immunized sharks by phage display. *Mol Immunol* **40**: 25–33.

Dooley H,. Flajnik M.F. (2006a). Antibody repertoire development in cartilaginous fish. *Dev Comp Immunol* **30**: 43–56.

Dooley H., Stanfield R.L., Brady R.A., Flajnik M.F. (2006b). First molecular and biochemical analysis of in vivo affinity maturation in an ectothermic vertebrate. *Proc Natl Acad Sci USA* **103**: 1846–51.

Dunussi-Joannopoulos K., Tchistiakova L., Seth N., Olland S., Zollner R., Brady W., Mohler K.M., Herber D., Vugmeyster Y., Wensel D., Gill D., Collins M., Nickerson-Nutter C. (2009). Kinetics of B cell depletion and reconstitution in pereferal blood and lymphoid tissues after a single dose of SBI-087, a humanized anti-CD20 SMIP, in cynomolgus monkeys. EULAR poster.

Ewert S., Honegger A., Plückthun A. (2004). Stability improvement of antibodies for extracellular and intracellular applications: CDR grafting to stable frameworks and structure-based framework engineering. *Methods* **34**: 184–99.

Finlay W.J., deVore N.C., Dobrovolskaia E.N., Gam A., Goodyear C.S., Slater J.E. (2005). Exploiting the avian immunoglobulin system to simplify the generation of recombinant antibodies to allergenic proteins. *Clin Exp Allergy* **35**: 1040–8.

Finlay W.J., Shaw I., Reilly J.P., Kane M. (2006). Generation of high-affinity chicken single-chain Fv antibody fragments for measurement of the Pseudonitzschia pungens toxin domoic acid. *Appl Environ Microbiol* **72**: 3343–9.

Frenken L.G., van der Linden R.H., Hermans P.W., Bos J.W., Ruuls R.C., de Geus B., Verrips C.T. (2000). Isolation of antigen specific llama V_HH antibody fragments and their high level secretion by Saccharomyces cerevisiae. *J Biotechnol* **78**: 11–21.

Gassmann M., Thommes P., Weiser T., Hubscher U. (1990). Efficient production of chicken egg yolk antibodies against a conserved mammalian protein. *FASEB J* **4**: 2528–32.

Greenberg A.S., Avila D., Hughes M., Hughes A., McKinney E.C., Flajnik M.F. (1995). A new antigen receptor gene family that undergoes rearrangement and extensive somatic diversification in sharks. *Nature* **374**: 168–73.

Hamers-Casterman C., Atarhouch T., Muyldermans S., Robinson G., Hamers C., Songa E.B., Bendahman N., Hamers R. (1993). Naturally occurring antibodies devoid of light chains. *Nature* **363**: 446–8.

Harmsen M.M., Ruuls R.C., Nijman I.J., Niewold T.A., Frenken L.G., de Geus B. (2000). Llama heavy-chain V regions consist of at least four distinct subfamilies revealing novel sequence features. *Mol Immunol* **37**: 579–90.

Harmsen M.M., De Haard H.J. (2007a). Properties, production, and applications of camelid single-domain antibody fragments. *Appl Microbiol Biotechnol* **77**: 13–22.

Harmsen M.M., van Solt C.B., Fijten H.P., van Keulen L., Rosalia R.A., Weerdmeester K., Cornelissen A.H., De Bruin M.G., Eble P.L., Dekker A. (2007b). Passive immunization of guinea pigs with llama single-domain antibody fragments against foot-and-mouth disease. *Vet Microbiol* **120**: 193–206.

Hatta H., Tsuda K., Akachi S., Kim M., Yamamoto T. (1993). Productivity and some properties of egg yolk antibody (IgY) against human rotavirus compared with rabbit IgG. *Biosci. Biotechnol Biochem* **57**: 450–4.

Hayden-Ledbetter M.S., Cerveny C.G., Espling E., Brady W.A., Grosmaire L.S., Tan P., Bader R., Slater S., Nilsson C.A., Barone D.S., Simon A., Bradley C., Thompson P.A., Wahl A.F., Ledbetter J.A. (2009). CD20-directed small modular immunopharmaceutical, TRU-015, depletes normal and malignant B cells. *Clin Cancer Res* **15**: 2739–46.

Hedges S.B. (2002). The origin and evolution of model organisms. *Nat Rev Genet* **3**: 838–49.

Henderson K.A., Streltsov V.A., Coley A.M., Dolezal O., Hudson P.J., Batchelor A.H., Gupta A., Bai T., Murphy V.J., Anders R.F., Foley M., Nuttall S.D. (2007). Structure of an IgNAR-AMA1 complex: targeting a conserved hydrophobic cleft broadens malarial strain recognition. *Structure* **15**: 1452–66.

Hof D., Hoeke M.O., Raats J.M. (2008). Multiple-antigen immunization of chickens facilitates the generation of recombinant antibodies to autoantigens. *Clin Exp Immunol* **151**: 367–77.

Ikemori Y., Peralta R.C., Kuroki M., Yokoyama H., Kodama Y. (1993). Research note: avidity of chicken yolk antibodies to enterotoxigenic Escherichia coli fimbriae. *Poult Sci* **72**: 2361–5.

Joosten V., Roelofs M.S., van den Dries N., Goosen T., Verrips C.T., van den Hondel C.A., Lokman B.C. (2005). Production of bifunctional proteins by Aspergillus awamori: llama variable heavy chain antibody fragment (V(HH) R9 coupled to Arthromyces ramosus peroxidase (ARP). *J Biotechnol* **120**: 347–59.

Kettleborough C.A., Saldanha J., Heath V.J., Morrison C.J., Bendig M.M. (1991). Humanization of a mouse monoclonal antibody by CDR-grafting: the importance of framework residues on loop conformation. *Protein Eng* **4**: 773–83.

Kohler G., Milstein C. (1975). Continuous cultures of fused cells secreting antibody of predefined specificity. *Nature* **256**: 495–7.

Kontermann R.E. (2010). Alternative antibody formats. *Curr Opin Mol Ther* **12**: 176–83.

Kopsidas G., Roberts A.S., Coia G., Streltsov V.A., Nuttall S.D. (2006). *In vitro* improvement of a shark IgNAR antibody by Qbeta replicase mutation and ribosome display mimics in vivo affinity maturation. *Immunol Lett* **107**: 163–8.

Krebber A., Bornhauser S., Burmester J., Honegger A., Willuda J., Bosshard H.R., Pluckthun A. (1997). Reliable cloning of functional antibody variable domains from hybridomas and spleen cell repertoires employing a reengineered phage display system. *J Immunol Methods* **201**: 35–55.

Larsson A., Karlsson-Parra A., Sjoquist J. (1991). Use of chicken antibodies in enzyme immunoassays to avoid interference by rheumatoid factors. *Clin Chem* **37**:411–14.

Laskowski R.A., Luscombe N.M., Swindells M.B., Thornton, J.M. (1996). Protein clefts in molecular recognition and function. *Protein Sci* **5**: 2438–52.

Lee C.V., Hymowitz S.G., Wallweber H.J., Gordon N.C., Billeci K.L., Tsai S.P., Compaan D.M., Yin J., Gong Q., Kelley R.F., DeForge L.E., Martin F., Starovasnik M.A., Fuh G. (2006). Synthetic anti-BR3 antibodies that mimic BAFF binding and target both human and murine B cells. *Blood* **108**: 3103–11.

Lemamy G.J., Roger P., Mani J.C., Robert M., Rochefort H., Brouillet J.P. (1999). High-affinity antibodies from hen's-egg yolks against human mannose-6-phosphate/insulin-like growth-factor-II receptor (M6P/IGFII-R): characterization and potential use in clinical cancer studies. *Int J Cancer* **80**: 896–902.

Leonard P., Safsten P., Hearty S., McDonnell B., Finlay W., O'Kennedy R. (2007). High throughput ranking of recombinant avian scFv antibody fragments from crude lysates using the Biacore A100. *J Immunol Methods* **323**: 172–9.

Li Y., Cockburn W., Kilpatrick J.B., Whitelam G.C. (2000). High affinity ScFvs from a single rabbit immunized with multiple haptens. *Biochem Biophys Res Commun* **268**: 398–404.

Litman G.W., Berger L., Murphy K., Litman R., Hinds K., Erickson, B.W. (1985). Immunoglobulin VH gene structure and diversity in Heterodontus, a phylogenetically primitive shark. *Proc Natl Acad Sci USA* **82**: 2082–6.

Liu J.L., Anderson G.P., Delehanty J.B., Baumann R., Hayhurst A., Goldman, E.R. (2007a). Selection of cholera toxin specific IgNAR single-domain antibodies from a naive shark library. *Mol Immunol* **44**: 1775–83.

Liu J.L., Anderson G.P., Goldman, E.R. (2007b). Isolation of anti-toxin single domain antibodies from a semi-synthetic spiny dogfish shark display library. *BMC Biotechnol* **7**: 78.

McCormack W.T., Tjoelker L.W., Thompson C.B. (1993). Immunoglobulin gene diversification by gene conversion. *Prog Nucleic Acid Res Mol Biol* **45**: 27–45.

Michael N., Accavitt, M.A., Masteller E., Thompson C.B. (1998). The antigen-binding characteristics of mAbs derived from in vivo priming of avian B cells. *Proc Natl Acad Sci USA* **95**: 1166–71.

Michaelson J.S., Demarest S.J., Miller B., Amatucci A., Snyder W.B., Wu X., Huang F., Phan S., Gao S., Doern A., Farrington G.K., Lugovskoy A., Joseph I., Bailly V., Wang X.,

Garber E., Browning J., Glaser, S.M. (2009). Anti-tumor activity of stability-engineered IgG-like bispecific antibodies targeting TRAIL-R2 and LTbetaR. *mAbs* **1**: 128–41.

Muyldermans S., Atarhouch T., Saldanha J., Barbosa J.A., Hamers, R. (1994). Sequence and structure of VH domain from naturally occurring camel heavy chain immunoglobulins lacking light chains. *Protein Eng* **7**: 1129–35.

Muyldermans S., Lauwereys M. (1999). Unique single-domain antigen binding fragments derived from naturally occurring camel heavy-chain antibodies. *J Mol Recognit* **12**: 131–40.

Nakamura N., Aoki Y., Horiuchi H., Furusawa S., Yamanaka H.I., Kitamoto T., Matsuda, H. (2000). Construction of recombinant monoclonal antibodies from a chicken hybridoma line secreting specific antibody. *Cytotechnology* **32**: 191–8.

Nakamura N., Shimokawa M., Miyamoto K., Hojyo S., Horiuchi H., Furusawa S., Matsuda, H. (2003). Two expression vectors for the phage-displayed chicken monoclonal antibody. *J Immunol Methods* **280**: 157–64.

Nakamura N., Shuyama A., Hojyo S., Shimokawa M., Miyamoto K., Kawashima T., Aosasa M., Horiuchi H., Furusawa S., Matsuda H. (2004). Establishment of a chicken monoclonal antibody panel against mammalian prion protein. *J Vet Med Sci* **66**: 807–14.

Neumann P.J. (2009). Are mAbs different?: a comment on Cohen and Wilson. *mAbs* **1**: 29–30.

Nguyen V.K., Muyldermans S., Hamers R. (1998). The specific variable domain of camel heavy-chain antibodies is encoded in the germline. *J Mol Biol* **275**: 413–18.

Nguyen V.K., Hamers R., Wyns L., Muyldermans S. (1999). Loss of splice consensus signal is responsible for the removal of the entire C(H)1 domain of the functional camel IGG2A heavy-chain antibodies. *Mol Immunol* **36**: 515–24.

Nguyen V.K., Hamers R., Wyns L., Muyldermans S. (2000). Camel heavy-chain antibodies: diverse germline V(H)H and specific mechanisms enlarge the antigen-binding repertoire. *EMBO J* **19**: 921–30.

Nguyen V.K., Desmyter A., Muyldermans S. (2001). Functional heavy-chain antibodies in Camelidae. *Adv Immunol* **79**: 261–96.

Nguyen V.K., Su C., Muyldermans S., van der Loo W. (2002). Heavy-chain antibodies in Camelidae; a case of evolutionary innovation. *Immunogenetics* **54**: 39–47.

Nickerson-Nutter C., Tchistiakova L., Seth N., Olland S., Zollner R., Brady W.A., Mohler K., Baum P., Wahl A., Herber D., Vugmeyster Y., Wensel D., Gill D., Collins M., Dunussi-Joannopoulos K. 2LM20-4, a humanized CD20 Small Modular ImmunoPharmaceutical (SMIP™), but not mutant 2LM20-4 P331S with no *in vitro* CDC activity, has potent and sustained lymph node B cell depletion activity in cynomolgus monkeys. Manuscript submitted.

Nishibori N., Horiuchi H., Furusawa S., Matsuda H. (2006). Humanization of chicken monoclonal antibody using phage-display system. *Mol Immunol* **43**: 634–42.

Nuttall S.D., Krishnan U.V., Hattarki M., De Gori R., Irving R.A., Hudson, P.J. (2001). Isolation of the new antigen receptor from wobbegong sharks, and use as a scaffold for the display of protein loop libraries. *Mol Immunol* **38**: 313–26.

Nuttall S.D., Krishnan U.V., Doughty L., Pearson K., Ryan M.T., Hoogenraad N.J., Hattarki M., Carmichael J.A., Irving R.A., Hudson P.J. (2003). Isolation and characterization of an IgNAR variable domain specific for the human mitochondrial translocase receptor Tom70. *Eur J Biochem* **270**: 3543–54.

Nuttall S.D., Humberstone K.S., Krishnan U.V., Carmichael J.A., Doughty L., Hattarki M., Coley A.M., Casey J.L., Anders R.F., Foley M., Irving R.A., Hudson, P.J. (2004). Selection and affinity maturation of IgNAR variable domains targeting Plasmodium falciparum AMA1. *Proteins* **55**: 187–97.

O'Brien P.M., Aitken R., O'Neil B.W., Campo, M.S. (1999). Generation of native bovine mAbs by phage display. *Proc Natl Acad Sci USA* **96**: 640–5.

Omidfar K., Rasaee M.J., Kashanian S., Paknejad M., Bathaie, Z. (2007). Studies of thermostability in Camelus bactrianus (Bactrian camel) single-domain antibody specific for the mutant epidermal-growth-factor receptor expressed by Pichia. *Biotechnol Appl Biochem* **46**: 41–9.

Pant N., Hultberg A., Zhao Y., Svensson L., Pan-Hammarstrom Q., Johansen K., Pouwels P.H., Ruggeri F.M., Hermans P., Frenken L., Boren T., Marcotte H., Hammarstrom L. (2006). Lactobacilli expressing variable domain of llama heavy-chain antibody fragments (lactobodies) confer protection against rotavirus-induced diarrhea. *J Infect Dis* **194**: 1580–8.

Polson A., Coetzer T., Kruger J., von Maltzahn E., van der Merwe K.J. (1985). Improvements in the isolation of IgY from the yolks of eggs laid by immunized hens. *Immunol Invest* **14**: 323–7.

Popkov M., Mage R.G., Alexander C.B., Thundivalappil S., Barbas C.F. III, Rader C. (2003). Rabbit immune repertoires as sources for therapeutic monoclonal antibodies: the impact of kappa allotype-correlated variation in cysteine content on antibody libraries selected by phage display. *J Mol Biol* **325**: 325–35.

Qian W.J., Kaleta D.T., Petritis B.O., Jiang H., Liu T., Zhang X., Mottaz H.M., Varnum S.M., Camp D.G., Huang L., Fang X., Zhang W.W., Smith, R.D. (2008). Enhanced detection of low abundance human plasma proteins using a tandem IgY12-SuperMix immunoaffinity separation strategy. *Mol Cell Proteomics* **7**: 1963–73.

Riechmann L., Muyldermans S. (1999). Single domain antibodies: comparison of camel VH and camelised human VH domains. *J Immunol Methods* **231**: 25–38.

Robak T., Robak P., Smolewski P. (2009). TRU-016, a humanized anti-CD37 IgG fusion protein for the potential treatment of B-cell malignancies. *Curr Opin Investig Drugs* **10**: 1383–90.

Roux K.H., Greenberg A.S., Greene L., Strelets L., Avila D., McKinney E.C., Flajnik M.F. (1998). Structural analysis of the nurse shark (new) antigen receptor (NAR): molecular convergence of NAR and unusual mammalian immunoglobulins. *Proc Natl Acad Sci USA* **95**: 11804–9.

Sandin S., Ofverstedt L.G., Wikstrom A.C., Wrange O., Skoglund U. (2004). Structure and flexibility of individual immunoglobulin G molecules in solution. *Structure* **12**: 409–15.

Shimizu M., Nagashima H., Sano K., Hashimoto K., Ozeki M., Tsuda K., Hatta H. (1992). Molecular stability of chicken and rabbit immunoglobulin G. *Biosci Biotechnol Biochem* **56**: 270–4.

Simmons D.P., Abregu F.A., Krishnan U.V., Proll D.F., Streltsov V.A., Doughty L., Hattarki M.K., Nuttall, S.D. (2006). Dimerisation strategies for shark IgNAR single domain antibody fragments. *J Immunol Methods* **315**: 171–84.

Stanfield R.L., Dooley H., Flajnik M.F., Wilson I.A. (2004). Crystal structure of a shark single-domain antibody V region in complex with lysozyme. *Science* **305**:1770–3.

Stanfield R.L., Dooley H., Verdino P., Flajnik M.F., Wilson I.A. (2007). Maturation of shark single-domain (IgNAR) antibodies: evidence for induced-fit binding. *J Mol Biol* **367**: 358–72.

Streltsov V.A., Varghese J.N., Carmichael J.A., Irving R.A., Hudson P.J., Nuttall, S.D. (2004). Structural evidence for evolution of shark Ig new antigen receptor variable domain antibodies from a cell-surface receptor. *Proc Natl Acad Sci USA* **101**: 12444–9.

Stuart C.A., Pietrzyk R.A., Furlanetto R.W., Green A. (1988). High affinity antibody from hen's eggs directed against the human insulin receptor and the human IGF-I receptor. *Anal Biochem* **173**: 142–50.

Tateishi Y., Nishimichi N., Horiuchi H., Furusawa S., Matsuda H. (2008). Construction of chicken-mouse chimeric antibody and immunogenicity in mice. *J Vet Med Sci* **70**: 397–400.

Tracey D., Klareskog L., Sasso E.H., Salfeld J.G., Tak P.P. (2008). Tumor necrosis factor antagonist mechanism of action: A comprehensive review. *Pharm & Therap* **117**: 244–79.

Tsen Y.C., Kao G.Y., Chang C.L., Lai F.Y., Huang C.H., Ouyang S., Yu M.H., Wang C.P., Chiou Y.N. (2003). Evaluation and validation of a duck IgY antibody-based immunoassay for high-sensitivity C-reactive protein: avian antibody application in clinical diagnostics. *Clin Chem* **49**: 810–13.

Tsurushita N., Park M., Pakabunto K., Ong K., Avdalovic A., Fu H., Jia A., Vasquez M., Kumar S. (2004). Humanization of a chicken anti-IL-12 monoclonal antibody. *J Immunol Methods* **295**: 9–19.

Van Bockstaele F., Holz J.B., Revets H. (2009). The development of nanobodies for therapeutic applications. *Curr Opin Investig Drugs* **10**: 1212–24.

Vandenbroucke K., de Haard H., Beirnaert E., Dreier T., Lauwereys M., Huyck L., Van Huysse J., Demetter P., Steidler L., Remaut E., Cuvelier C., Rottiers P. (2010). Orally administered L. lactis secreting an anti-TNF Nanobody demonstrate efficacy in chronic colitis. *Mucosal Immunol* **3**: 49–56.

Wallis J.W., Aerts J., Groenen M.A., Crooijmans R.P., Layman D., Graves T.A., Scheer D.E., Kremitzki C., Fedele M.J., Mudd N.K., Cardenas M., Higginbotham J., Carter J., McGrane R., Gaige T., Mead K., Walker J., Albracht D., Davito J., Yang S.P., Leong S., Chinwalla A., Sekhon M., Wylie K., Dodgson J., Romanov M.N., Cheng H., de Jong P.J., Osoegawa K., Nefedov M., Zhang H., McPherson J.D., Krzywinski M., Schein J., Hillier L., Mardis E.R., Wilson R.K., Warren W.C. (2004). A physical map of the chicken genome. *Nature* **432**: 761–4.

Wilson P.C., de Bouteiller O. (1998). Somatic hypermutation introduces insertions and deletions into immunoglobulin V genes. *J Exp Med* **187**(1): 59–70.

Wu A.M., Tan G.J., Sherman M.A., Clarke P., Olafsen T., Forman S.J., Raubitschek A.A. (2001). Multimerization of a chimeric anti-CD20 single-chain Fv-Fc fusion protein is mediated through variable domain exchange. *Protein Eng* **14**: 1025–33.

Yamanaka H.I., Inoue T., Ikeda-Tanaka O. (1996). Chicken monoclonal antibody isolated by a phage display system. *J Immunol* **157**: 1156–62.

Editor Biography

Clive R. Wood, Ph.D., joined Bayer as Senior Vice President and Head of GDD-Global Biologics, Bayer HealthCare AG, Germany, in July 2009. Formerly, he was Chief Scientific Officer and Executive Vice President at Dyax Corp., Cambridge, MA. Dr. Wood joined Dyax in 2003 after 17 years at Genetics Institute/Wyeth Research in Cambridge, MA. Prior to Genetics Institute, he spent four years at Celltech Ltd., UK, where between 1982 and 1985 he contributed to some of the earliest work in recombinant antibody expression. His research has covered each of the key areas of antibody technology, including antibody humanization, transgenic mouse and phage display-based approaches for the discovery of treatments for inflammation and cancer. Dr. Wood also serves as an Adjunct Professor in the Department of Pharmacology and Experimental Therapeutics of Boston University School of Medicine. He received his Biochemistry B.Sc. and Ph.D. degrees from Imperial College, University of London.

Index